Lecture Notes in Physics

Springer-Verlag Berlin Heidelberg GmbH

N. Bretón J. L. Cervantes-Cota M. Salgado (Eds.)

The Early Universe and Observational Cosmology

Springer

Editors

Nora Bretón
Centro de Investigación
y de Estudios Avanzados del I.P.N
Departamento de Física
A.P. 14-740, 07000 México D.F.

Marcelo Salgado
Universidad Nacional
Autónoma de México
Instituto de Ciencias Nucleares
A.P. 70-543, 04510 México D.F.

Jorge Luis Cervantes-Cota
Instituto Nacional
de Investigaciones Nucleares (ININ)
Departamento de Física
A.P. 18-1027, Col. Escandón
11801 México D.F.

N. Bretón, J. L. Cervantes-Cota, M. Salgado (Eds.), *The Early Universe and Observational Cosmology*, Lect. Notes Phys. **646** (Springer, Berlin Heidelberg 2004), DOI 10.1007/b97189

Library of Congress Control Number: 2004103823

ISSN 0075-8450
ISBN 978-3-642-06004-5 ISBN 978-3-540-40918-2 (eBook)
DOI 10.1007/978-3-540-40918-2

springeronline.com

Originally published by Springer-Verlag Berlin Heidelberg New York in 2004
Softcover reprint of the hardcover 1st edition 2004

Data conversion: PTP-Berlin Protago-TeX-Production GmbH
Cover design: *design & production*, Heidelberg

Printed on acid-free paper
54/3141/ts - 5 4 3 2 1 0

Preface

The Mexican School on Gravitation and Mathematical Physics, sponsored by the Mexican Physical Society, is a conference that started 10 years ago. The aim of the School is to cover different topics on the frontiers of gravitation, field theory and mathematical physics. It is held every two years and a different theme is chosen for each occasion. The School, which is oriented towards advanced graduate students and beyond, is gaining a reputation for the quality of lectures given by leaders in the field. In our previous Schools the subjects covered have been *Supergravity and Mathematical Physics*, *Branes*, *Black Holes* and the speakers have included A. Ashtekar, B. Carter, G. Gibbons, M. Heusler, W. Israel, F. Müller–Hoisen, Y. Neeman, R. Myers, L. Randall, R. Sorkin, P. Van Niewenhuizen, R. Wald, among other top ranked physicists.

Over the past few years remarkable discoveries in physics and astronomy have been achieved with enormous implications for cosmology. In particular, the recent experiments measuring anisotropies on the cosmic microwave background (CMB) and the distance–red–shift relation in type Ia supernovae (SNIa) have opened a new era in cosmology, sometimes called *the golden years* or *the high–precision era of cosmology.*

Such discoveries have not only corroborated several theoretical predictions and put stringent bounds on many cosmological models, but also renew some ancient paradigms like the origin of a cosmological constant.

In view of the primary importance of such a hot topic today, it was clear that a convenient theme for the Fifth Mexican School was *The Early Universe and Observational Cosmology.* We considered that subjects like *Inflation*, *Structure Formation*, *Cosmological Perturbations*, *Braneworld Cosmologies*, *Quintessence*, and *Dark Matter* would give the participants a good picture of the current status of modern cosmology.

Like in past Schools the topics were covered by leaders in the field, and the general perception by the participants was that the goals were well accomplished; of course, the beautiful setting of Playa del Carmen in the Mexican Caribbean did not hurt. About 80 people participated from all over the world and we are indebted to all of them.

Undoubtedly, the School would have not been possible without the main courses and plenary lectures. Therefore, we extend our deep gratitude to the invited speakers. The School was complemented with more specialized topics presented in parallel sessions, some of which are included in these Lecture Notes.

Finally, the goals of the School would certainly be unmet if there were not some hard–copy record of the ideas presented during that week of November 2002. To that end, we warmly thank all the contributors who made possible the publication of this book.

Mexico City,
January 2004

Nora Breton
Jorge L. Cervantes–Cota
Marcelo Salgado

Contents

List of Contributors

Miguel Alcubierre
Universidad Nacional
Autónoma de México,
Instituto de Ciencias Nucleares,
A.P. 70-543, 04510 México D.F.
malcubi@nuclecu.unam.mx

Alejandro Ayala
Universidad Nacional
Autónoma de México,
Instituto de Ciencias Nucleares,
A.P. 70-543, 04510 México D.F.
ayala@nuclecu.unam.mx

Ricardo Becerril
Universidad Michoacana,
Instituto de Física y Matemáticas,
Edif. C-3, Ciudad Universitaria
58040 Morelia, Michoacán, México.
becerril@ifm1.ifm.umich.mx

Martin Bojowald
The Pennsylvania State University,
Center for Gravitational Physics
and Geometry,
University Park, PA 16802, USA.
bojowald@gravity.psu.edu

Robert H. Brandenberger
Brown University,
Physics Department,
Providence, RI 02912, USA.
rhb@het.brown.edu

Nora Bretón
Centro de Investigación
y de Estudios Avanzados del I.P.N,
A.P. 14-740, 07000 México D.F.
Nora.Breton@fis.cinvestav.mx

Luis G. Cabral–Rosetti
Universidad Nacional
Autónoma de México,
Instituto de Ciencias Nucleares,
A.P. 70-543,04510 México D.F.
luis@nuclecu.unam.mx

Jorge L. Cervantes–Cota
Instituto Nacional
de Investigaciones Nucleares (ININ),
Departamento de Física,
A.P. 18-1027, Col. Escandón, 11801
México D.F.
jorge@nuclear.inin.mx

Edmund J. Copeland
University of Sussex,
Department of Physics
and Astronomy,
Brighton, BN1 9QJ, U.K.
e.j.copeland@susx.ac.uk

Rubén Cordero
Escuela Superior de Física
y Matemáticas del I.P.N.,
Unidad Adolfo López Mateos,
Edificio 9, 07738 México, D.F.
cordero@fis.cinvestav.mx

Michael Doran
Dartmouth College,
Dept. of Physics and Astronomy,
6127 Wilder Laboratory,
Hanover, NH 03755, USA.
Michael.Doran@dartmouth.edu

Alexei V. Filippenko
University of California,
Department of Astronomy,
Berkeley, CA,
94720-3411, USA.
alex@astro.berkeley.edu

Gabriel Germán
Universidad Nacional
Autónoma de México,
Centro de Ciencias Físicas,
Apartado Postal 48-3,
62251 Cuernavaca, Morelos, México.
gabriel@fis.unam.mx

Francisco S. Guzmán
Max Planck Institut
für Gravitationsphysik,
Albert Einstein Institut,
Am Müehlenberg 1, 14476 Golm,
Germany.
guzman@aei-potsdam.mpg.de

Xavier Hernández
Universidad Nacional
Autónoma de México,
Instituto de Astronomía,
A.P. 70264, 04510 México D.F.
xavier@astroscu.unam.mx

Joerg Jaeckel
Universität Heidelberg
Institut für Theoretische Physik,
Philosophenweg 16,
D-69120 Heidelberg, Germany.
jaeckel@thphys.uni-heidelberg.de

James E. Lidsey
Queen Mary, University of London,
Astronomy Unit,
School of Mathematical Sciences,
Mile End Road, London, E1 4NS,
UK.
J.E.Lidsey@qmul.ac.uk

Axel de la Macorra
Universidad Nacional
Autónoma de México,
Instituto de Física,
Apartado Postal 20-364,
01000 México D.F.
macorra@fisica.unam.mx

Kei-ichi Maeda
Waseda University,
Dept. of Physics,
Shinjuku, Tokyo 169-8555, Japan.
maeda@waseda.jp

Tonatiuh Matos
Centro de Investigación
y de Estudios Avanzados del I.P.N,
A.P. 14-740, 07000 México D.F.
tmatos@fis.cinvestav.mx

Hugo A. Morales–Técotl
Universidad Autónoma
Metropolitana–Iztapalapa,
Departamento de Física,
A.P. 55-534, 09340 México D.F.
hugo@xanum.uam.mx

Darío Núñez
Universidad Nacional
Autónoma de México,
Instituto de Ciencias Nucleares,
A.P. 70-543, 04510 México D.F.
nunez@nuclecu.unam.mx

Gabriella Piccinelli
Universidad Nacional
Autónoma de México,
Centro Tecnológico Aragón,
Av. Rancho Seco S/N, Bosques de
Aragón, Nezahualcóyotl,
57130 Estado de México, México.
gabriela@astroscu.unam.mx

Efraín Rojas
Universidad Veracruzana,
Facultad de Física
e Inteligencia Artificial,
Sebastián Camacho 5,
Xalapa, Veracruz, 91000 México.
efrojas@uv.mx

Marcelo Salgado
Universidad Nacional
Autónoma de México,
Instituto de Ciencias Nucleares,
A.P. 70-543, 04510 México D.F.
marcelo@nuclecu.unam.mx

Roberto A. Sussman
Universidad Nacional
Autónoma de México,
Instituto de Ciencias Nucleares,
A.P. 70-543, 04510 México D.F.
sussman@nuclecu.unam.mx

Max Tegmark
University of Pennsylvania,
Department of Physics,
Philadelphia, PA 19104, USA.
max@physics.upenn.edu

César A. Terrero–Escalante
Universidad Nacional
Autónoma de México,
Instituto de Física,
A.P. 20-364, 01000 México D.F.
cterrero@fis.cinvestav.mx

Luis Arturo Ureña–López
Universidad de Guanajuato,
Instituto de Física,
A.P. E-143, 37150 León,
Guanajuato, México.
lurena@fisica.ugto.mx

Introduction

Nora Bretón[1], Jorge L. Cervantes–Cota[2], and Marcelo Salgado[3]

[1] Departamento de Física, Cinvestav-IPN, Apdo. Postal 14-740, 07000 México, D.F. Nora.Breton@fis.cinvestav.mx
[2] Departamento de Física, ININ, Apdo. Postal 18-1027, Col. Escandón, 11801 México D.F. jorge@nuclear.inin.mx
[3] Instituto de Ciencias Nucleares, Universidad Nacional Autónoma de México Apdo. Postal 70-543, 04510 México D.F. marcelo@nuclecu.unam.mx

Abstract. We summarize the topics presented in this book, and when appropriate we enlighten cross–references among the different topics: high precision and observational cosmology, standard Big Bang model, inflation, reheating, baryogenesis, quintessence, strings, braneworlds, and loop quantum cosmology.

The last two decades have been glorious for modern cosmology. So glorious that the experimental advances in this direction have risen cosmology to the status of a *genuine* science: many speculative theoretical issues have found an almost direct verification and also many experiments can be performed now with better precision. Perhaps the experiment that started this new era was the one performed by the COBE satellite team in the early 1990's. This experiment, which was a modern version of that performed by Penzias and Wilson, for the first time revealed that the Universe was almost, but not completely homogeneous and isotropic. The small quantum fluctuations generated in the early Universe were imprinted in the tiny anisotropies that COBE detected in the Cosmic Microwave Background Radiation (CMBR). This and other more recent cosmological probes (CP), like BOOMERANG, MAXIMA, and WMAP, not only confirmed with a great accuracy some of the theoretical predictions of the standard *Big Bang* model (SBB), but also opened the possibility of testing theories and scenarios of the very early Universe, namely, the theory of inflation.

Inflation, in its many versions, tries, one way or the other, to solve the paradigms that emerge when confronting the SBB with current observations; that is, inflation solves the horizon and flatness problems, and provides a causal origin of density fluctuations. In the most standard version, the dominating vacuum energy of a hypothetical fundamental scalar field, the *inflaton*, is responsible for an exponential expansion of the Universe.

During inflation, the matter–energy content within a Hubble radius homogenize and isotropize, and hence, a flat Universe is achieved, like the one observed nowadays. This huge expansion rate, which is guaranteed by cosmic no–hair theorems, also made unimportant the shape of the Universe's initial conditions. The effective potentials (cosmological constants), which permit

N. Bretón, J.L. Cervantes–Cota, and M. Salgado, Introduction, Lect. Notes Phys. **646**, 1–4 (2004)
http://www.springerlink.com/

the application of these theorems, have their origin in particle physics's theories, whose modern descedents are the braneworld scenarios. In this way, the isotropy and homogeneity, assumed in the SBB, are within the inflationary scenario validated in a rather quantitative way.

This exponential expansion caused quantum fluctuations of the inflaton field to cross the Hubble radius, and later after inflation they reentered into a Hubble size. In this way, quantum fluctuations were imprinted in fluctuations of matter and radiation to yield the seeds of structure formation. Matter density fluctuations ultimately evolved in the form of galaxies and larger structures, whereas CMBR fluctuations decoupled from matter at the *last scattering surface*, and remain so until now, up to possible reionization processes. These topics, being known for some time, are reviewed with a modern view in the first part of the book. The lecture by J. L. Cervantes–Cota presents the SBB, emphasizing its long–standing problems. Then, a solution to these problems is found in the inflationary theory. In a complementary manner, E. J. Copeland covers some aspects of the Big Bang theory, namely its problems and solutions within inflationary string cosmology scenarios. C. A. Terrero–Escalante complements these lectures by analyzing the general properties of scalar fields in Friedman–Robertson–Walker (FRW) backgrounds to achieve a successful cosmological model, from the inflationary era to the present.

Perhaps one of the key predictions of the standard inflationary scenario is the average value of the total energy density of the Universe $\Omega = 1$. Until some years, prior to BOOMERANG and MAXIMA results, the assumption $\Omega = 1$ in many constructed–theoretical cosmological models was somehow seen as a prejudice of inflation due to the fact that the observations only showed that $\Omega_{obs} \sim 0.3$ at the best. Most of this density was attached to an unknown form of matter, *dark matter*, responsible for the clustering of galaxies and only a small fraction $\Omega_{\rm bar} \sim 0.04$, the primordial nucleosynthesis contribution, due to ordinary visible matter (baryons). The inflationary prediction $\Omega = 1$, if not completely confirmed, has been to a large extent validated by the very recent CP, whose results together with the simplest standard cosmological scenarios become consistent only if $\Omega \sim 1$. Of course this would be compatible with $\Omega_{obs} \sim 0.3$ only if another contribution of energy density is added. The current and simplest scenario appeals to a *cosmological constant*, or *dark energy*, contributing with $\Omega_\Lambda \sim 0.7$.

In addition to those provided by the CP there have been other cosmological observations of primary importance. One of the oldest but no less remarkable ones started at the time of E. Hubble, who was the first in revealing that the Universe was not static but in continuous expansion. Limited by the technology of that time, the observation of relatively close objects showed that galaxies were expanded at a rate H_0 of $50 - 100$ km s^{-1}Mpc^{-1}, known today as the *Hubble constant*. Recent direct observations have narrowed that range, but which is more important for modern cosmology are the recent observations of far SNIa. These observations which allow one to draw a modern Hubble diagram show that the Universe is not only expanding but expanding in an accelerated way, contrary to what would be expected if the Universe were filled only with a matter in the form of pressureless particles. This astonishing result, of an independent nature from the CP observations, when applied to the standard cosmological model is only compatible with $\Omega \sim 1$ if dark energy producing *negative pressure* is considered. Remarkably, the independent CP and the SNIa observations coincide to a large extent with the values of several cosmological parameters.

The CP and SNIa have not only confirmed the inflationary prediction $\Omega \sim 1$, their precision in measuring small scale temperature anisotropies also confirms some of the predictions of the sophisticated theory of cosmological perturbations. In this way the correlation of temperature fluctuations with angular scales can be well understood. The small room that the current CP leave in the space of cosmological parameters of theoretical models open the possibility to differentiate between competing models. The lectures by R. Brandenberger, M. Tegmark and A. Filippenko give a detailed review of the above ideas, and they complete the first part of the book: The Very Early Universe and High Precision Cosmology. R. Brandenberger provides a theoretical view of the current status of the theory of fluctuations in the early Universe around the FRW background. M. Tegmark provides a survey of recent measurements of spacetime, from local to cosmological, covering a factor of 10^{22} orders of magnitude in scale! The lecture of A. Filippenko gives a detailed account of the SNIa projects, their confrontation with the standard cosmological models, and the future expectations for ruling out some of them. These three authors confront the measurements of CP and SNIa with the competing cosmological models.

As it turns out in many branches of physics, the theoretical models together with precision experiments open new paradigms. This has not been the exception in cosmology. As we mentioned, while the insertion of a cosmological constant in the theoretical models allowed one to explain the requirement $\Omega \sim 1$ in a simple fashion, it also raised a paradigm known as the *coincidence problem*, that we can rephrase as follows: *how is it possible that after billions of years of evolution we live in a time of our Universe where the energy density of the different components of matter (which evolve in time) is of the same order of magnitude of the constant energy density attributed to the cosmological constant ?* For instance in the time of the radiation dominated epoch, the cosmological constant contribution was several orders of magnitude smaller than the photon contribution. Nowadays, however, the matter density contribution *almost* coincides with that of dark energy.

This coincidence problem has been intended to be solved in several fashions. Perhaps the most popular models in this direction are the ones that replace the cosmological constant with a scalar field: the *quintessence* or the *k–essence* field. As expected, the new proposals have to be submitted to all possible tests. In this regard, the lectures of M. Tegmark and A. Filippenko also show the constraints on the equation of state w in quintessential models imposed by the CP and SNIa experiments. Furthermore, the lectures of the second part of the book by A. de la Macorra, G. German and J. Jaeckel treat different aspects of quintessence scenarios: initial conditions, various potentials, and quantum corrections, among other topics. Two other contributions complement this part, namely, one on electroweak phase transitions by G. Piccinelli & A. Ayala, and one on a new method to infer which type of neutralinos could make up galactic dark halos by L. G. Cabral–Rosetti et al.

The last part of the book is dedicated to the most recent theoretical developments in the cosmological theory. Apart from the paradigms of dark–energy, dark–matter and their possible correlation, the dimensionality of nature and the quantum version of gravity are perhaps two of the most fundamental questions in physics, which presumably are interconnected with each other. In this regard, the braneworld scenarios offer the possibility that the Universe we observe lives in a brane embedded in a larger dimensional manifold (the bulk). These scenarios, some of them motivated by string theory, can provide alternative explanations to the origin of inflation and quintessence, or combine them in several fashions. The lecture by K. Maeda gives a thorough account of cosmologies in braneworld scenarios, whereas the lecture by J. Lidsey deals with specific issues (inflation, density perturbations, gravity waves) related to the Randall–Sundrum type II braneworld and extensions of it. The lecture by R. Cordero & E. Rojas covers the creation of brane Universes, whereby the probability of nucleation of the brane is computed, and its cosmological consequences are explored. T. Matos et al present a braneworld realization of their dark matter and dark energy models. Finally, the contribution by M. Bojowald and H. A. Morales–Técotl accounts for cosmological realizations of the theory of loop quantum gravity, in which the basic formalism of this new approach is included.

Part I

The Very Early Universe and High Precision Cosmology

An Introduction to Standard Cosmology

Jorge L. Cervantes–Cota

Instituto Nacional de Investigaciones Nucleares (ININ)
Departamento de Física, Apartado Postal 18-1027, Col. Escandón, 11801 México D.F. jorge@nuclear.inin.mx

Abstract. A short introduction to the Standard Big Bang model is provided, presenting its physical model, and emphasizing its long–standing problems such as the horizon, flatness, baryon asymmetry, among others. Next, an introduction to the inflationary cosmology is presented to elucidate a solution to some of the above–mentioned problems. It is shown that the inflationary scenario succeeds in explaining what the standard Big Bang model cannot, passing the tests of the high precision experimental constraints which have been performed since last decade. This contribution should serve as an introduction to the standard ideas and scenarios which will be used in the forthcoming lectures of this book.

1 On the Standard Big Bang Model

We would like to begin our study by reviewing some basic aspects of the the standard hot Big Bang model (SBB), paying attention to what particle physics theories would bring about in the very early Universe. Our primary focus is to present the achievements of the SBB, but also some difficulties or conundrums that cannot be understood without the incorporation of other concepts, such as extensions to both gravity and particle physics theories, which will give rise to an inflationary scenario.

1.1 FRW Models

The SBB is based on Einstein's general relativity (GR) theory, which can be derived from the Einstein–Hilbert Lagrangian:

$$\mathcal{L} = \frac{1}{16\pi G} R\sqrt{-g} \ , \tag{1}$$

where R is the Ricci scalar, G the Newton constant, and $g = |g_{\mu\nu}|$ the determinant of the metric tensor; for our geometric conventions see the table provided in [68] (cover page), here we have used "-" for the metric **g**, "+" for Riemann, and "-" for Einstein.

By performing the metric variation of this equation, one obtains the Einstein's well known field equations

$$R_{\mu\nu} - \frac{1}{2} R g_{\mu\nu} = -8\pi G T_{\mu\nu} \ , \tag{2}$$

J.L. Cervantes–Cota, An Introduction to Standard Cosmology, Lect. Notes Phys. **646**, 7–51 (2004)
http://www.springerlink.com/

where $R_{\mu\nu}$ is the Ricci tensor and $T_{\mu\nu}$ is the energy–momentum stress tensor. The left hand side (l.h.s.) of this equation represents the geometry, whereas the right hand side (r.h.s.) accounts for the fluid(s) present. In GR the space–time is four dimensional (three spatial dimensions plus time), and since both tensors are symmetric, (2) represents a collection of ten coupled, partial differential equations.

Once one is provided with the gravity theory, one should introduce a symmetry through the metric tensor. In cosmology one assumes a simple metric tensor according to the *cosmological principle* which states that the Universe is both homogeneous and isotropic. This turns out to be in very good agreement with the observed very–large–scale structure of the Universe. This homogeneous and isotropic space–time symmetry was originally studied by Friedmann, Robertson, and Walker (FRW); see [38, 81, 96]. The symmetry is encoded in the special form of the following line element:

$$ds^2 = -dt^2 + a^2(t)\left[\frac{dr^2}{1-kr^2} + r^2\left(d\theta^2 + \sin^2\theta\, d\phi^2\right)\right] , \tag{3}$$

where t is the time variable, r–θ–ϕ are polar coordinates, which can be adjusted so that the constant curvature takes the values $k = 0, +1$, or -1 for a flat, closed, or open space, respectively. $a(t)$ is the scale factor of the Universe.

The FRW solutions to the Einstein equations (2) represent a cornerstone in the development of modern cosmology, since with them it is possible to understand the expansion of Universe, as was realized in the late 20s through Hubble's law of expansion [53]. With this metric, the GR cosmological field equations are,

$$H^2 \equiv \left(\frac{\dot{a}}{a}\right)^2 = \frac{8\pi G}{3}\rho - \frac{k}{a^2} \tag{4}$$

and

$$\frac{\ddot{a}}{a} = -\frac{4\pi G}{3}(\rho + 3p) , \tag{5}$$

where H is the Hubble parameter; ρ and p are the density and pressure of the perfect fluid considered; that is, $T_{\mu\nu} = \rho u_\mu u_\nu + p(u_\mu u_\nu - g_{\mu\nu})$, where $u_\mu = \delta^0_\mu$ is the four–velocity of the fluid in co–moving coordinates, i.e. in coordinates that are moving with the expansion. Equations (4–5) can also be deduced within Newtonian cosmology, but there the pressure is not a source of gravitation; see the contribution of E. Copeland in this book.

The energy–momentum tensor conservation, ${T_\mu}^\nu{}_{;\nu} = 0$, is valid and from it one obtains that

$$\dot{\rho} + 3H(\rho + p) = 0 . \tag{6}$$

If one assumes further a barotropic equation of state for the fluid, $\omega =$ const.,

$$\frac{p}{\rho} = \omega = \begin{cases} \frac{1}{3} & \text{for} \quad \text{radiation and/or ultra} - \text{relativistic matter} \\ 0 & \text{for} \quad \text{dust} \\ 1 & \text{for} \quad \text{stiff fluid} \\ -1 & \text{for} \quad \text{vacuum energy} \end{cases} \tag{7}$$

to integrate (6), it yields

$$\rho = \frac{M_\omega}{a^{3(1+\omega)}} \ , \tag{8}$$

where M_ω is the integration constant and is differently dimensioned by considering different ω–fluids. Equations (4), (5), and (6) are not linearly independent, only two of them are. That is, one can derive, e.g., (5) from (4) and (6). Note that these equations are time symmetric, the interchange $t \to -t$ leaves the equations the same.

Let us very briefly recall which ω–values are needed to describe the different epochs of the Universe's evolution. At very early times, the Universe is believed to have experienced a huge expansion due to some cosmological constant ($\Lambda = 8\pi G\rho$, where $\rho =$ const.) or vacuum energy. This epoch, to be fully described later on, is roughly characterized by an equation of state $\omega = -1$. After inflation, the vacuum energy decays in some particle content, a process called *reheating* [3, 33, 1], after which the Universe is filled with a "fluid" of radiation or of ultra–relativistic matter where the material content of the Universe consisted of photons, neutrinos, electrons, and other massive particles with very high kinetic energy. During this epoch the assumption $\omega = 1/3$ is valid. After some Universe cooling, some massive particles decayed and others survived (protons, neutrons, electrons) and their masses eventually surpassed the radiation components (photons, neutrinos). From that epoch until very recent times, the matter content dominated and effectively produced no pressure on the expansion and, therefore, one accepts a model filled with dust, i.e. $\omega = 0$. Until the mid 90s we thought that a dust model would be representative for the current energy content of the Universe. Recent measurements (see contribution of A. Filippenko in this book), however, indicate that as of recently the Universe is again experiencing a huge expansion rate. It is believed that a kind of cosmological constant, or vacuum energy, is the largest energy contribution to the expansion of the Universe at present. Thus the cosmological constant is the generic factor of an inflationary solution, see the $k = 0$ solution below, (10), which is believed to be characteristic of both the very early inflationary epoch and today.

Finally, a stiff model, $\omega = 1$, is sometimes considered in order to describe very dense matter under very high pressures.

The ordinary differential equations system described above needs a set of either initial conditions or boundary conditions to be integrated. One can assume a set of two initial values, say, $(\rho(t_*), \dot{a}(t_*)) \equiv (\rho_*, \dot{a}_*)$ at some (initial) time t_*, in order to determine its evolution. Its full analysis has been reviewed by many authors [97, 68]. Here, in order to show some early Universe consequences we take $k = 0$, justified as follows: From (4) and (8) one notes that the expansion rate, given by the Hubble parameter, is dominated by the density term as $a(t) \to 0$, since $\rho \sim 1/a^{3(1+\omega)} > k/a^2$ for $\omega > -1/3$; that is, the flat solution is very well fitted at the very beginning of time. Furthermore, recent Cosmic Microwave Background Radiation (CMBR) [1] measurements [26, 6, 48, 13] are consistent with $k \sim 0$. Therefore, assuming $k = 0$, (4) implies

$$a(t) = [6\pi G M_\omega (1+\omega)^2]^{\frac{1}{3(1+\omega)}} (t - t_*)^{\frac{2}{3(1+\omega)}}$$

$$= \begin{cases} (\frac{32}{3}\pi G M_{\frac{1}{3}})^{1/4} \ (t-t_*)^{1/2} & \text{for} \quad \omega = \frac{1}{3} \text{ radiation} \\ (6\pi G M_0)^{1/3} \quad (t-t_*)^{2/3} & \text{for} \quad \omega = 0 \text{ dust} \\ (24\pi G M_1)^{1/6} \ (t-t_*)^{1/3} & \text{for} \quad \omega = 1 \text{ stiff fluid} \end{cases} \tag{9}$$

and

$$a(t) = a_* e^{Ht} \qquad \text{for} \qquad \omega = -1 \text{ vacuum energy} \tag{10}$$

where the letters with a subindex "$*$" are integration constants, representing quantities evaluated at the beginning of times, $t = t_*$.

From (9) one can immediately see that at $t = t_*$, $a_* = 0$ and from (8), $\rho_* = \infty$; that is, the solution has a singularity at that time, presumably at the Universe's beginning; this initial cosmological singularity is also called *Big Bang* singularity. As the Universe expands the Hubble parameter evolves as $H \sim 1/t$, i.e. the expansion rate decreases; whereas the matter–energy content acts as an expanding agent, cf. (4), it also decelerates the expansion in an asymptotically decreasing manner, cf. (5) and (8). In that way, H^{-1} represents an upper limit to the age of the Universe; for instance, $H^{-1} = 2t$ for $\omega = 1/3$ and $H^{-1} = 3t/2$ for $\omega = 0$, t being the Universe's age.

The solution (10) is inflationary and has no singularity. This solution is such that the Hubble parameter is indeed a constant. A fundamental ingredient of inflation is that the r.h.s. of (5) remains positive, $\ddot{a} > 0$. This is performed when the *inflation pressure* is negative [18], $\rho + 3p < 0$. In this way, one does not have necessarily to impose the strong condition $\omega = -1$, but it suffices that $\omega < -1/3$, in order to have a moderate inflationary solution; for example, $\omega = -2/3$ it implies $a = a_* t^2$, a mild power-law inflation. The issue of inflation will be discussed in Sect. 2.

1.2 The Physical Scenario

So far we have obtained some exact solutions for Einstein's cosmology. Now, to achieve a more physical scenario one considers the Universe filled with a plasma of particles and their antiparticles. This was originally done by G. Gamow [40], who first considered a hot Big Bang model for the Universe's

[1] The CMBR is also sometimes referred to in this book as Cosmic Microwave Background (CMB) or Cosmic Background Radiation(CBR).

beginning, which was later qualitatively confirmed by Penzias and Wilson [73] and interpreted by Dicke et. al. [29]. Furthermore, with the development of modern particle physics theories in the 70s it was unavoidable to think about a physical scenario for the early Universe which should include even the "new" physics. It was also realized that the physics described by GR should not be applied beyond Planck (Pl) initial conditions, because there the quantum corrections to the metric tensor become very important, a theory which is still in progress. Thus, we make some assumption at some early time, $t \gtrsim t_{Pl}$: the Universe was filled with a plasma of relativistic particles, including quarks, leptons, and gauge and Higgs bosons, all in thermal equilibrium at a very high temperature, T, with some gauge symmetry dictated by a particle physics theory.

Now, in order to work in that direction one introduces some thermodynamic considerations necessary for the description of the physical content of the Universe, which we would like to present here. Assuming an ideal-gas approximation, the number density n_i of the particles of type i, with a momentum q, is given by a Fermi or Bose distribution [60]:

$$n_i = \frac{g_i}{2\pi^2} \int \frac{q^2 dq}{e^{(E_i - \mu_i)/T} \pm 1} , \tag{11}$$

where $E_i = \sqrt{m_i^2 + q^2}$ is the particle energy, μ_i is the chemical potential, the sign $(+)$ applies for fermions and $(-)$ for bosons, and g_i is the number of spin states. One has that $g_i = 2$ for photons, quarks, baryons, electrons, muons, taus, and their antiparticles, but $g_i = 1$ for neutrinos because they are only left-handed. For the particles existing in the early Universe one usually assumes that $\mu_i = 0$: one expects that in any particle reaction the μ_i are conserved, just as the charge, energy, spin, and lepton and baryon number are conserved as well. The number density of photons (n_γ), which can be created and/or annihilated after some particle collisions, must not be conserved and its distribution with $\mu_\gamma = 0$, $E = q = h\nu$, reduces to the Planck one. For other constituents, in order to determine the μ_i, one needs n_i; one notes from (11) that for large $\mu_i > 0$, n_i is large too. One does not know n_i, but from nucleosynthesis that [72]

$$\eta \equiv \frac{n_B}{n_\gamma} \equiv \frac{n_{\text{baryons}} - n_{\text{anti-baryons}}}{n_\gamma} \approx (3-4) \times 10^{-10} . \tag{12}$$

The smallness of the baryon number density, n_B, relative to the photon's, suggests that n_{leptons} may also be small compared with n_γ. Therefore, one takes for granted that $\mu_i = 0$ for all particles. Why the ratio n_B/n_γ is so small, but not zero, is one of the puzzles of the SBB. This ratio is also often called $\eta \equiv n_B/n_\gamma$.

The above approximation allows one to treat the density and pressure of all particles as a function of the temperature only. According to the second law of thermodynamics, one has [97]

$$dS(V,T) = \frac{1}{T}[d(\rho V) + pdV] \ , \tag{13}$$

where S is the entropy in a volume $V \sim a^3(t)$ with $\rho = \rho(T), \ p = p(T)$ in equilibrium. Furthermore, the integrability condition $\frac{\partial^2 S}{\partial T \partial V} = \frac{\partial^2 S}{\partial V \partial T}$ is also valid, which turns out to be

$$\frac{dp}{dT} = \frac{\rho + p}{T} \ . \tag{14}$$

Additionally, the energy conservation law equation (6) leads to

$$a^3(t)\frac{dp}{dt} = \frac{d}{dt}[a^3(t)(\rho + p)] \tag{15}$$

and using (14), the latter takes the form

$$\frac{d}{dt}[\frac{a^3(t)}{T}(\rho + p)] = 0 \ . \tag{16}$$

Using (14), (13) can be written as

$$dS(V,T) = \frac{1}{T}d[(\rho + p)V] - \frac{V}{T^2}(\rho + p)dT \ . \tag{17}$$

Then, (16) together with (17) imply that the entropy

$$S = \frac{a^3}{T}[\rho + p] = \text{const.} \tag{18}$$

is a constant of motion.

The density and pressure are given by

$$\rho \equiv \int E_i n_i dq \quad \text{and} \quad p \equiv \int \frac{q^2}{3E_i} n_i dq \ . \tag{19}$$

For photons or ultra–relativistic fluids, $E = q$, these equations become such that

$$p = \frac{1}{3}\rho \ , \tag{20}$$

confirming (7), and after integrating (14), it results that

$$\rho = bT^4 \tag{21}$$

with the constant of integration, b. In a real scenario there are many relativistic particles present, each of which contributes as in (21). Summing up all of them, $\rho = \sum_i \rho_i$ and $p = \sum_i p_i$ over all relativistic species, it results that $b(T) = \frac{\pi^2}{30}(N_B + \frac{7}{8}N_F)$, which depends on the number of effective relativistic degrees of freedom of bosons (N_B) and fermions (N_F). Therefore, this quantity varies with the temperature; different i–species remain relativistic until some characteristic temperature $T \approx m_i$, after which the value N_{F_i} (or N_{B_i}) no longer contributes to $b(T)$. The factor $7/8$ accounts for the different statistics the particles have, see (11). In the standard model of particles physics $b \approx 1$ for $T \ll 1$ MeV and $b \approx 35$ for $T > 300$ GeV. Additionally, for relativistic particles one obtains from (11) that

$$n = cT^3, \quad \text{with } c = \frac{\zeta(3)}{\pi^2}(N_B + \frac{3}{4}N_F) \ . \tag{22}$$

where $\zeta(3) \approx 1.2$ is the Riemann zeta function of 3. Currently, $n_\gamma \approx \frac{422}{\text{cm}^3} T_{2.75}^3$, where $T_{2.75} \equiv \frac{T_{\gamma 0}}{2.75°K}$; the subscript "0" refers to quantities evaluated at present time.

From (18), using (20) and (21), one concludes that $T \sim 1/a(t)$ and from the $\omega = 1/3$ solution in (9) one arrives at the result

$$T = \sqrt[4]{\frac{M_{\frac{1}{3}}}{b}}\frac{1}{a(t)} = \sqrt[4]{\frac{3}{32\pi G b}}\frac{1}{(t-t_*)^{\frac{1}{2}}} \ , \tag{23}$$

a decreasing temperature behavior as the Universe expands. Therefore, initially at the Big Bang $t = t_*$ implies $T_* = \infty$, the Universe was very hot.

The entropy for an effective relativistic fluid is given by (18) together with (20) and (21):

$$S = \frac{4}{3}\ b\ (a\ T)^3 = \text{const.} \tag{24}$$

Combining this with (23), one can compute the value of $M_{\frac{1}{3}}$ to be $M_{\frac{1}{3}} = (\frac{3}{4}S)^{4/3}/b^{1/3} \approx 10^{116}$, since $b \approx 35$ and the photon entropy $S_0 = \frac{4}{3}\ b\ (a_0\ T_0)^3 \approx 10^{88}$ for the currently evaluated quantities $a_0 = d_H(t_0) = 10^{28}$cm and $T_{\gamma 0} = 2.7$ °K. For later convenience, we define the entropy per unit volume, *entropy density*, to be $s \equiv S/V = \frac{4}{3}\frac{\pi^2}{30}(N_B + \frac{7}{8}N_F)T^3$; thus, currently $s \approx 7n_\gamma$. The nucleosynthesis bound on η, (12), implies that $n_B/s \approx (4-6) \times 10^{-11}$.

Now we consider particles in their non–relativistic limit ($m \gg T$). From (11) one obtains for both bosons and fermions that

$$n = g\left(\frac{mT}{2\pi}\right)^{3/2} e^{-m/T} \ . \tag{25}$$

The abundance of equilibrium massive particles decreases exponentially once they become non–relativistic; this situation is referred as *in equilibrium annihilation.* Their density and pressure are given through (19) and (25) by

$$\begin{aligned} \rho &= nm \\ p &= nT \ll \rho \ . \end{aligned} \tag{26}$$

Therefore, the entropy given by (18) for non–relativistic particles through (25) and (26) also diminishes exponentially during their in equilibrium annihilation. The entropy of these particles is transferred to that of relativistic components by augmenting their temperature. Hence, the constant total entropy is essentially the same as that given by (24), but the i–species contributing to it are just those which are in equilibrium and maintain their relativistic behavior, that is, particles without mass such as photons.

Having introduced the abundances of the different particle types, we would like to comment on the equilibrium conditions for the constituents of the Universe as it evolves. This is especially of importance in order to have an idea whether or not a given i–species disappears or decouples from the primordial brew. To see this, let us consider n_i when the Universe's temperature, T, is such that (a) $T \gg m_i$, during the ultra–relativistic stage of some particles of type i and (b) $T \ll m_i$, when the particles i are non–relativistic; both cases originally in thermal equilibrium. From (22) one obtains for case (a) that $n_i \sim T^3$; the total number of particles, $\sim n_i a^3$, remains constant. Whereas for case (b), from (25), $n_i \sim T^{3/2} e^{-m_i/T}$, i.e. when the Universe temperature goes down below m_i, the number density of the i–species significantly diminishes; an "in equilibrium annihilation" occurs. Let us take as an example the neutron–proton annihilation: one then has

$$\frac{n_n}{n_p} \sim e^{\frac{m_p - m_n}{T}} = e^{-\frac{1.5\times 10^{10}\ ^\circ K}{T}} \tag{27}$$

which drops with the temperature, from nearly 1 at $T \geq 10^{12}\ ^\circ K$ to about 5/6 at $T \approx 10^{11}\ ^\circ K$, and 3/5 at $T \approx 3 \times 10^{10}\ ^\circ K$ [70]. If this is forever valid, one ends up without massive particles, meaning that our Universe should have consisted only of radiative components; our own existence contradicts that! Therefore, the in–equilibrium annihilation eventually stopped. The quest is now to freeze out this ratio (to be $n_n/n_p \approx 1/6$) [2] in order to leave some hadrons for posteriorly achieving successful nucleosynthesis. The answer comes by comparing the Universe expansion rate, H, with particle physics reaction rates, Γ. Hence, for $H < \Gamma$, the particles interact with each other faster than the Universe expansion rate and then equilibrium is established. For $H > \Gamma$ the particles cease to interact effectively and then thermal equilibrium drops out [3]. In this way, the more interacting the particles are, the longer they remain in equilibrium annihilation and, therefore, the lower their number densities are after some time; e.g., baryons vanish first, then charged leptons,

[2] Due to neutron decays, until the time when nucleosynthesis begins, n_n/n_p reduces to 1/7.

[3] This is only approximately true; a proper account of this involves a Boltzmann equation analysis. In doing so a numerical integration should be carried out in which annihilation rates are balanced with inverse processes; see for example [90, 60].

neutral leptons, etc. Finally, the particle numbers of (massless) photons and neutrinos remain constant, as it was mentioned above; see Fig. 1. Note that if interactions of an i−species freeze out when it is still relativistic, then its abundance can be significant at present.

It is worth mentioning that if the Universe were to expand faster, then the temperature of decoupling at $H \sim \Gamma$ would be higher, thus the fixed ratio n_n/n_p would be greater, thus leading to profound implications in the nucleosynthesis of the light elements. For instance the Helium, ^{4}He, abundance should be higher. Therefore, the expansion of the Universe cannot arbitrarily be augmented during the equilibrium era of some particles. Furthermore, if a particle species is still highly relativistic ($T \gg m_i$) or highly non–relativistic ($T \ll m_i$) when decoupling from primordial plasma occurs, it maintains an equilibrium distribution; the former characterized by $T_r a$ =const. and the latter by $T_m a^2$ =const., cf. (30).

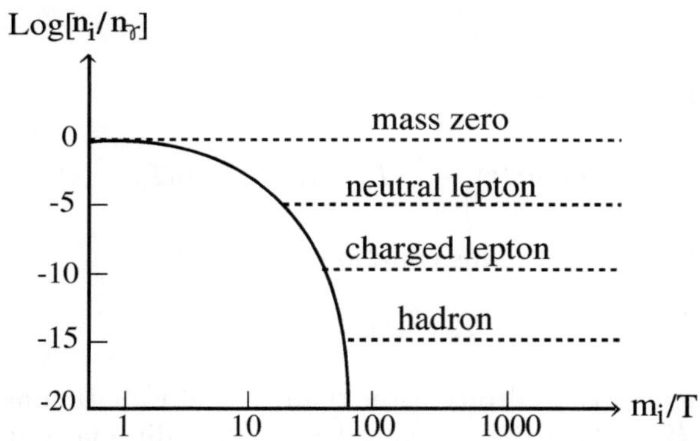

Fig. 1. The evolution of the particle density of different i−species. If an i−species is in equilibrium its abundance diminishes exponentially after the particle becomes non–relativistic (solid line). However, interactions of an i−species can freeze out, causing the particle species to decouple from equilibrium and maintain its abundance (dashed line). (Figure adapted from Kolb and Turner 1990).

There are also some other examples of decoupling, like the neutrino decoupling: during nucleosynthesis there exist reactions like $\nu\bar{\nu} \longleftrightarrow e^+e^-$, which maintain neutrinos efficiently coupled to the original plasma ($\Gamma > H$) until about 1 MeV, since $\frac{\Gamma}{H} \approx \left(\frac{T}{\mathrm{MeV}}\right)^3$. Below 1 MeV reactions are no longer efficient and neutrinos decouple and continue evolving with a temperature $T_\nu \sim 1/a$. Then, at $T \gtrsim m_e = 0.51$MeV the particles in equilibrium are photons (with $N_B = 2$) and electron and positron pairs (with $N_F = 4$) which contribute to the entropy with $b(T) = \frac{\pi^2}{30} \cdot (11/2)$. Later, when the temperature drops to $T \ll m_e$, the reactions are no longer efficient ($\Gamma < H$) and after the $e^\pm$ pair annihilation there are only photons in equilibrium with

$b(T) = \frac{\pi^2}{30} \cdot (2)$. Since the total entropy, $S = \frac{4}{3} b(aT)^3$, must be conserved, the decrease in $b(T)$ must be balanced with an increase in the radiation temperature; this gives a result of $\frac{T_\gamma}{T_\nu} = \left(\frac{11}{4}\right)^{1/3}$, which should remain to the present day, implying the existence of a cosmic background of neutrinos with a temperature today of $T_{\nu_0} = 1.96$ °K.

Another example of this is the gravitation decoupling, which should also be present if gravitons were in thermal equilibrium at the Planck time and then decoupled. The present–day background of temperature should be characterized at most by $T_{\text{grav.}} = \left(\frac{4}{107}\right)^{1/3} \approx 0.91$ °K.

For the matter dominated era we have stressed that effectively $p = 0$; next we will see the reason for this: First consider an ideal gas (like atomic Hydrogen) with mass m, then $\rho = nm + \frac{3}{2} n T_m$ and $p = n T_m$. From (15) one obtains, equivalently, that

$$\frac{d}{da}(\rho a^3(t)) = -3 p a^2(t) \tag{28}$$

and substituting the above ρ and p, one obtains

$$\frac{d}{da}(nma^3(t) + \frac{3}{2} n T_m a^3(t)) = -3 n T_m a^2(t) \tag{29}$$

where $nma^3(t)$ is a const. This equation yields

$$T_m a^2(t) = \text{const.} \ , \tag{30}$$

that matter temperature drops faster than that of radiation as the Universe expands; see (23). Now, if one considers both radiation and matter, it is valid that $\rho = nm + \frac{3}{2} n T_m + b T_r^4$ and $p = n T_m + \frac{1}{3} b T_r^4$; the source of the Universe's expansion is proportional to $\rho + p = nm + \frac{5}{2} n T_m + \frac{4}{3} b T_r^4$; the first term dominates the second, precisely because T_m decreases very rapidly. The third term diminishes as $\sim 1/a^4$, whereas the first as $\sim 1/a^3$, and after the time of densities equality (eq.), $\rho_m = \rho_r$, the matter density term is greater than the others, which is why one assumes no pressure for that era.

From now on, when we refer to the temperature, T, it should be related to the radiation temperature.

The detailed description of the thermal evolution of the Universe for the different particle types, depending on their masses, cross-sections, etc., is well described in many textbooks, going from the physics known in the early 70s [97] to the late 80s [60], or late 90s [62], and therefore it will not be presented here. However, we notice that as the Universe cools down, a series of spontaneous symmetry–breaking phase transitions are expected to occur. The type and/or nature of these transitions depend on the specific particle physics theory considered. Among the most popular are Grand Unification theories (GUT) which bring together all known interactions except for gravity. One could also be more modest and just consider the standard model of particle physics or some extensions of it. Ultimately, one should decide, in constructing a cosmological theory, according to which energy scale one wants to use to describe physics. For instance, at a temperature between 10^{14} GeV to 10^{16} GeV the transition of the $SU(5)$ GUT should take place -if this theory is valid- in which a Higgs field breaks this symmetry to $SU(3)_C \times SU(2)_W \times U(1)_{HC}$, a process through which some bosons acquire their masses. Due to the gauge symmetry, there are color (C), weak (W) and hypercharge (HC) conservation, as the subindexes indicate. Later on, when the Universe evolves to about a few hundred GeV, the electroweak phase transition takes place, in which a second Higgs field breaks the symmetry $SU(3)_C \times SU(2)_W \times U(1)_{HC}$ to $SU(3)_C \times U(1)_{EM}$; through this second breaking the fermions also acquire their masses. At this stage, there are only color and electromagnetic (EM) charge conservation, due to the gauge symmetry. Afterwards, at a temperature of about 100 to 300 MeV the Universe should undergo a transition associated to the chiral symmetry–breaking and color confinement, from which baryons and mesons are formed out of quarks. Subsequently, at approximately 10 MeV the synthesis of light elements (nucleosynthesis) begins, producing most of the observed Hydrogen and Helium observed in the present day, along with abundances of some other light elements. The nucleosynthesis represents the earliest scenario tested in the SBB. After some time, matter dominates, over radiation components, in the Universe, and the large scale structure (galaxies, clusters, superclusters, voids, etc.) begins to form. At about 1 eV the *recombination* takes place; that is, the Hydrogen ions and electrons combine to compose neutral Hydrogen atoms, then matter and EM radiation decouple from each other. At this moment the surface of last scattering (ls) of the CMBR evolves as an imprint of the Universe at that time. In Fig. 2 the main events of the SSB are sketched.

Let us go back to the FRW cosmological equations. In observing the two terms involved in (4), the matter term $8\pi G\rho/3$ and the curvature term k/a^2, one should be aware of the validity of the approximation $8\pi G\rho/3 > k/a^2$. Let us for the moment elucidate that k is tiny but different from zero. Then, eventually when the energy density has diminished enough due to the expansion, $8\pi G\rho/3 \sim k/a^2$, and further on the Universe will be dominated by its curvature. Let us consider this case, but for both $k = \pm 1$ separately. First, take $k = -1$, then $H = 1/a$ and the solution is $a \sim t$, that is, the Universe expands forever. Otherwise, for $k = +1$, at the moment of maximum expansion, say[4] $\tau_c/2$, $8\pi G\rho/3 = k/a^2$, the Universe stops its expansion and then the scale factor begins to decrease. The solution given by the negative square root of (4) again ends with a singularity but now at $t = \tau_c > t_*$, where

[4] τ_c stands for τ_{collapse}. The lifetime of such an Universe, the time of a cycle, is just twice the time of maximal expansion, because the solution is time symmetric.

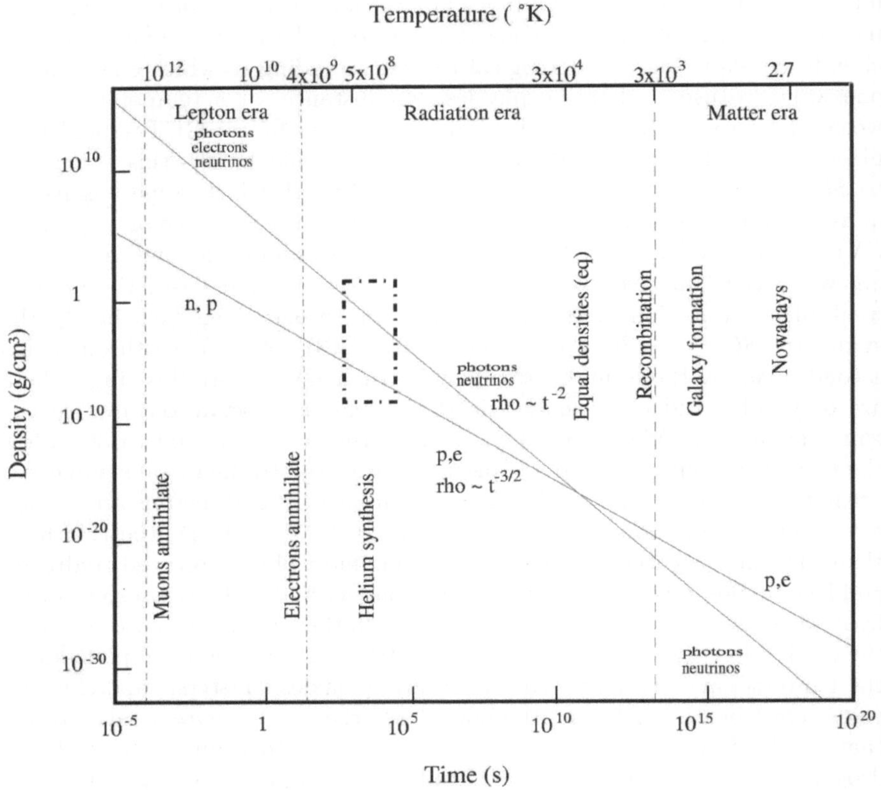

Fig. 2. The thermal history of the SBB (Figure adapted from Harrison 1970).

$\rho = \infty$, $T = \infty$, and $a = 0$; this is the so-called Big Crunch. For instance, the lifetime τ_c for a model filled with cold dark matter is

$$\tau_c \approx MG \approx \frac{M}{M_{Pl}} 10^{-43} \ \text{s} \ ; \tag{31}$$

that is, if our Universe were dark–matter dominated and with a closed curvature, then it must presently have (10^{17} s) a mass $M > 10^{60} M_{Pl} \approx 10^{55}$gr.

For a radiation model, in terms of its entropy, from (18) and (23), one obtains that

$$\tau_c \approx \frac{S^{2/3}}{M_{Pl}} \approx S^{2/3} 10^{-43} \ \text{s} \ , \tag{32}$$

in this case $S > 10^{90}$, a huge entropy! In trying to understand such big numbers one is forced to recognize some problems with the SBB. Next, we present some of them.

1.3 Problems of the Standard Big Bang Model

In considering a theory of the Universe one is open to think about an Universe's "arena" as general as possible. In doing so one finds a large list of problems to be understood. However, not all of them are of the same nature. For instance, some problems arise as a result of computations, others by implementing a physical scenario for GUT or from the conception itself of how the Universe should have begun; that is, apart from the choice of initial conditions, $(\rho_*, \dot{a}_*)$, are the number of dimensions or the global topology. In this section we list some of these problems, emphasizing those for which the inflationary Universe offers an explanation:

Dimensionality

Why should the Universe have four space–time dimensions, at least locally in our surroundings? A first attempt to consider theories in more dimensions was carried out by Kaluza and Klein [54, 56], who tried, unsuccessfully, to unify gravity with the electromagnetic interaction. However, from that we learned to use more than four dimensions for unifying meanings.

Other theories such as fundamental strings are conceivable in $D-$dimensions, but by demanding Lorentz invariance of the quantized bosonic string theory one has to choose $D = 26$, or in fermionic strings $D = 10$. Yet, there arises the problem of compactifying the $D-4$ dimensions, to a compact space whose size is of the order of the Planck length (l_{Pl}). There is no unique procedure. The compactification can be achieved in a number of ways, many of them casting different particle content in their low energy effective Lagrangians. In addition, there exists no compelling principle which would determine the space–time dimension to be four. All dimensions below D seem to be on an equal footing [66].

Euclidicity

What is the global geometry of the Universe? The space geometry is almost perfectly Euclidean on large scales, but on very small scales -say, slightly smaller than the Planckian- GR is not any more tractable, as quantum fluctuations of the metric make it impossible to extend a classical formalism. Within GR one understands a large–scale euclidicity, but not at the very small scale, even though the only natural length in GR is $l_{Pl} = \sqrt{G}$. Why this? Naturally, it is tempting to go beyond GR, a theory which is not yet completed.

Singularity

As we have already mentioned, at $t = t_*$ the scale factor is $a = 0$, the density $\rho = \infty$ and $T = \infty$, see (9), (8) and (23). It can also be shown that the curvature tensor $R^{\mu}{}_{\nu\gamma\delta} = \infty$ at that time. There exists no such a theory to explain gravity as $a(t)$ approaches zero. In fact, one expects GR to be valid as far as $a(t) \rightarrow l_{Pl}$; in going beyond this limit the problems mentioned in the Euclidicity item appear.

Homogeneity and Isotropy

The large–scale structure of the Universe seems to be very homogeneous and isotropic. However, looking on small scales, the isotropy and homogeneity break down: there exist planets, stars, compact objects, galaxies, clusters...a large–scale structure. Hence, it is tempting to consider more general inhomogeneous and anisotropic models, which should explain, as a consequence of their evolution, the currently observed large–scale structure along with the isotropy limits observed in the CMBR, in x–ray backgrounds (e.g. quasars at high redshift), and in number counts in faint radio sources.

In GR, without the aid of a cosmological constant or inflation, Collins and Hawking [25] examined the question in terms of an "initial conditions" analysis. They obtained that the set of spatially homogeneous cosmological models approaching isotropy in the limit of infinite times is of measure zero in the space of all spatially homogeneous models. This in turn implies that the isotropy of the models is unstable to homogeneous and anisotropic perturbations. However, their definition of isotropization demands asymptotic stability of the isotropic solution. An asymptotic stability analysis of Bianchi models in GR [10] shows, e.g., that in the Bianchi type VII$_h$ the anisotropy will not exactly vanish but can be bounded. In this sense, the open FRW model may be stable. Attempts to understand this question in other gravity theories, such as Brans–Dicke theory, shed some light on the solution [20].

Horizon

The region of space which can be connected to some other region by causal physical processes, at most through the propagation of light with $ds^2 = 0$, defines the *causal* or *particle horizon*, d_H. For the FRW equation (3), in spherical coordinates with $\theta, \phi =$const. and after redefining r, this means that [82, 97]:

$$\int_0^t \frac{dt}{a(t)} = \int_0^{r_H} \frac{dr}{\sqrt{1-kr^2}}$$

$$d_H(t) \equiv a(t) \int_0^{r_H} \frac{dr}{\sqrt{1-kr^2}} \quad . \tag{33}$$

In order to analyze the whole horizon evolution, from the present (t_0) to the Planck time (t_{Pl}), we first compute the horizon for the matter dominated era $t_{\rm eq.} \leq t \leq t_0$ and secondly for the radiation era $t \leq t_{\rm eq.}$, because they are differently determined by (9), where we set $t_* = 0$ for convenience. For the matter epoch one has $a(t) = a_0(t/t_0)^{2/3}$, then the first equation above gives $r_H = \frac{3}{a_0}(t_0^2\, t)^{1/3}$; from the second equation one obtains the horizon $d_H(t) = 3t = 2H^{-1}$. For the radiation period, one finds that $r_H = \frac{2}{a_{\rm eq.}}(t_{\rm eq.}\, t)^{1/2}$ and $d_H(t) = 2t = H^{-1}$. We see for the matter dominated era that the causal horizon is twice the Hubble distance, H^{-1}, and that they are equal to each other during the radiation dominated era; therefore, one uses them interchangeably. It is clearly seen for both eras that as $t \to 0$, the Universe is causally disconnected, being $a(t) > d_H(t)$.

The evolution of a typical co–moving distance scale, L, due to the Universe expansion is given by $L(t) = L_0 \frac{a(t)}{a_0}$. Next, let us compare the past evolution of that scale with the corresponding traced by the horizon, $d_H(t) = d_{H_0}(t/t_0)$, where $d_{H_0} = 3t_0$, for the matter dominated era. Then, one finds that

$$\frac{d_H}{L} = \frac{d_{H_0}}{L_0}\left(\frac{t}{t_0}\right)^{1/3} \qquad \text{for} \qquad t_{\rm eq.} \le t \le t_0 \; . \tag{34}$$

Now consider the typical scale, L_0, to be the present observed particle horizon, $L_0 = d_{H_0}$. Then, the amount by which the three dimensional horizon was smaller than the "volume" $L^3(t)$ is determined by the following relation:

$$\left(\frac{d_H}{L}\right)^3 = \frac{t}{t_0} = \left(\frac{T_0}{T}\right)^{3/2} \qquad \text{for} \qquad t_{\rm eq.} \le t \le t_0 \; , \tag{35}$$

in which we have made use of (9) and (24). At the time when the CMBR was last scattered (ls) one has then that $\left(\frac{d_H}{L}\right)^3_{\rm ls} \approx 10^{-5}$; that is, there were approximately one hundred thousand small horizon regions without causal connection! But, on the other hand, by that time the CMBR was already highly isotropic. Thus, one has to take for granted that the initial conditions for all the 10^5 volume horizons were fine tuned so as to account for the present observed large angle CMBR levels of isotropy, with $\delta T/T \sim 10^{-5}$. This is the horizon problem.

One can go further and compute the number of disconnected regions up to the Planck epoch. But first, one needs to evaluate the ratio d_H/L when the radiation and matter densities equal (eq.) each other; this is $\left(\frac{d_H}{L}\right)^3_{\rm eq.} \approx 10^{-6}$. Up until this time, one has to use the radiation solution given by

$$\frac{d_H}{L} = \frac{d_{H_{\rm eq.}}}{L_{\rm eq.}}\left(\frac{t}{t_{\rm eq.}}\right)^{1/2} \qquad \text{for} \qquad t \le t_{\rm eq.} \; , \tag{36}$$

again taking as the typical scale that of the horizon at that time, which is given by $\left(\frac{d_H}{L}\right)^3_{\rm eq.} \approx 10^{-6}$. Then, one finds that

$$\left(\frac{d_H}{L}\right)^3 = 10^{-6}\left(\frac{t}{t_{\rm eq.}}\right)^{3/2} = 10^{-6}\left(\frac{T_{\rm eq.}}{T}\right)^3 \qquad \text{for} \qquad t \le t_{\rm eq.} \; , \tag{37}$$

which at the time of nucleosynthesis (ns) is $\left(\frac{d_H}{L}\right)^3_{\rm ns} \approx 10^{-24}$; then one has to tune the initial conditions even finer (than at $t = t_{eq}$) to explain the homogeneous Universe element composition. Further, at Planck time it yields $\left(\frac{d_H}{L}\right)^3_{Pl} \approx 10^{-89}$, that is, $\frac{L}{d_H} \approx 10^{29} \approx e^{68}$; such large numbers will later be explained in a successful inflationary model.

Now, let us try to link this issue with the big numbers encountered in (31) and (32). To do that, next we compute the entropy per horizon, S_H, using (24), finding

$$S_H = \frac{4}{3}\, b\, (d_H\, T)^3 \ , \tag{38}$$

now using (34), (24), and (9) for the matter dominated era and (36), (24), and (23) for the radiation era, we obtain the following results:

$$S_H = S\left(\frac{T_0}{T}\right)^{3/2} \qquad \text{for} \qquad t_{\rm eq.} \le t \le t_0 \tag{39}$$

and

$$S_H = S\left(\frac{d_{H_{\rm eq.}}}{L_{\rm eq.}}\right)^3 \left(\frac{T_{\rm eq.}}{T}\right)^3 = S \times 10^{-6}\left(\frac{T_{\rm eq.}}{T}\right)^3 \qquad \text{for} \qquad t \le t_{\rm eq.} \ , \tag{40}$$

where $S = 10^{88}$ and should be a constant of motion; see (18). From these equations one obtains at $t = t_{\rm ls}$ that $S_{H_{\rm ls}} = 10^{83}$. At a typical time during the nucleosynthesis one finds $S_{H_{\rm ns}} = 10^{63}$, and so on, until the Planck time, where $S_{H_{Pl}} \approx 1$. That is to say that the horizon problem is related to the increase of the horizon entropy as the Universe expands: this increase should be such that currently $S_{H_0} \gtrsim 10^{88}$ can explain the Universe's age, cf. (32). The evolution of horizon entropy in the standard Big Bang model is depicted in Fig. 3. Within the context of the *anthropic principle*, the existence of such big numbers invites us to reflect on our own existence; why are they so big (or so small)? The anthropic principle states that only in this way can life exist to account for it!

Flatness

Why is our Universe today nearly flat, and why was it almost identically flat at the very beginning? [30]. From (4) and (8) one finds that

$$\begin{aligned} \Omega(t) - 1 = \frac{\rho - \rho_c}{\rho_c} &= \frac{k}{a^2 H^2} \\ &= \frac{k}{\frac{8\pi G}{3} M_\omega a^{-(1+3\omega)} - k} \ , \end{aligned} \tag{41}$$

where the density parameter is defined as $\Omega(t) \equiv \rho(t)/\rho_c(t)$ and the critical density as $\rho_c(t) \equiv 3H^2(t)/8\pi G$. From (41) one can see that $k = 0$, $\Omega = 1$

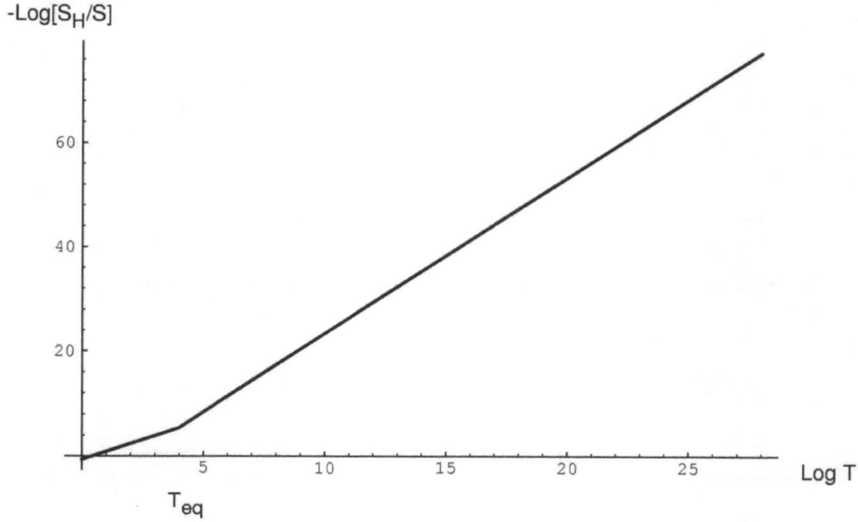

Fig. 3. The entropy per horizon is shown as the Universe cools. For the matter era the solution is given by (39) and for the radiation era by (40). The entropy per horizon presently is $S_{H_0} = S \sim 10^{88}$ at $T = 2.7\ °K$.

is an unstable point. Consider the limit $a \to 0$, then $\Omega \to 1$ for $\omega > -1/3$. Now, if $k = -1$, as $a \to \infty$ then $\Omega \to 0$; while for $k = +1$, as $a \to a_{\text{max.}}$ then $\Omega \to \infty$. That is, unless $k = 0$ and exactly $\Omega = 1$, the spatially flat Universe is unstable [72]; see Fig. 4.

Let us analyze in greater detail the first limit taken above. In order to compare the presently observed $\Omega_0 = \mathcal{O}(1)$ with that in the past, we first consider the evolution during the matter dominated era, given by (9) with $\omega = 0$. It implies that

$$\Omega(t) - 1 = k \left(\frac{H_0^{-1}}{a_0}\right)^2 \left(\frac{t}{t_0}\right)^{2/3} \qquad \text{for} \qquad t_{\text{eq.}} \le t \le t_0 \tag{42}$$

which at $t = t_0$ implies $\Omega - 1 \approx k$, but at $t = t_{\text{ls}}$, $\Omega - 1 = k\ 10^{-4}$. Therefore, in order to explain the present $\Omega_0 = \mathcal{O}(1)$ one has to fine tune the density value at $t = t_{\text{ls}}$ to be very similar to the critical value, the difference being of the order of only one part in ten thousand. This is the flatness problem.

For $t < t_{\text{eq.}}$, we use the radiation solution, given by (9) with $\omega = 1/3$, to have

$$\Omega(t) - 1 = k \left(\frac{H_{\text{eq.}}^{-1}}{a_{\text{eq.}}}\right)^2 \frac{t}{t_{\text{eq.}}} \qquad \text{for} \qquad t \le t_{\text{eq.}}\ , \tag{43}$$

at $t = t_{Pl}$, $\Omega - 1 = k\ 10^{-59}$! Thus, considering the entire evolution of the Universe beginning with Planckian initial conditions, one needs again to fine–tune the initial density value to be $\rho = (1 \pm 10^{-59})\rho_c$ in order to explain the

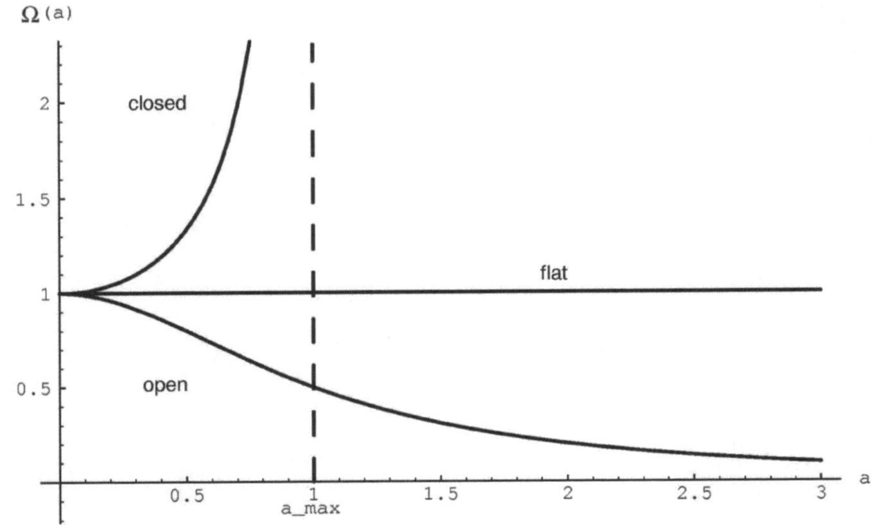

Fig. 4. The parameter Ω as a function of the scale factor, a, in a radiation dominated Universe. For closed models, with $k = +1$, Ω diverges as the scale factor approaches its maximum value, whereas for open models, $k = -1$, Ω asymptotically approaches to zero as the Universe expands. Finally, for a flat metric, $k = 0$, Ω is always equal to one. The behavior for a dust model is similar.

currently observed energy content of the Universe, i.e. to explain our own existence. The anthropic principle would just restate that the Universe has chosen those initial conditions necessary for us to be here! Nevertheless, this is no explanation but more a philosophical posture.

Let us try again to relate this issue to the aforementioned big numbers, (31) and (32). To do that, we express the above quantities in terms of the entropy within the horizon, (39) and (40). Since (24) is always valid, one obtains for both eras that

$$\Omega(t) - 1 = k\left(\frac{S_H}{S}\right)^{2/3} \qquad \text{for all times;} \tag{44}$$

again, at $t = t_0$, $\Omega(t) - 1 \approx k$. At $t = t_{\rm ls}$, $S_{H_{\rm ls}} = 10^{83}$ implies that $\Omega(t) - 1 \approx k10^{-4}$, whereas at the Planck time $S_H \approx 1$, one once again obtains that $\Omega(t) - 1 \approx k10^{-59}$. Very similar to the horizon problem, here one finds that the very small numbers come from the vast entropy increase within the horizon, which is the entropy necessary to fit the Universe's age, cf. (32).

Thus, the last two puzzles can be restated as: Why was the horizon entropy at the Planck time $S_{H_{Pl}} \approx 1$, but now $S_{H_0} \approx 10^{88}$?

Baryon Asymmetry

We observe that our Universe is apparently made of matter but not of antimatter. Why is this? Furthermore, the present different types of matter (fermions, bosons) are not in equal proportion. As we have already mentioned, nucleosynthesis restricts the value of η to be[5] $\eta \approx (3-4)\times 10^{-10}$, cf. (12); this fact tells us that the Universe is far more filled with photons than with baryons and if the baryon number is conserved η must also be conserved since the beginning of nucleosynthesis. In the standard model of cosmology one has to assume this as a given input. Let us explain this in more detail:

As far as observations show, within the solar system and our galaxy there is no evidence of primordial anti–baryons; if there were, some amount of gamma rays would be detected because of their annihilation with their baryon counterparts, something which has been not observed [89]. In going beyond galaxy scales, antimatter in galaxy clusters is ruled out by simple arguments that in fact are related to the horizon problem: one can imagine a baryon symmetric early Universe, whereby baryons and anti–baryons coexist in equilibrium. Their particle numbers in a co–moving volume should remain constant only until they become non–relativistic, when (25) begin to be valid; after that their particle abundances decrease exponentially. The particles remain in equilibrium annihilation until the temperature $T \approx 22$ MeV, when the annihilation rate, Γ, falls below the expansion rate. Then the ratio n_B/s is fixed to be $n_b/s = n_{\bar{b}}/s = 7\times 10^{-20}$ [89, 60], nine orders of magnitude smaller than the currently observed ratio n_B/s! In order to avoid this annihilation catastrophe one can try in some manner to stop the annihilation mechanism some time before, at about $T \approx 38$ MeV when $n_b/s = n_{\bar{b}}/s = 10^{-10} - 10^{-11}$, by separating baryons from anti–baryons. Even so the horizon at that time contained the following amount of matter:

$$M_d = \rho d_H^3 = (\rho a^3)_0 \left(\frac{d_H}{a_0}\right)^3 = M_0 \times 10^{-6}\left(\frac{T_{\rm eq.}}{T}\right)^3 , \qquad (45)$$

where we have used (37); at $T = 38$ MeV, $M_d = 5\times 10^{26}\text{gr} = 2.7\times 10^{-7} M_\odot$, which is clearly very much smaller compared to galaxy cluster mass scale; again, very fine tuning must be done. Instead of appealing to rare initial conditions, an alternative is to explain the baryon asymmetry by means of the Universe's evolution. Accordingly, at some high temperature $T \gtrsim 1$ GeV,

[5] In fact, η cannot be directly determined, nor can n_B/s. They are fitted to the currently observed values of light element composition in the Universe, i.e. $0.22 \lesssim Y_{^4\mathrm{He}} \lesssim 0.26$, $\mathrm{D/H} \gtrsim (1-2)\times 10^{-5}$, $(\mathrm{D} + {}^3\mathrm{He})/\mathrm{H} \lesssim 10^{-4}$ and $({}^7\mathrm{Li/H}) \lesssim 2\times 10^{-10}$. In this way, one can relate η with the baryon content of the Universe. Accordingly, for baryons $n_B = \rho_B/m_B = 1.13\times 10^{-5}\Omega_B h^2/cm^3$ and $n_\gamma = \frac{2\zeta(3)}{\pi^2}T^3 \approx \frac{422}{cm^3}T_{2.75}^3$, therefore, $\Omega_B = 3.6\times 10^7 \eta T_{2.75}^3/h^2$, and from the above mentioned value of η one finds that $0.01 \le \Omega_B \le 0.10$; that is, the Universe cannot be closed by baryons alone.

when $n_q \approx n_{\bar{q}} \approx n_\gamma$, a tiny asymmetry was already present and it prevented the total annihilation of quarks (q) and anti–quarks ($\bar{q}$), in such a manner that $\frac{n_q - n_{\bar{q}}}{n_q} = 3 \times 10^{-8}$ in order to explain the n_B/s needed for successful nucleosynthesis to take place [60].

The particle physics implementation in the early Universe by which that tiny asymmetry could be solved is called *baryogenesis*. The first attempt to address this problem was made by Sakharov [84], who pointed out three ingredients necessary to attain a baryon asymmetry in the Big Bang model. Let us review these: (*i*) baryon number violation, otherwise the baryon asymmetry can only reflect asymmetric initial conditions; (*ii*) violation of charge conjugation (C) and charge conjugation combined with parity (CP) are necessary to achieve different production rates for baryon and anti–baryons, otherwise a net zero baryon number is maintained; and (*iii*) non–equilibrium conditions, otherwise the same Fermi distribution of baryons and anti–baryons would guarantee the same phase space for them, i.e. $n_b = n_{\bar{b}}$.

It is curious that these three conditions were pointed out before there was a theory which could accomplish them. Indeed, first GUT appeared in the 70s and when one realized them in an early Universe scenario, they met the three ingredients: the first two are fulfilled because, by construction, strong and electroweak interactions are unified; this implies that quarks and leptons are members of a common irreducible representation of the GUT gauge group. In that way, gauge bosons mediate interactions in which baryons can decay into leptons, or the inverse, giving rise to a baryon number violation. C is violated by weak interactions and CP violation is observed in Kaon K^0 (meson) interaction. Thus, one also expects that the massive X-bosons decay into quarks/leptons, with a branching ratio of, say, r, and $\bar{X}$ with $\bar{\text{r}}$, such that $\text{r} \neq \bar{\text{r}}$. The third condition is attained due to Universe expansion, which evolves as $H \sim T^2/M_{Pl}$ in the radiation dominated era. For that to happen, one takes the reaction rates (decay, annihilation, and inverse processes) $\Gamma_X > H$. Then, through the *out-of-equilibrium decay* mechanism the $X-$bosons have a long enough lifetime so that their inverse decays go out of equilibrium as they are still abundant. In this way, the baryon number is produced by the X free decay, whereas the inverse rates are turned off.

Nevertheless, GUT have their own problems. For instance, precisely because of the first two ingredients above, the proton should decay too; in the minimal SU(5) GUT its lifetime is $\tau_p \approx 10^{29\pm1}$ years, but the experimental limit is greater, $\tau_p \approx 10^{31-32}$ years. Thus, something is wrong with this theory.

Another problem of GUT is that unless the model is B-L conserving, any net baryon number generated might be brought to zero by efficient anomalous electroweak processes, at temperatures of about $T \sim 100$ GeV. Though this seems not possible within the standard model electroweak baryogenesis, there are model extensions where this can be possible; see contribution of Piccinelly and Ayala in this book. This possibility represents a serious problem to GUT, and it opens new windows for low energy physics. Here, we depict briefly the idea of how electroweak baryogenesis works[31]: the vacuum manifold of the electroweak model, the so-called $\theta-$vacuum, has degenerated minima separated by energy barriers in the field configuration space as a result of non–trivial vacuum gauge configurations ($A^a{}_\mu \neq 0$) of non–Abelian gauge theories. Different minima have different baryon and lepton numbers, with the net difference between two adjacent minima being given by the number of families. Thus, for the standard model, jumps between these minima imply the creation of three baryons and leptons, hence, there is B-L conservation and B+L violation. At $T = 0$, tunneling (jumps) between two adjacent minima is mediated by *instantons* and the tunneling rate is exponentially suppressed [92] $\Gamma \sim e^{-\frac{1}{\alpha_{EW}}}$, where $\alpha_{EW} = 1/170$ is the electroweak coupling constant; this is why the proton is stable. However, at finite temperature, $T \approx 100$ GeV, one can go over the energy barrier to achieve a baryon number violation, as first described in [61]. The height of the barrier is a solution of an unstable static configuration called *sphaleron*, whose rate is $\Gamma \sim e^{-\frac{E_S}{T}}$, with its associated energy $E_S \approx M_W/\alpha_{EW}$. For temperatures above the critical temperature of electroweak symmetry restoration, the rate is no more strongly suppressed, but $\Gamma \sim (\alpha_{EW} T)^4$, indeed making possible baryon number violation. The other two ingredients to achieve baryogenesis could also be present, but a detailed analysis is in order; for a short review see [32, 44] and for an extended one see [31].

Monopole and Other Relics

Another problem of GUT is the production of magnetic monopoles [91, 77] as a consequence of GUT symmetry–breaking to some semi simple group U(1). In the course of the phase transition, bubbles of the new phase are produced and on scales greater than d_H one expects different Higgs field alignments. Because of this randomness, topological knots are present and they are the magnetic *monopoles.* It has been proved that their number density should be comparable to the baryon density, but their mass is 10^{16} times greater than that of the protons; in this case, the Universe should have recollapsed long before [55, 102, 78].

Additionally, some theories predict primordial cosmological particles (or structures) that could be present currently, also as a result of some spontaneous symmetry–breaking process. Among these cosmological relics are massive neutrinos, gravitinos, domain walls, cosmic strings, axions, etc.

Cosmological Constant

Another problem that arises as a consequence of theories of grand unification (or theories of everything, including gravity) is that the vacuum energy associated with these, $< 0|T_{\mu\nu}|0 > = < \rho > g_{\mu\nu}$, turns out to be very large.

Summing the zero-point energies of all normal modes of some field of mass m, one obtains $< \rho >\approx M^4/(16\pi^2)$, where M represents some cutoff in the integration, $M \gg m$. Then, assuming GR is valid up to the Planck scale, one should take $M \approx 1/\sqrt{8\pi G}$, which gives $< \rho >= 10^{71}$ GeV^4. This term plays the role of an effective cosmological constant of $\Lambda = 8\pi G < \rho >\approx M_{Pl}^2 \sim 10^{38}$ GeV^2 which must be added to the l.h.s. of Einstein equations (2) and yields an inflationary solution (10). However, if the cosmological constant is at present of the order of magnitude of the material content of the Universe, one has that

$$\Lambda \sim 8\pi G \rho_0 = 3H_0^2 \sim 10^{-83} \mathrm{GeV}^2, \tag{46}$$

which is very small compared with the value derived above on dimensional grounds. Thus, the cosmological constraint and theoretical expectations are rather dissimilar, by about 121 orders of magnitude! Even if one considers symmetries at lower energy scales, the theoretical Λ is indeed smaller, but never as small as the cosmological constraint. One finds that $\Lambda_{GUT} \sim 10^{21}$ GeV^2 and $\Lambda_{SU(2)} \sim 10^{-29}$ GeV^2 in contrast to (46). For an analysis of this problem in terms of longitude scales (not of mass square scales), see the contribution by E. Copeland in this book. This problem has been reviewed in [98, 19].

Large–Scale Structure

The problem of explaining structure formation in the Universe is most fascinating. There exist stars, galaxies, clusters of galaxies, superclusters, voids, and a variety of large–scale structures in the currently observed Universe, whose origin one hopes to understand within the framework of Newtonian or GR physics. Such systems represent complicated problems, for which one needs a deep understanding of both the initial conditions of the relevant physical quantities and their evolution: among them are the Universe composition (accounted in the density of the different i-species, Ω_i) and the type of perturbation the Universe experienced, i.e. adiabatic or isocurvature (isothermal).

Imagine an early Universe filled with a radiation fluid (i.e. effective relativistic) and some non–relativistic components. Let us consider the following density contrast:

$$\delta(\boldsymbol{x}) \equiv \frac{\delta\rho(\boldsymbol{x})}{\bar{\rho}} = \frac{\rho(\boldsymbol{x}) - \bar{\rho}}{\bar{\rho}} \ , \tag{47}$$

where $\bar{\rho}$ is the average density of the Universe. If in the past there were small density perturbations that grew as time went on, the formation of some structure will be favored. This density contrast is commonly expanded into a Fourier expansion:

$$\delta(\boldsymbol{x}) = \frac{V}{(2\pi)^3}\int_{\text{vol.}} \delta_k e^{-i\boldsymbol{k}\cdot\boldsymbol{x}} d^3k \qquad \text{and}$$

$$\delta_k = \frac{1}{V}\int_{\text{vol.}} \delta(\boldsymbol{x}) e^{i\boldsymbol{k}\cdot\boldsymbol{x}} d^3x \ , \tag{48}$$

where for $\boldsymbol{x}$ are chosen co–moving coordinates. A given "perturbation" mode λ is associated with its wave number $k = 2\pi/\lambda$. The physical mode is given by $\lambda_{\text{phys.}} = \lambda\frac{a(t)}{a_*}$, then $k_{\text{phys.}} = k\frac{a_*}{a(t)}$, when the expansion begins at $t = t_*$, $a(t_*)$. One can also relate to a given λ a mass defined as the rest mass contained in a radius $\lambda_{\text{phys.}}/2$, $M \equiv \frac{\pi}{6}\rho_m\lambda^3_{\text{phys.}} = 1.5\times 10^{11}M_\odot(\Omega_0 h^2)\lambda^3_{\text{Mpc}}$. For a galaxy, for instance this corresponds to $\lambda_{\text{gal.}} \approx \frac{1.9\text{Mpc}}{(\Omega_0 h^2)^{1/3}}$; this is the physical scale that would contain today a galaxy mass (of approximately $\sim 10^{12}M_\odot$) and after its non–linear regime would give rise to a typical galaxy size of approximately 30 kpc.

The fundamental quantity $|\delta_k|^2$, called the power spectrum, $\mathcal{P}(k)$, determines any statistical quantity for gaussian random fluctuations. In the absence of a fundamental theory of structure origin, one admits a power spectrum of the type

$$\mathcal{P}(k) \equiv |\delta_k|^2 = \text{const.}\ k^{n_s} \ , \tag{49}$$

where an isotropic wave number $|\boldsymbol{k}| = k$ has been assumed which is allowed in an FRW Universe symmetry; n_s is a constant called the *spectral index*[6]. At first, the Cosmic Background Explorer (COBE) satellite DMR results[86] suggested that $n_s \sim 1$ [11, 12, 100, 7]. Recently, the WMAP satellite measurements of the CMBR concluded more precisely that $n_s = 0.93 \pm 0.03$ [13].

The evolution of the density contrast determines whether and when the perturbation grows to arrive at its non–linear stage, when $|\delta_k|^2 > 1$, it starts to develop structure formation. This comes out by analyzing the Jeans equations in an expanding Universe. One finds that effectively there exists growing modes solutions (for $k < k_J$), which open, in principle, the possibility of describing the presently observed large–scale structure.

A particular perturbation is given through a Fourier component and is characterized by its amplitude and its co–moving wave number, in terms of which one can write the root-mean-square (*rms*) density fluctuation $\delta\rho/\rho$ as,

$$\left(\frac{\delta\rho}{\rho}\right)^2 \equiv \langle\delta(\boldsymbol{x})\delta(\boldsymbol{x})\rangle = \frac{1}{V}\int_0^\infty \frac{k^3|\delta_k|^2}{2\pi^2}\frac{dk}{k} \ , \tag{50}$$

where $\langle\cdots\rangle$ stands for spatial average. An *rms* mass fluctuation, δM, corresponds to a density contrast such that

[6] The spectral index, n_s, is sometimes referred as n, cf. contribution of E. Copeland in this book.

$$\left(\frac{\delta M}{M}\right)^2_\lambda = \frac{1}{(2\pi)^3}\frac{1}{VV_W^2}\int \mathcal{P}(k)|W(k)|^2 d^3k \ , \tag{51}$$

where $W(k)$ is a window function, typically Gaussian, i.e. $W(k) = V_W e^{-k^2 r_o^2/2}$, where r_o is the radius within which the mass M is contained; $V_W = (2\pi)^{3/2} r_o^3$ being its volume. Then, one has for the *rms* mass perturbation at a given λ that

$$\left(\frac{\delta M}{M}\right)^2_\lambda \sim k^3 \mathcal{P}(k) \sim k^{3+n_s} \ , \tag{52}$$

where the overall normalization amplitude, for all λ, has not yet been specified, according to (49). Note that $\left(\frac{\delta\rho}{\rho}\right)_k \approx \left(\frac{\delta M}{M}\right)_\lambda$, where the subindex k in the density contrast means $\frac{\delta\rho}{\rho}$ in a given logarithmic interval $\frac{dk}{k} \sim 1$.

Because of the existence of a causal horizon, there should be some $\lambda-$ modes that were once super-horizon sized and that some time later enter inside the horizon. These modes begin to grow posteriorly at $t > t_{\rm eq.}$. Once the perturbation enters the horizon, the Universe is well described with Newtonian physics and the distinction between adiabatic and isothermal perturbations becomes irrelevant; see Fig. 5.

The density and matter contrasts evolve for superhorizon modes as $\frac{\delta M}{M} \sim a^m(t) \sim t^{\frac{2m}{3(1+\omega)}}$, see (9); during the radiation era $\omega = 1/3$, $m = 2$ and for the matter dominated era $\omega = 0$, $m = 1$. During the time the physical mode is superhorizon sized, it scales as $\lambda_{\rm phys.} = \lambda\, a(t) \sim \lambda\, t^{\frac{2}{3(1+\omega)}}$ and at the moment this mode enters the horizon it is valid that $\lambda_{\rm phys.} = d_H \sim t_H$, therefore, $\lambda^{\frac{3(1+\omega)}{1+3\omega}} \sim k^{-\frac{3(1+\omega)}{1+3\omega}} \sim t_H$; since $\frac{3(1+\omega)}{1+3\omega} > 0$, then the larger the initial perturbation wavelength is, the later it enters the horizon. This means larger perturbation wavelengths begin later to develop to their non–linear regime, thus, one expects large–scale perturbations in the present to be smaller than the small scale perturbations. This is in accordance with the fact that $\delta\rho/\rho \sim 10^{30}$ for stars, 10^5 for galaxies, 10^{1-3} for cluster of galaxies, and $\mathcal{O}(1)$ for superclusters.

One has at the moment of horizon entering for every λ–scale that

$$\delta_{\rm H}(k) \equiv \frac{\delta M}{M}(k, t_H) \sim t_H^{\frac{2m}{3(1+\omega)}} \cdot \frac{\delta M}{M}(k,t) = k^{-\frac{2m}{1+3\omega}} \cdot k^{\frac{3+n_s}{2}} = k^{\frac{n_s-1}{2}} \ . \tag{53}$$

This is valid for both radiation and matter dominated horizon entering modes, since $\frac{2m}{1+3\omega} = 2$ for both cases. Harrison [49] and Zel'dovich [101] have argued that at the time perturbations enter the horizon they should have equal amplitude, that is, a *scale invariant spectrum*, which is achieved by choosing $n_s = 1$. This value is preferred by observations, as mentioned above. For instance, if $n_s > 1$, the perturbations are too strong and tend to close up and form island Universes; for $n_s < 1$ they are too weak to form galaxies [72]. Furthermore, the amplitude is required to be $\mathcal{O}(10^{-5})$ in magnitude when the perturbations start to grow at $t_{eq.}$.

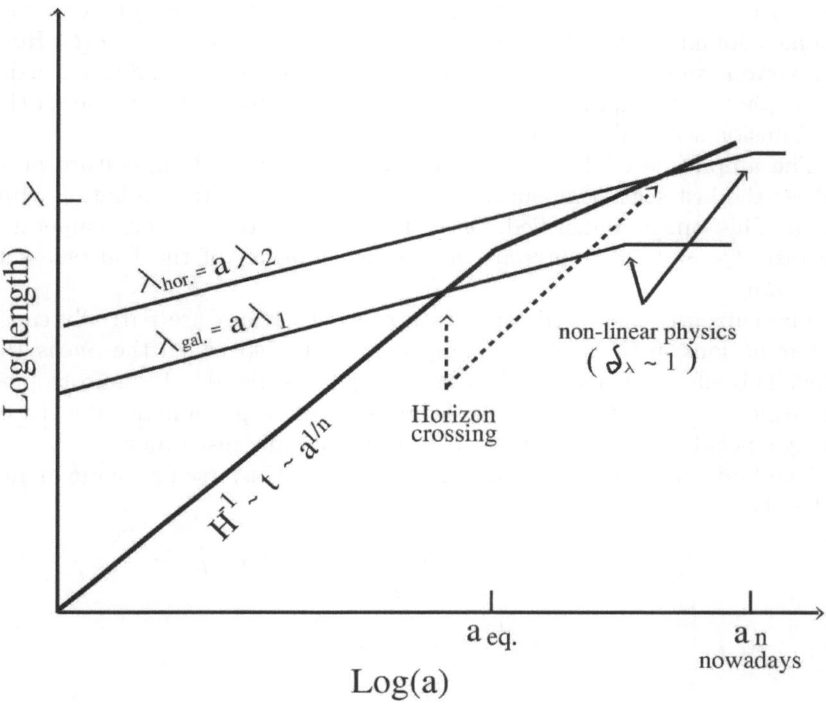

Fig. 5. At some early times there were superhorizon sized perturbations characterized by their wavelength and amplitude. As the Universe evolves these perturbations grow with the scale factor and eventually they cross inside the horizon, H^{-1}. After that, these modes first have a linear regime ($\delta_\lambda < 1$) and afterwards they start to develop the sky structure we see nowadays. In the figure we show two typical modes, one corresponding to galaxy scales and the other to horizon scales. The horizon evolves as $H^{-1} \sim t \sim a^{1/n}$, where here n denotes $n = 3(1+\omega)/2$, from solution (9). (Figure adapted from Kolb and Turner 1990).

During its radiation dominated phase, the Universe will not significantly develop non–linearities inside the horizon. Furthermore, some λ–scales are forbidden because of damping phenomena due to collisionless phase mixing of relativistic particles, known as Landau damping or *free–streaming* (fs). This is caused because relativistic particles move freely away from overdense regions and their velocity dispersion dissolves the compression regions. In this way, only those wavelengths greater than some free streaming scale [7] (λ_{fs}) are

[7] That is, wavelengths corresponding to scales greater than the horizon size when the particles become non–relativistic. $\lambda_{fs} = \mathcal{O}(1) \times \left(\frac{t}{a}\right)_{nr} \approx 1\mathrm{Mpc}\frac{\mathrm{keV}}{m_X}\frac{\mathrm{T_X}}{T_\gamma}$, where nr refers to when the X–component, with mass m_X and temperature T_X, becomes non–relativistic.

allowed to maintain overdense regions. Once a non–relativistic component becomes dominant, the growing modes increase typically as $\delta \sim a(t)$. However, baryons suffer from collisional damping due to photon diffusion, mainly during photon decoupling. This allows growing $\lambda-$modes that are larger than the diffusion scale, also known as the *Silk scale*[8] (λ_S).

The amplitude of density perturbations affects the temperature profile and, at the last scattering surface, the last stage of matter-radiation equilibrium. This can be quantified, since the present–day observed temperature contrast, $\frac{\delta T}{T} \sim 10^{-5}$, represents a "cosmic imprint" of the Universe since decoupling.

The currently measured temperature perturbations are partially caused by fluctuations in the gravitational potential, the so–called the *Sachs-Wolf effect*. This effect is responsible for large angular–scale ($\theta > 1°$) anisotropies[9]. The other known effect, the dipole anisotropy, is presumably due to our galaxy's peculiar velocity with respect to the cosmic rest frame.

One finds in the synchronous gauge for a flat Universe and large angular scales that

$$\begin{aligned}\left.\frac{\delta T(\boldsymbol{x})}{T}\right|_{\mathrm{ls}} &= -\frac{a_0^2 H_0^2}{2(2\pi)^3}\int k^{-2}\delta_{\boldsymbol{k}} e^{-i\boldsymbol{k}\cdot\boldsymbol{x}} d^3k \\ &= \frac{1}{3}\Delta\phi(\boldsymbol{x},t_0) \approx \left(\frac{\delta\rho}{\rho}\right)_{\lambda\sim H_0^{-1}} ,\end{aligned} \tag{54}$$

where $\Delta\phi(\boldsymbol{x},t_0)$ is the perturbed Newtonian potential and $\boldsymbol{x}$ points to the last scattering surface and has a length of $2H_0^{-1}$. It is convenient to expand $\delta T/T$ in spherical harmonics,

$$\frac{\delta T(\boldsymbol{x})}{T} = \sum_{l=2}^{\infty}\sum_{m=-l}^{m=+l} \mathbf{a}_{lm} Y_{lm}(\theta,\phi) \ , \tag{55}$$

where θ and ϕ are the spherical angles in the sky ($\theta > 1°$ corresponds to $l < 100$). The coefficients $\mathbf{a}_{lm}$ can be computed for the power spectrum of (49) resulting in

$$\begin{aligned}C_l \equiv \langle|\mathbf{a}_{lm}|^2\rangle &= \frac{H_0^4}{2\pi V}\int_0^\infty |\delta_k|^2 [j_l(kx)]^2 \frac{dk}{k^2} \\ &\sim H_0^{n_s+3} \approx \left(\frac{\delta\rho}{\rho}\right)^2_{H_0^{-1}}\end{aligned} \tag{56}$$

---

[8] $\lambda_S \approx \left(\frac{\Omega_0}{\Omega_B}\right)^{1/2} \frac{3.5\mathrm{Mpc}}{(\Omega_0 h^2)^{3/4}}$, where Ω_B is the baryonic contribution to the total density, i.e. $\Omega_0 = \Omega_B + \Omega_\gamma + \cdots$.

[9] $1°$ because the horizon scale at decoupling subtended approximately that angle. Then, $\theta > 1°$ corresponds to a $\lambda-$perturbation that was super-horizon sized. The existence of such super-horizon scales represents an initial density spectrum beyond causality!

where j_l is the spherical Bessel function of order l. Hence, $\delta T/T$ depends upon the present horizon scale, H_0^{-1}. For a Harrison-Zel'dovich form, $n_s = 1$, the spectrum of density fluctuations with a λ–scale greater than $\lambda_{\rm eq.} \approx 13(\Omega_0 h^2)^{-1}$ Mpc should conserve its initial form, and is related to an *rms* mass perturbation as follows:

$$C_l \approx (H_0 L)^4 \left(\frac{\delta M}{M}\right)^2_L . \tag{57}$$

Evaluating it in $L = 30h^{-1}$ Mpc and $(\delta M/M)_L = 1/4$, as inferred from measurements, one gets $C_l^{1/2} \approx 2 \times 10^{-5}$ which is experimentally confirmed from the CMBR anisotropies measurements; a more general discussion is found, e.g., in [60].

Note that, on the one hand, one has that $\left(\frac{\delta\rho}{\rho}\right)_{\rm ls} = {\rm const.} \left(\frac{\delta T}{T}\right)_{\rm ls}$ and, on the other hand[10], $\delta\rho/\rho \sim a(t) \sim 1/(1+z)$; thus, the maximal amplitude grow from last scattering until now is given by $1 + z_{\rm ls} \equiv \frac{a_0}{a_{\rm ls}} = 1200$ times 10^{-5}, which is not sufficient to form the non–linear structures we observe today. Therefore, other (dark) components must have played a role in the growing of perturbations. It seems that considering dark matter together with a quintessence field the correct grow can be achieved, as is suggested in the forthcoming lectures in this book.

One more intriguing aspect of the large–scale structure arises by observing the autocorrelation function of galaxies, clusters, etc.; $\xi(r) \equiv \langle \delta n(\boldsymbol{x}+\boldsymbol{r})\delta(\boldsymbol{x})\rangle$, which is proportional to the probability of finding an emitting object at a distance r from a given object, i.e. $\delta P = n\delta V[1 + \xi(r)]$; for instance, in a totally uniform distribution $\xi(r) = 0$, whereas $\xi(r) > 0$ indicates an enhancement of density near a given object. This has been measured for galaxies, clusters, and superclusters, showing that approximately [5, 9, 74]

$$\begin{aligned}
\xi_g &\approx 20\, r^{-\gamma} && \text{for} \quad 2 \lesssim r_{\rm Mpc} \lesssim 10 \ , \\
\xi_c &\approx 360\, r^{-\gamma} && \text{for} \quad 15 \lesssim r_{\rm Mpc} \lesssim 100 \ , \\
\xi_{sc} &\approx 1500\, r^{-\gamma} && \text{for} \quad 100 \lesssim r_{\rm Mpc} \lesssim 200 \ ,
\end{aligned} \tag{58}$$

where $\gamma \approx 1.7 - 1.8$ and the distance $r_{\rm Mpc}$ is given in Mpc. Again, due to the existence of a causal horizon, if one goes back in time one finds out that the co–moving separation between two emitting objects, e.g. cluster of galaxies, is larger than the light cone of causality at $t = t_{\rm eq}$. Therefore, if the initial density perturbations, $\delta(\boldsymbol{x})$, were produced before, or at, $t_{\rm eq}$, the correlations cannot be explained by causal mechanisms. Hence, why do have approximately the same slope, but different magnitudes or correlation lengths, is a mysterious aspect that should be explained by the structure formation theory,

[10] By considering the evolution of the Universe, one often uses, instead of the time parameter, the redshift (z), defined as $1 + z \equiv a_0/a$.

still in development. The different magnitudes of $\xi(r)$s suggest the existence of some dark matter present in greater quantities on bigger scales. This is in agreement with the determination of Ω_0 from dynamical computations, according to which one needs more dark matter as one goes from galaxy to cluster scales, and so forth.

In order to explain these issues, investigators have developed some numerical codes in which they include Hot Dark Matter (HDM) like light neutrinos in order to achieve a *top–down* scenario, first performing large structures that should fragment to give rise to smaller ones. One finds that typical large–scale filamentary structures and voids are well reproduced, but smaller scales are underweighted and, for instance, galaxy formation should have taken place rather recently, at $z \lesssim 1$, in order to obtain the observed galaxy correlation function mentioned above. Besides, the predicted limits for $\Delta T/T$ are near the upper limit measured. Numerical simulations also include Cold Dark Matter (CDM) with WIMPs (weakly interacting massive particles) to have now a *bottom–up* scenario, where first the smaller structures are formed and, later on, the larger ones. These simulations are in good agreement with galaxy correlation functions for acceptable redshifts, say $z \sim$ few. Nevertheless, some negative correlation function is expected for galaxies that have not been observed. Furthermore, the cluster correlation function is predicted to be about three times less than the measured value, that is, large scales are underweighted; the temperature contrast in this model can be as much as one order of magnitude below that observed. One notes that HDM and CDM play opposite roles for structure formation, because of their different free streaming ranges. Therefore, one considers mixed models (MDM) which include both CDM and HDM, and additionally some smooth components as a cosmological constant term (ΛDM). Up until now, it is not known for sure which particles have participated as the relevant building blocks that eventually brought about the formation of the large–scale structure with the right spectrum of anisotropies we observe today; computations hint there may be mixture of them: cold, hot, and a Λ–term; see the contribution of Cabral–Rosetti et al in this book. A review of these topics can be found in [4].

Summarizing, the problem of density perturbations lies in the understanding of its growth during its linear era in such away that on large scales $\delta\rho/\rho \sim \delta T/T \sim 10^{-5}$, but on small scales $\delta\rho/\rho > 1$ in order to reproduce the structures we see in the sky nowadays.

We have mentioned above some important aspects to be considered in order to achieve a better understanding of our present Universe. One should mention that these problems do not imply logical inconsistencies with the SBB. Nevertheless, for their explanation one is forced to appeal to very special initial conditions, a thing that a physicist would hardly accept. Moreover, solutions like the anthropic principle result to be dissatisfying. Therefore, extensions of the SBB are required, because by nature some of the presented problems, but not all, already come from its extensions, as by incorporating high energy ($\gg$ MeV) physics; ergo, the solutions to the puzzles should hopefully come from correct implementations of such extensions. Some proposed scenarios are already known, such as the inflationary one, which is the best candidate and is the subject of the next section. In order to achieve a further understanding of the early Universe, it is tempting to move from energies in MeVs to the Planck energy scale, that is, 22 orders of magnitude greater!

2 Beyond the Standard Big Bang Model: Inflation

In this section we show a way to solve the problems of the SBB presented in Sect. 1. We explain the general inflationary scenario that took place in the very early Universe and that gave rise to a successful cosmological model; that is, a Universe with its right causal size, age, temperature, and the perturbations spectrum that originated structure formation. At the end of this section we point out some remarks on inflationary models.

Inflation was accomplished as a natural extension of the "new" physics of the 70s being incorporated into the SBB. With the advent of GUT it was natural to study their cosmological consequences. In the late 70s and the beginning of the 80s some publications appeared about effects of GUT phase transitions in the very early Universe; even in this respect some authors considered exponential solutions, see [65, 72]. But the cornerstone paper was that of A. Guth [46], where he stressed that due to these phase transitions, the Universe could have experienced an exponential expansion of approximately e^{65} foldings, and in this way one could solve the horizon, flatness, and monopole problems, all at once. This was the first model of inflation. Although the original model suffered from some problems, this has shown that it is, in principle, possible to tackle the problems of the SBB by considering some vacuum energy or scalar fields to be present at the very beginning of time. Next, we explain how inflation addresses this.

2.1 Inflation: The General Idea

As we mentioned in Sect. 1, the FRW cosmological (4)-(6) admit very rapid expanding solutions for the scale factor. This is achieved when the inflation pressure, $\rho + 3p$, is negative, i.e. when the equation of state admits negative pressure such that $\omega < -1/3$, to have $\ddot{a} > 0$. For instance, if $\omega = -2/3$, one has that $a \sim t^2$ and $\rho \sim 1/a$, that is, the source of rapid expansion decreases inversely proportional to the expansion. Of special interest is the case when $\omega = -1$, $\rho =$ const., because this guarantees that the expansion rate will not diminish. Thus, if $\rho =$ const. is valid for a period of time, τ, then the Universe will experience an expansion of $N = \tau H$ $e-$foldings, given by $a = a_* e^N$, (10).

This is the well known de Sitter cosmological solution[11] [27], achieved here only for a $\tau-$stage in an FRW model.[12]

We shall now see how an inflationary stage helps to solve the problems of the SBB. First consider the particle (causal) horizon, given by (33), during inflation; with $k = 0$, again one obtains

$$d_H = H^{-1}(e^{Ht} - 1) \ ; \tag{59}$$

the causal horizon grows exponentially, whereas H^{-1} remains constant. Since $d_H \neq H^{-1}$, we call H^{-1} the *Hubble horizon* to distinguish it from the causal horizon. We again compare, in analogy to (34) and (36), the horizon distance with that of any physical length scale, $L(t) = L_* \frac{a(t)}{a_*} = L_* e^{Ht}$, to get

$$\frac{d_H}{L} = \frac{H^{-1}(e^{Ht} - 1)}{L_* e^{Ht}} \gtrsim 1 - e^{-Ht} \ , \tag{60}$$

for initial length scales $L_* \lesssim H^{-1}$. After a few e-fold times the causal horizon is as big as any length scale which initially was subhorizon sized. Therefore, if the original patch before inflation is causally connected, and presumably in equilibrium[13], after inflation this region of causality is exponentially bigger than it was; thus all the present observed (apparent) Universe can stem from it. Therefore, at some later time, say, at the time of last scattering (photon decoupling) the Universe has all the mentioned 10^5 regions (and more than that!) causally connected, then solving the horizon problem. In fact, if the inflation stage is sufficiently long, there can presently exist regions which are still so distant from each other that they are no longer in contact, even though originally they come from the same causal patch in existence before inflation; they will be in contact again when light reaches these distant points.

From (59) one can observe that if the initial physical length scale is greater than the Hubble distance, $L_* > H_*^{-1}$, then $L > d_H$ during inflation. Events initially outside the Hubble horizon remain acausal. This is better observed by considering the *event horizon*, d_e, that determines the region of space

[11] The de Sitter model contained no matter, $p = \rho = 0$. Instead de Sitter considered a cosmological constant such $H^2 = \Lambda/3$. In this sense, this was an anti–Machian solution, since matter was not needed to produce inertia. Alternatively, by choosing $p = -\rho$, $\Lambda = 0$ the dynamic is the same and the solution is Machian, but the price paid is that of having such exotic matter, $p = -\rho$.

[12] From now on, *inflation* will refer to the period of exponential expansion. However, an *inflationary scenario* also implies some other physical processes such generation of density fluctuations, reheating, etc.

[13] One can imagine the initial stage of the Universe to possess some inhomogeneities, anisotropies, and a rather chaotic distribution of particles. At some time after the Planck scale, $10^3 t_{Pl}$, one expects the dampening of the anisotropies and inhomogeneities in the metric, and due to statistical processes, the Universe should thermalize in some local scale ($< d_H$), which we now call the *original patch* [46].

which will remain in causal contact after some time; that is, it delimits the region from which one can ever receive (up to some time $t_{\max}$) information about events taking place now (at the time t):

$$d_e(t) = a(t) \int_t^{t_{\max}} \frac{dt'}{a(t')} \ . \tag{61}$$

For a flat model during its matter dominated era ($a \sim t^{2/3}$), like our Universe in the present, $d_e \to \infty$ as $t_{\max} \to \infty$. However, during inflation one finds that

$$d_e = H^{-1}(1 - e^{-(t_{\max}-t)H}) \approx H^{-1}, \tag{62}$$

which implies that any observer sees only those events that take place within a distance $\leq H^{-1}$. In this respect, there is an analogy with black holes, from whose surface no information can escape. Here, in an exponential expanding Universe, observers encounter themselves in a region which apparently was surrounded by a black hole [43, 65], since they receive no information located further than H^{-1}.

So far we have seen how the exponential solution can offer means to understand the horizon problem of the SBB. Now, we analyze it quantitatively. First, consider the evolution of a co–moving length since the very beginning. Before inflation starts the Universe is characterized by some initial values of temperature, Hubble horizon, etc., related by

$$\begin{aligned} \rho_* &= b\, T_*^4 \ , \\ H_*^2 &= \frac{8\pi G}{3}\, b\, T_*^4 \ , \\ S_* &= \frac{4}{3}\, b\, (a_* T_*)^3 \ . \end{aligned} \tag{63}$$

If inflation does not take place, an initial horizon–sized length scale, $L_* \leq H_*^{-1}$, grows only proportionally to $t^{1/2}$ (radiation era) and, later, to $t^{2/3}$ (matter era). Therefore its size would currently be much smaller than our apparent Universe; in others words, this is the horizon problem. Hence, suppose that inflation indeed takes place during a time period $\tau = NH_*^{-1}$ and, afterwards, a usual Friedmann Universe follows. Then, a typical co–moving scale evolves as

$$L = L_* e^N \left(\frac{t_{\rm eq}}{t_f}\right)^{1/2} \left(\frac{t}{t_{\rm eq}}\right)^{2/3} \ . \tag{64}$$

The first term accounts for the inflationary stage (until the final time $t_f \approx \tau = NH_*^{-1}$), the second for the radiation era that lasts until the time of equal densities ($\rho_r = \rho_m$ at $t = t_{\rm eq}$), and the third term for the matter dominated era; all of them being solutions of the FRW cosmological equations characterized by different equations of state, cf. (9) and (10). Consequently, the number of e-folds (N) of growth in the initial co–moving scale L_*, to achieve at present (t_0) a size L_0, is given by

$$N = \frac{1}{\log e}\left(\log\frac{L_0}{L_*} - \frac{1}{2}\log\frac{t_{eq}}{t_f} - \frac{2}{3}\log\frac{t_0}{t_{eq}}\right) \equiv N_{\min 1} . \qquad (65)$$

In principle, one could substitute the value $t_f \approx NH_*^{-1}$ into (65) and try to solve it for N, but one finds no analytical solution. As a matter of fact, one typically obtains $t_f \approx 10^2 H_*^{-1}$ and this turns out to be a very good approximation. For definiteness, let us assume this and, furthermore, that the initial co–moving scale is the horizon at the beginning, i.e. our original patch; then from (63), $L_* = H_*^{-1} \approx 10^{-1} M_{Pl}/T_*^2$. T_* is the temperature that characterizes some phase transition and is typically $T_* = 10^{14}$ GeV. Then, $H_*^{-1} = 10^{-24}$ cm, $t_* = 10^{-35}$ s, $\frac{t_{eq}}{t_f} = \frac{10^{12}\text{s}}{10^{-33}\text{s}} = 10^{45}$, $\frac{t_0}{t_{eq}} = \frac{10^{17}\text{s}}{10^{12}\text{s}} = 10^5$, and $\frac{L_0}{L_*} = \frac{10^{28}\text{cm}}{10^{-24}\text{cm}} = 10^{52}$ to yield $N = 60.2$. In fact, this value represents the minimal number of e-folds of inflation necessary for an initial horizon sized co–moving length scale to grow as big as our presently observed Universe, $L_0 = H_0^{-1} \approx 10^{28}$ cm. If the original patch is horizon sized, then during inflation it remains within the causal horizon, d_H, according to (59). After inflation the causal horizon grows as $d_H \sim t$, whereas the co–moving scale expands only as the scale factor does, in conformity with (9). Therefore, the co–moving length scale, L, remains always within the causal horizon.

For the inflationary Universe the currently apparent horizon comes from a region delimited by the original patch H^{-1}, which during inflation remains almost constant and, after, evolves as $H^{-1} \sim t$. At the end of inflation $a(t) \gg H^{-1}(t)$. Subsequently, the scale factor expands only with the power law solution $t^{1/2}$ (or $t^{2/3}$), whereas the Hubble horizon evolves faster, $H^{-1} \sim t$. Then, at some later time the Hubble horizon is as large as the scale factor, $H^{-1} \sim a(t)$. Accordingly, the value $N_{\min 1}$ defines the minimal number of e-folds of inflation necessarily to have this equality at present; that is, the original patch grown until now is as big as our apparent Hubble horizon. Hence, some time ago, say, at the last scattering surface (photon decoupling), the Universe consisted of 10^5 Hubble horizon regions, yet all these regions stem from one original patch of size H_*^{-1} just at the start of inflation. Recall that the causal horizon, d_H, is always bigger than H^{-1}, except at the outset of inflation. Naturally, our original patch could have experienced a longer period of inflation. In this case, $N > N_{\min 1}$ and the Universe is bigger than we observe it to be today: our event horizon will enable us to explore the Universe beyond.

Let us see how the flatness comes out of inflation. Consider originally a Hubble patch, H^{-1}, that might even possess some curvature different from zero. If there were the above conditions for such a gigantic expansion to take place, then in a short time the original Hubble patch will become very flat, since $H = \sqrt{\frac{8\pi G}{3}\rho}$ stays constant during that τ-stage. On the other hand, a typical scale $L_* \leq H^{-1}$ will exponentially increase in size as $L(t) = L_* \frac{a(t)}{a_*} = L_* e^N$. That is, all physical inhomogeneities, anisotropies and/or 'perturbations' of any kind (including particles!) will be diluted away. Their density becomes insignificant, thus solving the monopole (and other relics) problem.

When we discussed the flatness problem it was pointed out that Ω closely approaches unity as one goes back in time (see 42 and 43), in a way such that one must choose very special initial density values for explaining our flatness today, i.e. $\Omega_0 \approx \mathcal{O}(1)$. Imagine the Universe with initial conditions taken above, $a_* \sim H_*^{-1}$, then $\Omega_* - 1 \approx k$. Now, if the exponential expansion occurs, $\Omega(t = \tau)$ evolves to

$$\Omega(\tau) - 1 = \frac{\rho - \rho_c}{\rho_c} = \frac{k}{a^2 H^2} = k e^{-2N} \ . \tag{66}$$

If N is sufficiently large, which will be case since typically $N > N_{\rm min1}$, after a de Sitter stage the Universe looks like an almost perfectly flat model. Therefore, the initial density plays almost no role; if the exponential expansion occurs the Universe becomes effectively flat; see Fig. 6. In this way, instead of appealing to very special initial conditions, one starts with a Universe with more normal -that is, not very fine tuned- conditions which permit the Universe to evolve to an inflationary stage, after which it looks like it would have very special conditions, i.e. with $\Omega \approx 1$ with exponential accuracy. Thereupon, the flatness problem is no longer present.

One can also observe this by geometric means. The *radius of curvature* of the Universe is defined as

$$R_{\rm curv} \equiv \frac{a(t)}{|k|^{1/2}} = \frac{H^{-1}}{|\Omega - 1|^{1/2}} \ , \tag{67}$$

where we have used (4). Initially one may find that $\Omega \approx \mathcal{O}(1)$, implying that $R_{\rm curv} \sim H^{-1}$, but after inflation Ω is very close to unity; thus the radius of curvature is exponentially larger than the Hubble distance, making the latter look very flat.

We have argued that the Universe becomes very flat. However, after inflation the r.h.s. of (66) starts to grow linearly with time during the radiation dominated era, as was derived in (43), and during the matter era it grows as $\sim t^{2/3}$, cf. (42). Therefore, the time evolution of $\Omega(t)$ is given by

$$\Omega(t) - 1 = \frac{k}{a^2 H^2} = k \ e^{-2N} \ \frac{t_{\rm eq}}{t_f} \left(\frac{t}{t_{\rm eq}}\right)^{2/3} \quad \text{for} \quad t_{\rm eq} \leq t \ . \tag{68}$$

If the number of e-folds of inflation is not sufficient, at some t, Ω will be very different from one. Accordingly, from (68) one can compute the minimal number of e-folds such that currently $\Omega_0 \approx \mathcal{O}(1)$, as suggested from observations. One gets,

$$N > \frac{1}{2 \log e} \left(\log\frac{t_{\rm eq}}{t_f} + \frac{2}{3}\log\frac{t_0}{t_{\rm eq}} \right) \equiv N_{\rm min2} \ , \tag{69}$$

which with the values used above implies that $N > 56 = N_{\rm min2}$. However, the value of N chosen greater than (65), $N > N_{\rm min1}$, already fulfills our new requirement, since $N_{\rm min1} \gtrsim N_{\rm min2}$. Therefore, in most models of inflation $N \gg N_{\rm min1}$ to predict Ω_0 to be unity to high accuracy, in accordance with the recent WMAP measurements [13] of the CMBR. However, if Ω is close, but different than the unity, then there will come a future time when the value of Ω is arbitrarily close to zero (for $k = -1$) or arbitrarily large (for $k = 1$), the same as happens in the SBB; compare Figs. 4 and 6.

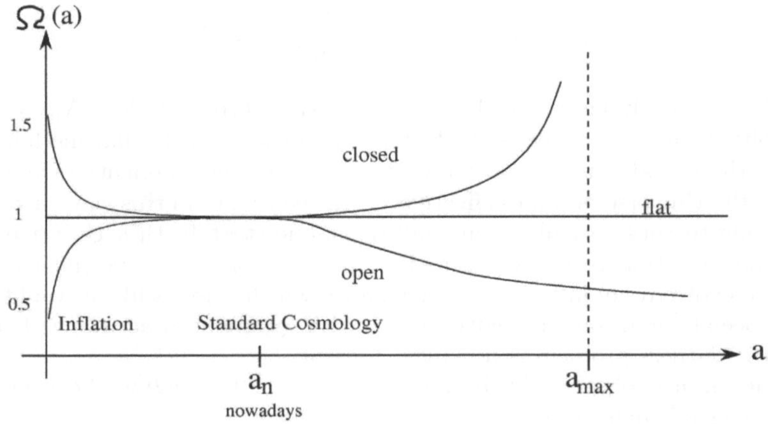

Fig. 6. The parameter Ω as a function of the scale factor, a, during inflation and thereafter in a radiation dominated Universe. Inflation makes the space seem almost flat. Having enough e-folds of inflation to solve the horizon problem implies that the Universe still looks currently very flat. Later on, the behavior is as in Fig. 4.

2.2 Transition to the Physical Universe

Provided that the Universe underwent a period of $N(\geq N_{\rm min1} \approx 60)$ e-folds of inflation, it seems that the horizon and flatness problems are no longer present: all physical events are, or were, in causal contact. However, as we shall see now, this is only a necessary, but not a sufficient, condition to assure that the original patch, H_*^{-1}, contains the properties of our Universe today.

In Sect. 1 we saw that both problems are related to the increase of entropy per horizon and that the age of the Universe is related to the total entropy, cf. (32): the Universe is too old because the entropy is too large. Then, within the inflationary Universe there arises the question of how the $N_{\rm min's}$ are related to the entropy enhancement.

In this section we have remarked that with the aid of inflation any co-moving length scale remains within causal horizon. Therefore, in the present case the entropy per causal horizon remains constant if the Universe evolves adiabatically ($T \sim 1/a$), and therefore it is no longer necessary to distinguish between horizon entropy and the total entropy. Thus, it suffices to compute the entropy at any time to know how large it is. Accordingly, at the very beginning $S_* = \frac{4}{3} b(a_* T_*)^3 \sim 10^{13}$ with initial conditions taken from above. Clearly, this value is much smaller than the observed photon entropy today, $S_0 \sim 10^{88}$. Once the entropy augments to 10^{88}, the flat patch can explain the currently observed Universe with a big mass and age; see (31) and (32). In this way, to fully solve the horizon and flatness problems, one has to find an entropy production mechanism such that the increase factor grows over 10^{75} orders of magnitude! Would this entropy enhancement not exist, the original patch must contain a huge entropy (10^{88}), and therefore, the Universe would consist at the very beginning of too many disconnected causal horizons; ergo, the horizon problem would stay unsolved. The natural solution is to obtain after inflation a mechanism by which the entropy increases from some initial value to $S_0 \sim 10^{88}$. To see how this happens, consider first a Universe model filled with some relativistic components with an energy density given by

$$\rho = bT^4 + V(0) \ , \qquad (70)$$

where $V(0) \equiv M^4$ is a constant associated with the vacuum energy density of some GUT; M is some mass term. As the Universe cools the energy density diminishes until certain time, say, $t = t_c$, at which the constant dominates the dynamics over the radiative components. At that moment the entropy within the horizon is $S_{d_H} = \frac{4}{3} b(a_c T_c)^3 = \text{const.}$, assuming adiabatic processes. At the moment when the constant $V(0)$ begins to dominate the dynamics, the solution is given by (10). Then, the original patch, $a_c \sim H_c^{-1}$, expands exponentially and the Universe supercools $T = T_c e^{-H\tau}$, since $aT = \text{const.}$ during the τ-stage of inflation. Note that the entrance to an inflationary era is natural as a consequence of the Universe cooling and, of course, of the presence of the constant $V(0)$. Typically, $M \sim T_c \sim 10^{14}$ GeV is related to the critical temperature of a spontaneous symmetry–breaking process, whereas $H_c^{-1} \approx 10^{-1} M_{Pl}/M^2 \sim 10^{-34}$s, that is, the values we have chosen above to yield $S_{d_H} \sim 10^{13} \ll 10^{88}$. Furthermore, after inflation the Universe contains a very low particle density and is very cold, even as cold as it is today! The transition to a radiation–(or matter–)dominated era with sufficient entropy and particle content comes from the 'decaying' or transformation of the energy source of inflation, $\rho = V(0)$, into heat, a process called *reheating* (RH). In his original model of inflation A. Guth [46] showed that if the Universe super–cools sufficiently its temperature is $T_s \sim e^{-N} T_c$ and a phase transition proceeds releasing latent heat of characteristic temperature $\sim T_c$. Then, the Universe is reheated to some temperature (T_{RH}) of the order of T_c. Through this mechanism the entropy increases from the initial value $S_* = \frac{4}{3} b(a_c T_c)^3$ to the final, after reheating, $S_f = \frac{4}{3} b(a_{RH} T_{RH})^3 \approx \frac{4}{3} b(e^N a_c T_c)^3$, that is, by a factor of $(T_{RH}/T_s)^3 \sim e^{3N}$, achieving the desired numbers.

2.3 Density Perturbations

The vacuum energy density responsible for inflation is associated with some scalar field that experiences quantum fluctuations around some vacuum expectation value. The theory of quantum fluctuations in a de Sitter space was developed by Bunch and Davies [17] and was applied by several authors [50, 47, 88, 8] to the inflationary universe in order to compute its contribution to $\delta\rho/\rho$.

It is our intention in this subsection to depict qualitatively how these fluctuations are responsible for a density perturbation spectrum, as required to understand structure formation; see the problem of structure formation above. For a detailed treatment of this topic see the lecture by R. Brandenberger in this book and his review article [69] .

We begin by noting that the event horizon during a de Sitter stage is $d_e \approx H^{-1}$, cf. (62). This means that microphysics can only operate coherently within distances at most as big as the Hubble horizon, H^{-1}. Recall that the causal horizon, d_H, expands exponentially and it is very large compared to the almost constant H^{-1} during inflation, see (59). Hence, during the de Sitter stage the generation of perturbations, which is a causal microphysical process, is localized in regions of the order of H^{-1}. That is, in all regions of size H^{-1} that comprise the Universe during inflation there should be such generation of perturbations.

Furthermore, it has been shown that the amplitude of inhomogeneities produced corresponds to the Hawking temperature in the de Sitter space, $T_H = H/(2\pi)$. In turn, this means that perturbations with a fixed physical wavelength of size H^{-1} are produced throughout the inflationary era. Accordingly, a physical scale associated with a quantum fluctuation, $\lambda_{\mathbf{phys}} = \lambda a(t)$, expands exponentially and once it leaves the (Hubble) horizon it behaves as a metric perturbation; its description is then classical, general relativistic. If inflation lasts for enough time, the physical scale can grow as much as a galaxy– or horizon–sized perturbation. The field fluctuation always expands with the scale factor and after inflation it evolves according to t^n ($n = 1/2$ radiation or $n = 2/3$ matter). On the other hand, the Hubble horizon evolves after inflation as $H^{-1} \sim t$. This means that there will come a time at which field fluctuations cross inside the Hubble horizon and re-enter as density fluctuations. Thus, inflation produces a large spectrum of perturbations, of which the largest originated at the start of inflation with a size H_i^{-1}, and the smallest with H_f^{-1} at the end of inflation; see Fig. 7.

Finally, analogous to the density perturbations spectrum created by the scalar field during inflation, any massless (or very light $H \gg M$) field is excited in the de Sitter space. Once the excited modes re–enter the Hubble horizon they will propagate as particles [94]. In this way, gravitational wave (GW) perturbations re–enter the Hubble horizon during the radiation do-

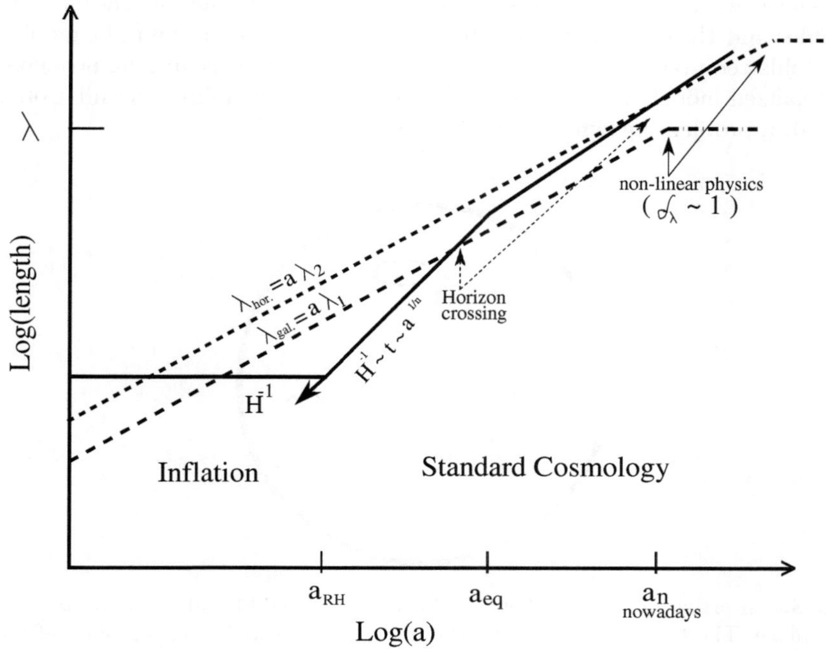

Fig. 7. Quantum perturbations were initially subhorizon–sized. During inflation they grow exponentially ($\lambda_{\mathrm{phys.}} = \lambda a(t)$), whereas the Hubble horizon remains almost constant. Then they eventually cross outside H^{-1} and evolve as classical perturbations. Later on, they re-enter the Hubble horizon to produce an almost scale invariant, Harrison-Zel'dovich density perturbation spectrum; in this way, its origin is no longer a mystery. In the figure there are two physical perturbations scales depicted, galaxy and horizon sized. (Figure adapted from Kolb and Turner 1990).

minated era [45, 87, 83, 35]. The amplitude of these perturbations must be $\leq 10^{-5}$ in order to be consistent with the isotropy levels of the CMBR.

2.4 Final Remarks on Inflationary Models

The descriptions presented above generically describe the inflationary scenario without referring to the many specific models that have been proposed in the course of more than twenty years of inflationary cosmology. Below we mention the primary features of the first inflationary models that permitted the understanding of the physical properties of a general inflationary scenario.

The first inflationary model was due to A. Guth, who in 1981 published [46] a model which later became known as *old inflation*. This model proposed the formation of bubbles of scalar fields obeying a first–order GUT phase transition in the early Universe; see Fig. 8. However, the nucleation of the bubbles and Hubble expansion rates could not encompass the right numbers of e-folds required ($\geq N_{\rm min}$) and, at the same time, achieve a homogeneous thermalized model. The model also had difficulties in ending the inflationary period, a problem known as *graceful exit.*

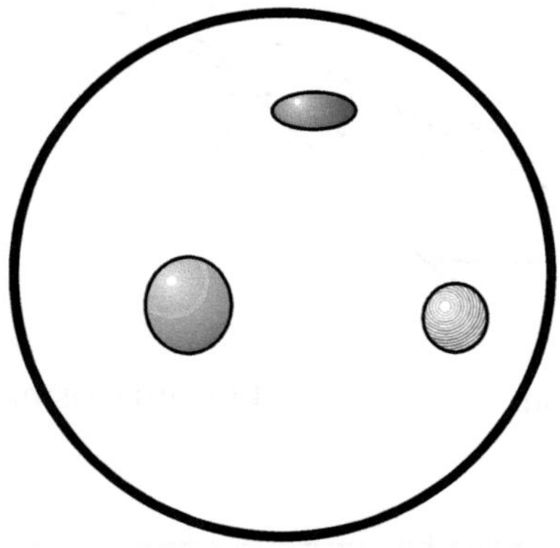

Fig. 8. An artistic picture of how the formation of bubbles after inflation should take place. The big circle represents the universe filled with a background of false vacuum, $\phi = 0$. Thus, some bubbles are forming in a sea of false vacuum. The different bubble ϕ-field values are represented by different gray tones.

Shortly afterwards, a model called *new inflation* was proposed [63, 2], in which the scalar field experiences a second–order phase transition. In this model the whole Universe stems from an initially uniform single bubble or fluctuation region ($\leq d_H$), which after its exponential expansion can be as large as our apparent horizon. In this way the Universe does not possess the problem of bubble nucleation, nor that of the presence of unwanted relics, because they are produced at bubble boundaries, which in this case are exponentially far away from our apparent horizon. The source of exponential expansion is achieved by permitting the scalar field to slowly evolve from its symmetric state ($\phi = 0$) to its ground state, $\phi = v$, of a typical potential $V(\phi) = \lambda(\phi^2 - v^2)^2$. During this time the potential associated with the scalar field is almost constant, thus providing an effective cosmological constant to the FRW equations; see Fig. 9. The process of *slow rolling down* of ϕ along its potential curvature is the main new ingredient of this model of achieving an inflationary stage.

After inflation the ϕ-field begins to oscillate around its stable minimum, $\phi = v$. The energy stored in $V(0)$ decays to reheat the Universe to have the required entropy. One must point out that over the course of the years, important steps to consolidate the theory of reheating have been made; see for example [93]. But, qualitative new ideas have been introduced only since the

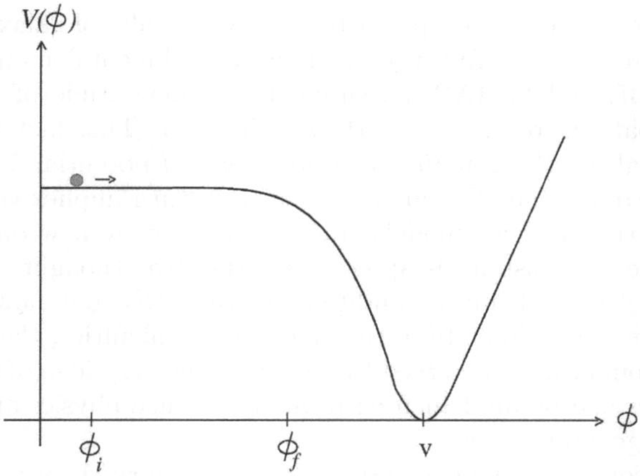

Fig. 9. A sketch of the new inflationary potential is shown. The potential curvature is very flat in order to permit the field to slow roll down the hill to yield enough e-folds of inflation during that time. Inflation begins at some ϕ_i and ends at ϕ_f when the field begins to evolve rapidly to its stable symmetry–breaking state $\phi = v$, around which the field oscillates until reheating.

mid 90s [57, 58]. Accordingly, the process of reheating should consist of three different stages. At the first phase, the ϕ-field decays into massive bosons (fermions) due to parametric resonance given through a Mathieu equation that determines the regions of stability and instability (particle production) in the quantum fluctuations of the created particles. These can be ϕ-particles or other bosons (fermions) coupled to the ϕ-field. This process is very efficient, even explosive, and many bosons can be created in this stage. Note that the original theory is based upon the decay of the ϕ-particles, whereas in the present theory the ϕ-field decays into ϕ-particles, and perhaps others, and only after this process does the decay of ϕ-particles proceed. Then, to distinguish this explosive process from the normal stage of particle decay, the authors of [57] have called it *preheating*. Bosons produced at this stage are far from thermal equilibrium and have very big occupational numbers. The second stage of this scenario describes the decay of the already produced particles. This phase is described as in the original theory. Thus, the methods developed for the original theory are now applied to the product particles, but not to the decay of the ϕ-field itself. The third stage is the thermalization by which the system reaches equilibrium; for review of this topic see [59].

A very important result found in the context of the "new" inflationary model is that perturbations of the scalar field can explain the required initial conditions of structure formation, i.e. an almost scale invariant, Harrison-Zel'dovich spectrum. This spectrum results from the original quantum fluctuations of the ϕ-field. These field fluctuations cross outside the Hubble horizon during inflation, evolve classically and, eventually, return back to re–enter the Hubble horizon as density perturbations with scales of galaxy or present–horizon size, see Sect. 2.3. But its correct accomplishment demands, according with the COBE and WMAP measurements, a magnitude of $\delta\rho/\rho \sim 10^{-5}$ when perturbations re–enter the Hubble horizon. This fact demands the above potential, or the generic toy *chaotic model* potential $V = \lambda\phi^n$ [64], to have an extremely small value for λ which in turn implies very particular choices of particle physics models (fine–tuning), or even wrong models. In this way, some inconsistencies appear. Inflation was thought to circumvent the choosing of special initial conditions of the SBB, but now we see that it encounters its own. In spite of this and other difficulties, the new and the chaotic inflationary models served to show how the very idea of field slow rollover dynamics can be implemented in many particle physics and/or gravity theories with general success.

Over the course of two decades of the inflationary theory, many related scenarios and models have been proposed with concrete physical mechanisms to achieve inflation, reheating, baryogenesis and a causal perturbation spectrum, among others. It has been found that many models generally suffer from "unnatural" fine–tuning of parameters. Nevertheless, some of these models have interesting properties, and the relation among theoretical and observational cosmology and particle physics has become tighter than ever. For instance, typical unification theories have different scalar fields which have been used to have one or more inflationary stages. Additionally, they have been used to produce the correct density perturbation spectrum together with a sufficient reheating temperature. That is, it is tempting to use the various fields for achieving different cosmological tasks, as in the case of *hybrid inflation*. Then, to distinguish among the different fields, the field responsible for the period of exponential expansion is generically called the *inflaton*. The modern view is that this inflaton is a primary ingredient in offering a solution to the above–mentioned problems. A general description of the scalar field (inflaton) dynamics, as well as some of its quantitative parameters, are found in the contributions of E. Copeland and C. A. Terrero–Escalante in this book.

3 Overview

Finally, we are going to review the topics that were covered in this contribution. We have presented a general view of the SBB, its problems, and the main ideas involved in the inflationary Universe. We have shown how inflation achieves an explanation of the horizon and flatness problems. In doing so, a period exponential expansion of about 60 e-folds in the scale factor is necessary. However, this condition is not sufficient at all to yield a Universe like the one we live in: after inflation the region of the Universe which will give rise to our present apparent horizon is almost devoid of particles and, because of the adiabatic exponential expansion, it is also very cold. Therefore, a process of reheating is mandatory. GUT offer through phase transition phenomena an appealing scenario for creating both a constant energy density in the very early Universe, to have inflation, and its subsequent preheating and reheating. After inflation Baryogenesis can take place in the context of GUT, a scenario which can attain the Sakharov required conditions.

We have also shown that the inflationary theory provides means to solve the monopole and other relics problems within the new inflationary scenario. Furthermore, there are inflationary models with no singularity, because they begin with finite initial conditions, for instance, those attached to the potential energy. In this way, one excludes singularities by appealing to the physical limitations of the classical theory; this is the case for chaotic inflation. Other inflationary models without initial singularity have also been proposed [95].

The homogeneity and isotropy of the large–scale structure of the Universe are also a consequence of a long period of exponential expansion. Recall that all inhomogeneities are shifted away from the Hubble horizon. Thus, inflation makes the space very homogeneous and isotropic. This assertion is known as the *cosmic no hair* conjecture [43, 15]. The question remains whether the initial patch was sufficiently smooth to enable inflation to start. That is, in some sense, the Universe should be, at some level homogeneous and isotropic to consider it as an original patch with which to begin inflation. It turns out that scalar fields, initially without considering potential terms, can bring an initially homogeneous, anisotropic patch to an almost FRW symmetry [20, 67]. That is, an anisotropic Universe can begin with such initial conditions that the potential term is yet not the primary contribution to the cosmological field equations and, after some time, it can become nearly isotropic. Then, inflation occurs when the potential term begins to dominate [21, 22, 23].

The cosmological constant, $\Lambda = V(0)$, provides the energy density to have an exponential expansion. However, in the inflationary theory this constant must be assumed, and inflation provides no explanation of this. From the particle physics point of view, it is also intriguing to consider why this constant should or should not exist; there is no known principle that demands it to vanish. Thus, for convenience one usually assumes $V(0) \neq 0$ in order to be able to achieve inflation and to have today $V(v) = 0$ with $\phi = v$. In this way, one avoids choosing the tuned value $V(v) \lesssim \rho_{c_0} \approx 10^{-47}$ GeV^4 today. However, this last possibility is very interesting in the context of the present huge expansion rate; see below.

It is worthwhile to notice that because inflation predicts $\Omega_0 \approx 1$ and since our observed baryonic Universe only contributes $\Omega_B \sim 0.05$ to the total energy density, then inflation predicts some amount of dark components, namely, $\Omega_{\rm dark} \sim 0.95$! The motivation for one of the dark candidates, Λ, came from different cosmological measurements which without a cosmological constant (or function) are rather difficult to explain. For instance, W. Priester et al. [51, 52, 14] pointed out that with a present cosmological constant one can explain the absorption lines of quasars, the so called *Ly* α-forest spectrum, assuming a Hubble constant of $H_0 = 90$ km s^{-1} Mpc^{-1}. Further, measurements of the Hubble constant employing a variety of techniques suggested a rather high value for it: $H_0 = 80 \pm 5$ km s^{-1} Mpc^{-1} [39]. Also, the *Hubble Space Telescope* (HST) measurements of Cepheid variable stars in the Virgo cluster evidence such high values as $H_0 = 80 \pm 17$ km s^{-1} Mpc^{-1} [36, 71]. If this evidence is correct, it turns out that the age of a flat Big Bang Universe (suggested by inflation) is too small, $t_0 = \frac{2}{3} H_0^{-1} = 8 - 10$ Gyr, to explain the oldest globular clusters estimated to be 16 ± 3 Gyr [85] or 9.3 ± 2 Gyr [99]. This means that some other contribution to the FRW cosmological equations should be present to let the Universe be older. This effect is carried out precisely by a cosmological constant, or function term. This is so because Λ corresponds to a negative pressure (repulsive force) so that the expansion rate first decreases more slowly (than if $\Lambda = 0$) and eventually decreases faster, yielding a larger expansion age.

Astonishingly, recent, independent observational data measured in the CMBR on various angular scales [26, 13], in type Ia supernovae[14] [79, 76, 80], as well as in the 2dF Galaxy Redshift survey [75, 34], suggest that $\Omega = \Omega_\Lambda + \Omega_{\rm m} \approx 1$, or $\Omega_\Lambda \approx 0.7$ and $\Omega_{\rm m} \approx 0.3$, implying the existence of dark energy and dark matter, respectively. One particular candidate for dark energy is a scalar field usually called quintessence [24]. Naturally, particular inflationary scenarios motivated from different particle physics theories have their own dark matter candidates, as such the Axion, neutralino, Higgs particle, etc, and additionally a quintessence field; see the contributions of E. Copeland and A. de la Macorra in this book.

Additionally, three–dimensional numerical simulations of structure formation have incorporated cold, warm, or hot particles into their analyses and, up to now, the best fittings with sky surveys turn out to be a mixture of the different dark matter ingredients with $\Omega_{\rm matter} \sim 0.3$, also including a cosmological constant with $\Omega_\Lambda \sim 0.7$ [4].

Next we comment on the first two problems listed in Sect. 1: the dimensionality and euclidicity problems. They seem to go beyond the scope of the inflationary theory. They touch the foundations of a theory of everything, including gravity. However, some cosmological solutions in Kaluza Klein theories have been found that are related to modern particle physics and gravity theories [37]. With the advent of fundamental string theory, some cosmological solutions have been found that compactify the $D - 4$ dimensions to four [28], and there are inflationary solutions stemming from effective string theories in which various fields exist; one of them, the dilaton, plays the role of the inflaton [41, 42, 16]. The issue of string cosmology has become of much interest in recent years, and modern implementations are accomplished within braneworld scenarios; this topic is extensively explained in the contributions of K. Maeda and J. Lidsey in this book.

[14] See the contribution of A. Filippenko in this book.

Inflation turns out to be a possible, natural, cosmological realization of high energy physics with qualitatively outstanding results. However, a better implementation of it must be achieved, perhaps within new theories or as extensions of the known ones, such as the ones presented in the forthcoming chapters of this book.

Acknowledgements

This work has been supported by CONACYT grant number 33278-E.

References

1. Abbott L F, Farhi E and Wise M B 1982 *Phys. Lett. B* **117** 29.
2. Albrecht A and Steinhardt P J 1982 *Phys. Rev. Lett.* **48** 1220.
3. Albrecht A, Steinhardt P J, Turner M S and Wilczek F 1982 *Phys. Rev. Lett.* **48** 1437.
4. Varios Contributions in 2003 *Rev Mex A A (SC)*, **17**, eds. Avila-Reese V Firmani C, Frenk C S and Allen C.
5. Bahcall N A 1988 *Ann. Rev. Astron. Astrophys.* **26** 631.
6. Balbi A et al. 2000, *Astrophys. J.* **545** L1.
7. Banday A J *et al* 1994 *Astrophys. J.* **436** L99.
8. Bardeen J M, Steinhardt P J and Turner M S 1983 *Phys. Rev. D* **28** 679.
9. Barrow J D 1993 *Phys. Rev. D* **47** 5329.
10. Barrow J D and Sonoda D H 1986 *Phys. Reports* **139** 1.
11. Bennett C L *et al* 1994 *Astrophys. J.* **436** 423.
12. Bennett C L *et al* 1996 *Astrophys. J.* **464** L1.
13. Bennett C L *et al* 2003 *Astrophys. J. Suppl.* **148** 1.
14. Blome H J and Priester W 1991 *Astron. Astrophys.* **250** 43.
15. Boucher W and Gibbons G W 1983 *The very early universe* Editors: G.W. Gibbons and S.W. Hawking (Cambridge Univ. Press) 273.
16. Brustein R, Gasperini M, Giovannini M, Mukhanov V F and Veneziano G 1995 *Phys. Rev. D* **51** 6744.
17. Bunch T S and Davies P C W 1978 *Proc. Roy. Soc. London A* **360** 117.
18. Cahill K and Podolský J 1994 *J. Phys. G* **20** 571.
19. Carrol S, Press W and Turner E 1992 *Ann. Rev. Astron. Astrophys.* **30** 499.
20. Chauvet P and Cervantes-Cota J L 1995 *Phys. Rev. D* **52** 3416.
21. Cervantes-Cota J L 1996 *Ph.D. thesis: Induced gravity and Cosmology* (University of Konstanz, Hartung Verlag)
22. Cervantes-Cota J L and Chauvet P 1999 *Phys. Rev. D* **59** 043501-1.
23. Cervantes-Cota J L 1999 *Class. Quant. Grav.* **16** 3903.
24. Caldwell R R Dave R and Steinhard P J 1998 *Phys. Rev. Lett.*, **80** 1582.
25. Collins C B and Hawking S W 1973 *Astrophys. J.* **180** 317.
26. de Bernardis P. et al, 2000, Nat, 404, 995.
27. de Sitter W 1917 *Proc. Kon. Ned. Akad. Wet.* **19** 1217; ibid **20** 229; 1916 *Mon. Not. R. Astron. Soc.* **76** 699; ibid **77** 155; 1917 ibid **78** 3.

28. de Vega H J and Sánchez N 1995 *Current Topics in Astrofundamental Physics: The Early Universe.* Editors: N. Sánchez and A. Zichichi (Kluwer Acad. Pu.) 99.
29. Dicke R H, Peebles P J E, Roll P G and Willkinson D T 1965 *Astrophys. J.* **142** 414.
30. Dicke R H and Peebles P J E 1979 *General Relativity: An Einstein centenary survey* Editors: S.W. Hawking and W. Israel (Cambridge Univ. Press) 504.
31. Dolgov A D 1992 *Phys. Reports* **222** 309.
32. Dolgov A D 1994 *Nucl. Phys. B* (Proc. Suppl.) **35** 28.
33. Dolgov A D and Linde A D 1982 *Phys. Lett. B* **116** 329.
34. Efstathiou G 2002 *MNRAS* **330** L29.
35. Fabri R and Pollock M D 1983 *Phys. Lett. B* **125** 445.
36. Freedman W L *et al* 1994 *Nature* **371** 757.
37. Freund P G O 1982 *Nucl. Phys. B***209** 146.
38. Friedmann A 1922 Zeit. f. Phys. **10** 377; 1924 ibid **21** 326.
39. Fukugita M, Hogan C J and Peebles P J E 1993 *Nature* **366** 309.
40. Gamov G 1946 *Phys. Rev.* **70** 572; 1948 ibib **74** 505.
41. Gasperini M and Veneziano G 1992 Phys. Lett. B **277** 256;
42. Gasperini M and Veneziano G 1994 *Phys. Rev. D* **50** 2519.
43. Gibbons G W and Hawking S W 1977 *Phys. Rev. D* **15** 2738.
44. Gleiser M, Lect. Notes Phys. 455, 387 (1995). Preprint: hep-ph/9407369.
45. Grishchuk L P 1975 *Sov. Phys. JETP* **40** 409.
46. Guth A H 1981 *Phys. Rev. D* **23** 347.
47. Guth A H and Pi S Y 1982 *Phys. Rev. Lett.* **49** 1110.
48. Hanany S et al. 2000, ApJ, **545**, L5.
49. Harrison E R 1970 *Phys. Rev. D* **1** 2726.
50. Hawking S W 1982 *Phys. Lett. B* **115** 295.
51. Hoell J and Priester W 1991 *Comments Astrophys.* **15** 127.
52. Hoell J and Priester W 1991 *Astron. Astrophys.* **251** L23.
53. Hubble E P 1929 *Proc. Nat. Acad. Sci.* **15** 168.
54. Kaluza T 1921 *Sitzungsber. Preus. Akad. Wiss. Berlin* **1** 966.
55. Kibble T W B 1976 *J. Phys. A* **9** 1387.
56. Klein O 1926 Zeit. Phys. **37** 895; 1926 *Nature* **118** 516.
57. Kofman L, Linde A and Starobinsky A A 1994 *Phys. Rev. Lett.* **73** 3195.
58. Kofman L, Linde A and Starobinsky A A 1996 *Phys. Rev. Lett.* **76** 1011.
59. Kofman L, Linde A and Starobinsky A A 1997 *Phys. Rev. D* **56** 3258.
60. Kolb E W and Turner M S 1990 *The Early Universe* "Frontiers in Physics" no. 69 (Addison-Wesley).
61. Kuzmin V A, Rubakov V A and Shaposhnikov M E 1985 *Phys. Lett. B* **155** 36.
62. Liddle A R and Lyth D H; 2000 *Cosmological Inflation and Large-Scale Structure*, Cambridge U.P., Cambridge, UK.
63. Linde A D 1982 *Phys. Lett. B* **108** 389.
64. Linde A D 1983 *Phys. Lett. B* **129** 177.
65. Linde A D 1990 *Particle Physics and Inflationary Cosmology*, (Harwood Ac.).
66. Lüst D and Theisen S 1989 *Lectures on string theory* (Springer-Verlag).
67. Mimoso J P and Wands D 1995 *Phys. Rev. D* **52** 5612.
68. Misner C W, Thorne K S and Wheeler J A 1973 *Gravitation* (Freeman and Company).

69. Mukhanov V F, H.A. Feldman H A and Brandenberger R H 1992 *Phys. Reports* **215** 203.
70. Narlikar J V 1993 *Introduction to Cosmology* (Cambridge University press).
71. Pierce M *et al* 1994 *Nature* **371** 385.
72. Olive K A 1990 *Phys. Reports* **190** 307.
73. Penzias A A and Wilson R W 1965 *Astrophys. J.* **142** 419.
74. Peacock J A 1999 *Cosmological Physics* (Cambridge University press).
75. Peacock J A et al 2002 in *A New Era in Cosmology* (ASP Conference Proceedings), eds Shanks T and Metcalfe N. Preprint astro-ph/0204239.
76. Perlmutter S et al. 1999 *Astrophys. J.* **517** 565.
77. Polyakov A M 1974 *Sov. Phys. JETP* Lett. **20** 194.
78. Preskill J 1984 *Ann. Rev. Nucl. Part. Sci.* **34** 461.
79. Riess A G et al 1998 *Astronom. J.* **116** 1009.
80. Riess A G et al. 2001 *Astrophys. J.* **560** 49.
81. Robertson H P 1935 *Astrophys. J.* **82** 284; 1936 ibid **83** 187, 257.
82. Rindler W 1956 *Mon. Not. Roy. Astron. Soc.* **116** 663.
83. Rubakov V A, Sazhin M V and Veryaskin A V 1982 *Phys. Lett. B* **115** 189.
84. Sakharov A D 1967 *Sov. Phys. JEPT Lett.* **5** 24.
85. Sarajedini A and King C R 1989 *Astron. J.* **98** 1624.
86. Smoot G et al. 1992 *Astrophys. J.* **396** L1.
87. Starobinsky A A 1979 *Sov. Phys. JETP* Lett. **30** 682.
88. Starobinsky A A 1982 *Phys. Lett. B* **117** 175.
89. Steigman G 1976 *Ann. Rev. Astr. Astrophys.* **14** 339.
90. Steigman G 1979 *Ann. Rev. Nucl Part. Sci.* **29** 313.
91. 't Hooft G 1974 *Nucl. Phys. B***79** 276.
92. 't Hooft G 1976 Phys. Rev. Lett **37** 8; 1976 *Phys. Rev. D* **14** 3432.
93. Traschen J H and Brandenberger R H 1990 *Phys. Rev. D* **42** 2491.
94. Turner M S and Widrow L M 1988 *Phys. Rev. D* **37** 3428.
95. Vilenkin A 1982 *Phys. Lett. B* **117** 25.
96. Walker A G 1937 *Proc. Lond. Math. Soc. (2)* **42** 90.
97. Weinberg S 1972 *Gravitation and Cosmology: principles and applications of the general theory of relativity* (John Wiley & Sons).
98. Weinberg S 1989 *Rev. Mod. Phys.* **61** 1.
99. Winget D E *et al* 1987 *Astrophys. J.* **315** L77.
100. Wright E l *et al* 1994 *Astrophys. J.* **436** 443;
101. Zel'dovich Y B 1972 *Mon. Not. R. Astron. Soc.* **160** 1p.
102. Zel'dovich Y B and Khlopov M Y 1978 *Phys. Lett. B* **79** 239.

Inflation – In the Early Universe and Today

Edmund J. Copeland

Department of Physics and Astronomy, University of Sussex, Brighton, BN1 9QJ, U.K., e.j.copeland@susx.ac.uk

Abstract. In these lectures we will be giving a basic introduction to modern ideas in cosmology. Beginning with a review of the Standard Big Bang (SBB) scenario, we will introduce the observed cosmological parameters and indicate the features that the SBB can not explain, such as the initial conditions. This will lead to an introduction to the inflationary cosmology, which postulates a period of accelerated expansion during the Universe's earliest stages [1, 2, 3, 4, 5]. We will provide some examples of inflationary solutions and demonstrate how they can be used to make distinctive predictions which in principle can be tested with current observations. In particular it provides a possible model for the origin of structure in the Universe. The state of these observations will also be discussed with particular attention being given to the most recent experiments which have detected anisotropies in the cosmic microwave background radiation. We will discuss some of the most exciting developments that have recently emerged in cosmology, arising from string and M-theory models. A particular example of inflation arising out of branes will be given to emphasise the potential new features these solutions have. Finally we will discuss models of Quintessence, scalar field models used to explain the exciting results that the Universe is undergoing a period of acceleration today.

1 The Standard Big Bang Model

The standard hot big bang (SBB) theory is an extremely successful one, and has been around for over 60 years, since Gamow originally proposed it. Remarkably, for such a simple idea, it provides us with an understanding of many of the basic features of our Universe. All that you require in the cooking pot, are initial conditions of an expanding scale factor, gravity, plus the standard particle physics we are used to, to provide the matter in the Universe. It can then pass a number of crucial observational tests.

- The expansion of the Universe – $t_{\mathrm{age}} \sim 10 - 20$ Billion years.
- The existence and spectrum of the cosmic microwave background radiation (CBR) – Planck Back Body spectrum with $T \sim 2.73$K.
- The abundance of light elements in the Universe (nucleosynthesis).
- Gravitational collapse – responsible for the formation of structure in the Universe, although it relies on the presence of initial irregularities being present in the CBR consistent with that detected by the COBE satellite.

E.J. Copeland, Inflation – In the Early Universe and Today, Lect. Notes Phys. **646**, 53–107 (2004)
http://www.springerlink.com/

However, as mentioned the hot big bang theory can successfully proceed only if the initial conditions are very carefully chosen, and even then it only really works at temperatures low enough, so that the underlying physics can be well understood. The very early Universe is out of bounds, yet there is a hope that accurate observations of the present state of the Universe may highlight the types of process occurring during these early stages, providing an insight on the nature of physical laws at energies which it would be inconceivable to explore by other means. Another unresolved issue is the cause of the apparent acceleration of the Universe today, as seen through the distribution of distant Type Ia Supernovae.

To begin with we will give a quick review of the big bang cosmology. Surprisingly, for a theory which is usually associated with solving highly non-linear Einstein equations, it is possible to obtain the key evolution equation (the Friedmann equation) simply from Newtonian cosmology. The hot big bang theory is based on the *cosmological principle*, which states that the Universe should look the same to all observers. That tells us that the Universe must be homogeneous and isotropic. With this in mind, imagine (you will need a healthy imagination throughout these lectures!) a uniform homogeneous 'dust' filled Universe of mass M, with a test particle at radius a. The acceleration experienced by the particle is (ignoring the mass outside of radius a),

$$\ddot{a} = -\frac{GM}{a^2},$$

where G is Newton's constant and $\dot{a} \equiv \frac{da}{dt}$. This can be integrated to give

$$\frac{\dot{a}^2}{2} - \frac{GM}{a} = -\frac{k}{2}, \tag{2}$$

where k is an integration constant. Equation (2) is simply the statement that energy is conserved. Now, for a uniform dust distribution we have

$$M = \frac{4\pi}{3}\rho a^3 = \text{constant},$$

where ρ is the energy density (i.e. mass per unit volume). Substituting for M in (2) we obtain the Friedmann equation,

$$H^2 \equiv \frac{\dot{a}^2}{a^2} = \frac{8\pi G}{3}\rho - \frac{k}{a^2}, \tag{4}$$

where H is the Hubble parameter. This derivation is perfectly adequate because of the assumption of homogeneity. Birkhoff's theorem allows us to consider a region of arbitrary small a, where we expect the Newtonian approximation to be a valid. Homogeneity allows us to then extend this to large a. The parameter $a(t)$ is an important one, it is the 'Scale factor' of the universe, so called because all length scales grow by the same factor $a(t)$ in a homogeneous Universe. It measures the physical size of the Universe. The constant k has a geometrical interpretation (although we need to return to the full General Relativity to see it), it measures the spatial curvature with k negative, zero or positive corresponding to open, flat and closed Universes respectively. The *cosmological principle* tells us which metric must be used to describe it. It is the Robertson–Walker metric, which is given by (3) in the contribution of J.L. Cervantes–Cota in this book.

In cosmology we often use comoving coordinates (r) which are rescaled to the physical coordinates (x) through $x = a(t)r$. Comoving coordinates are so useful, they allow for the expansion of the Universe to be removed from a problem. The crucial link that Einstein spotted was that the geometry of the Universe or its expansion is governed by the properties of material within it. This is specified by the energy density $\rho(t)$ and the pressure $p(t)$, usually related by an equation of state, $\rho = wp$ which gives p as a function of ρ. The key examples are

$$p = \frac{\rho}{3} \quad \text{Radiation}\,, \tag{5}$$

$$p = 0 \quad \text{Non-relativistic matter}\,, \tag{6}$$

$$p = -\rho \quad \text{Vacuum Dominated}\,. \tag{7}$$

In general though there need not be a simple equation of state for example when there is a combination of radiation and non-relativistic matter. This matter satisfies energy momentum conservation, also known as the Fluid equation,

$$\dot{\rho} + 3H(\rho + p) = 0 \tag{8}$$

In (8), $3H\rho$ is the reduction in density due to the increase in volume, and $3Hp$ is the reduction in energy caused by the thermodynamic work done by the pressure when this expansion occurs. Combining (4) and (8) we obtain the useful k-independent Acceleration equation

$$\frac{\ddot{a}}{a} = -\frac{4\pi G}{3}(\rho + 3p)\,. \tag{9}$$

1.1 Standard Big Bang Solutions

In most of what follows we will assume a flat Universe (although there will be exceptions which hopefully will be clear). When $k = 0$ (4) and (8) can be solved for the various equations of state to give the cosmological solutions

$$\text{Matter Domination} \quad p = 0: \quad \rho \propto a^{-3} \quad a(t) \propto t^{2/3} \tag{10}$$

$$\text{Radiation Domination} \quad p = \rho/3: \rho \propto a^{-4} \quad a(t) \propto t^{1/2} \tag{11}$$

$$\text{Vacuum Domination} \quad p = -\rho: \quad \rho \propto \rho_0 \quad a(t) \propto exp(Ht). \tag{12}$$

In both radiation and matter cases the density falls as t^{-2}, whereas in the vacuum (or *cosmological constant* case it remains constant (in general it will be constant or decrease less quickly than $1/t^2$). If there is both matter and radiation the Friedmann equation can be solved using conformal time $\tau = \int dt/a$, while, as we shall see, if there is matter and a non-zero curvature term the solution can be given either in parametric form using normal time t.

1.2 Characteristic Scales and Density Parameters

When the spatial geometry is flat, for a given H, (4) determines the critical density

$$\rho_c(t) = \frac{3H^2}{8\pi G} . \tag{13}$$

Densities are usually measured as fractions of ρ_c:

$$\Omega(t) \equiv \frac{\rho}{\rho_c} , \tag{14}$$

where Ω is known as the density parameter, and can be applied either to individual types of material or to the total density.

The present value of the Hubble parameter is not that well known, so it is parameterized as

$$H_0 = 100h \text{ km s}^{-1} \text{Mpc}^{-1} = \frac{h}{3000} \text{Mpc}^{-1} , \tag{15}$$

where h is normally assumed to lie in the range $0.5 \leq h \leq 0.8$. The current most popular value for h is around $h \simeq .7$ based on a number of different observations. Note, the subscript '0' refers to the present day and reflects the value a parameter has today. Having defined the Hubble parameter and curvature scale, it follows that these can be used to define two length scales: The Hubble time (or length) $H_0^{-1} = 9.8 \times 10^9 h^{-1}$ years gives an approximation to the actual age of the Universe, providing the typical time scale of evolution for $a(t)$. Of course, the Hubble parameter is not constant, varying in general as t^{-1}. The second scale is the curvature scale $a|k|^{-1/2}$ and gives the distance up to which space can be taken as having a flat (Euclidean) geometry. The present critical density is

$$\rho_c(t_0) = 1.88\, h^2 \times 10^{-29} \text{ g cm}^{-3}, \tag{16}$$

incredibly small bearing in mind what we are used to dealing with on earth.

Both the Hubble length and curvature length are physical scales; to obtain the corresponding comoving scale we must divide by $a(t)$. The ratio of these scales actually gives a measure of Ω; from the Friedmann equation we find

$$|\Omega - 1| = \frac{|k|}{H^2 a^2} . \tag{17}$$

A crucial property of the big bang Universe is that it possesses *horizons* which arise because light can only have traveled a finite distance since the start of the Universe t_*. To obtain the horizon, we simply use the fact that light travels on null geodesics ($ds^2 = 0$), see (3) in the contribution of J.L. Cervantes–Cota in this book, hence for fixed θ and ϕ, we obtain $dr = dt/a(t)$ which integrates to give the physical distance

$$d_{\rm H}(t) = a(t) \int_{t_*}^{t} \frac{dt}{a(t)} . \tag{18}$$

In a matter dominated Universe $d_{\rm H}(t) = 3t = 2H^{-1}$; see also Sect. 1 (horizon) in the contribution of Jorge L. Cervantes–Cota in this book.

1.3 Introducing the Cosmic Background Radiation

The redshift measures the expansion of the Universe via the stretching of light

$$1 + z = \frac{a(t_0)}{a(t_{\rm emit})} . \tag{19}$$

As a measure of time, the redshift refers to the time at which light would have to be emitted to have a present redshift z. As a measure of distance, it refers to the *present* distance to an object from which light is received with a redshift z. We can combine (4), (17) and (19) to solve for more general cosmologies involving the situation $k \neq 0$ (remember we are assuming that the cosmological constant vanishes here). Unfortunately it is generally too difficult to obtain explicit solutions in these cases for $a(t)$, rather we obtain $t(a)$. For example, in a matter dominated Universe we obtain

$$\begin{aligned} t_0 &= H_0^{-1} \frac{\Omega}{2(\Omega-1)^{3/2}} \left[\cos^{-1}(2\Omega^{-1}-1) - \frac{2}{\Omega}(\Omega-1)^{1/2} \right] , \quad \Omega > 1 \\ &= H_0^{-1} \frac{\Omega}{2(1-\Omega)^{3/2}} \left[\frac{2}{\Omega}(1-\Omega)^{1/2} - \cosh^{-1}(2\Omega^{-1}-1) \right] , \quad \Omega < 1. \end{aligned} \tag{20}$$

Expanding about $\Omega = 1$ we obtain

$$t_0 \simeq \frac{2}{3} H_0^{-1} \left[1 - \frac{1}{5}(\Omega - 1) + \cdot \right] , \tag{21}$$

implying that for $\Omega < 1$, the Universe is older for a given value of h.

The Universe is full of *radiation*, a remnant of the big bang. The detection of this primordial soup in 1965 by Penzias and Wilson provided one of the major breakthroughs for cosmologists trying to understand the nature of the Universe. Its existence is a prediction of the model and was first proposed in the 1940's by George Gamow. The radiation was emitted at a red shift $z \sim 1100$ the epoch known as the surface of last scattering, corresponding to a time $t \sim 180,000(\Omega h^2)^{(-1/2)}$ years, the time when the photons decoupled from the the electrons as they found their way into their ground state. The temperature then was about 2500 K, and was the moment the Universe went from being opaque to being transparent. Gamow argued that as the Universe expanded and cooled, the photons would be stretched and would today have a temperature of order 10 K and be close to a perfect blackbody spectrum. In 1990, based on just 9 minutes of data, the **CO**smic microwave **B**ackground **E**xplorer satellite (COBE) detected the radiation and showed that it was almost perfectly isotropic, with a Planck blackbody spectrum of $T = 2.735 \pm 0.01$ K. This corresponds to a photon density in the Universe today of $n_\gamma = 422\mathrm{cm}^{-3}$. The radiation is peaked at wavelength $\lambda = 2$ mm corresponding to a frequency $\nu = 150$GHz (i.e. in the microwave region of the electromagnetic spectrum). Of course, this temperature is today much lower than it was in the early Universe. This is because as the Universe expands, it cools. We can determine the rate that it cools through the following simple argument. At high density, because of the high interaction rate, any matter rapidly approaches thermal equilibrium. For radiation, Planck showed us that a quanta of frequency ν had energy $E = h\nu = h/\lambda \propto a(t)^{-1}$ where h is Planck's constant, and the scaling with the scale factor simply represents the fact that all length scales are stretched by the expansion of the Universe. The corresponding energy density in radiation evolves as $\rho_\gamma(t) = E_\gamma/V \propto a(t)^{-4}$. From the world of statistical mechanics we have Stefan's law which tells us[1] $\rho_\gamma \propto T^4$, from which we see that the Universe cools as it expands according to the law

$$T \propto \frac{1}{a} . \tag{22}$$

In its earliest stages the Universe may have been arbitrarily hot and dense, so although matter dominated since nucleosynthesis, far enough in the past it will have been dominated by radiation.

1.4 The Mass of the Universe

How much mass is there in the Universe and can we determine the answer? This is a crucial issue in cosmology. Earlier, we defined the density parameter $\Omega(t) \equiv \rho/\rho_c$. This is the parameter which is important if we are to determine the future evolution of the Universe. Current bounds on its value place it between $.3 < \Omega < 1.2$, but what is it composed of? One of the most significant and successful predictions of the SBB is Nucleosynthesis, the formation of the lightest elements in the Universe. The spectrum of these elements can be predicted and has been compared to observation through numerous experiments. Basically as the Universe expands and cools, it reaches a critical temperature (around 1 MeV) when the reversible reaction of the neutron decaying into protons ceases and the neutron freezes out. The neutron's then

[1] Compare to (21) in the contribution of J.L. Cervantes–Cota in this book.

decay solely into protons and the lightest elements begin to form. This is not the place to discuss nucleosynthesis in detail, it involves analysing rate equations to determine the net fractions of light elements formed, but it is worth discussing the main results. The actual fraction of the light elements formed depends sensitively on the density of ordinary matter (the baryons) at the time of nucleosynthesis. This can vary from $10^{-30} - 10^{-31}$gm cc^{-1}. The light elements formed are Hydrogen (75% by mass), Helium (24% by mass), Deuterium (10^{-5} compared to hydrogen), Helium3 (10^{-5} compared to hydrogen), Lithium 7 (10^{-10} compared to hydrogen). These incredibly small numbers are crucial as slight variations in the baryon density lead to large changes in the abundance, changes that can be ruled out by observation. The key feature though as far as we are concerned are bounds on the total amount of **baryonic matter** in the Universe. The current Nucleosynthesis constraints give

$$\Omega_{\text{baryon}} = (0.019 \pm .0012)h^{-2}. \tag{23}$$

So the maximum contribution to the total energy density arising from baryons is bounded by $\Omega_{\text{baryon}} < 0.08$. However, we just saw earlier that the lower bound on Ω from all sources is $\Omega > .3$. This arises from a number of observations including analysing the dynamics of clusters of galaxies, from the gravitational lensing of distant quasars by rich clusters of galaxies and by determining the baryon abundance in the centres of clusters of galaxies. All of these long distance observations provide bounds of

$$\Omega_{\text{matter}} = (0.3 \pm .05)h^{-(1/2)}. \tag{24}$$

The conclusion that emerges from comparing (23) and (24) are incredibly significant and one of the principle reasons why particle physicists should be interested in cosmology. Clearly, some of the matter in the Universe must be **non-baryonic** and it must be dark, we can not see it. This is clear simply from analysing the rotation curves of light emitted from neighbouring galaxies. These can be interpreted as placing constraints on the matter distribution in our own neighbourhood and points to the existence of large almost spherical dark halos around our visible galaxy.

The current boom in high precision cosmic microwave background experiments such as BOOMERANG, MAXIMA and DASI have enabled a new and exciting approach on the matter question. Since these lectures were given WMAP has come on the scene, and so we should really make use of their wonderful data (whilst bearing in mind the pioneering work of the other high precision experiments). Given a particular cosmological model, there is an associated distribution of peaks and troughs in the power spectrum associated with the anisotropies in the CBR, the position and height of which depend on the nature of the cosmological parameters. In particular the location in ℓ-space of the first 'doppler' peak depends on the quantity $\Omega_{\text{matter}} + \Omega_\Lambda$, where Ω_Λ is the density parameter associated with a cosmological constant (and will be discussed later), through $\ell \sim 220/\sqrt{\Omega_{\text{matter}} + \Omega_\Lambda}$. Based on parameter fits and combining their data with other astronomical data, the WMAP team find that the best fits are [6]: $\Omega_{\text{matter}} + \Omega_{\Lambda} = 1.02 \pm .002$, with $\Omega_{\text{matter}} h^2 = .135^{+.008}_{-.009}$ and $\Omega_{\text{baryons}} h^2 = .0224 \pm .0009$.

This all opens up the intriguing question, what could be the source of this dark matter? There are a number of particle physics candidates including cold (non-relativistic at decoupling) particles such as Axions, neutralinos, primordial black holes and supermassive non-thermal relics. There are also Hot particles possible such as massive neutrinos. At first sight the amazing discovery by the Super-Kamiokande team of evidence for massive neutrinos could be thought to point in this direction, however, the proposed masses for the light neutrinos appear to be too light for them to play a significant cosmological role. Another fascinating aspect of the matter question is the fact that even with dark matter present there is still too little matter to cause the Universe to be flat as can be seen from (17). This amount of matter would lead to an open Universe but as we shall shortly discuss the Universe appears to be flat today. Where then is the remaining contribution required to yield $\Omega = 1$? We shall see that it appears to be coming from an unusual source, namely something is providing an energy contribution through an effective cosmological constant which is dominating the Universe today – a dark energy contribution!

1.5 The Timetable for the Universe

Up until the mid 1990's, any cosmology book would state with some authority that the present Universe is dominated by non-relativistic matter which scales as $\rho \propto a^{-3}$. Since we know radiation reduces more quickly with the expansion, this implies that at earlier times the Universe was radiation dominated. We can estimate this period by simply relating the two contributions to obtain $1/a_{eq} = 2.4. \times 10^4 (\Omega h^2)$ where the two energy densities are equal when the scale factor is a_{eq}. During the radiation era, since $a \propto t^{1/2}$, temperature and time are related by

$$\frac{t}{1\,\text{sec}} \simeq \left(\frac{10^{10}\,\text{K}}{T}\right)^2 \simeq \left(\frac{\text{MeV}}{T}\right)^2 . \tag{25}$$

From this we quickly see that

$$\begin{aligned} T &\sim \text{MeV} \Longleftrightarrow t \sim \text{sec} \\ T &\sim \text{GeV} \Longleftrightarrow t \sim 10^{-6}\text{sec} \\ T &\sim 10^{15}\text{GeV} \Longleftrightarrow t \sim 10^{-35}\text{sec}. \end{aligned} \tag{26}$$

The highest energies accessible to terrestrial experiment, generated in particle accelerators, correspond to a temperature of about $10^{15}\,$K, which was attained when the Universe was about $10^{-10}\,$sec old. Earlier times rely on extrapolation of our known physics and possibly new mathematical insights (such as string theory may offer). Later times appear to be well understood, with a possible timetable being:

- 10^{-34} seconds: Grand unified phase transition, where strong force decouples from electroweak force.
- 10^{-10} seconds: Electroweak phase transition, where weak force decouples from electromagnetic force. Possible origin for the observed baryon asymmetry in the Universe.
- 10^{-4} seconds: Quarks condense to form protons and neutrons.
- 1 second: The Universe has cooled sufficiently that light nuclei are able to form – **nucleosynthesis**.
- 10^4 years: **Matter–radiation equality**. Subsequently the Universe is matter dominated.
- 10^5 years: Decoupling of radiation from matter leads to the formation of the microwave background. Similar time to recombination, when the up-to-now free electrons combine with the nuclei to form atoms.
- 10^{10} years: The present era where Beckham joins Real Madrid for a bargain of £20M – and the Universe is accelerating!.

We have seen that up to two years ago, this would have been the accepted folklore. However, it now appears that the Universe, rather than steadily decelerating is actually accelerating, going faster and faster. The evidence for this lies in the distribution of Type Ia Supernovae at very large scales, but the conclusion is dramatic. As we shall shortly discover, an accelerating universe requires an energy source which effectively has a negative pressure. There is something out there providing this constant source of energy density, and it is dominating over everything else present – there appears to be a cosmological constant present in the Universe today! We will return to this amazing issue later on.

2 Problems with the Big Bang

There are a number of issues that the SBB simply can not address and have to adopt as initial conditions. These provided the original motivation for the inflationary cosmology, and we now turn our attention to these issues.

2.1 The Flatness Problem

In the absence of a cosmological constant contribution, the Friedmann equation can be written in terms of the density parameter equation (17). During SBB evolution, a^2H^2 is decreasing, and so Ω moves away from one, for example

$$\text{Matter domination:} \quad |\Omega - 1| \propto t^{2/3} \tag{27}$$

$$\text{Radiation domination:} \quad |\Omega - 1| \propto t \tag{28}$$

where the solutions apply provided Ω is close to one. So $\Omega = 1$ is an *unstable* critical point. However, today Ω is certainly within an order of magnitude of one, so it must have been much closer in the past. Inserting the appropriate behaviours for the matter and radiation eras gives for example

$$\text{nucleosynthesis } (t \sim 1\,\text{sec}) : |\Omega - 1| < \mathcal{O}(10^{-16}) \qquad (29)$$

That is, hardly any choices of the initial density lead to a Universe like our own. Typically, the Universe will either swiftly recollapse, or will rapidly expand and cool below 3K within its first second of existence.

2.2 The Horizon Problem

The observation by COBE that all cosmic microwave photons appear to be in thermal equilibrium at almost the same temperature is a puzzle? Why is it so isotropic? It is not difficult to see that in the SBB the Universe has not had enough time for different regions to reach a state of thermal equilibrium by today. The regions could not have interacted before the photons were emitted because of the finite horizon size,

$$\int_{t_*}^{t_{\rm dec}} \frac{dt}{a(t)} \ll \int_{t_{\rm dec}}^{t_0} \frac{dt}{a(t)} \,. \qquad (30)$$

In other words, the distance light could travel before the microwave background was released is much smaller than the present horizon distance. In fact, any regions separated by more than about 2 degrees would be causally separated at decoupling in the hot big bang theory. In the big bang theory there is therefore no explanation of why the Universe appears so homogeneous.

The same argument that prevents the smoothing of the Universe also prevents the creation of irregularities. The COBE satellite has detected irregularities in the CMB on all large angular scales, too large to be accounted for as emerging in the period between the big bang and the time of decoupling, because the horizon size at decoupling subtends only a degree or so. Hence these perturbations must have been part of the initial conditions.

2.3 The Monopole Problem

Modern particle theories predict a variety of 'unwanted relics', which can not be present today as they would have dramatically altered the evolution of the Universe. These include magnetic monopoles, domain walls, gravitinos and moduli fields associated with the extra dimensions arising in superstring theories. They are all massive particles created in the very early Universe but are diluted less rapidly than radiation as the Universe expands. Hence they would rapidly come to dominate the dynamics, and lead to rapid closure of the Universe. We must eliminate them, while preserving the rest of the matter which we like.

2.4 The Cosmological Constant

When Einstein developed the theory of General Relativity, the consensus was that the Universe was static, after all there was no apparent movement of the galaxies in the night sky with respect to one another. However, the Friedmann equations were unstable to a static solution, not surprising since matter attracts. To resolve the problem, Einstein introduced a constant balancing term (Λ) which would allow for a static solution to exist. The modified Friedmann equation now became

$$H^2 = \frac{8\pi G}{3}\rho + \frac{\Lambda}{3} - \frac{k}{a^2}, \tag{31}$$

with the new *cosmological constant* appearing as a constant contribution to the Hubble parameter. As soon as the Universe was discovered to be expanding, the need for this term went away. Unfortunately, the equations suggested otherwise, there was no apriori reason to set $\Lambda = 0$. This term has haunted cosmologists and particle physicists ever since and could well have just come back to visit us again today. What is the problem then?

From (31), as we are not dominated by the curvature term and since the present energy density is close to the critical value, we see that today,

$$|\Lambda| \leq H_0^2. \tag{32}$$

Thus the length scale $\ell_\Lambda \equiv |\Lambda|^{-1/2}$ associated with the cosmological constant must be larger than $H_0^{-1} = h_0^{-1} \times 10^{26}$ m, a macroscopic distance. In a classical regime this is fine, it is simply saying the cosmological constant length scale is larger than the Hubble length. Problems arise when we combine gravity and quantum mechanics. At the quantum level the natural scales which emerge at the Planck epoch are the Planck mass and Planck length given by (where we have reinserted Planck constant and the speed of light)

$$m_P = \sqrt{\frac{\hbar c}{8\pi G}} = 2.4 \times 10^{18}\ \mathrm{GeV/c^2},$$

$$\ell_P = \frac{\hbar}{m_P c} = 8.1 \times 10^{-35}\ \mathrm{m}$$

The above constraint now reads :

$$\ell_\Lambda \equiv |\Lambda|^{-1/2} \geq \frac{1}{H_0} \sim 10^{60}\ \ell_P. \tag{33}$$

There are more than sixty orders of magnitude between the scale associated with the cosmological constant and the scale of quantum gravity. We could of course simply set $\Lambda = 0$, indeed this is generally what is done. Unfortunately when there is matter hanging around this is not such a good idea. The matter itself experiences quantum fluctuations (called zero-point fluctuations) and these can act like an effective cosmological constant ($\Lambda_{\mathrm{eff}} \sim \lambda^4/m_P^2$). In fact the natural value this constant should then have usually reflects the scale associated with these quantum fluctuations. So, λ would be typically of the order of 100 GeV in the case of the gauge symmetry breaking of the Standard Model or 1 TeV in the case of supersymmetry breaking. But the constraint (33) now reads:

$$\lambda \leq 10^{-30}\, m_P \sim 10^{-3} \text{ eV}. \tag{34}$$

It is this very unnatural fine-tuning of parameters that is referred to as the *cosmological constant problem*, or more accurately the vacuum energy problem.

3 Enter Inflation

Inflation is defined to be any epoch where $\ddot{a} > 0$, an accelerated expansion. From (9) this corresponds to a negative pressure ($p < -\frac{\rho}{3}$) and from the definition $H = \frac{\dot{a}}{a}$, we see that it also corresponds to $\frac{d(H^{-1}/a)}{dt} < 0$, i.e. the Hubble length as measured in comoving coordinates, *decreases* during inflation. At any other time, the comoving Hubble length increases. This is the key property of inflation; although typically the expansion of the Universe is very rapid, the crucial characteristic scale of the Universe is actually becoming smaller, when measured relative to that expansion.

We have already seen an example of an inflationary solution, the vacuum dominated regime $p = -\rho$, has a solution is

$$a(t) \propto \exp(Ht)\ . \tag{35}$$

There are many many more! Of course, we know the SBB has many successes, and it is none inflating, so inflation can not last for ever, it must terminate and enter the SBB regime smoothly at some epoch.

3.1 The Flatness Problem

Inflation solves the flatness problem by rapidly forcing Ω towards unity rather than away from it. This is clear from the fact that the comoving Hubble length H^{-1}/a is decreasing. We require enough inflation to force Ω extremely close to unity to ensure that it will remain close to it today. Remember, as soon as we enter the SBB phase, $\Omega = 1$ is an unstable point. Including a possible cosmological constant contribution, modifies the Friedmann equation to

$$|\Omega + \Omega_\Lambda - 1| = \frac{|k|}{a^2 H^2}\,, \tag{36}$$

and so it is $\Omega + \Omega_\Lambda$ which is forced to one. In general, it is spatial flatness ($k \simeq 0$) that we are driven towards, not a critical matter density.

3.2 Relic Abundances

The rapid expansion of the inflationary stage rapidly dilutes the unwanted relic particles, because the energy density during inflation falls off more slowly than the relic particle density. Very quickly their density becomes negligible. Of course they do not disappear totally and will one day re-enter the horizon – the ultimate in sweeping something under the carpet.

We need to ensure that after inflation, the energy density of the Universe can be turned into conventional matter without recreating the unwanted relics. This *reheating* period must have a temperature that never gets hot enough to allow their thermal recreation. It will then allow for the particles we want to create and lead naturally into the SSB period, vital for the success of nucleosynthesis and the CMB.

3.3 The Horizon Problem and Homogeneity

Inflation rapidly increases the size of any region of the Universe, but it keeps its characteristic scale, the Hubble scale fixed. So, a small patch of the Universe, small enough for thermalisation before inflation, can expand to a patch much larger than the size of our presently observable Universe. This ensures that all the cosmic microwave radiation are in thermal equilibrium. Moreover, it also allows for irregularities to be generated in the CMB, irregularities which would then evolve to form structures. We can rephrase the horizon solution by saying that because of inflation, light can travel much further before decoupling than it can afterwards.

3.4 The Cosmological Constant

Unfortunately, a period of inflation says nothing about why the present value of the cosmological constant should be so small. In fact it should now be clear that inflation effectively relies on such a constant if only for a finite period of time.

4 Inflation out of Particle Physics

The most common framework in which inflation is obtained is based on the existence of scalar fields, in particular scalar field potentials. Needless to say, as far as particle physics is concerned they remain elusive – yet we really need them! They represent spin zero particles, transforming as a scalar (that is, it is unchanged) under coordinate transformations. In a homogeneous Universe, the scalar field is a function of time alone.

The traditional starting point for particle physics models is the action, which is an integral of the Lagrange density over space and time and from which the equations of motion can be obtained. A scalar field Lagrangian is like one for a particle, the difference between the kinetic energy and the potential energy of the field

$$L = \frac{1}{2}(\partial_\mu \phi)(\partial^\mu \phi) - V(\phi). \tag{37}$$

The stress energy tensor is defined in terms of the Lagrangian

$$T_{\mu\nu} = (\partial_\mu \phi)(\partial_\nu \phi) - L g_{\mu\nu}, \tag{38}$$

where $g_{\mu\nu}$ is the metric tensor. If ϕ represents an isotropic fluid then we can write down the pressure and energy density from the definition

$$T^\mu_\nu = \mathrm{diag}(-\rho, p, p, p), \tag{39}$$

from which we obtain for a homogeneous field

$$\rho_\phi = \frac{1}{2}\dot{\phi}^2 + V(\phi) \tag{40}$$

$$p_\phi = \frac{1}{2}\dot{\phi}^2 - V(\phi)\,. \tag{41}$$

The potential energy $V(\phi)$ measures how much internal energy is associated with a particular field value. Normally, like all systems, scalar fields try to minimize this energy; however, a crucial ingredient which allows inflation is that scalar fields are not always very efficient at reaching this minimum energy state. In a given theory, there would be a specific form for the potential $V(\phi)$. However, we are not presently in a position where there is a well established fundamental theory that one can use, so, in the absence of such a theory, inflation workers tend to regard $V(\phi)$ as a function to be chosen arbitrarily, with different choices corresponding to different models of inflation. Some example potentials are

$$V(\phi) = \lambda\left(\phi^2 - M^2\right)^2 \quad \text{Higgs potential} \tag{42}$$

$$V(\phi) = \tfrac{1}{2}m^2\phi^2 \quad \text{Massive scalar field} \tag{43}$$

$$V(\phi) = \lambda\phi^4 \quad \text{Self-interacting scalar field} \tag{44}$$

4.1 Inflation Dynamics

The equations for an expanding Universe containing a homogeneous scalar field are easily obtained by substituting (40) and (41) into the Friedmann and fluid equations, giving

$$H^2 = \frac{8\pi G}{3}\left[V(\phi) + \frac{1}{2}\dot{\phi}^2\right], \tag{45}$$

$$\ddot{\phi} + 3H\dot{\phi} = -V'(\phi)\,, \tag{46}$$

where prime indicates $d/d\phi$. Here we have ignored the curvature term k, since we know that by definition it will quickly become negligible once inflation starts. Since

$$\ddot{a} > 0 \Longleftrightarrow p < -\frac{\rho}{3} \Longleftrightarrow \dot{\phi}^2 < V(\phi) \tag{47}$$

we will have inflation whenever the potential energy dominates. This should be possible provided the potential is flat enough, as the scalar field would then be expected to roll slowly. The potential should also have a minimum or some other feature which would allow inflation to end.

To solve these equations we use the **slow-roll approximation** (SRA), which assumes that a term can be neglected in each of the equations of motion to leave the simpler set

$$H^2 \simeq \frac{8\pi G}{3} V \tag{48}$$

$$3H\dot{\phi} \simeq -V' \tag{49}$$

The **slow-roll parameters** fist introduced by Liddle and Lyth [14]

$$\epsilon(\phi) \equiv \frac{1}{16\pi G}\left(\frac{V'}{V}\right)^2 \quad ; \quad \eta(\phi) \equiv \frac{1}{8\pi G}\frac{V''}{V}, \tag{50}$$

measures the slope of the potential (ϵ), and the curvature (η), and the necessary conditions for the slow-roll approximation to hold are

$$\epsilon \ll 1 \quad ; \quad |\eta| \ll 1\,. \tag{51}$$

4.2 The Amount of Inflation

The amount of inflation is normally specified by the *the number of e-foldings* N, given by

$$N \equiv \ln \frac{a(t_{\text{end}})}{a(t_{\text{initial}})} = \int_{t_i}^{t_e} H\,dt\,, \tag{52}$$

$$\simeq -8\pi G \int_{\phi_i}^{\phi_e} \frac{V}{V'}\,d\phi\,, \tag{53}$$

where the final step uses the SRA. Notice that the amount of inflation between two scalar field values can be calculated without needing to solve the equations of motion, and also that it is unchanged if one multiplies $V(\phi)$ by a constant. We can estimate the amount of inflation required to solve the various cosmological problems. Consider the flatness problem. First we make a few plausible assumptions to ease the situation: inflation is of the exponential form ending at $t = 10^{-34}$ sec, with the Universe immediately entering a radiation era which persists until today some 3×10^{17} sec later. Imagine also that today $|\Omega - 1| \leq 0.01$, a reasonable constraint on the value of Ω. Now during the radiation era, from (17), $|\Omega-1| \propto t$, hence $|\Omega(10^{-34}\,\mathrm{sec})-1| \leq 3\times 10^{-54}$. During inflation H is constant, so $|\Omega - 1| \propto \frac{1}{a^2}$. From this it follows that in order to satisfy the constraint by the end of inflation, the scale factor has to grow during inflation by an amount

$$\frac{a_{\mathrm{tend}}}{a_{\mathrm{tbegin}}} \sim 10^{27} \sim exp(62), \tag{54}$$

corresponding to around 62 e-foldings. Although this looks large, inflation is typically so rapid that most inflation models give much more.

4.3 Some Examples of Inflation: Polynomial Chaotic Inflation

A particularly nice example of an inflaton potential is a simple polynomial potential first introduced by Linde (for a review see [7]). It could be a massive non-interacting field, $V(\phi) = m^2\phi^2/2$ where m is the mass of the scalar field, or it could be a massless self-interacting field, $V(\phi) = \lambda\phi^4$, where λ is the self coupling of the field. Consider the first case. The slow-roll equations are

$$3H\dot{\phi} + m^2\phi = 0 \quad ; \quad H^2 = \frac{4\pi G m^2\phi^2}{3}, \tag{55}$$

and the slow-roll parameters are

$$\epsilon = \eta = \frac{1}{4\pi G\phi^2}, \tag{56}$$

implying that inflation can proceed provided $|\phi| > 1/\sqrt{4\pi G}$, i.e. away from the minimum.

The solutions to the equations give

$$\phi(t) = \phi_{\mathrm{i}} - \frac{m}{\sqrt{12\pi G}}\,t, \tag{57}$$

$$a(t) = a_{\mathrm{i}} \exp\left[\sqrt{\frac{4\pi G}{3}}\,m\left(\phi_{\mathrm{i}} t - \frac{m}{\sqrt{48\pi G}} t^2\right)\right], \tag{58}$$

(where $\phi = \phi_{\mathrm{i}}$ and $a = a_{\mathrm{i}}$ at $t = 0$) and the total amount of inflation is

$$N_{\mathrm{tot}} = 2\pi G\,\phi_{\mathrm{i}}^2 - \frac{1}{2}. \tag{59}$$

An important thing to bear in mind is that we need to ensure that we are in a position where classical physics remains a valid approximation. This is simply the requirement $V \ll G^{-2}$, but it is still easy to get enough inflation provided m is small enough. In fact, m is required to be small from observational limits on the size of density perturbations produced.

As an exercise the reader may want to try and repeat the exercise for potential $V(\phi) = \lambda\phi^4$, assuming the field starts at $t = 0$ from rest rolling towards $\phi = 0$ from the positive side of the potential. Show that the SR equations give

$$\phi(t) = \phi_{\rm i}\, exp\left(-\sqrt{\frac{2\lambda}{3\pi G}}\,t\right), \tag{60}$$

$$a(t) = a_{\rm i}\exp\left\{\phi_{\rm i}^2\pi G\left[1 - exp\left(-\sqrt{\frac{4\lambda}{3\pi G}}\,t\right)\right]\right\}, \tag{61}$$

(where $\phi = \phi_{\rm i}$ and $a = a_{\rm i}$ at $t = 0$) and the total amount of inflation is

$$N_{\rm tot} = \pi G\,\phi_{\rm i}^2 - 1\,. \tag{62}$$

Since these lectures were given, the WMAP results are now beginning to place constraints on the viability of polynomial inflation models, and it appears that the $\lambda\phi^4$ marks the boundary between viable models (powers less than 4) and un-viable models (powers greater than 4)[8]. However, a word of caution. The 'real' inflaton potential is likely to be a bit more complicated than a simple single scalar field power law model, so lets not get too excited yet about ruling out large classes of potentials.

4.4 From Inflation to the SBB – Reheating

During inflation, all matter except the inflaton scalar field is redshifted to extremely low densities. **Reheating** is the process whereby the inflaton's energy density is converted back into conventional matter after inflation, re-entering the standard big bang theory.

As the slow-roll conditions break down, ϕ evolves from being overdamped to being underdamped, moving rapidly on the Hubble timescale and oscillating at the bottom of the potential, where it decays into conventional matter. This is an active and technically demanding area of research and there has recently been something of a revolution in the way we think reheating takes place. Traditional treatments (e.g. as given in Kolb & Turner[9]) added a phenomenological decay term; this was constrained to be very small with reheating being inefficient. In particular there was a long time delay (redshifting) between the end of inflation and the Universe returning to thermal equilibrium; hence a low reheat temperature compared to the energy density at the end of inflation.

In **preheating** [10], this picture is turned on its head. Kofman et al have shown that the decay can initially proceed through broad parametric resonance, with extremely efficient transfer of energy from the coherent oscillations of the inflaton field. The result is a very short reheating period, with most of the inflaton energy density at the end of inflation available for conversion into thermalized form. A higher reheat temperature is possible with some amazing possibilities, such as non-thermal phase transitions [11] and baryogenesis occurring at the electroweak scale[12, 13].

4.5 Inflation Models

There are a number of models on offer, some better motivated than others. **Chaotic inflation models** are the generic type found in a number of situations because they just require a single scalar field, rolling in a potential $V(\phi)$, which in some regions satisfies the slow-roll conditions, while also possessing a minimum with zero potential in which inflation is to end. the initial conditions place the field well up the potential, and could be due to large fluctuations at the Planck era. Examples include (see [14])

Polynomial chaotic inflation	$V(\phi) = \frac{1}{2}m^2\phi^2$
	$V(\phi) = \lambda\phi^4$
Power-law inflation	$V(\phi) = V_0 \exp(\sqrt{\frac{16\pi G}{p}}\,\phi)$
'Natural' inflation	$V(\phi) = V_0[1 + \cos\frac{\phi}{f}]$
Intermediate inflation	$V(\phi) \propto \phi^{-\beta}$

Hybrid inflation models are a very interesting class as they have more than one scalar field and appear to offer the possibility of occurring in particle physics contexts. An example is one with a potential

$$V(\phi, \psi) = \frac{\lambda}{4}\left(\psi^2 - M^2\right)^2 + \frac{1}{2}m^2\phi^2 + \frac{1}{2}\lambda'\phi^2\psi^2 \,. \tag{63}$$

When ϕ^2 is large, the minimum of the potential is in the ψ-direction is at $\psi = 0$. The field rolls down this 'valley' until it reaches $\phi^2_{\rm inst} = \lambda M^2/\lambda'$, where $\psi = 0$ becomes unstable and the field rolls into one of the true minima at $\phi = 0$ and $\psi = \pm M$. Note for suitable choices of the potential, topological defects could form at the end of a period of inflation.

While in the 'valley', it is like a single field model with an effective potential for ϕ of the form

$$V_{\rm eff}(\phi) = \frac{\lambda}{4}M^4 + \frac{1}{2}m^2\phi^2 \,. \tag{64}$$

The constant term would not normally be allowed as it would give a present-day cosmological constant. When it dominates, it allows both for the energy density during inflation to be much lower than normal while still giving suitably large density perturbations, and for ϕ to roll very slowly.

Models of inflation can also be found in scalar-tensor theories of gravity where the gravitational constant may vary. One interesting case arises from the low energy string action, where two scalar fields, the dilaton and moduli field lead to a period of inflation driven not by the potential energy of the fields (in fact the potential vanishes), rather by the kinetic energy of the fields. This interesting possibility is known as the **pre big bang model**, so called because this evolution occurs before the usual big bang singularity is met. These will be discussed in Sect. 5. There are also fascinating models which lead to Open universes, but likewise we do not have time to discuss them here.

4.6 Density Perturbations and Gravitational Waves

Perhaps the most important property of inflationary cosmology is that it produces spectra of both density perturbations and gravitational waves. The former would be responsible for the formation and clustering of galaxies, as well as creating anisotropies in the microwave background radiation. The gravitational waves do not affect the formation of galaxies, but may contribute extra microwave anisotropies on the large angular scales sampled by the COBE satellite.

The beauty of having models like inflation, or topological defects (which we do not discuss here) is that they are predictive. We can predict the form of the initial perturbation spectra, as opposed to simply assuming it as is often done in studies of large-scale structure. For example, the gravitational waves may be assumed not to be present, and the density perturbations to take on a simple form such as the scale-invariant Harrison–Zel'dovich spectrum, or a scale-free power-law spectrum. In his lectures Robert Brandenberger has gone into a great deal of detail describing the cosmological perturbations theory [15]. Here we will just be picking out the bits useful for inflation without deriving any of the formalism.

4.7 Perturbations Produced During Inflation

Inflation generates perturbations on large scales because the comoving Hubble length decreases during inflation, where as in the SBB the comoving Hubble length is always increasing, all scales are initially much larger than it, and hence unable to be affected by causal physics. Once they become smaller than the Hubble length, they remain so for all time. The fact that COBE sees perturbations on large scales at a time when they were much bigger than the Hubble length, means that in the standard picture no mechanism could have created them.

During inflation a given comoving scale could be well inside the Hubble length, and hence be affected by causal physics, thereby enabling it to generate homogeneity to solve the horizon problem and to superimpose small quantum perturbations. Before inflation ends, as the comoving Hubble length decreases, the given scale crosses outside the Hubble radius rendering causal physics ineffective. Any perturbations generated become imprinted, or, 'frozen in'. Long after inflation is over, as the comoving Hubble length increases the scales cross inside the Hubble radius again. Perturbations are created on a very wide range of scales, but the most readily observed ones range from about the size of the present Hubble radius (i.e. the size of the presently observable Universe) down to a few orders of magnitude less. Thus inflation allows perturbations to be generated causally. These quantum fluctuations are present simply from the Uncertainty principle, they have to be and can explain the initial inhomogeneities that later grow by gravitational collapse to the structures we see today.

The size of the irregularities depends on the energy scale at which inflation takes place. It is outside the scope of these lectures to describe in detail how this calculation is performed. We will just briefly outline the necessary steps and then quote the result, which we can go on to apply (see Liddle and Lyth [14] for details).

(a) Perturb the scalar field	$\phi = \phi(t) + \delta\phi(\mathbf{x}, t)$
(b) Expand in comoving wavenumbers	$\delta\phi = \sum(\delta\phi)_{\mathbf{k}} e^{i\mathbf{k}.\mathbf{x}}$
(c) Linearized equation for classical evolution	
(d) Quantize theory	
(e) Find solution with initial condition giving flat space quantum theory ($k \gg aH$)	
(f) Find asymptotic value for $k \ll aH$	$\langle\lvert\delta\phi_{\mathbf{k}}\rvert^2\rangle = H^2/2k^3$
(g) Relate field perturbation to metric or curvature perturbation	$\mathcal{R}_{\mathbf{k}} = H\,\delta\phi_{\mathbf{k}}/\dot{\phi}$

Unfortunately, analytic results are not known for general inflation models. The results given below are lowest-order in the SRA. There are results known to second-order in slow-roll for arbitrary inflaton potentials. Power-law inflation is the only standard model for which exact results are known.

The curvature perturbation $\mathcal{R}_{\mathbf{k}}$ is so important because it provides the vital bridge which allows us to link the primordial fluctuations in ϕ to the physically observed matter fluctuations present today. The reason is that it remains effectively constant for scales much greater than the co-moving Hubble length, hence the scales that 'freeze-in' as they leave the Hubble length during inflation, remain unaffected (apart from stretching due to the Universe expanding) until they re-enter the Hubble radius much later in the standard big bang era. Given the definition of the power spectra

$$\mathcal{P}_g(k) = \frac{k^3}{2\pi^2}|\delta g_k|^2 , \tag{65}$$

then the amplitude of the density perturbation $\delta_{\rm H}^2(k) = \frac{4}{25}\mathcal{P}_{\mathcal{R}}$ is given by

$$\begin{aligned} \delta_{\rm H}^2(k) &= \frac{4}{25}\frac{k^3}{2\pi^2}\left(\frac{H\delta\phi_{\mathbf{k}}}{\dot{\phi}}\right)^2 \\ &= \left|\cdot\frac{4}{25}\left(\frac{H}{\dot{\phi}}\right)\left(\frac{H}{2\pi}\right)^2\right|_{k=aH} . \end{aligned} \tag{66}$$

Using the SRA this then gives

$$\delta_{\rm H}(k) = \sqrt{\frac{512\pi}{75}}\left.\frac{V^{3/2}G^{3/2}}{|V'|}\right|_{k=aH} . \tag{67}$$

A similar calculation gives the amplitude of the gravitational waves

$$A_{\rm G}(k) = \sqrt{\frac{32}{75}}\, V^{1/2}G\Bigg|_{k=aH} , \tag{68}$$

where $A_{\rm G}^2(k) = \frac{1}{25}\mathcal{P}_g(k)$.

The right-hand sides of the above equations are to be evaluated at the time when the comoving wave number $k = aH$ during inflation, which for a given k corresponds to some particular value of ϕ. We see that the amplitude of perturbations depends on the properties of the inflaton potential at the time the scale crossed the Hubble radius during inflation. The relevant number of e-foldings from the end of inflation is approximately given by

$$N \simeq 62 - \ln \frac{k}{a_0 H_0} \tag{69}$$

Approximating this to to say that the scales of interest to us crossed outside the Hubble radius 60 e-foldings before the end of inflation then gives

$$N \simeq -8\pi G \int_{\phi}^{\phi_{\rm end}} \frac{V}{V'}\, d\phi\,, \tag{70}$$

which tells us the value of ϕ to be substituted into (67) and (68).

We can apply this formalism to the specific example of the $m^2\phi^2/2$ potential. Inflation ends when $\epsilon = 1$, so $\phi_{\rm end} \simeq 1/\sqrt{4\pi G}$. 60 e-foldings before this, gives $\phi_{60} \simeq \frac{3}{\sqrt{G}}$ from (59), which in turn upon substitution yields

$$\delta_{\rm H} \simeq 12\, m\sqrt{G} \quad ; \quad A_{\rm G} \simeq 1.4\, m\sqrt{G}.$$

The COBE result corresponds to $\delta_{\rm H} \simeq 2 \times 10^{-5}$ (provided $A_{\rm G} \ll \delta_{\rm H}$), hence $m\sqrt{G} \simeq 10^{-6}$ and we have an inflaton mass of $m = 10^{13}$ GeV. Such a small mass satisfies the condition $VG^2 < 1$, which implies that $\phi < 1/(mG) \simeq 10^6(1/\sqrt{G})$. Since $N_{\rm tot} \simeq 2\pi\phi^2 G$, we can get up to about 10^{13} e-foldings in principle, compared with the 70 or so actually required.

4.8 Observational Consequences

The current high precision CMB experiments like BOOMERANG, MAXIMA I and now WMAP are beginning to probe key features of the spectra, such as the scale-dependence and the relative size of the two spectra. Again the slow-roll parameters ϵ and η can be used to estimate these quantities for any given inflation potential. The standard approximation used to describe the spectra is the **power-law approximation**, where we take[2]

$$\delta_{\rm H}^2(k) \propto k^{n-1} \quad ; \quad A_{\rm G}^2(k) \propto k^{n_{\rm G}}\,, \tag{71}$$

[2] Cf. (53) in the contribution of J.L. Cervantes–Cota in this book.

where the spectral indices[3] n and n_{G} are given by

$$n-1=\frac{d\ln\delta_{\mathrm{H}}^2}{d\ln k} \quad ; \quad n_{\mathrm{G}}=\frac{d\ln A_{\mathrm{G}}^2}{d\ln k}\,. \tag{72}$$

The power-law approximation is usually valid because only a limited range of scales are observable, with the range 1 Mpc to 10^4 Mpc corresponding to $\Delta\ln k \simeq 9$.

The crucial equation we need is that relating ϕ values to when a scale k crosses the Hubble radius,

$$\frac{d\ln k}{d\phi}=8\pi G\frac{V}{V'}\,. \tag{73}$$

This comes from the noticing the right hand side of the amplitude equations are evaluated when $k=aH$, and during inflation $\dot{H}$ is very small compared to the rate of change of a. Hence we can take $d\ln\, k = Hdt$, from which it follows $k\simeq\exp N$. Then make use of (70). Direct differentiation then yields[14]

$$n = 1-6\epsilon+2\eta\,, \tag{74}$$
$$n_{\mathrm{G}} = -2\epsilon\,, \tag{75}$$

where now ϵ and η are to be evaluated on the appropriate part of the potential.

A measure of the relevant importance of density perturbations and gravitational waves is seen in the microwave background which gives gives

$$R\equiv\frac{C_\ell^{\mathrm{GW}}}{C_\ell^{\mathrm{DP}}}\simeq 4\pi\epsilon\,. \tag{76}$$

where the C_ℓ are the contributions to the microwave multipoles. Briefly, the temperature difference between two regions of the sky separated by (θ,ϕ) is given in terms of spherical harmonics Y_m^ℓ as $\Delta T/T=\sum a_{\ell m}Y_m^\ell(\theta,\phi)$ where $C_\ell=\langle|a_{\ell m}|^2\rangle$; see (56) in the contribution of Jorge L. Cervantes–Cota in this book.

From n, n_G and R, it follows that if and only if $\epsilon \ll 1$ and $|\eta| \ll 1$ do we get $n\simeq 1$ and $R\simeq 0$ whereas gravitational waves can have a significant effect even if ϵ is quite a bit smaller than one.

Different models predict different things which implies that large-scale structure observations, and especially microwave background observations, can strongly discriminate between inflationary models. When they are made, most existing inflation models will be ruled out. As an example the recent WMAP data appears to be placing the $\lambda\phi^4$ model under some pressure [8]. Fortunately, inflation as an idea has one very useful and hopefully unique test, which will allow it to be verified or ruled out, independent of the particular

[3] The scalar spectral index, n, is sometimes referred as n_s.

Table 1. The spectral index and gravitational wave contribution for a range of inflation models – taken from Liddle - astro-ph/9910110.

MODEL	POTENTIAL	n	R
Polynomial	ϕ^2	0.97	0.1
chaotic inflation	ϕ^4	0.95	0.2
Power-law inflation	$\exp(-\lambda\phi)$	any $n < 1$	$2\pi(1-n)$
'Natural' inflation	$1+\cos(\phi/f)$	any $n < 1$	0
Hybrid inflation (standard)	$1+B\phi^2$	1	0
Hybrid inflation (extreme)	$1+B\phi^2$	$1 < n < 1.15$	~ 0

model being investigated. There exists a *consistency equation*

$$R = -2\pi n_{\mathrm{G}}\,, \tag{77}$$

independent of the choice of inflationary model (though it does rely on the slow-roll and power-law approximations). There are no other models that produce such a relation, unfortunately as we have already seen it may turn out that the gravitational wave contribution is so small that the consistency equation can never be verified!

4.9 The Cosmological Parameters

Cosmologists are aiming to fully understand and explain the origin and contents of our Universe, and this includes all the parameters that make it up. So far, we have discussed three primordial ones, δ_H, n and R which describe the initial perturbations laid down in the first 10^{-34} sec or so. Most of the perturbations except the largest ones just re-entering the horizon today, have been heavily processed by real astrophysics to give the non-linear features we observe. We can break the parameters up into cosmological and inflationary:

Inflationary parameters: $\delta_H\,, n\,, n_G\,, R, dn/d\ln k\,\cdot$

Cosmological parameters: $h\,, \Omega_{\mathrm{baryon}}\,, \Omega_{\mathrm{CDM}}\,, \Omega_{\mathrm{HDM}}\,, \Omega_{\mathrm{Lambda}}\,, k\,, g_*\,, \tau\,\cdot,$

where g_* is the number of massless species of particles and τ is the reionisation optical depth. As we mentioned earlier, through a combination of observations and parameter fitting techniques these parameters are already being constrained The recent WMAP data coupled with other astronomical data have led to the following published constraints: $n = 1.13 \pm 0.08$, $dn/d\ln k = -0.055^{+0.028}_{-0.029}$, $\Omega_{\mathrm{matter}} + \Omega_\Lambda = 1.02 \pm .002$, with $\Omega_{\mathrm{matter}} h^2 = .135^{+.008}_{-.009}$ and $\Omega_{\mathrm{baryons}} h^2 = .0224 \pm .0009$[6, 8]. Note there may be tentative evidence for a running of the spectral index, something that would be highly significant if it holds.

The COBE normalization allows the energy scale associated with inflation to be determined, since it is probing perturbations still in their primordial form, dependent only on the initial seed perturbations. Using the present Hubble scale, $\delta_{\mathrm{H}} \equiv \delta_{\mathrm{H}}(k = a_0 H_0)$, to be given by the COBE normalisation

$$\delta_{\mathrm{H}} \simeq 2 \times 10^{-5}\,, \tag{78}$$

then since

$$\delta_{\mathrm{H}}^2 = \frac{32}{75} G^2\, V\, \frac{1}{\epsilon}\,, \tag{79}$$

this implies

$$V^{1/4} \simeq 10^{-3}/\sqrt{G} \simeq 10^{16}\,\mathrm{GeV}\,, \tag{80}$$

at the time when observable scales crossed outside the horizon. A scale consistent with many GUT models.

5 String Cosmology

String theory, and its most recent incarnation, that of M-theory, has been accepted by many as the most likely candidate theory to unify the forces of nature as it includes General Relativity in a consistent quantum theory. If it is to play such a pivotal role in particle physics, it should also include in it all of cosmology. It should provide the initial conditions for the Universe, perhaps even explain away the singularity associated with the standard big bang. It should also provide a mechanism for explaining the observed density fluctuations, perhaps by providing the inflaton field or some other mechanism which would lead to inflation. Should the observations survive the test of time, string theory should be able to provide a mechanism to explain the current accelerated expansion of the Universe. In other words, even though it is strictly a theory which can unify gravity with the other forces in the very early Universe, for consistency, as a theory of everything it will have a great deal more to explain. In this article, we will introduce some of the developments that have occurred in string cosmology over the past decade or so, initially basing the discussion on an analyse of the low energy limit of string theory, and then later extending it to include branes arising in Heterotic M-theory.

5.1 Dilaton-Moduli Cosmology (Pre-Big Bang)

Strings live in 4+d spacetime dimensions, with the extra d dimensions being compactified. For homogeneous, four–dimensional cosmologies, where all fields are uniform on the surfaces of homogeneity, we can consider the compactification of the $(4+d)$–dimensional theory on an isotropic d–torus. The radius, or 'breathing mode' of the internal space, is then parameterized by a modulus field, β, and determines the volume of the internal dimensions. We can then assume that the $(4+d)$–dimensional metric is of the form

$$ds^2 = -dt^2 + g_{ij}dx^i dx^j + e^{\sqrt{2/d}\beta}\delta_{ab}dX^a dX^b \tag{81}$$

where indices run from $(i,j) = (1,2,3)$ and $(a,b) = (4,\ldots,3+d)$ and δ_{ab} is the d–dimensional Kronecker delta. The modulus field β is normalized in such a way that it becomes minimally coupled to gravity in the Einstein frame.

The low energy action that is commonly used as a starting point for string cosmology is the four dimensional effective Neveu-Schwarz- Neveu-Schwarz (NS-NS) action given by:

$$S_* = \int d^4x\sqrt{|g|}e^{-\varphi}\left[R + (\nabla\varphi)^2 - \frac{1}{2}(\nabla\beta)^2 - \frac{1}{2}e^{2\varphi}(\nabla\sigma)^2\right], \tag{82}$$

where φ is the effective dilaton in four dimensions, and σ is the pseudo–scalar axion field which is dual to the fundamental NS–NS three–form field strength present in string theory, the duality being given by

$$H^{\mu\nu\lambda} = \epsilon^{\mu\nu\lambda\kappa}\, e^{\varphi}\, \nabla_\kappa\sigma. \tag{83}$$

The dimensionally reduced action (82) may be viewed as the prototype action for string cosmology because it contains many of the key features common to more general actions. Cosmological solutions to these actions have been extensively discussed in the literature – for a review see [16]. Some of them play a central role in the pre–big bang inflationary scenario, first proposed by Veneziano [17, 18]. An important point can be seen immediately in (82) where there is a non-trivial coupling of the dilaton to the axion field, a coupling which will play a key role later on when we are investigating the density perturbations arising in this scenario.

All homogeneous and isotropic external four–dimensional spacetimes can be described by the Friedmann-Robertson-Walker (FRW) metric. The general line element in the string frame can be written as

$$ds_4^2 = a^2(\eta)\left\{-d\eta^2 + d\Omega_\kappa^2\right\}, \tag{84}$$

where $a(\eta)$ is the scale factor of the universe, η is the conformal time and $d\Omega_\kappa^2$ is the line element on a 3-space with constant curvature κ:

$$d\Omega_\kappa^2 = d\psi^2 + \left(\frac{\sin\sqrt{\kappa}\psi}{\sqrt{\kappa}}\right)^2 (d\theta^2 + \sin^2\theta d\varphi^2) \tag{85}$$

To be compatible with a homogeneous and isotropic metric, all fields, including the pseudo–scalar axion field, must be spatially homogeneous.

The models with vanishing form fields, but time-dependent dilaton and moduli fields, are known as *dilaton-moduli-vacuum* solutions. In the Einstein–frame, these solutions may be interpreted as FRW cosmologies for a stiff perfect fluid, where the speed of sound equals the speed of light. The dilaton and moduli fields behave collectively as a massless, minimally coupled scalar field, and the scale factor in the Einstein frame is given by

$$\tilde{a} = \tilde{a}_* \sqrt{\frac{\tau}{1+\kappa\tau^2}} \tag{86}$$

where $\tilde{a} \equiv e^{-\varphi/2}a$, $\tilde{a}_*$ is a constant and we have defined a new time variable:

$$\tau \equiv \begin{cases} \kappa^{-1/2}|\tan(\kappa^{1/2}\eta)| & \text{for } \kappa > 0 \\ |\eta| & \text{for } \kappa = 0 \\ |\kappa|^{-1/2}|\tanh(|\kappa|^{1/2}\eta)| & \text{for } \kappa < 0 \end{cases} . \tag{87}$$

The time coordinate τ diverges at both early and late times in models which have $\kappa \geq 0$, but $\tau \to |\kappa|^{-1/2}$ in negatively curved models. There is a curvature singularity at $\eta = 0$ with $\tilde{a} = 0$ and the model expands away from it for $\eta > 0$ or collapses towards it for $\eta < 0$. The expanding, closed models recollapse at $\eta = \pm\pi/2$ and there are no bouncing solutions in this frame.

The corresponding string frame scale factor, dilaton and modulus fields are given by the 'rolling radii' solutions [19]

$$a = a_* \sqrt{\frac{\tau^{1+\sqrt{3}\cos\xi_*}}{1+\kappa\tau^2}} , \tag{88}$$

$$e^{\varphi} = e^{\varphi_*}\tau^{\sqrt{3}\cos\xi_*} , \tag{89}$$

$$e^{\beta} = e^{\beta_*}\tau^{\sqrt{3}\sin\xi_*} \tag{90}$$

The integration constant ξ_* determines the rate of change of the effective dilaton relative to the volume of the internal dimensions. Figures 1 and 2 show the dilaton-vacuum solutions in flat FRW models when stable compactification has occurred, so that the volume of the internal space is fixed, with $\xi_* \bmod \pi = 0$.

The solutions just presented have a scale factor duality which when applied simultaneously with time reversal implies that the Hubble expansion parameter $H \equiv d(\ln a)/dt$ remains invariant, $H(-t) \to H(t)$, whilst its first derivative changes sign, $\dot{H}(-t) \to -\dot{H}(t)$. A decelerating, post–big bang solution – characterized by $\dot{a} > 0$, $\ddot{a} < 0$ and $\dot{H} < 0$ – is mapped onto a pre–big bang phase of inflationary expansion, since $\ddot{a}/a = \dot{H} + H^2 > 0$. The Hubble radius H^{-1} decreases with increasing time and the expansion is therefore super-inflationary. Thus, the pre-big bang cosmology ($\kappa = 0$ case in (88–90)) is one that has a period of super-inflation driven simply by the kinetic energy of the dilaton and moduli fields [17, 18]. This is related by duality to the usual FRW post–big bang phase. The two branches are separated by a curvature singularity, however, and it is not clear how the transition between the pre– and post–big bang phases might proceed. This will be the focus of attention in Sect. 5.

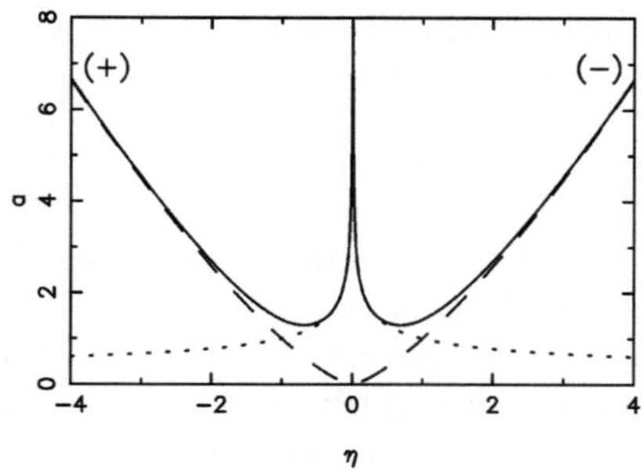

Fig. 1. String frame scale factor, a, as a function of conformal time, η, for flat $\kappa = 0$ FRW cosmology in dilaton-vacuum solution in (88) with $\xi_* = 0$ (dashed-line), $\xi_* = \pi$ (dotted line) and dilaton-axion solution in (93) with $r = \sqrt{3}$ (solid line). The (+) and (−) branches are defined in the text.

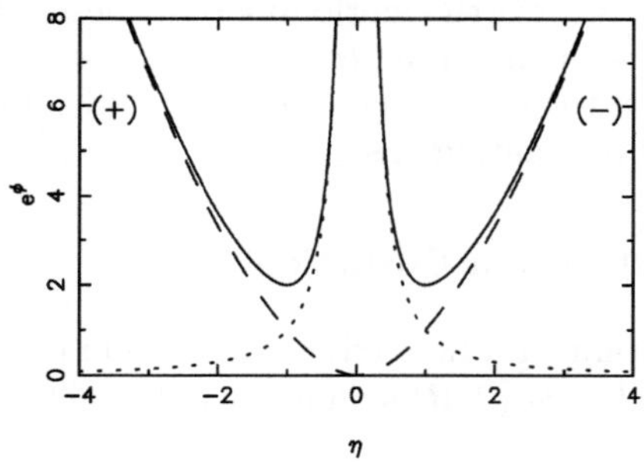

Fig. 2. Dilaton, e^{φ}, as a function of conformal time, η, for flat $\kappa = 0$ FRW cosmology in dilaton-vacuum solution in (89) with $\xi_* = 0$ (dashed-line), $\xi_* = \pi$ (dotted line) and dilaton-axion solution in (92) with $r = \sqrt{3}$ (solid line).

The solution for a flat ($\kappa = 0$) FRW universe corresponds to the well–known monotonic power-law, or 'rolling radii', solutions. For $\cos \xi_* < -1/\sqrt{3}$ there is accelerated expansion, i.e., inflation, in the string frame for $\eta < 0$ and $e^{\varphi} \to 0$ as $t \to -\infty$, corresponding to the weak coupling regime. The expansion is an example of 'pole–law' inflation [20, 21].

The solutions have semi-infinite proper lifetimes. Those starting from a singularity at $t = 0$ for $t \geq 0$ are denoted as the (−) branch in [22], while those which approach a singularity at $t = 0$ for $t \leq 0$ are referred to as the (+) branch (see Figs. 1–2). These (+/−) branches do *not* refer to the choice of sign for $\cos \xi_*$. On either the (+) or (−) branches of the dilaton-moduli-vacuum cosmologies we have a one-parameter family of solutions corresponding to the choice of ξ_*, which determines whether e^{φ} goes to zero or infinity as $t \to 0$. These solutions become singular as the conformally invariant time parameter $\eta \equiv \int dt/a(t) \to 0$ and there is no way of naively connecting the two branches based simply on these solutions [22].

In the Einstein frame, where the dilaton field is minimally coupled to gravity, the scale factor given in (86), becomes

$$\tilde{a} = \tilde{a}_* |\eta|^{1/2} \tag{91}$$

As $\eta \to 0$ on the $(+)$ branch, the universe is collapsing with $\tilde{a} \to 0$, and the comoving Hubble length $|d(\ln \tilde{a})/d\eta|^{-1} = 2|\eta|$ is decreasing with time. Thus, in both frames there is inflation taking place in the sense that a given comoving scale, which starts arbitrarily far within the Hubble radius in either conformal frame as $\eta \to -\infty$, inevitably becomes larger than the Hubble radius in that frame as $\eta \to 0$. The significance of this is that it means that perturbations can be produced in the dilaton, graviton and other matter fields on scales much larger than the present Hubble radius from quantum fluctuations in flat spacetime at earlier times – this is a vital property of any inflationary scenario.

For completeness, it is worth mentioning that these solutions can be extended to include a time-dependent axion field, $\sigma(t)$, by exploiting the SL$(2,R)$ S-duality invariance of the four–dimensional, NS-NS action [19]. We now turn our attention to this fascinating case.

5.2 Dilaton-Moduli-Axion Cosmologies

The cosmologies containing a non–trivial axion field can be generated immediately due to the global SL$(2,R)$ symmetry of the action (82). The resultant solutions are [19]:

$$e^{\varphi} = \frac{e^{\varphi_*}}{2}\left\{\left(\frac{\tau}{\tau_*}\right)^{-r} + \left(\frac{\tau}{\tau_*}\right)^{r}\right\}, \tag{92}$$

$$a^2 = \frac{a_*^2}{2(1+\kappa\tau^2)}\left\{\left(\frac{\tau}{\tau_*}\right)^{1-r} + \left(\frac{\tau}{\tau_*}\right)^{1+r}\right\}, \tag{93}$$

$$e^{\beta} = e^{\beta_*}\tau^s , \tag{94}$$

$$\sigma = \sigma_* \pm e^{-\varphi_*}\left\{\frac{(\tau/\tau_*)^{-r} - (\tau/\tau_*)^{r}}{(\tau/\tau_*)^{-r} + (\tau/\tau_*)^{r}}\right\}, \tag{95}$$

where the exponents are related via

$$r^2 + s^2 = 3 , \tag{96}$$

and without loss of generality we may take $r \geq 0$.

In all cases, the dynamics of the axion field places a *lower* bound on the value of the dilaton field, $\varphi \geq \varphi_*$. In so doing, the axion smoothly interpolates between two dilaton–moduli–vacuum solutions, where its dynamical influence asymptotically becomes negligible. The effects of time–dependent axion solutions for the scale-factor and dilaton are plotted in Figs. 1 and 2 for the flat FRW model when the modulus field is trivial ($s = 0$). When the internal space is static, it is seen that the string frame scale factors exhibit a bounce. However we still have a curvature singularity in the Einstein frame as $\tau \to 0$. The actual time-dependent axion solutions is shown in Fig. 3.

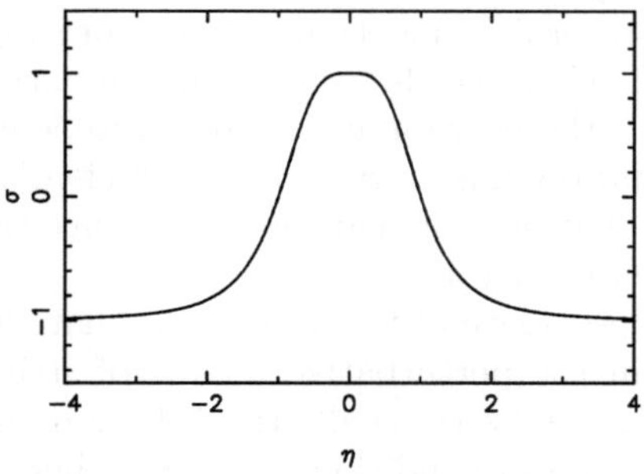

Fig. 3. Axion, σ, as a function of conformal time, η, for flat $\kappa = 0$ FRW cosmology in dilaton-axion solution in (95) with $r = \sqrt{3}$ (solid line).

The spatially flat solutions reduce to the power law, dilaton–moduli–vacuum solution given in (88–90) at early and late times. When $\eta \to \pm\infty$ the solution approaches the vacuum solution with $\sqrt{3}\cos\xi_* = +r$, while as $\eta \to 0$ the solution approaches the $\sqrt{3}\cos\xi_* = -r$ solution. Thus, the axion solution interpolates between two vacuum solutions related by an S-duality transformation $\varphi \to -\varphi$. When the internal space is static the scale factor in the string frame is of the form $a \propto t^{1/\sqrt{3}}$ as $\eta \to \pm\infty$, while as $\eta \to 0$ the solution becomes $a \propto t^{-1/\sqrt{3}}$. These two vacuum solutions are thus related by a scale factor duality that inverts the spatial volume of the universe. This asymptotic approach to dilaton–moduli–vacuum solutions at early and late times will lead to a particularly simple form for the semi-classical perturbation spectra that is independent of the intermediate evolution. However, there is a down side to these solutions from the standpoint of pre big bang cosmologies. As $\eta \to \pm\infty$ and as $\eta \to 0$ the solution approaches the strong coupling regime where $e^{\varphi} \to \infty$. Thus there is no weak coupling limit, the axion interpolates between two strong coupling vacuum solutions. We will shortly see how a similar affect arises when we include a moving brane in the dilaton-moduli picture, as it too mimics the behaviour of a non-minimally coupled axion field.

The overall dynamical effect of the axion field is negligible except near $\tau \approx \tau_*$, when it leads to a bounce in the dilaton field. Within the context of M–theory cosmology, the radius of the eleventh dimension is related to the dilaton by $r_{11} \propto e^{\varphi/3}$ when the modulus field is fixed. This bound on the dilaton may therefore be reinterpreted as a lower bound on the size of the eleventh dimension.

5.3 Fine Tuning Issues

The question over the viability of the initial conditions required in the pre Big Bang scenario has been a cause for many an argument both in print and in person. Since both $\dot{H}$ and $\dot{\varphi}$ are positive in the pre–big bang phase, the initial values for these parameters must be *very small.* This raises a number of important issues concerning fine–tuning in the pre–big bang scenario [23, 24, 25, 26, 27, 28, 29]. There needs to be enough inflation in a homogeneous patch in order to solve the horizon and flatness problems which means that the dilaton driven inflation must survive for a sufficiently long period of time. This is not as trivial as it may appear, however, since the period of inflation is limited by a number of factors.

The fundamental postulate of the scenario is that the initial data for inflation lies well within the perturbative regime of string theory, where the curvature and coupling are very small [18]. Inflation then proceeds for sufficiently homogeneous initial conditions [27, 28], where time derivatives are dominant with respect to spatial gradients, and the universe evolves into a high curvature and strongly–coupled regime. Thus, the pre–big bang initial state should correspond to a cold, empty and flat vacuum state. Initial the universe would have been huge relative to the quantum scale and hence should have been well described by classical solutions to the string effective action. This should be compared to the initial state which describes the standard hot big bang, namely a dense, hot, and highly curved region of spacetime. This is quite a contrast and a primary goal of pre–big bang cosmology must be to develop a mechanism for smoothly connecting these two regions, since we believe that the standard big bang model provides a very good representation of the current evolution of the universe.

Our present observable universe appears very nearly homogeneous on sufficiently large scales. In the standard, hot big bang model, it corresponded to a region at the Planck time that was 10^{30} times larger than the horizon size, $l_{\rm Pl}$. This may be viewed as an initial condition in the big bang model or as a final condition for inflation. It implies that the comoving Hubble radius, $1/(aH)$, must decrease during inflation by a factor of at least 10^{30} if the horizon problem is to be solved. For a power law expansion, this implies that

$$\left|\frac{\eta_f}{\eta_i}\right| \leq 10^{-30} \tag{97}$$

where subscripts i and f denote values at the onset and end of inflation, respectively. In the pre–big bang scenario, (89) implies that the dilaton grows as $e^{\varphi} \propto |\eta|^{-\sqrt{3}}$, and since at the start of the post–big bang epoch, the string coupling, $g_s = e^{\varphi/2}$, should be of order unity, the bound (97) implies that the initial value of the string coupling is strongly constrained, $g_{s,i} \leq 10^{-26}$. Turner and Weinberg interpret this constraint as a severe fine–tuning problem in the scenario, because inflation in the string frame can be delayed by the effects of spatial curvature [23]. It was shown by Clancy, Lidsey and Tavakol that the bounds are further tightened when spatial anisotropy is introduced, actually preventing pre–big bang inflation from occurring [24]. Moreover, as we have seen the dynamics of the NS–NS axion field also places a lower bound on the allowed range of values that the string coupling may take [19]. In the standard inflationary scenario, where the expansion is quasi–exponential, the Hubble radius is approximately constant and $a \propto (-\eta)^{-1}$. Thus, the homogeneous region grows by a factor of $|\eta_i/\eta_f|$ as inflation proceeds. During a pre–big bang epoch, however, $a \propto (-\eta)^{-1/1+\sqrt{3}}$ and the increase in the size of a homogeneous region is reduced by a factor of at least $10^{30\sqrt{3}/(1+\sqrt{3})} \approx 10^{19}$ relative to that of the standard inflation scenario. This implies that the initial size of the homogeneous region should exceed 10^{19} in string units if pre–big bang inflation is to be successful in solving the problems of the big bang model [17, 25]. The occurrence of such a large number was cited by Kaloper, Linde and Bousso as a serious problem of the pre–big bang scenario, because it implies that the universe must already have been large and smooth by the time inflation began [25].

On the other hand, Gasperini has emphasized that the initial homogeneous region of the pre–big bang universe is not larger than the horizon even though it is large relative to the string/Planck scale [30]. The question that then arises when discussing the naturalness, or otherwise, of the above initial conditions is what is the basic unit of length that should be employed. At present, this question has not been addressed in detail.

Veneziano and collaborators conjectured that pre–big bang inflation generically evolves out of an initial state that approaches the Milne universe in the semi–infinite past, $t \to -\infty$ [27, 28]. The Milne universe may be mapped onto the future (or past) light cone of the origin of Minkowski spacetime and therefore corresponds to a non–standard representation of the string perturbative vacuum. The proposal was that the Milne background represents an early time attractor, with a large measure in the space of initial data. If so, this would provide strong justification for the postulate that inflation begins in the weak coupling and curvature regimes and would render the pre-big bang assumptions regarding the initial states as 'natural'. However, Clancy *et al.* took a critical look at this conjecture and argued that the Milne universe is an unlikely past attractor for the pre–big bang scenario [31]. They suggested that plane wave backgrounds represent a more generic initial state for the universe [24]. Buonanno, Damour and Veneziano have subsequently proposed that the initial state of the pre–big bang universe should correspond to an ensemble of gravitational and dilatonic waves [29]. They refer to this as the state of 'asymptotic past triviality'. When viewed in the Einstein frame these waves undergo collapse when certain conditions are satisfied. In the string frame, these gravitationally unstable areas expand into homogeneous regions on large scales.

To conclude this section, it is clear that the question of initial conditions in the pre–big bang scenario is currently unresolved. We turn our attention now to another unresolved problem for the scenario – the Graceful Exit.

5.4 The Graceful Exit

We have seen how in the pre Big Bang scenario, the Universe expands from a weak coupling, low curvature regime in the infinite past, enters a period of inflation driven by the kinetic energy associated with the massless fields present, before approaching the strong coupling regime as the string scale is reached. There is then a branch change to a new class of solutions, corresponding to a post big bang decelerating Friedman-Robertson-Walker era. In such a scenario, the Universe appears to emerge because of the gravitational instability of the generic string vacua – a very appealing picture, the weak coupling, low curvature regime is a natural starting point to use the low energy string effective action. However, how is the branch change achieved without hitting the inevitable looking curvature singularity associated with the strong coupling regime? The simplest version of the evolution of the Universe in the pre-big bang scenario inevitably leads to a period characterised by an unbounded curvature. The current philosophy is to include higher-order corrections to the string effective action. These include both classical finite size effects of the strings (α' corrections arising in higher order derivatives), and quantum string loop corrections (g_s corrections). The list of authors who have worked in this area is too great to mention here, for a detailed list see [16, 32]. A series of key papers were written by Brustein and Madden, in which they demonstrated that it is possible to include such terms and successfully have an exit from one branch to the other [33, 34]. More recently this approach has been generalised by including combinations of classical and quantum corrections [35]. Brustein and Madden [33, 34] made use of the result that classical corrections can stabilize a high curvature string phase while the evolution is still in the weakly coupled regime[36]. The crucial new ingredient that they added was the inclusion of terms of the type that may result from quantum corrections to the string effective action and which induce violation of the null energy condition (NEC – The Null Energy Condition is satisfied if $\rho + p \geq 0$, where ρ and p represent the effective energy density and pressure of the additional sources). Such extra terms mean that evolution towards a decelerated FRW phase is possible. Of course this violation of the null energy condition can not continue indefinitely, and eventually it needs to be turned off in order to stabilise the dilaton at a fixed value, perhaps by capture in a potential minimum or by radiation production – another problem for string theory!

The analysis of [33] resulted in a set of necessary conditions on the evolution in terms of the Hubble parameters H_S in the string frame, H_E in the Einstein frame and the dilaton φ, where they are related by $H_E = e^{\varphi/2}(H_S - \frac{1}{2}\dot{\varphi})$. The conditions were:

– Initial conditions of a (+) branch and $H_S, \dot{\varphi} > 0$ require $H_E < 0$.
– A branch change from (+) to (−) has to occur while $H_E < 0$.
– A successful escape and exit completion requires NEC violation accompanied by a bounce in the Einstein frame after the branch change has occurred, ending up with $H_E > 0$.
– Further evolution is required to bring about a radiation dominated era in which the dilaton effectively decouples from the "matter" sources.

There is as yet no definitive calculation of the full loop expansion of string theory. This is of course a big problem if we want to try and include quantum effects in analysing the graceful exit issue. The best we can do, is to propose plausible terms that we hope are representative of the actual terms that will eventually make up the loop corrections. We believe that the string coupling g_S actually controls the importance of string-loop corrections, so as a first approximation to the loop corrections we multiplied each term of the classical correction by a suitable power of the string coupling [33, 34].

Not surprisingly the field equations need to be solved numerically, but this can be done and the solutions are very encouraging as they show there exists a large class of parameters for which successful graceful exits are obtained [35]. One such example is shown in Fig. 4.

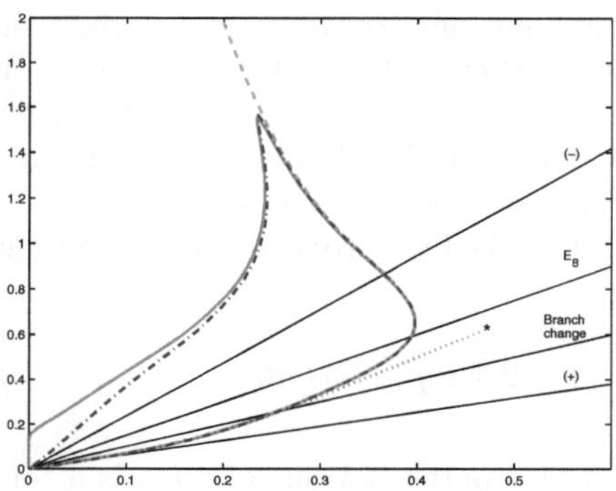

Fig. 4. Hubble expansion in the S-frame as a function of the dilaton for a successful exit. The y-axis corresponds to H, and the x-axis to $2\dot{\varphi}/3$. The initial conditions for the simulations have been set with respect to the lowest-order analytical solutions at $t_S = -1000$. For details see [35]

We should point out though, that although it is possible to have a successful exit, it is not so easy to ensure that the exit takes place in a weakly coupled regime, and typically we found that as the exit was approached $\varphi_{\rm final} \sim 0.1 - -0.3$. Thus it is fair to say that although great progress has been made on the question of Graceful Exit in string cosmology, it remains a problem in search of the full solution. It is a fascinating problem, and not surprisingly alternative prescriptions which aim to address this issue have recently been proposed, involving colliding branes [37] and Cyclic universes [39]. We now turn our attention to the observational consequences of string cosmology, in particular the generation of the observed cosmic microwave background radiation.

5.5 Density Perturbations in String Cosmology

We have to consider inhomogeneous perturbations that may be generated due to vacuum fluctuations, and follow the formalism pioneered by Mukhanov and collaborators [40, 41]. During a period of accelerated expansion the comoving Hubble length, $|d(\ln a)/d\eta|^{-1}$, decreases and vacuum fluctuations which are assumed to start in the flat-spacetime vacuum state may be stretched up to exponentially large scales. The precise form of the spectrum depends on the expansion of the homogeneous background and the couplings between the fields. The comoving Hubble length, $|d(\ln \tilde{a})/d\eta|^{-1} = 2|\eta|$, does indeed decrease in the Einstein frame during the contracting phase when $\eta < 0$. Because the dilaton, moduli fields and graviton are minimally coupled to this metric, this ensures that small-scale vacuum fluctuations will eventually be stretched beyond the comoving Hubble scale during this epoch.

As we remarked earlier, the axion field is taken to be a constant in the classical pre-big bang solutions. However, even when the background axion field is set to a constant, there will inevitably be quantum fluctuations in this field. We will see that these fluctuations can not be neglected and, moreover, that they are vital if the pre-big bang scenario is to have any chance of generating the observed density perturbations.

In the Einstein frame, the first-order perturbed line element can be written as

$$d\widetilde{s}^2 = \widetilde{a}^2(\eta)\left\{-(1+2\widetilde{A})d\eta^2 + 2\widetilde{B}_{,i}d\eta dx^i + [\delta_{ij}+h_{ij}]\,dx^i dx^j\right\}, \tag{98}$$

where $\widetilde{A}$ and $\widetilde{B}$ are scalar perturbations and h_{ij} is a tensor perturbation.

5.6 Scalar Metric Perturbations

First of all we consider the evolution of linear metric perturbations about the four-dimensional spatially flat dilaton-moduli-vacuum solutions given in (88–90). Considering a single Fourier mode, with comoving wavenumber k, the perturbed Einstein equations yield the evolution equation

$$\widetilde{A}'' + 2\widetilde{h}\widetilde{A}' + k^2\widetilde{A} = 0\,, \tag{99}$$

plus the constraint

$$\widetilde{A} = -(\widetilde{B}' + 2\widetilde{h}\widetilde{B})\,, \tag{100}$$

where $\widetilde{h}$ is the Hubble parameter in the Einstein frame derived from (91), and $\widetilde{A}' \equiv \frac{d\widetilde{A}}{d\eta}$. In the spatially flat gauge we have the simplification that the evolution equation for the scalar metric perturbation, (99), is independent of the evolution of the different massless scalar fields (dilaton, axion and moduli), although they will still be related by the constraint

$$\widetilde{A} = \frac{\varphi'}{4\widetilde{h}}\,\delta\varphi + \frac{\beta'}{4\widetilde{h}}\,\delta\beta\,, \tag{101}$$

where $\delta\varphi$ and $\delta\beta$ are the perturbations in φ and β respectively. To first-order, the metric perturbation, $\widetilde{A}$, is determined solely by the dilaton and moduli field perturbations, although its evolution is dependent only upon the Einstein frame scale factor, $\widetilde{a}(\eta)$, given by (91), which in turn is determined solely by the stiff fluid equation of state for the homogeneous fields in the Einstein frame.

One of the most useful quantities we can calculate is the curvature perturbation on uniform energy density hypersurfaces (as $k\eta \to 0$). It is commonly denoted by ζ [42]and in the Einstein frame, we obtain

$$\zeta = \frac{\widetilde{A}}{3}\,, \tag{102}$$

in any dilaton–moduli–vacuum or dilaton–moduli–axion cosmology [43, 46].

The significance of ζ is that in an expanding universe it becomes constant on scales much larger than the Hubble scale ($|k\eta| \ll 1$) for purely adiabatic perturbations. In single-field inflation models this allows one to compute the density perturbation at late times, during the matter or radiation dominated eras, by equating ζ at "re-entry" ($k = \widetilde{a}\widetilde{H}$) with that at horizon crossing during inflation. To calculate ζ, hence the density perturbations induced in the pre-big bang scenario we can either use the vacuum fluctuations for the canonically normalised field at early times/small scales (as $k\eta \to -\infty$) or use the amplitude of the scalar field perturbation spectra to normalise the solution for $\widetilde{A}$. This yields, (after some work), the curvature perturbation spectrum on large scales/late times (as $k\eta \to 0$):

$$\mathcal{P}_\zeta = \frac{8}{\pi^2} l_{\mathrm{Pl}}^2 \widetilde{H}^2 (-k\eta)^3 [\ln(-k\eta)]^2\,, \tag{103}$$

where l_{Pl} is the Planck length in the Einstein frame and remains fixed throughout. The scalar metric perturbations become large on superhorizon scales ($|k\eta| < 1$) only near the Planck era, $\widetilde{H}^2 \sim l_{\mathrm{Pl}}^{-2}$.

The spectral index of the curvature perturbation spectrum is conventionally given as [44]

$$n \equiv 1 + \frac{d\ln \mathcal{P}_\zeta}{d\ln k} \tag{104}$$

where $n = 1$ corresponds to the classic Harrison-Zel'dovich spectrum for adiabatic density perturbations favoured by most models of structure formation in our universe. By contrast the pre–big bang era leads to a spectrum of curvature perturbations with $n = 4$. Such a steeply tilted spectrum of metric perturbations implies that there would be effectively no primordial metric perturbations on large (super-galactic) scales in our present universe if the post-Big bang era began close to the Planck scale. Fortunately, as we shall see later, the presence of the axion field could provide an alternative spectrum of perturbations more suitable as a source of large-scale structure. The pre-big bang scenario is not so straightforward as in the single field inflation case, because the full low-energy string effective action possesses many fields which can lead to non-adiabatic perturbations. This implies that density perturbations at late times may not be simply related to ζ alone, but may also be dependent upon fluctuations in other fields.

5.7 Tensor Metric Perturbations

The gravitational wave perturbations, h_{ij}, are both gauge and conformally invariant. They decouple from the scalar perturbations in the Einstein frame to give a simple evolution equation for each Fourier mode

$$h_k'' + 2\widetilde{h}\, h_k' + k^2 h_k = 0\,. \tag{105}$$

This is exactly the same as the equation of motion for the scalar perturbation given in (99) and has the same growing mode in the long wavelength ($|k\eta| \to 0$) limit given by (103). The spectrum depends solely on the dynamics of the scale factor in the Einstein frame given in (91), which remains the same regardless of the time-dependence of the different dilaton, moduli or axion fields. It leads to a spectrum of primordial gravitational waves steeply growing on short scales, with a spectral index $n_T = 3$ [18], in contrast to conventional inflation models which require $n_T < 0$ [44]. The graviton spectrum appears to be a robust and distinctive prediction of any pre-big bang type evolution based on the low-energy string effective action, although recently in the non-singular model of Sect. 5, we have demonstrated how passing through the string phase could lead to a slight shift in the tilt closer to $n_T \sim 2$ [45]

5.8 Dilaton–Moduli–Axion Perturbation Spectra

We will now consider inhomogeneous linear perturbations in the fields about a homogeneous background given by [46, 47]

$$\varphi = \varphi(\eta) + \delta\varphi(\mathbf{x},\eta), \quad \sigma = \sigma(\eta) + \delta\sigma(\mathbf{x},\eta), \quad \beta = \beta(\eta) + \delta\beta(\mathbf{x},\eta)\,. \tag{106}$$

The perturbations can be re-expressed as a Fourier series in terms of Fourier modes with comoving wavenumber k. Considering the production of dilaton, moduli and axion perturbations during a pre-big bang evolution where the background axion field is constant, $\sigma' = 0$, the evolution of the homogeneous background fields are given in (89–90). The dilaton and moduli fields both evolve as minimally coupled massless fields in the Einstein frame. In particular, the dilaton perturbations are decoupled from the axion perturbations and the equations of motion in the spatially flat gauge become

$$\delta\varphi'' + 2\widetilde{h}\delta\varphi' + k^2\delta\varphi = 0\,, \tag{107}$$
$$\delta\beta'' + 2\widetilde{h}\delta\beta' + k^2\delta\beta = 0\,, \tag{108}$$
$$\delta\sigma'' + 2\widetilde{h}\delta\sigma' + k^2\delta\sigma = -2\varphi'\delta\sigma'\,, \tag{109}$$

Note that these evolution equations for the scalar field perturbations defined in the spatially flat gauge are automatically decoupled from the metric perturbations, although as we have said they are still related to the scalar metric perturbation, $\widetilde{A}$ through (101).

On the (+) branch, i.e., when $\eta < 0$, we can normalise modes at early times, $\eta \to -\infty$, where all the modes are far inside the Hubble scale, $k \gg |\eta|^{-1}$, and can be assumed to be in the flat-spacetime vacuum. Whereas in conventional inflation where we have to assume that this result for a quantum field in a classical background holds at the Planck scale, in this case the normalisation is done in the zero-curvature limit in the infinite past. Just as in conventional inflation, this produces perturbations on scales far outside the horizon, $k \ll |\eta|^{-1}$, at late times, $\eta \to 0^-$.

Conversely, the solution for the (−) branch with $\eta > 0$ is dependent upon the initial state of modes far outside the horizon, $k \ll |\eta|^{-1}$, at early times where $\eta \to 0$. The role of a period of inflation, or of the pre-big bang (+) branch, is precisely to set up this initial state which otherwise appears as a mysterious initial condition in the conventional (non-inflationary) big bang model.

The power spectrum for perturbations is commonly denoted by

$$\mathcal{P}_{\delta x} \equiv \frac{k^3}{2\pi^2}|\delta x|^2\,, \tag{110}$$

and thus for modes far outside the horizon ($k\eta \to 0$) we have

$$\mathcal{P}_{\delta\varphi} = \frac{32}{\pi^2}l_{\mathrm{Pl}}^2\widetilde{H}^2(-k\eta)^3[\ln(-k\eta)]^2\,, \tag{111}$$
$$\mathcal{P}_{\delta\beta} = \frac{32}{\pi^2}l_{\mathrm{Pl}}^2\widetilde{H}^2(-k\eta)^3[\ln(-k\eta)]^2\,, \tag{112}$$

where $\widetilde{H} \equiv \widetilde{a}'/\widetilde{a}^2 = 1/(2\widetilde{a}\eta)$ is the Hubble rate in the Einstein frame. The amplitude of the perturbations grows towards small scales, but only becomes large for modes outside the horizon ($|k\eta| < 1$) when $\widetilde{H}^2 \sim l_{\mathrm{Pl}}^{-2}$, i.e., the Planck scale in the Einstein frame. The spectral tilt of the perturbation spectra is given by

$$n - 1 \equiv \Delta n_x = \frac{d\ln\mathcal{P}_{\delta x}}{d\ln k} \tag{113}$$

which from (111) and (112) gives $\Delta n_\varphi = \Delta n_\beta = 3$ (where we neglect the logarithmic dependence). This of course is the same steep blue spectra we obtained earlier for the metric perturbations, which of course is far from the observed near H-Z scale invariant spectrum. We have recently examined the case of the evolution of the field perturbations in the non-singular cosmologies of Sect. 5 and as with the metric-perturbation case, amongst a number of new features that emerge there is a slight shift produced in the spectral index [48].

While the dilaton and moduli fields evolve as massless minimally coupled scalar fields in the Einstein frame, the axion field's kinetic term still has a non-minimal coupling to the dilaton field. This is evident in the equation of motion, (109), for the axion field perturbations $\delta\sigma$. The non-minimal coupling of the axion to the dilaton leads to a significantly different evolution to that of the dilaton and moduli perturbations.

After some algebra, we find that the late time evolution in this case is logarithmic with respect to $-k\eta$, (for $\mu \neq 0$)

$$\mathcal{P}_{\delta\sigma} = 64\pi l_{\rm Pl}^2 C^2(\mu) \left(\frac{e^{-\varphi}\widetilde{H}}{2\pi} \right)^2 (-k\eta)^{3-2\mu} , \tag{114}$$

where $\mu \equiv |\sqrt{3}\cos\xi_*|$ and the numerical coefficient

$$C(\mu) \equiv \frac{2^{\mu}\Gamma(\mu)}{2^{3/2}\Gamma(3/2)} , \tag{115}$$

approaches unity for $\mu \to 3/2$.

The key result is that the spectral index can differ significantly from the steep blue spectra obtained for the dilaton and moduli fields that are minimally coupled in the Einstein frame. The spectral index for the axion perturbations is given by [46, 47]

$$\Delta n_\sigma = 3 - 2\sqrt{3}|\cos\xi_*| \tag{116}$$

and depends crucially upon the evolution of the dilaton, parameterised by the value of the integration constant ξ_*. The spectrum becomes scale-invariant as $\sqrt{3}|\cos\xi_*| \to 3/2$, which if we return to the higher-dimensional underlying theory corresponds to a fixed dilaton field in ten-dimensions. The lowest possible value of the spectral tilt Δn_σ is $3 - 2\sqrt{3} \simeq -0.46$ which is obtained when stable compactification has occurred and the moduli field β is fixed. The more rapidly the internal dimensions evolve, the steeper the resulting axion spectrum until for $\cos\xi_* = 0$ we have $\Delta n_\sigma = 3$ just like the dilaton and moduli spectra.

When the background axion field is constant these perturbations, unlike the dilaton or moduli perturbations, do not affect the scalar metric perturbations. Axion fluctuations correspond to isocurvature perturbations to first-order. However, if the axion field does affect the energy density of the universe at later times (for instance, by acquiring a mass) then the spectrum of density perturbations need not have a steeply tilted blue spectrum such as that exhibited by the dilaton or moduli perturbations. Rather, it could have a nearly scale-invariant spectrum as required for large-scale structure formation. Such an exciting possibility has received a great deal of attention recently, notably in [49, 50, 51, 52, 53], and could be a source for the 'curvaton' field recently introduced by Lyth and Wands as a way of converting isocurvature into adiabatic perturbations [54]. Time will tell if the axion has any role to play in cosmological density perturbations although already it is beginning to look as the curvaton route is an interesting one to follow in this context [55, 56].

5.9 Smoking Guns?

Are there any distinctive features that we should be looking out for which would act as an indicator that the early Universe underwent a period of kinetic driven inflation? We have already mentioned the possibility of observing the presence of axion fluctuations in the cosmic microwave background anisotropies. Some of the other smoking guns include:

- The spectrum of primordial gravitational waves steeply growing on short scales, with a spectral index $n_T = 3$, although of no interest on large scales, such a spectrum could be observed by the next generation of gravitational wave detectors such as the Laser Interferometric Gravitational Wave Observatory (LIGO) if they are on the right scale [57, 58, 45]. The current frequency of these waves depends on the cosmological model, and in general we would require either an intermediate epoch of stringy inflation, or a low re-heating temperature at the start of the post-big bang era [59] to place the peak of the gravitational wave spectrum at the right scale. Nonetheless, the possible production of high amplitude gravitational waves on detector scales in the pre–big bang scenario is in marked contrast to conventional inflation models in which the Hubble parameter decreases during inflation.
- Because the scalar and tensor metric perturbations obey the same evolution equation, their amplitude is directly related. The amplitude of gravitational waves with a given wavelength is commonly described in terms of their energy density at the present epoch. For the simplest pre–big bang models this is given in terms of the amplitude of the scalar perturbations as

$$\Omega_{\mathrm{gw}} = \frac{2}{z_{\mathrm{eq}}} \mathcal{P}_\zeta \tag{117}$$

where $z_{\mathrm{eq}} = 24000 \Omega_o h^2$ is the red-shift of matter-radiation equality. The advanced LIGO configuration will be sensitive to $\Omega_{\mathrm{gw}} \approx 10^{-9}$ over a range of scales around 100Hz. However, the maximum amplitude of gravitational waves on these scales is constrained by limits on the amplitude of primordial scalar metric perturbations on the same scale [59]. In particular, if the fractional over-density when a scalar mode re-enters the horizon during the radiation dominated era is greater than about 1/3, then that horizon volume is liable to collapse to form a black hole with a lifetime of the order the Hubble time and this would be evaporating today! If we find PBH's and gravitational waves together then this would indeed be an exciting result for string cosmology!

– Evidence of a primordial magnetic field could have an interpretation in terms of string cosmology. In string theory the dilaton is automatically coupled to the electromagnetic field strength, for example in the heterotic string effective action the photon field Lagrangian is of the form

$$\mathcal{L} = e^{-\varphi} F_{\mu\nu} F^{\mu\nu} , \tag{118}$$

where the field strength is derived from the vector potential, $F_{\mu\nu} = \nabla_{[\mu} A_{\nu]}$.

Now in an isotropic FRW cosmology the magnetic field must vanish to zeroth-order, and thus the vector field perturbations are gauge-invariant and we can neglect the metric back-reaction to first-order. In the radiation gauge ($A^0 = 0$, $A^i_{|i} = 0$) then the field perturbations can be treated as vector perturbations on the spatial hypersurfaces. The field perturbation A_i turns out to have a clear unique dependence on the dilaton field. In fact the time dependence of the dilaton (rather than the scale factor) leads to particle production during the pre–big bang from an initial vacuum state [60, 61, 62]. Using the pre–big bang solutions given in (88)–(90), we find that the associated Power spectrum of the gauge fields have a minimum tilt for the spectral index for $\xi_* = 0$ when $\mu = (1+\sqrt{3})/2$ with a spectral tilt $\Delta n_{\mathrm{em}} = 4 - \sqrt{3} \approx 2.3$. This is still strongly tilted towards smaller scales, which currently is too steep to be observably acceptable.

6 Dilaton-Moduli Cosmology Including a Moving Five Brane

We turn our attention briefly to M-theory, and in particular to cosmological solutions of four-dimensional effective heterotic M-theory with a moving five-brane, evolving dilaton and T modulus [63]. It turns out that the five-brane generates a transition between two asymptotic rolling-radii solutions, in a manner analogous to the case of the NS-NS axion discussed in Sect. 3. Moreover, the five-brane motion generally drives the solutions towards strong coupling asymptotically. The analogous solutions to those presented in the pre-big-bang involves a negative-time branch solution which ends in a brane collision accompanied by a small-instanton transition. Such an exact solution should be of interest bearing in mind the recent excitement that has been generated over the Ekpyrotic Universe scenario, which involves solving for the collision of two branes [37, 38].

The four-dimensional low-energy effective theory we will be using is related to the underlying heterotic M-theory. Of particular importance for the interpretation of the results is the relation to heterotic M-theory in five dimensions, obtained from the 11-dimensional theory by compactification on a Calabi-Yau three-fold. This five-dimensional theory provides an explicit realisation of a brane-world. The compactification of 11 dimensional Horava-Witten theory, that is 11-dimensional supergravity on the orbifold $S^1/Z_2 \times M_{10}$, to five dimensions on a Calabi-Yau three fold, leads to the appearance of extra three-branes in the five-dimensional effective theory. Unlike the "boundary" three-branes which are stuck to the orbifold fix points, however, these three-branes are free to move in the orbifold direction, and this leads to a fascinating new cosmology.

Our starting point is the four dimensional action

$$S = -\frac{1}{2\kappa_P^2}\int d^4x\sqrt{-g}\left[\frac{1}{2}R + \frac{1}{4}(\nabla\varphi)^2 + \frac{3}{4}(\nabla\beta)^2 + \frac{q_5}{2}e^{(\beta-\varphi)}(\nabla z)^2\right] , \tag{119}$$

where φ is the effective dilaton in four dimensions, β is the size of the orbifold, z is the modulus representing the position of the five brane and satisfies $0 < z < 1$, and q_5 is the five brane charge. Due to the non-trivial kinetic term for z, solutions with exactly constant φ or β do not exist as soon as the five-brane moves. Therefore, the evolution of all three fields is linked and (except for setting $z =$ const) cannot be truncated consistently any further. Looking for cosmological solutions for simplicity, we assume the three-dimensional spatial space to be flat. Our Ansatz then reads

$$ds^2 = -e^{2\nu}d\tau^2 + e^{2\alpha}d\mathbf{x}^2 \tag{120}$$

$$\varphi = \varphi(\tau) \tag{121}$$

$$\alpha = \alpha(\tau) \tag{122}$$

$$\beta = \beta(\tau) \tag{123}$$

$$z = z(\tau) \tag{124}$$

The cosmological solutions are given by [63]

$$\alpha = \frac{1}{3}\ln\left|\frac{t-t_0}{T}\right| + \alpha_0 \tag{125}$$

$$\beta = p_{\beta,i}\ln\left|\frac{t-t_0}{T}\right| + (p_{\beta,f}-p_{\beta,i})\ln\left(\left|\frac{t-t_0}{T}\right|^{-\delta}+1\right)^{-\frac{1}{\delta}} + \beta_0 \tag{126}$$

$$\varphi = p_{\varphi,i}\ln\left|\frac{t-t_0}{T}\right| + (p_{\varphi,f}-p_{\varphi,i})\ln\left(\left|\frac{t-t_0}{T}\right|^{-\delta}+1\right)^{-\frac{1}{\delta}} + \varphi_0 \tag{127}$$

$$z = d\left(1+\left|\frac{T}{t-t_0}\right|^{-\delta}\right)^{-1} + z_0 \, . \tag{128}$$

where t is the proper time, the time-scales t_0 and T are arbitrary constants as are the constants d and z_0 which parameterise the motion of the five-brane. For $-\infty < t < t_0$ we are in the positive branch of the solutions and for $t_0 < t < \infty$ we are in the negative branch.

We see that both expansion powers for the scale factor α are given by 1/3, a fact which is expected in the Einstein frame. The initial and final expansion powers for β and φ are less trivial and are subject to the constraint

$$3p_{\beta,n}^2 + p_{\varphi,n}^2 = \frac{4}{3} \tag{129}$$

for $n = i, f$. These are mapped into one another by

$$\begin{pmatrix} p_{\beta,f} \\ p_{\varphi,f} \end{pmatrix} = P \begin{pmatrix} p_{\beta,i} \\ p_{\varphi,i} \end{pmatrix} , \qquad P = \frac{1}{2}\begin{pmatrix} 1 & 1 \\ 3 & -1 \end{pmatrix} . \tag{130}$$

This map is its own inverse, that is $P^2 = 1$, which is a simple consequence of time reversal symmetry. The power δ is explicitly given by

$$\delta = p_{\beta,i} - p_{\varphi,i} . \tag{131}$$

For $\delta < 0$ we are in the negative branch and for $\delta > 0$ we are in the positive time branch. Finally, we have

$$\varphi_0 - \beta_0 = \ln\left(\frac{2q_5 d^2}{3}\right) . \tag{132}$$

The solutions have the following interpretation: at early times, the system starts in the rolling radii solution characterised by the initial expansion powers p_i while the five-brane is practically at rest. When the time approaches $|t - t_0| \sim |T|$ the five-brane starts to move significantly which leads to an intermediate period with a more complicated evolution of the system. Then, after a finite comoving time, in the late asymptotic region, the five-brane comes to a rest and the scale factors evolve according to another rolling radii solution with final expansion powers p_f. Hence the five-brane generates a transition from one rolling radii solution into another one. While there are perfectly viable rolling radii solutions which become weakly coupled in at least one of the asymptotic regions, the presence of a moving five-brane always leads to strong coupling asymptotically, a phenomenon similar to what we observed in the dilaton-moduli-axion dynamics (see Fig. 2).

These general results can be illustrated by an explicit example. Focusing on the negative-time branch and considering the solutions with an approximately static orbifold at early time, Fig. 5 shows the evolution of β and φ, whereas Fig. 6 shows the evolution of the dynamical brane.

At early times, $|t - t_0| \gg |T|$, the evolution is basically of power-law type with powers p_i, because at early time the five-brane is effectively frozen at $z \simeq d + z_0$ and does not contribute a substantial amount of kinetic energy.

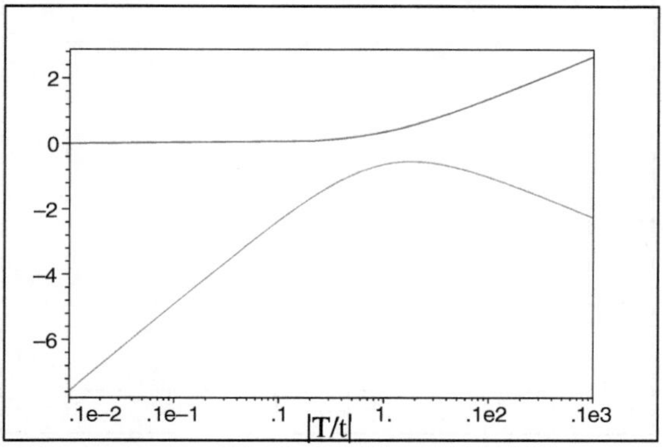

Fig. 5. Time-behaviour of β (upper curve) and φ (lower curve).

This changes dramatically once we approach the time $|t - t_0| \sim |T|$. In a transition period around this time, the brane moves from its original position by a total distance d and ends up at $z \simeq z_0$. At the same time, this changes the behaviour of the moduli β and φ until, at late time $|t| \ll |T|$, they correspond to another rolling radii solution with powers controlled by p_f. Concretely, the orbifold size described by β turns from being approximately constant at early time to expanding at late time, while the Calabi-Yau size controlled by φ undergoes a transition from expansion to contraction. We also find that as with the axion case discussed earlier, the solution runs into strong coupling in both asymptotic regions $t - t_0 \to -\infty$ and $t - t_0 \to 0$ which illustrates our general result.

In Fig. 6 we have shown a particular case which leads to brane collision. The five-brane is initially located at $d + z_0 \simeq 0.9$ and moves a total distance of $d = 1.5$ colliding with the boundary at $z = 0$ at the time $|t - t_0|/|T| \simeq 1$.

This represents an explicit example of a negative-time branch solution which ends in a small-instanton brane-collision. Solving for these systems has only just the begun, but already interesting features have emerged including a new mechanism for baryogenesis arising from the collision of two branes [64], and a more detailed understanding of the vacuum transitions associated with brane collisions [65].

7 Inflation Today – Quintessence

Now we will look at the general form Quintessence scenarios take. They are of course attempts to account for the observed accelerated expansion of the universe [66, 67], but are based on the evolution of as yet unobserved time dependent scalar fields. In particular they are not: a true cosmological constant; a time-dependent cosmological constant or solid dark energy such as arising from frustrated network of domain walls.

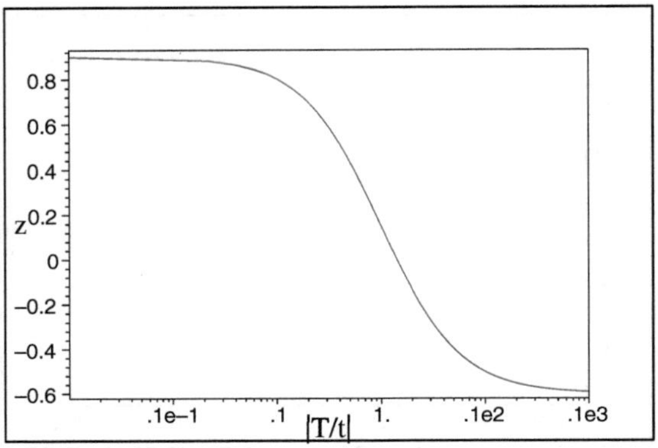

Fig. 6. Time-behaviour of the five-brane position modulus z for the example specified in the text. The boundaries are located at $z = 0, 1$ and the five-brane collides with the $z = 0$ boundary at $|t/T| \simeq 1$.

In Quintessence, the time dependent solutions arising out of scalar field potentials usually involves some form of tracking behaviour, where the energy density in the scalar field evolves so as to mimic that of the background fluid density for a period of time [68]. As we approach a redshift between $0.5 < z < 1$ the potential energy of the Quintessence field becomes the dominant contribution to the energy density and the Universe begins to accelerate [69, 70]. We will not go into details of the solutions in these lectures, rather we will discuss the general behaviour one expects from Quintessence scenarios. A nice review of the rich structure present in these models is presented in [71, 72], and Axel de la Macorra has given some detailed lectures here at the meeting [73].

Using a particular potential $V(\phi) = exp(0.3e^{0.3\phi})$ as an example, Fig. 7 shows the generic behaviour that is expected to be followed in Quintessence models.

Region 1 corresponds to the period where the initial potential energy in the scalar field is converted into kinetic energy as the field begins to roll down its potential. This scalar field kinetic energy soon comes to dominate the energy density of the scalar field as $\rho_\phi \propto a(t)^{-6}$ where $a(t)$ is the scale factor [region 2]. As the kinetic energy decreases rapidly, the system slows down again [region 3] leading to a constant field regime. This is then followed by the crucial period where the kinetic energy in the scalar field scales in proportion to its potential energy [region 4]. This is an attractor regime and as can be seen from Fig. 7 it corresponds to an extended period in which the energy density tracks that of the background energy density. These attractor properties are very useful because they make the reliance on initial conditions of the scalar field less important. Finally in region 5, we see the specific property of the scalar field potential coming into its own, as it determines when the scalar field potential energy density comes to dominate over the

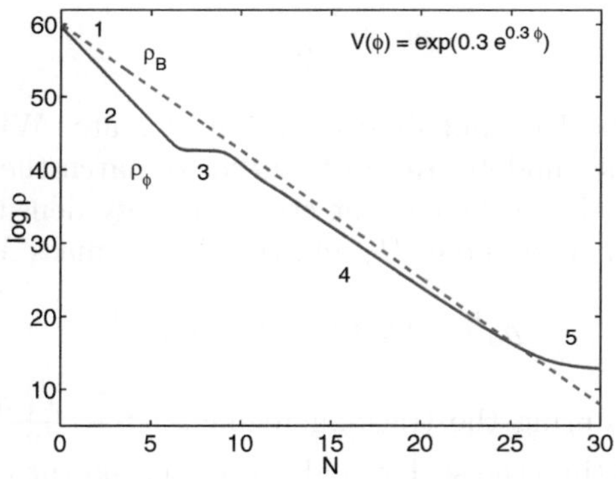

Fig. 7. Typical scaling properties of Quintessence potentials. ρ_B, ρ_ϕ are the energy densities in the background fluid and scalar field respectively[72].

background fluid energy density leading to the observed acceleration of the Universe. Fig. 7 gives a flavour for some of the fine tuning issues that arises in Quintessence. There are two obvious ones, the first is that the value of the energy density today must be very close to the critical density $10^{-3}eV^4$, the second is that domination had to occur very recently $z \leq 1$ in order to account for the fact that galaxy formation is not affected too much by the Quintessence field. There are also tight constraints on the energy density in the Quintessence field at the time of nucleosynthesis, as the field acts like an extra light degree of freedom and we already know that there are tight constraints on the number of families from nucleosynthesis. We will now go on to look at some individual models.

7.1 Specific Quintessence Models

The original Quintessence model [69, 70] has an inverse power law type of potential,

$$V(\phi) = \frac{M^{4+\alpha}}{\phi^\alpha}, \tag{133}$$

where α is thought of as a positive number (it could actually also be negative) and M is constant.

Most models of Quintessence are analysed through their effective equation of state,

$$w_\phi \rho_\phi,$$

where p_ϕ is the pressure in the field and ρ_ϕ is the energy density in the field. We know from Einstein's acceleration equation that for the Quintessence field to lead to acceleration of the Universe we require $\rho_\phi + 3p_\phi < 0$ or $w_\phi < -\frac{1}{3}$. Applying this to the inverse power case we find

$$w_\phi = \frac{\alpha w_B - 2}{2 + \alpha},$$

where w_B is the background fluid equation of state. Where does the fine tuning arise in these models? Recall we need to match the energy density in the Quintessence field to the current critical energy density, which in terms of the Hubble parameter today H_0 and the Planck mass M_{pl} is given by

$$\rho_\phi < M_{pl}^2 H_0^2 \sim 10^{-47} GeV^4.$$

It turns out that during the tracking regime, $H^2 \sim \frac{V(\phi)}{\phi^2} \sim \frac{\rho_\phi}{M_{pl}^2}$, hence it follows that at the time the scalar field is dominating the energy density and leading to acceleration today, we must have $\phi_0 \sim M_{pl}$, the value of the scalar field today has to be of order the Planck scale. This is typical of virtually all Quintessence models. The real fine tuning now becomes clear, substituting for the value of ϕ_0 in to the bound on the energy density today ρ_ϕ^0, we see:

$$M = (\rho_\phi^0 M_{pl}^\alpha)^{\frac{1}{4+\alpha}}.$$

This then constrains the allowed combination of α, M. For example for $\alpha = 2$ the constraint implies $M = 1GeV$ etc... Within the class of parameters which satisfy the coincidence problem the inverse power law potentials suffer in that their predicted equation of state w_ϕ is only marginally compatible with the values emerging from observations. At the 1σ confidence level in the $\Omega_{\mathrm{M}} - w_\phi$ plane, the data prefer $w_\phi < -0.8$ with possibly a favoured cosmological constant $w_Q = -1$ whereas the values permitted by these tracker potentials (for $\alpha \geq 1$, have $w_Q > -0.8$. A general problem we will always have to tackle is finding such Quintessence models in particle physics. For an interesting attempt at this in the context of Supersymmetric QCD see the model proposed by Binetruy [74].

Multiple exponential potentials also offer interesting possibilities for a successful Quintessence scenario [75]. Such potentials are expected to arise as a result of compactifications in superstring models, hence are well motivated. Unfortunately we still have not obtained what one would call a 'natural' model for reasons we will discuss below. Nevertheless it remains a model with some potential for success in it as it delivers Quintessence scenarios for a wide range of initial conditions.

It has been known for some time that single exponential potentials lead to scaling solutions[68, 76, 77]. Consider the case of $V(\phi) = V_0 \exp(\alpha\kappa\phi)$, where $\kappa^2 \equiv 8\pi/M_{pl}^2$. The two late time attractor solutions depend on the values of α and the background's equation of state w_B:

(1) $\alpha^2 > 3(w_B+1)$: the scalar field mimics the evolution of the barotropic fluid with $w_\phi = w_B$, and the relation $\Omega_\phi = 3(w_B + 1)/\alpha^2$ holds.

(2) $\alpha^2 < 3(w_B + 1)$. The late time attractor is the scalar field dominated solution ($\Omega_\phi = 1$) with $w_Q = -1 + \alpha^2/3$.

By including two exponential terms it allows for the possibility of the system entering two scaling regimes which depend on the value of the slope of the two terms: one tracks radiation and matter, while the second one dominates at end. To be specific we can consider

$$V(\phi) = V_0 \left(e^{\alpha\kappa\phi} + e^{\beta\kappa\phi}\right), \tag{134}$$

where for convenience we assume α to be positive (the case $\alpha < 0$ can always be obtained taking $\phi \to -\phi$). Figures 8 and 9 show the results of a typical run with such a potential leading to potential domination today and acceleration.

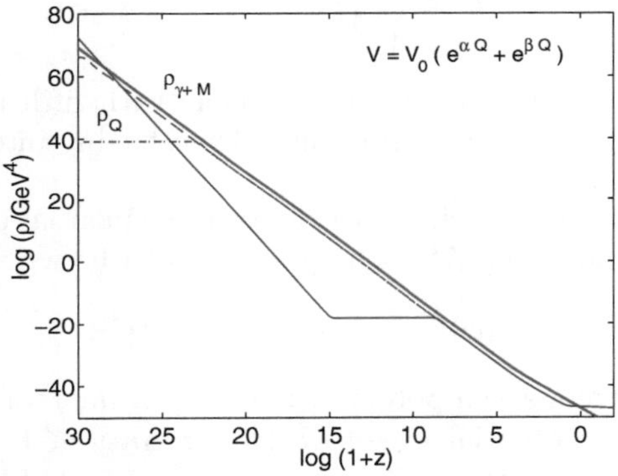

Fig. 8. Plot of the energy density, ρ_Q, for the 2EXP model with $\alpha = 20$, $\beta = 0.5$ and several initial conditions admitting an $\Omega_Q = 0.7$ flat universe today. The line labeled by $\rho_{\gamma+\mathrm{M}}$ is the evolution of radiation and matter.[72].

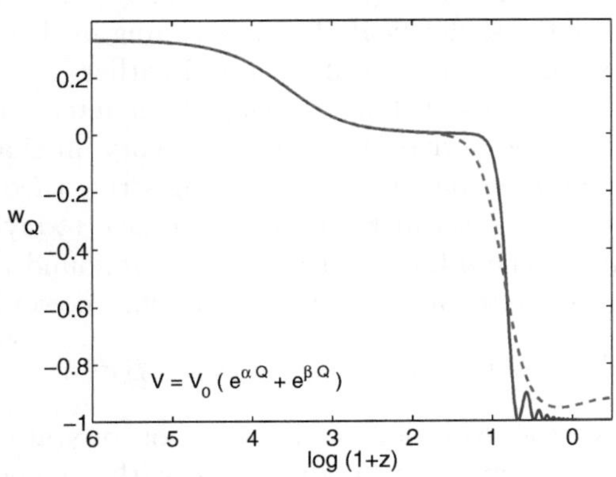

Fig. 9. The late time evolution of the equation of state, in the 2EXP model, for parameters $(\alpha, \beta) = (20,0.5)$ dashed line; $(20,-20)$ solid line for $\Omega_Q = 0.7$.[72].

It is clear from Fig. 9, where the evolution of the equation of state is shown and compared to the case with $\beta/\alpha > 0$, that the field mimics the radiation ($w_Q = 1/3$) and matter ($w_Q = 0$) evolution before settling in an accelerating ($w_Q < -1/3$) expansion. As a result of the scaling behaviour of attractor (1), it is clear that there exists a wide range of initial conditions that provide realistic results.

Where in this case is the fine tuning to be found then? Demanding the energy density in the field matches the critical density today, places the bound $V_0 \sim \rho_\phi^0 \sim 10^{-47} GeV^4 \sim (10^{-3} eV)^4$. This very low energy density converts into an extremely light scalar field, in particular its mass is given by

$$m \simeq \sqrt{\frac{V_0}{M_{pl}^2}} \sim 10^{-33} eV.$$

Such a tiny mass is very difficult to reconcile with fifth force experiments, unless there is a mechanism to prevent ϕ from having interactions with the other matter fields!

A model which can be related to the two exponential case has been suggested by Sahni and Wang [78]. The potential can be written as:

$$V(\phi) = V_0 \left[\cosh(\alpha\kappa\phi) - 1\right]^n . \tag{135}$$

It behaves as an exponential potential $V \to \exp(n\alpha\kappa\phi)$ for $|\alpha\kappa\phi| \gg 1$ and as a power law type of potential $V \to (\alpha\kappa\phi)^{2n}$ for $|\alpha\kappa\phi| \ll 1$. It follows that the evolution scales as radiation and matter when dominated by the exponential form and later enters into an oscillatory regime when the minimum is reached. In this regime the time average equation of state is

$$\langle w_\phi \rangle = \frac{n-1}{n+1} . \tag{136}$$

We see that for $n < 1/2$ then $w_\phi < -1/3$, implying late times accelerated expansion driven by the scalar field. The fine tuning in this case is similar to that of the two exponential potential discussed earlier.

Albrecht and Skiordis [79] have developed an interesting model which they have argued can be derived from String theory, in that they claim the parameters are all of order one in the underlying string theory. The potential has a local minimum which can be adjusted to have today's critical energy density value (this is where the fine tuning is to be found by the way). The actual potential is a combination of exponential and power-law terms:

$$V(\phi) = V_0 e^{-\alpha\kappa\phi} \left[A + (\kappa\phi - B)^2\right] . \tag{137}$$

In Fig. 10 we show the evolution of the equation of state. For early times the exponential term dominates the dynamics, with the energy density of ϕ scaling as radiation and matter. For suitable choices of the parameters the field gets trapped in the local minimum because the kinetic energy during

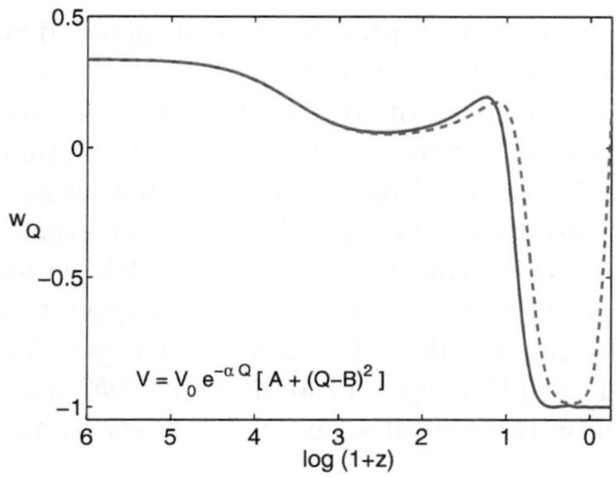

Fig. 10. The evolution of the equation of state of quintessence when the Albrecht-Skiordis potential has a local minimum (solid) and when it does not (dashed). In this case $\alpha = 10$, $V_0 = \kappa^{-4}$, $A = 0.9/\alpha^2$ and $B = 27.2$, for the former case and $\alpha = 6$, $V_0 = \kappa^{-4}$, $A = 1.1/\alpha^2$ and $B = 45.5$, for the latter.[72].

scaling is small. The field then enters a regime of damped oscillations leading to $w_\phi \to -1$ and an eternally expanding universe.

There are many other models which we could describe: coupled quintessence, extended quintessence, tracking oscillatory quintessence to name but three. They all have similar properties to those described above, but rather than concentrate on them we will turn our attention finally to the case of Quintessential Inflation, developed by Peebles and Vilenkin [80]. One of the major drawbacks often used to attack models of Quintessence is that it introduces yet another weakly interacting scalar field. Why can't we use one of those scalars already 'existing' in cosmology, to also act as the Quintessence field? This is precisely what Peebles and Vilenkin set about doing. They introduced a potential for the field ϕ which allowed it to play the role of the inflaton in the early Universe and later to play the role of the Quintessence field. To do this it was important that the potential did not have a minimum in which the inflaton field would completely decay at the end of the initial period of inflation. The potential they proposed was:

$$\begin{aligned} V(\phi) &= \lambda(\phi^4 + M^4) \quad \text{for } \phi < 0 \\ &= \frac{\lambda M^4}{(1 + (\frac{\phi}{M})^\alpha)} \quad \text{for } \phi \geq 0 \end{aligned}$$

For $\phi < 0$ we have ordinary chaotic inflation. When this ends the Universe is reheated via gravitational particle production. Much later on, for $\phi > 0$ the Universe once again begins to inflate but this time at the lower energy scale associated with Quintessence. Needless to say, Quintessential Inflation also requires a degree of fine tuning, in fact perhaps even more than before as there are no tracker solutions we can rely on for the initial conditions. The initial period of inflation must produce the observed density fluctuations, hence constrains $\lambda \sim 10^{-14}$. Demanding that $\Omega_\phi^0 \sim 0.7$, we find we can constrain the parameter space of (α, M). For example, for $\alpha = 4$, we have $M = 10^5$ GeV. Time does not permit us to elaborate further on this aspect of Quintessence, but it is worth at least mentioning that there are some very nice resolutions of Quintessential Inflation in Brane world scenarios (for details see [81, 82, 83]. Neither have we time to go into the wealth of Quintessence models that have been proposed within the context of supergravity, apart from giving a brief flavour of the general idea. Brax and Martin [84] demonstrated that a supergravity model with Superpotential $W = \Lambda^{3+\alpha}\Phi^\alpha$ and Kahlar potential $K = \Phi\Phi^*$ (where Φ is the Chiral scalar field) leads to an associated scalar potential

$$V(\phi) = \frac{\Lambda^{6+\alpha}}{\phi^{2\alpha+2}} e^{\frac{\kappa^2}{2}\phi^2},$$

under the rather strict assumption that $< W >= 0$. A working example is the case $\alpha = 11$ which has an associated equation of state $w_\phi^0 = -0.8$. There are more models that have been investigated [85, 86, 87]. A word of caution though about Quintessence in supergravity. Kolda and Lyth [88], have argued that all current supergravity inspired models suffer from the fact that loop corrections will always couple the Quintessence field to other sources of matter so as to lift the potential thereby breaking the flatness criteria required for Quintessence today.

7.2 Evidence for Quintessence?

If there is a scalar field responsible for the current acceleration of the Universe how might we see it? In this conference there have been many talks addressing this issue, so we will not go into great details here, other than remind the reader of some of the attempts that are under way and have been proposed recently. Ideally we would look for evidence of evolution in the equation of state, w_ϕ as a function of redshift. These include

- Precision CMB anisotropies – lots of models are currently compatible.
- Combined LSS, SN1a and CMB data tend to give $w_\phi < -0.8$, which is difficult to tell from a true cosmological constant.
- Look for more supernova of the type SN1a. The proposed satellite, SNAP will find over 2000 which may then enable us to start constraining the equation of state.
- Constraining the equation of state with Sunayaev-Zeldovich cluster surveys from which we can compute the number of clusters for a given set of cosmological parameters.
- Probing the Dark Energy with Quasar clustering in which redshift distortions constrain cosmological parameters.

– Reconstruct the equation of state from observations – this approach at least offers the hope of developing a method independent of potentials – an example is the Statefinder method developed by Sahni et al. [89].
– Look for evidence in the variation of the fine structure constant.

We finish off the lectures by discussing in a bit more detail one of the items just mentioned. Finding a suitable parameterisation of the equation of state an issue of importance for those interested in reconstructing w_ϕ from observation, such as those working on SNAP [90, 91]. Two approaches suggested to date involve a polynomial expansion either in terms of the red-shift, z (i.e. $w_\phi(z) = \sum_{i=0}^{N} w_i z^i$) [92] or in terms of the logarithm of the red-shift (i.e. $w_\phi(z) = \sum_{i=0}^{N} w_i \ln(1+z)^i$) [93]. A third approach has recently been developed by Corasaniti and Copeland [94]. It allows for tracker solutions in which there is a rapid evolution in the equation of state, something that the more conventional power-law behaviour can not accommodate. This has has some nice features in that it allows for a broad class of Quintessence models to be accurately reconstructed and it opens up the possibility finding evidence of quintessence in the CMB both through its contribution to the Integrated Sachs Wolfe Effect [95] and as a way of using the normalisation of the dark energy power spectrum on cluster scales, σ_8, to discriminate between dynamical models of dark energy (Quintessence models) and a conventional cosmological constant model[96].

8 Summary

In these lectures we have addressed a number of issues relating to inflationary cosmology, both in the early Universe and today. We have seen how inflation arises in both potential dominated cases and as a result of rolling radii solutions associated with the low energy string action. We have also seen how hard it is to relate inflation to realistic particle physics inspired models. This area is one of intense interest at the moment. In our attempt to bridge this gap, we have related these solutions to the exciting new solutions arising in M-theory cosmology, and showed how a moving five brane could act in a manner similar to the axion field in the pre Big Bang case. This is an exciting time for string and M-theory cosmology, the subject is developing at a very fast rate, and no doubt there will be new breakthroughs emerging over the next few years. Hopefully out of these we will be in a position to address a number of the issues we have raised in this article, as well as other key ones such as stabilising the dilaton and explaining the current observation of an accelerating Universe. We have investigated a number of Quintessence models and tried to argue why Quintessence offers a plausible explanation for the observational fact that the Universe is accelerating today. We have also tried to emphasise the issues that Quintessence as a model simply fails to answer naturally, requiring some form of fine tuning in order to do so. These include:

- Why is there a Λ type term dominating today?
- Why are the matter and Λ contributions comparable today – 'coincidence' problem?
- Why is Λ so small compared to typical particle physics scale?
- Is there any need for a quintessence field? Is it simply a cosmological constant?

There is little doubt that this very exciting field is being driven by observations, especially in the CMBR and LSS. They are constraining the cosmological parameters, even before Map or Planck arrives on the scene. Yet we do not know why the universe inflating today and through Quintessence we are hoping that particle physics provides an answer. The existence of scaling solutions and tracker behaviour may yet show up through time varying constants [97]. There is much going on in Brane inspired cosmology and it may provide important clues to the nature of dark energy. In general as we have seen, there are many models of Quintessence but they may yet prove too difficult to separate from a cosmological constant. We need to try though – it is too exciting a prospect not to!

Acknowledgements

I am very grateful to the organisors for inviting me to this wonderful meeting in one of the most beautiful places I have ever been.

References

1. A. H. Guth, Phys. Rev. D**23**, 347 (1981).
2. A.D. Linde, Phys. Lett. B**108**, 389 (1982).
3. A. Albrecht and P.J. Steinhardt, Phys. Rev. Lett.**48**, 1220 (1982).
4. S.W. Hawking, I.G. Moss and J.M. Stewart, Phys. Rev. D**26**, 2681 (1982).
5. A. A. Starobinsky, Phys. Lett. B **91**, 99 (1980).
6. D. N. Spergel et al., Astrophys. J. Suppl. **148**, 175 (2003) [arXiv:astro-ph/0302209].
7. A. D. Linde: *Particle Physics and Inflationary Cosmology* (Harwood Academic, Chur, Switzerland, 1990).
8. H. V. Peiris et al., Astrophys. J. Suppl. **148**, 213 (2003) [arXiv:astro-ph/0302225].
9. E.W. Kolb and M.S. Turner: *The Early Universe* (Addison-Wesley, Redwood City, 1994).
10. L. Kofman, A. D. Linde and A. A. Starobinsky, Phys. Rev. Lett. **73**, 3195 (1994) [arXiv:hep-th/9405187].
11. L. Kofman, A. D. Linde and A. A. Starobinsky, Phys. Rev. Lett. **76**, 1011 (1996) [arXiv:hep-th/9510119].
12. E. J. Copeland, D. Lyth, A. Rajantie and M. Trodden, Phys. Rev. D **64**, 043506 (2001) [arXiv:hep-ph/0103231].

13. E. J. Copeland, S. Pascoli and A. Rajantie, Phys. Rev. D **65**, 103517 (2002) [arXiv:hep-ph/0202031].
14. A.R. Liddle and D.H. Lyth, Phys. Rep **231**, 1 (1993), [astro-ph/9303019]; A.R. Liddle and D.H. Lyth: *Cosmological inflation and large scale structure*, (Cambridge Univ. Press, 2000)
15. R. H. Brandenberger, Lectures on the Theory of Cosmological Perturbations, Lect. Notes Phys. 646, 127 (2004) [arXiv:hep-th/0306071].
16. J. H. Lidsey, D. Wands and E. J. Copeland, Phys. Rep. **337**, 343 (2000).
17. G. Veneziano, Phys. Lett. B**265**, 287 (1991).
18. M. Gasperini and G. Veneziano. Astropart. Phys. **1**, 317 (1993).
19. E. J. Copeland, A. Lahiri, and D. Wands, Phys. Rev. D**50**, 4868 (1994).
20. M. D. Pollock and D. Sahdev, Phys. Lett. B**222**, 12 (1989).
21. J. J. Levin and K. Freese, Phys. Rev. D**47**, 4282 (1993).
22. R. Brustein and G. Veneziano, Phys. Lett. B**329**, 429 (1994).
23. M. S. Turner and E. J. Weinberg, Phys. Rev. D**56**, 4604 (1997).
24. D. Clancy, J. E. Lidsey, and R. Tavakol, Phys. Rev. D **58**, 044017 (1998).
25. N. Kaloper, A. D. Linde, and R. Bousso, Phys. Rev. D**59**, 043508 (1999).
26. J. Maharana, E. Onofri, and G. Veneziano, J. High Energy Phys. **01**, 004 (1998).
27. G. Veneziano, Phys. Lett. B**406**, 297 (1997).
28. A. Buonanno, K. A. Meissner, C. Ungarelli, and G. Veneziano, Phys. Rev. D**57**, 2543 (1998).
29. A. Buonanno, T. Damour, and G. Veneziano, Nucl. Phys. B**543**, 275 (1999).
30. M. Gasperini, Phys. Rev. D**61** 087301 (2000).
31. D. Clancy, J. E. Lidsey, and R. Tavakol, Phys. Rev. D**59**, 063511 (1999).
32. M. Gasperini's web page, http://www.to.infn.it/ gasperin/
33. R. Brustein and R. Madden, Phys. Lett. B**410**, 110 (1997). R. Brustein and R. Madden, Phys. Lett. B**410**, 110 (1997).
34. R. Brustein and R. Madden, Phys. Rev. D**57**, 712 (1998). R. Brustein and R. Madden, Phys. Rev. D**57**, 712 (1998).
35. C. Cartier, E. J. Copeland and R. Madden, JHEP **0001**, 035 (2000).
36. M. Gasperini, M. Maggiore, and G. Veneziano, Nucl. Phys. B**494**, 315 (1997).
37. J. Khoury, B. A. Ovrut, P. J. Steinhardt and N. Turok, Phys. Rev. D**64**, 123522 (2001) and hep-th/0108187
38. R. Kallosh, L. Kofman and A. Linde, Phys. Rev. **D64**, 123523 (2001).
39. P. J. Steinhardt and N. Turok, Phys. Rev. D**65** 126003 (2002) hep-th/0111030 and hep-th/0111098
40. V. F. Mukhanov, Sov. Phys. JETP **68**, 1297 (1988).
41. V. F. Mukhanov, H. A. Feldman, and R. H. Brandenberger, Phys. Rep. **215**, 203 (1992).
42. J. M. Bardeen, P. J. Steinhardt, and M. S. Turner, Phys. Rev. D**28**, 679 (1983).
43. R. Brustein, M. Gasperini, M. Giovannini, V. F. Mukhanov, and G. Veneziano, Phys. Rev. D **51**, 6744 (1995).
44. A. R. Liddle and D. H. Lyth, Phys. Rep. **231**, 1 (1993).
45. C. Cartier, E. J. Copeland and M. Gasperini, Nucl. Phys. B**607**, 406 (2001).
46. E. J. Copeland, R. Easther, and D. Wands, Phys. Rev. D**56**, 874 (1997).
47. E. J. Copeland, J. E. Lidsey, and D. Wands, Nucl. Phys. B**506**, 407 (1997).
48. C. Cartier, J. Hwang and E. J. Copeland, Phys. Rev. D**64**, 103504 (2001).
49. R. Durrer, M. Gasperini, M. Sakellariadou, and G. Veneziano, Phys. Lett. B**436**, 66 (1998).

50. R. Durrer, M. Gasperini, M. Sakellariadou, and G. Veneziano, Phys. Rev. D**59**, 043511 (1999).
51. A. Melchiorri, F. Vernizzi, R. Durrer and G. Veneziano, Phys. Rev. Lett. **83**, 4464 (1999).
52. F. Vernizzi, A. Melchiorri and R. Durrer, Phys. Rev. D **63**, 063501 (2001).
53. K. Enqvist and M. S. Sloth, Nucl. Phys. B**626**, 395 (2002) [hep-ph/0109214].
54. D. Lyth and D. Wands, Phys. Lett. B**524**, 5 (2002).
55. V. Bozza, M. Gasperini, M. Giovannini and G. Veneziano, Phys. Lett. B **543**, 14 (2002).
56. V. Bozza, M. Gasperini, M. Giovannini and G. Veneziano, Phys. Rev. D**67**, 063514 (2003) [arXiv: hep-ph/0212112].
57. B. Allen and R. Brustein, Phys. Rev. D **55**, 3260 (1997).
58. M. Maggiore, Phys. Rept. **331**, 283 (2000).
59. E. J. Copeland, A. R. Liddle, J. E. Lidsey, and D. Wands, Phys. Rev. D**58**, 063508 (1998).
60. D. Lemoine and M. Lemoine, Phys. Rev. D**52**, 1955 (1995).
61. M. Gasperini, M. Giovannini, and G. Veneziano, Phys. Rev. Lett. **75**, 3796 (1995).
62. M. Gasperini, M. Giovannini, and G. Veneziano, Phys. Rev. D**52**, 6651 (1995).
63. E.J. Copeland, J. Gray and A. Lukas, Phys. Rev. D**64**, 126003 (2001).
64. M. Bastero-Gill, E.J. Copeland, J. Gray, A. Lukas, M. Plumacher, Phys. Rev. **D66**, 066005 (2002)
65. N. Antunes, E.J. Copeland, M. Hindmarsh and A. Lukas, Phys. Rev. D**68**, 066005 (2003) [hep-th/0208219].
66. S. Perlmutter et al., Astrophys. J **517**, 565 (1999).
67. A. Riess, et al., Astrophys. J,**117**, 707 (1999).
68. C. Wetterich, Nucl. Phys. B**302**, 668 (1998).
69. B. Ratra, and P.J.E. Peebles, Phys. Rev. D**37**, 3406 (1988).
70. R.R. Caldwell, R. Dave, and P.J. Steinhardt, Phys. Rev. Lett. **80**, 1582 (1998). I. Zlatev, L. Wang, and P.J. Steinhardt, Phys. Rev. Lett. **82**, 896 (1999).
71. C. Ng, N. Nelson, and F. Rosati, Phys. Rev. D **64** 083510 (2001).
72. N. Nelson. D. Phil thesis, University of Sussex (2001).
73. A. de la Macorra, Quintessence and Dark Energy, Lect. Notes Phys. 646, 225 (2004).
74. P. Binetruy, Phys. Rev. D**60**, 063502 (1999).
75. T. Barreiro, E.J. Copeland, and N. Nunes, Phys. Rev. D**61**, 127301 (2000).
76. E.J. Copeland, , A.R. Liddle, and D. Wands, Phys. Rev. D **57**, 4686 (1998).
77. P. Ferreira and M. Joyce, Phys. Rev. D**58**, 023503 (1998).
78. V. Sahni and L. Wang, Phys. Rev. D**62**, 103517 (2000).
79. A. Albrecht and C. Skordis, Phys. Rev. Lett. **84**, 2076 (1999).
80. P.J.E. Peebles and A. Vilenkin, Phys. Rev. D**59**, 063505 (1999).
81. E.J. Copeland, A. R. Liddle and J.E. Lidsey, Phys. Rev. D **64**, 023509 (2001)
82. G. Huey and J.E. Lidsey, Phys. Lett. B**514**, 217 (2001).
83. N. Nunes and E.J. Copeland, Phys. Rev. D**66**, 043524 (2002).
84. P. Brax and J. Martin, Phys. Lett. B**468**, 40 (1999).
85. K. Choi, Phys. Rev. D**62**, 043509 (2000).
86. A. Masiero, M. Pietroni and F. Rosati, Phys. Rev. D**61**, 023504 (2000).
87. E. J. Copeland, N. Nunes and F. Rosati, Phys. Rev. D**62**, 123503 (2000).
88. C. F. Kolda and D. H. Lyth, Phys. Lett. B**458**, 197 (1999).

89. V. Sahni et al., JETP Lett. **77**, 201 (2003) [astro-ph/0201498]; [astro-ph/0211084].
90. D. Huterer and M. Turner, Phys. Rev. D**60**, 081301 (1999).
91. D. Huterer and G. Starkman, Phys. Rev. Lett. **90**, 031301 (2003) [astro-ph/0207517].
92. J. Weller and A. Albrecht, Phys. Rev. Lett. **86**, 1939 (2001); Phys. Rev. D**65** 103512 (2002).
93. B. F. Gerke and G. Efstathiou, Mon. Not. Roy. Astron. Soc. **335**, 33 (2002).
94. P. S. Corasaniti and E. J. Copeland, Phys. Rev. D **67**, 063521 (2003) [arXiv:astro-ph/0205544].
95. P. S. Corasaniti, B. A. Bassett, C. Ungarelli and E. J. Copeland, Phys. Rev. Lett. **90**, 091303 (2003) [arXiv:astro-ph/0210209].
96. M. Kunz, P. S. Corasaniti, D. Parkinson and E. J. Copeland, arXiv:astro-ph/0307346.
97. E. J. Copeland, N. J. Nunes and M. Pospelov, Phys. Rev. D **69**, 023501 (2004), [arXiv:hep-ph/0307299].

Cosmic Acceleration, Scalar Fields, and Observations

César A. Terrero-Escalante

Instituto de Física, UNAM, Apdo. Postal 20-364, 01000, México D.F., México.
cterrero@fis.cinvestav.mx

Abstract. Conditions for accelerated expansion of Friedmann–Robertson–Walker space–time are analyzed. Connection of this scenario with present–day observations are reviewed. It is explained how a scalar field could be responsible for cosmic acceleration observed in present times and predicted for the very early Universe. Ideas aimed at answering whether is that the actual case for our Universe are described.

1 Introduction

The Standard Big Bang Model (SBB), based on a Friedmann–Robertson–Walker (FRW) Universe, evolves in time with essentially two phases. In the first one the energy related to relativistic matter (known as radiation) dominates over any other form of energy. During that period phase transitions described by particle physics took place to give rise to hadron formation, baryogenesis, nucleosynthesis, Quintessence fields and so on. Because in an expanding Universe radiation energy dilutes faster than the energy of pressureless matter, in the second phase the latter becomes dominant. Large scale structures (LSS) like galaxies and galaxy clusters formed during that period. In both phases the Universe expands in a decelerated fashion (for a short review of the SBB see the contribution to this book by J. L. Cervantes–Cota[1], and for more details the book [2]).

The SBB can easily accommodate phases of accelerated expansion of the Universe. According to cosmological observations, such a phase could correspond to the present state of the observable Universe and seems to be necessary in the very early Universe in order to solve several problems inherent to the SBB, particularly those problems related to the initial conditions.

For the Universe to expand acceleratingly, a very special kind of energy density is required to dominate over the remaining contributions to the total energy budget. This kind of energy is related to a negative pressure. One of the outstanding problems in modern cosmology is to find out what exactly this kind of matter is. It could be the case that there are different explanations of what causes the Universe to undergo cosmic acceleration in the present and in the very early phases of its evolution. As to the present era, a dominating vacuum energy is good enough to explain the observations but it introduces

C.A. Terrero-Escalante, Cosmic Acceleration, Scalar Fields, and Observations, Lect. Notes Phys. **646**, 109–125 (2004)
http://www.springerlink.com/

other problems which seem to be very difficult to solve. A good candidate is, instead, a single scalar field with dynamics dominated by its potential energy. This is also the favorite candidate to explain an era of cosmic acceleration in the very early Universe. In both cases, the problem is then to determine what the high-energy physics framework is where such scalar fields arise.

In the next section the conditions for an accelerated FRW Universe will be derived. Then, in Sect. 3 a scalar field will be described from a cosmological point of view, along with how it could be used to induce an accelerated expansion. Section 4 is devoted to explaining some ideas aimed at finding observational signatures in the data, allowing us to distinguish the origin of cosmic acceleration. If scalar fields are responsible for the observed and predicted eras of accelerated expansion, the observations should say something about high-energy physics that is outside the scope of Earth-based laboratories. Finally, conclusions are presented in Sect. 5.

Throughout this contribution natural units are used, i.e. $c = \hbar = 1$.

2 Accelerated Friedmann–Robertson–Walker Universe

Models in physics are based on a set of principles derived from observations and assumptions that make computations simpler without wiping out crucial features of the phenomenon to be modeled. The SBB is not an exception. The first assumption is that there exists a *cosmic scale*, quite a bit larger than the galaxy clusters scale, $100 - -200$ Mpc. After averaging on those scales, everything we observe in the sky dilutes into an isotropic picture. This means that the Universe is the same when seen by a terrestrial observer in any direction. But, according to scientific history, the human being does not seem to be such a special being. Thus, it is assumed that we live on an ordinary planet orbiting an ordinary star in a ordinary galaxy which is an ordinary member of an ordinary galaxy cluster. It implies that any observer located at any other point in the observable Universe will see exactly the same picture of the sky, averaged in cosmic scales, that a human observer does. In other words, the Universe is assumed to be isotropic and homogeneous. This is known as the *cosmological principle.*

The physical distance X_{phys} between two observers will change with time, even if they are in relative rest each with respect to the other,

$$\boldsymbol{X_{phys}} = a(t) \times \boldsymbol{X_{com}} \, . \tag{1}$$

The physical distance between observers will be equal to the distance between them if space does not expand (or contract), i.e. X_{com}, times factor $a(t)$ which describes the change in size of the expanding (contracting) homogeneous Universe after given amount of time. The coordinate system where X_{com} is defined is called the comoving frame. Since the age of the Universe is one of the quantities that can be inferred from observations, the homogeneity of the Universe must be defined on a surface of constant proper time since the Big Bang. Time dilation causes the proper time measured by an observer to depend on the velocity of the observer, hence the time variable t is actually the proper time for comoving observers since the beginning of the cosmological evolution. Distances in such a space–time are given by the FRW metric; see (3) in [1]. There the FRW equations are deduced (see (4) and (5) in [1]), and solutions are found for a flat Universe (7), (8) and (9), (10) in [1]). The Friedmann equation ((4) in [1]) that determines the Hubble parameter can be rewritten as

$$\Omega - 1 = \frac{k}{\dot{a}^2} = \frac{k}{a^2 H^2} = k d_{Hcom}^2 , \tag{2}$$

where a new definition is introduced, namely the *comoving Hubble radius* $d_{Hcom} \equiv d_H/a$, with $d_H \equiv H^{-1}$ being the *physical Hubble radius*[1]. Hence, $\Omega(\equiv \rho/\rho_c)$ can take values less, equal or greater than unity in open, flat and closed Universes, respectively. From this equation important conclusions can already be drawn about the differences between accelerated and decelerated cosmologies. In the case of cosmological evolution with $\ddot{a} > 0$ ($\ddot{a} < 0$) the comoving Hubble radius decreases (increases) with time and Ω converges to (diverges from) unity, implying that, with time, the corresponding spatial hypersurfaces look more and more (less and less) flat. The general condition for a Universe to expand or to contract acceleratingly can be drawn from (5) in [1].

$$p < -\frac{\rho}{3} , \tag{3}$$

i.e. it must be filled with a fluid having a sufficiently large negative pressure.

From the continuity equation ((6) in [1]) is not difficult to see that if the energy density is constant, then either the Universe is static (i.e. the scale factor a does not change in time) or the fluid satisfies the equation of state $p = -\rho$. Since a particular value of this constant energy density can be $\rho = 0$, this energy is commonly associated with the (scalable) energy of the cosmic vacuum. According to condition (3), the case of the non-static Universe with equation of state $p_\Lambda = -\rho_\Lambda = -(8\pi G)^{-1}\Lambda =$ constant, implies that, whenever the weak condition $\rho_\Lambda > 0$ is satisfied, the Universe is described by an accelerated expansion (or contraction) of the spatial hypersurfaces.

Vacuum is a particular case of barotropic fluids with equation of state, $p = \omega\rho$, where ω is, in general, a function of time. Condition (3) now reads, $\omega < -\frac{1}{3}$. For the case of $\omega =$ const., the continuity equation yields $\rho = \rho_0 a^{-3(1+\omega)}$.

The SBB includes two evolutionary stages. One extends from the Big Bang until nearly the beginning of the epoch of galaxies formation. To match several observations (the more important being the abundances of light elements as predicted by nucleosynthesis), the period had to be dominated

[1] Note that this definition does not coincide with the one for the causal horizon given in [1, 5] for $d_H(t) \equiv a(t)\int_{t_*}^{t} \frac{dt}{a(t)}$; They are different during an inflationary era but are proportional to each other when the expansion is of the power–law type.

by relativistic matter known, in general, as radiation with $\omega_R = 1/3$. After that period, the formation of large–scale cosmological structure requires non-relativistic pressureless matter ($\omega_M = 0$) to dominate over radiation. Therefore, according to $\omega < -\frac{1}{3}$, the SBB describes a Universe that expands non-acceleratingly from the very beginning through the far future.

The SBB provides us a picture of an expanding Universe that evolves from the initial singularity until today, passing through the above–described w-epochs. It is successful in explaining the formation of light elements (nucleosynthesis) and provides a general framework to understand the evolution of perturbations that eventually gave rise to the formation of LSS. However, as it is pointed out in the contributions by J. L. Cervantes–Cota [1] and E. Copeland [5] to this book, the SBB has unavoidable problems (horizon, flatness, causal origin of primordial perturbations, etc) that cannot be understood without the incorporation of new concepts and ideas. The main ingredient that particle physics has brought to the modern cosmological understanding is that of scalar field dynamics. The scalar field represents a generic matter field that evolves with the Universe expansion and should be responsible for an inflationary epoch at the very beginning of time, and perhaps should also be responsible for the present accelerating dynamics of the Universe. In the next section we study the dynamics of this generic field.

3 Scalar Fields

A real scalar field is a map $\phi : \mathrm{M} \to \mathbb{R}$, i.e. a real function that puts a point in the space–time M into relation with a point in the line $\mathbb{R}$. In Quantum Field Theory this function is used to represent a boson particle. If the boson lives in a Minkowski space–time $\mathcal{M}^4$, then the corresponding action is given by (the details of the calculations in this section can be found in [2]),

$$S = \int_{\mathcal{M}^4} dx^4 \mathcal{L} \,, \tag{4}$$

with Lagrangian density,

$$\mathcal{L} = -\frac{1}{2}\eta^{\mu\nu}\phi_{,\mu}\phi_{,\nu} - V'(\phi) \,, \tag{5}$$

where $\eta^{\mu\nu} = diag\{-1,1,1,1\}$ stands for a metric of $\mathcal{M}^4$, $\mu,\nu = 0,1,2,3$, $V(\phi)$ is the scalar field potential and a prime denotes the derivative with respect to ϕ. Varying the action, the equation of motion for the scalar field is obtained as,

$$\ddot{\phi} - \nabla^2\phi + V'(\phi) = 0 \,. \tag{6}$$

In the cosmological framework, $\mathcal{M}^4$ is substituted by the $\mathcal{FRW}$ space–time $\mathcal{RW}$ and the action is that of the Einstein–Hilbert,

$$S = \int_{\mathcal{FRW}} dx^4 \sqrt{-g}\mathcal{L}\,, \tag{7}$$

where g is the determinant of the FRW metric,

$$\mathcal{L} = \frac{1}{2} m_{\rm Pl}^2 R - \frac{1}{2} g^{\mu\nu} \phi_{,\mu} \phi_{,\nu} - V'(\phi)\,, \tag{8}$$

$m_{\rm Pl}$ is the Planck mass and R is the scalar curvature. Equation (6) becomes,

$$\ddot{\phi} + 3H\dot{\phi} + V'(\phi) = 0\,, \tag{9}$$

where a friction-like term arises due to the cosmic expansion and the gradient terms were omitted, consistent with the cosmological principle.

After comparison with the continuity equation ((6) in [1]), the scalar field becomes equivalent to a perfect fluid with energy density and pressure,

$$\rho = \frac{\dot{\phi}^2}{2} + V(\phi) \quad , \quad p = \frac{\dot{\phi}^2}{2} - V(\phi)\,. \tag{10}$$

Hence, for a Universe dominated by the energy density of a real scalar field, the condition (3) for accelerated expansion is rewritten as,

$$\frac{\dot{\phi}^2}{2} < V(\phi)\,. \tag{11}$$

With the aim of facilitating the analysis of the cosmic dynamics, it is convenient to define the horizon–flow functions that are given in terms of Hubble horizon; the latter is a primordial ingredient to understand the causal evolution of the cosmological dynamics; see the horizon problem in [1, 5]. Thus, horizon–flow functions are [6]:

$$\epsilon_0 \equiv \frac{d_{\rm H}(N)}{d_{\rm Hi}}, \qquad \epsilon_{m+1} \equiv \frac{{\rm d}\ln|\epsilon_m|}{{\rm d}N}, \qquad m \geq 0\,, \tag{12}$$

where $d_{\rm Hi} \equiv d_{\rm H}(t_{\rm i})$. Note that $\epsilon_1 = \dot{d_H}$ and $\epsilon_1\epsilon_2 = d_H \ddot{d_H}$. According to condition $\ddot{a} > 0 \Longrightarrow \frac{d\ln d_H}{dN} < 1$, for a positive energy density, $0 \leq \epsilon_1 < 1$ during inflation.

Further, in accordance with definitions (12),

$$H^2(N) = H_0^2 \exp\left(-2\int \epsilon_1(N) dN\right)\,, \tag{13}$$

where H_0 is an integration constant. Now, substituting ρ as given by (10) in the Friedmann equation ((4) in [1]) and using the definition (12) for ϵ_1 [2, 8],

$$\frac{d\phi}{dN} = \sqrt{\frac{2}{\kappa}}\sqrt{\epsilon_1}\,, \tag{14}$$

where $\kappa \equiv 8\pi G = 8\pi/m_{\rm Pl}^2$. Given a scalar field cosmology characterized by $\epsilon_1(N)$ the corresponding potential as a function of the field is given by,

$$V(\phi) = \begin{cases} \phi(N) = \sqrt{\frac{2}{\kappa}} \int \sqrt{\epsilon_1} dN - \phi_0 \,, \\ V(N) = \frac{H_0^2}{\kappa} \left[3 - \epsilon_1\right] \exp\left(-2 \int \epsilon_1 dN\right) \,, \end{cases} \tag{15}$$

where the potential as a function of N is derived from the Friedmann equation for scalar field cosmology using expressions (14) and (13) [2, 9].

4 Observations and Modeling

4.1 Present Day Acceleration

Recent observations of the celestial candles known as type Ia supernovae have been made that indicate a new feature of the present Universe composition. Currently the physics behind the peak light output from such supernovae seems to be well understood. Thus, by observing a type Ia supernova in a distant galaxy, measuring the peak light output, and comparing the relative intensity of light observed from the object with that expected from its absolute magnitude, the inverse square law for light intensity can be used to infer its distance. Because type Ia supernovae are very bright objects they are used to measure distances out to around 1000 Mpc, which is a significant fraction of the radius of the observable Universe. According to the analysis of the data collected for several type Ia supernovae, the observable Universe seems to be in a phase of accelerated expansion; for details on the type Ia supernovae and on the analysis of the redshift data see the contribution by A. Filippenko to this book [3].

Therefore, in accordance with (3), the Universe would be currently dominated not by pressureless matter but by some kind of fluid with negative pressure. Since this component of the cosmic energetic budget has eluded direct observation so far, it is generically known as *dark energy* (see the contribution to this book by de la Macorra [4]). The dark energy can be, in principle, the non-zero vacuum energy parametrized by the cosmological constant Λ. Adding such an energy does not strongly modify the cosmological picture as described by the SBB. In fact, the dark energy seems also to be necessary in order to match data from the observation of cosmic microwave background radiation and from large–scale structure formation. This fact is a very interesting confirmation of the existence of the dark energy.

First of all, if one would like to describe the present time cosmic acceleration as induced by a cosmological constant, the associated vacuum energy required to match the observations is $\rho_\Lambda^{Obs} \leq (10^{-12} GeV)^4$. On the other hand, from Quantum Field Theory with cutoff at the Planck scale it is expected that $\rho_{\Lambda}^{QFT} \sim (10^{18} GeV)^4$. The large disagreement between the two estimations of the vacuum energy is one of the hottest problems of modern cosmology. This is referred to as the *cosmological constant problem.*

A solution to this problem is to consider a cosmological "constant" decaying in such a way that its associated energy is currently the one required by observations. According to (10), the simplest candidate for such a dynamical vacuum energy could be a scalar field slowly changing over time. For a high enough potential energy, condition (11) is fulfilled and the Universe permeated by the scalar field potential energy undergoes accelerated expansion. This scalar field has been coined *quintessence.*

Many quintessence models have been devised and some other candidates for the dark energy have been proposed such as the *Cardassian expansion* [10] and the *Chaplygin gas* [11]. Among the problems that arise when modeling the dark energy, one of the outstanding ones is to find observational signatures differentiating between the candidates. One expects the observations to help in that task but often it is necessary to know exactly what to look for in the data. Since different dark energy candidates evolve in different ways, it could be useful to look for the imprint of these differences in data. With that aim, a model-independent parameterization of the dark energy evolution can be useful. In Fig. 1 it is shown how this parameterization can be done.

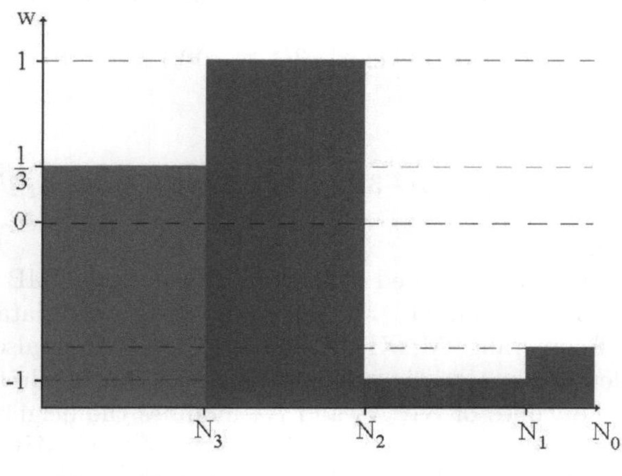

Fig. 1. Parameterization of the dark energy evolution

In the horizontal axis the e-folds numbers are calculated now with respect to the present time N_0, i.e. N increases leftwards. In the vertical axis is $\omega(N)$, which determines the equation of state $p = \omega\rho$. With dashed lines are denoted four cases where w is a constant: a stiff fluid ($\omega_s = 1$), radiation ($\omega_R = 1/3$), matter ($\omega_M = 0$), a tracker scalar field (which is a special case of quintessence models where w can be safely approximated to be constant with $-1 < \omega_{tracker} < -1/3$), and a cosmological constant ($\omega_\Lambda = -1$). Those cases that do not satisfy the condition $\omega < -\frac{1}{3}$ get thrown out of the game because they do not produce an accelerated expansion.

More generally, one can look for parameterizations where ω can be approximated by different constants for different ranges of e-folds numbers. In the figure the filled polygons represent each case,

$$\omega(N) = \begin{cases} \omega_4 = \frac{1}{3}, & N > N_3 \\ \omega_3 = 1, & N_3 \geq N > N_2 \\ \omega_2 = -1, & N_2 \geq N > N_1 \\ -1 < \omega_1 < -\frac{1}{3}, & N_1 \geq N > N_0 \end{cases} \tag{16}$$

where $-1 < \omega_1 < -1/3$. As it is explained in [4], this could be the case of some fluid which initially behaves like radiation, at some energy scale condensates into a scalar field with dynamics dominated by its kinetic energy, then undergoes a strong friction changing to a phase of totally potential-dominated dynamics and, finally, behaves like a tracker field.

It is interesting to check, for instance, how such a parameterization will modify the CMB spectrum and how it compares to a cosmological constant and to tracker fields. CAMB is a program which permits one to compute the CMB spectrum after specifying a number of parameters including $\omega_{tracker}$. To include the constant in sectors parameterization of $w(N)$ it was necessary to match the values of the energy densities ρ_{wi} at the borders of those sectors,

$$\rho_{wi}(N) = \rho_i^{eff} \exp\left[-3(1+\omega_i)N\right] , \tag{17}$$

where,

$$\rho_i^{eff} = \rho_0 \exp\left(-3\Delta\omega_{12}N_2\right) \exp\left(-3\Delta\omega_{23}N_3\right) \ldots \exp\left(-3\Delta\omega_{(i-1)i}N_i\right) \tag{18}$$

and $\Delta\omega_{(i-1)i} = \omega_{(i-1)} - \omega_i$.

In Fig. 2, examples are presented of the variations of the CMB spectrum after varying the parameters in (16). In this figure the binned data from experiments DASI, Boomerang, MAXIMA and COBE–DMR is also shown. Though a more detailed analysis is still in progress, it can be already noted that variation of the values of N_1, N_2 and N_3 modifies the heights and positions of the peaks and dips of the theoretical curve of the CMB spectrum. In fact, one can choose the values of these parameters in such a way that the fit to data might be improved compared to the cases of a cosmological constant and a tracker field, as shown in Fig. 3, where the blue curve represents the spectrum for a tracker field with $\omega_{tracker} = -0.86$, the green curve stands for a cosmological constant and the yellow one corresponds to $N_1 = 1.97$, $N_2 = 4.2$, $N_3 = 6.5$ and $\omega_1 = -0.86$. Certainly, the data error bars are still too large to make any strong conclusion in this direction but results are encouraging.

4.2 Cosmic Acceleration in the Very Early Universe

As in the case of the present-day Universe, if the acceleration in the very early times –required to solve the SBB problems– is proposed to be induced by the vacuum energy associated with a cosmological constant, then several problems are confronted. The theory of nucleosynthesis, which is so successful in accounting for the observed cosmic abundances of the lightest atoms, describes a radiation-dominated Universe. In this way, in order to reproduce the SBB success the inflationary period must end some time before nucleosynthesis started. Moreover, if the density perturbations leading to large–scale structure are wanted to be causally seeded by an inflationary period in the very early Universe, then the Hubble radius must start to increase, which implies the end of the accelerated expansion.

Once the energy of a cosmological constant begins to dominate over other non-exotic forms of energy, it will dominate forever. Thus, the inflationary epoch will have no end. Once more, a single scalar field seems to be the simplest candidate to solve all the problems associated with a cosmological constant while still producing enough inflation. In this framework, the origin of the density perturbations leading to LSS is thought to be the quantum fluctuations of the scalar field during inflation [2, 7]. A distinguishing feature of this scenario is that, along with the density perturbations associated with the inflaton field, tensor perturbations are also produced which are associated with space–time metric perturbations. All these fluctuations will be amplified by the accelerated expansion and seeded at the required time through the

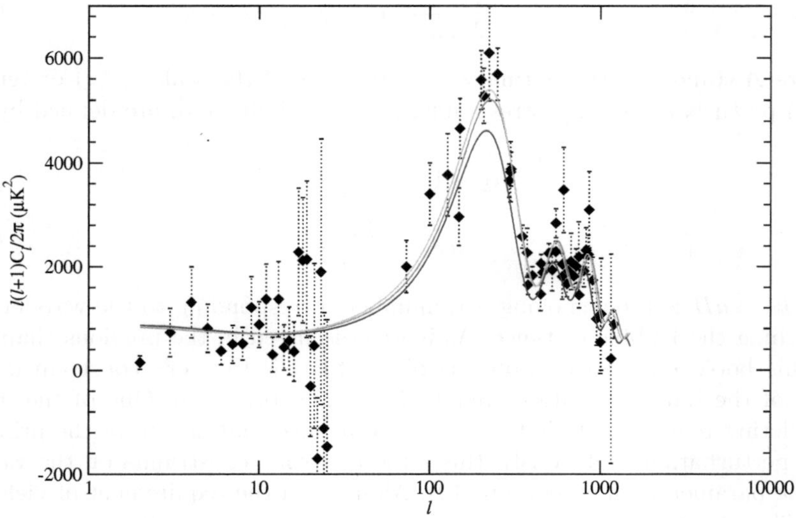

Fig. 2. The CMB spectra for different versions of the parametrization by sectors of the state parameter ω.

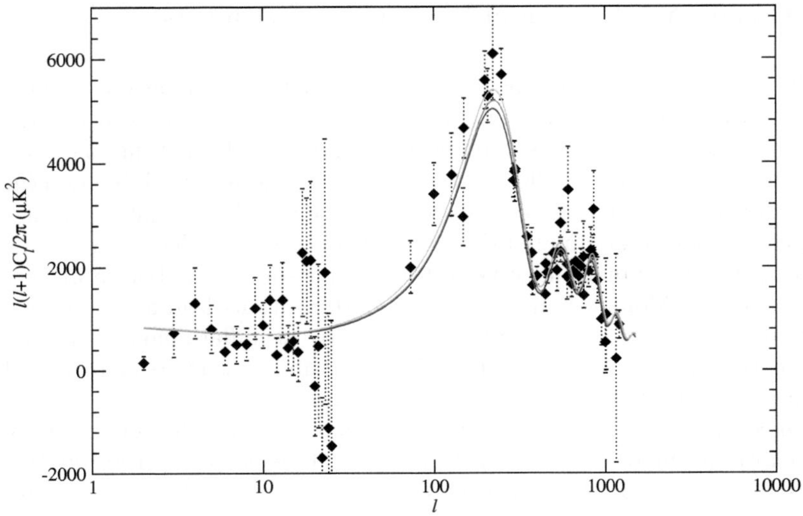

Fig. 3. Comparing a tracker field case (blue curve), the Λ case (green curve) and a case with $\omega(N)$ parametrized by sectors (yellow line).

mechanism of crossing out and crossing back into the Hubble horizon. The primordial perturbations are parameterized as,

$$\ln A^2(k) = \ln A^2(k_1) + n(k_1)\ln\frac{k}{k_1} + \frac{1}{2}\frac{d\,n(k)}{d\ln k}|_{k=k_1}\ln^2\frac{k}{k_1} + \cdots\,, \tag{19}$$

where A stands for the normalized amplitudes of the scalar (A_S) or tensor (A_T) perturbations, the corresponding spectral indices, n, are defined by,

$$n_S - 1 \equiv \frac{d\ln A_S^2}{d\ln k}\,, \tag{20}$$

$$n_T \equiv \frac{d\ln A_T^2}{d\ln k}\,, \tag{21}$$

and $k_1 = aH$ is the comoving wavenumber corresponding to the wavelength matching the Hubble distance. As it was discussed in the previous chapters of this book, one of the problems of the SBB is the very constrained nature of the initial conditions required for LSS formation. One of the most restricting requirements is the almost *scale invariant* nature of the primordial perturbations. Obviously, this imposes strong constraints on the values of the parameters in expansion (19). Along with the requirement of yielding a sufficiently large number of e-foldings, $n_S \approx 1$ and $n_T \approx 0$ are strong conditions in determining whether a proposed scalar field model is a successful inflationary model.

The horizon–flow functions (12) serve as link between observations and the inflationary dynamics. For most inflationary models, exact expressions for parameters in (19) are unknown. The usual way to calculate them is as an expansion in terms of the horizon–flow functions (see [12, 6] for details). To next–to–leading order in terms of the horizon–flow functions, indices (20) and (21) are written as [12, 13]

$$n_S - 1 = -2\epsilon_1 - \epsilon_2 - 2\epsilon_1^2 - (2C+3)\epsilon_1\epsilon_2 - C\epsilon_2\epsilon_3, \tag{22}$$

$$n_T = -2\epsilon_1 - 2\epsilon_1^2 - 2(C+1)\epsilon_1\epsilon_2\,, \tag{23}$$

where $C \approx -0.7293$. Dynamics described using the horizon-flow functions correspond to an inflationary potential. In this way, the corresponding spectral indices can be calculated and compared with the observational values.

The big problem here is that there exist a large number of single scalar field models in good agreement with observations, therefore it is difficult to determine which is the one corresponding to the actual physics of the very early Universe. With this aim, a more efficient approach seems to be to constrain and determine the main features of the inflaton potential according to observations, instead of building models and comparing their predictions with the measured data (see [5] and [14] for references on this approach).

Recalling that, according to (12), the definition of ϵ_{m+1} involves the derivative with respect to N of ϵ_m, (22) and (23) are therefore differential equations for ϵ_1. In this way, solving these equations and using expression (15), the inflaton potential can be determined from the information on the functional forms of the tensor and scalar spectral indices.

The strong limitation for this program to be useful is that the most that is known (and will be for a while) about the scale or time dependence of the spectral indices is the observed values of a very few parameters in expansions (19), together with the corresponding error bars. Taking this limitation into account, the best one can do is to look for generic features of the potential yielding values of the primordial parameters, in agreement with those derived from observations, using some "well-based" assumptions for the values of those parameters which have not been observed so far.

For instance, in [15] it was proved that if it is assumed $dn/d\ln k = 0$ for the scalar and the tensor perturbations, then the resulting potential is an exponential function of the inflaton field. This corresponds to the scenario known as *power-law inflation* because $a \sim t^p$ with $p \gg 1$ [16].

If $dn_S/d\ln k \ll 1$ is allowed to be non-zero while keeping $dn_T/d\ln k = 0$, it can be seen that power–law inflation is an attractor of the corresponding inflationary dynamics [15]. This implies that it is difficult to distinguish the actual potential from the exponential one using the observational information.

A similar result is obtained if both spectral indices are allowed to be scale–dependent but with this dependence being detectible up to the second order on the horizon–flow functions [8].

The role of the tensor perturbations deserves special attention when determining the best–fit values of the cosmological parameters from CMB and LSS spectra. This is motivated in part by the possibility of measuring the cosmic background polarization [2], allowing the tensorial contribution to be indirectly determined. This contribution can be parametrized in terms of the relative amplitudes of the tensor and scalar perturbations,

$$r \equiv \alpha \frac{A_T^2}{A_S^2}, \tag{24}$$

where α is a constant. The expectation is to measure a central value of r. Thus the question is what inflationary dynamics, yielding an almost constant ratio r, look like. The answer was given in [17] where it was shown that in the case of an exactly constant r, power–law inflation is a repulsor of the corresponding dynamics. Since it describes a quick and strong departure from scale–invariance, for the model to be successful, the perturbations must be produced in the quasi power–law regime, once more making it difficult to observationally distinguish between the two dynamics.

All of the above-discussed results imply a serious handicap for any program of reconstruction of the inflaton potential. A way to improve this situation could be to combine the information on Δn (the difference between the spectral indices) and the value of r, with the two first horizon–flow functions [18]. It follows from definitions (20), (21) and (24) that

$$\frac{d \ln r}{d \ln k} = \Delta n \equiv n_T - (n_S - 1), \tag{25}$$

this way, any information on the evolution of both spectral indices can be used as information on the scale dependence of the tensor to scalar ratio.

With regards to these, it becomes important to analyze in detail the case,

$$\ln \frac{r}{16} = a_0 + a_1(N - N_0), \tag{26}$$

where the corresponding solution for ϵ_1 is [18],

$$\epsilon_1(N) = \epsilon_{1(i)} \exp\left[B \exp\left(-\frac{N}{C}\right)\right] \exp(AN), \tag{27}$$

with $\epsilon_{1(i)} \equiv \epsilon_{1(0)} \exp\{[a_0 + (N_0 - C)\, a_1]\, C\}$ and $A \equiv a_1 C$. The asymptotes of this solution for $B \neq 0$ will be mainly determined by the value and sign of B. However, for $A = 0$ (i.e. $\ln r/16 = a_0$), if the model yielding ϵ_1 given by (27) is expected to be compatible with current data, B has to be chosen an extraordinarily small number. Once more, this will make it very difficult to observationally distinguish the corresponding scenario from power–law inflation. More interesting is the case $B = 0$ where the potential is,

$$V = V_0 \left(3 - \frac{A^2}{4}\psi^2\right) \exp\left(-\frac{A}{2}\psi^2\right), \tag{28}$$

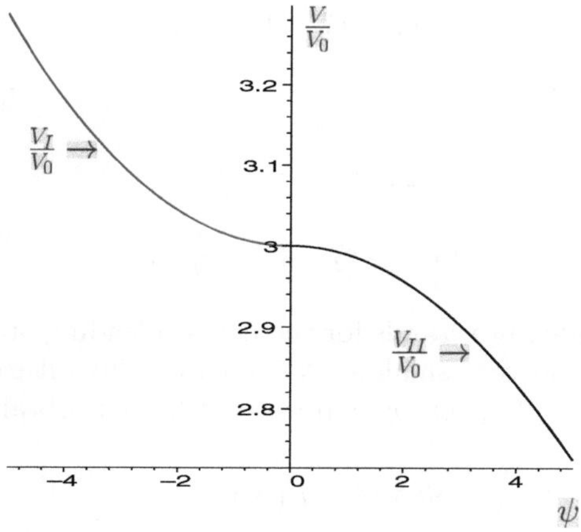

Fig. 4. Sectors of the inflaton potential given by (28) for $A = -0.0073$ (V_I) and $A = 0.0073$ (V_{II}).

with $\psi \equiv \sqrt{\kappa/2}(\phi + \phi_0)$. In Fig. 4 sectors of this potential are plotted for $A = -0.0073$ (V_I) and $A = 0.0073$ (V_{II}). For this model, information on the existence of extrema and on the curvature can be derived. As can be seen in Fig. 4, realizations of this potential resemble the cases of monomial potentials with even order (V_I, $\epsilon_2 < 0$), and inflation near a maximum (V_{II}, $\epsilon_2 > 0$) (see [2] for examples of such inflationary scenarios motivated by particle physics) allowing, therefore, one to observe features of the inflaton potential beyond the exponential form characteristic of power–law inflation. In [18] it was shown that for a large set of A and $\epsilon_{1(i)}$ values, the corresponding spectra agree with CMB and LSS observations.

On the Order of the Approximations

A crucial question in the analysis of the previous section is to what extent these results depend on the order of expressions underlying the calculations. It is widely believed that this is not of concern. Let us show that it must be.

To next-to-next-to-leading order of the tensor to scalar ratio is given by,

$$\ln\frac{r}{r_0} = \ln\epsilon_1 + C\epsilon_2 + \left(-\frac{\pi^2}{2} + 5 + C\right)\epsilon_1\epsilon_2 + \left(-\frac{\pi^2}{8} + 1\right)\epsilon_2^2 + \left(-\frac{\pi^2}{24} + \frac{C^2}{2}\right)\epsilon_2\epsilon_3 \,. \tag{29}$$

If the order of this expression was of little concern, then the corresponding solution for the case with r given by (26) must be very similar to potential (28). To check this, one can assume,

$$\epsilon_1 = \epsilon_1^{nlo} (1 + \delta) \,, \tag{30}$$

with $\delta << 1$, yielding

$$\epsilon_2 = \epsilon_2^{nlo} + \frac{d\delta}{dN} \tag{31}$$

$$\epsilon_2\epsilon_3 = \epsilon_2^{nlo}\epsilon_3^{nlo} + \frac{d^2\delta}{dN^2} \,, \tag{32}$$

where the super-index nlo stands for the next-to-leading order solution. $\delta(N)$ is expected to remain very small as N increases. Substituting (30), (31) and (32) into (29) with ϵ_1^{nlo} given by expression (27) it is obtained that

$$\frac{d^2\delta}{dN^2} + D(N)\frac{d\delta}{dN} + S(N)\delta = F(N) \tag{33}$$

$$D(N) = C_3 + C_2\epsilon_{1(i)}\, e^{Be^{NC_1}} e^{AN} \tag{34}$$

$$S(N) = -C_4 + C_2\epsilon_{1(i)}\, e^{Be^{NC_1}} e^{AN} \left(C_1 B e^{NC_1} + A\right) \tag{35}$$

$$F(N) = C_2\epsilon_{1(i)}\, e^{Be^{NC_1}} e^{AN} \left(C_1 B e^{NC_1} + A\right) + C_5 B e^{NC1} + C_6 A \,, \tag{36}$$

where non-linear terms of δ were neglected and C_i (with $i = 1 \ldots 6$) are fixed constants given in terms of C and π. Constants $\epsilon_{1(i)}$, A and B are those already given in the next-to-leading order solution (27). In Figs. 5, 6 and 7 the N-dependent parameters $D(N)$, $S(N)$ and $F(N)$ are plotted for the same numerical values used for the constants in [18].

At least for this case, the parameters can be safely approximated by constants. Therefore, a qualitative analysis of the phase space for the flow given by (33) can be carried out. It yields that there exists a saddle point at $(\delta = F/S, \dot{\delta} = 0)$, shown in Fig. 8. It means that solutions for $\delta(N)$ with $A = -0.0073$ and $\epsilon_{1(i)} = 0.05$ are likely unstable. A more complete analysis is indeed required, but this result would serve as a warning to pay more attention to the order of the approximations used when information on the inflationary dynamics is drawn from observational data.

5 Conclusions

Cosmic acceleration is a trivial solution to the Einstein equations for an isotropic and homogeneous Universe, as appears to be the one where we live. An accelerated expansion is typical of Universes filled with a kind of energy yielding a strong enough negative pressure. This is the case when the cosmic energy is dominated by the contribution of the vacuum.

Observational evidence strongly suggests that our Universe's evolution includes three well-defined epochs with regards to the increase of the cosmic volume. First, the very early Universe would undergo an accelerated expansion known as inflation. Then, a period of non-accelerated expansion

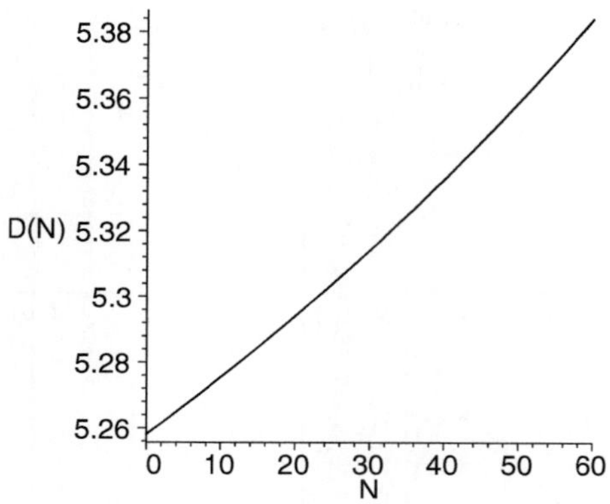

Fig. 5. The behavior of D(N) for $A = -0.0073$ and $\epsilon_{1(i)} = 0.05$.

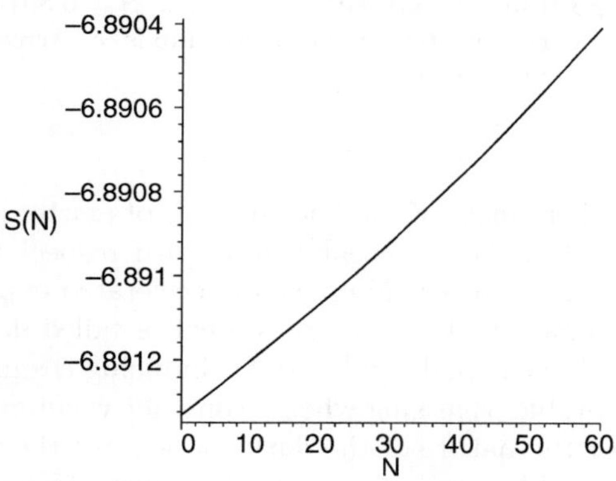

Fig. 6. The behavior of S(N) for $A = -0.0073$ and $\epsilon_{1(i)} = 0.05$.

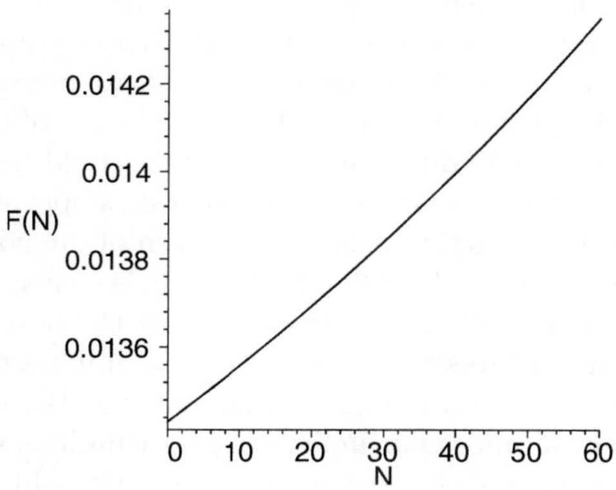

Fig. 7. The behavior of F(N) for $A = -0.0073$ and $\epsilon_{1(i)} = 0.05$.

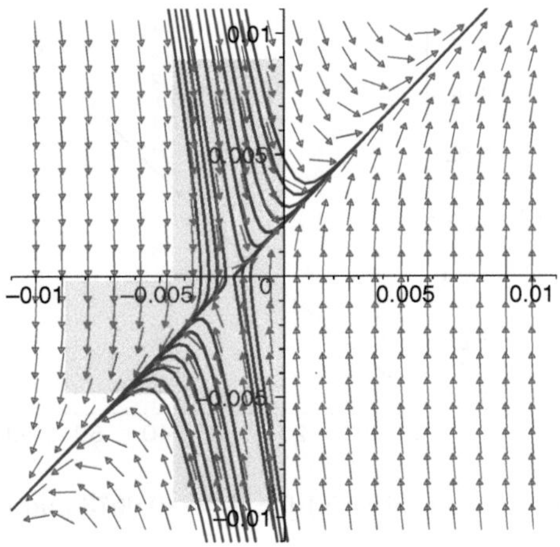

Fig. 8. The phase portrait for (33) with $D = -5.32$, $S = 6.891$ and $F = 0.014$. In the horizontal axis δ is represented; in the vertical axis, $\dot{\delta}$. Arrows show the vector field and lines show some trajectories.

would take place where most of the known kinds of matter and matter structures were formed. Finally, in recent times (with respect to cosmic scales) the Universe would enter a second epoch of accelerated expansion where the corresponding dominant matter–energy content is called dark energy.

A real scalar field is a good candidate for inducing cosmic acceleration. It may help to solve problems arising when a constant vacuum energy is used to explain inflation or the nature of the dark energy. For the required negative pressure, the scalar field dynamics must be dominated by its potential energy.

A hot question in cosmology is whether the observed (predicted) cosmic acceleration is (was) induced by a scalar field. If this is the case, the relevant question is to determine the origin and nature of the corresponding potential. This will open an important window into high energy physics.

Here, an idea was hinted at on how to differentiate between candidates for the dark energy. The proposal is to divide the evolution of the dark energy in periods where the corresponding equation of state could be approximated to be linear. The best–fit values for the corresponding slopes would indicate the favorite candidate. Encouraging results have been obtained in this direction.

It also explained some of the difficulties that arise when deriving the inflationary potential from observations. It seems that the best that can be done is to indicate generic features of the potentials yielding perturbations spectra matching the measured data. It was emphasized that the use of data on the difference of the tensor and scalar indices of perturbations yields information on the scale–dependence of the tensor to scalar ratio. This information may be very useful in classifying the inflationary potentials. Finally, a warning was issued about the possibility that the features of the inflaton potential drawn from the observational data could be biased by the order of the approximations used to derive the expressions underlying the calculations.

Acknowledgments

I would like to thank School organizers for inviting me to give this talk. I also thank A. García, A. de la Macorra and A. Coley for their solid support and helpful discussions. Supported in part by CONACyT grant 38495–E and SNI (Mexico).

References

1. J. L. Cervantes–Cota: An Introduction to Standard Cosmology, Lect. Notes Phys. 646, 7 (2004).
2. A. R. Liddle and D. H. Lyth: *Cosmological Inflation and Large-Scale Structure*, 1st edn (Cambridge University Press, Cambridge, 2000).
3. A. Filippenko: The Accelerating Universe and Dark Energy: Evidence from Type Ia Supernovae, Lect. Notes Phys. 646, 191 (2004).
4. A. de la Macorra: Quintessence and Dark Energy, Lect. Notes Phys. 646, 225 (2004).
5. E. Copeland: Inflation – In the Early Universe and Today, Lect. Notes Phys. 646, 53 (2004).
6. D. J. Schwarz, C. A. Terrero-Escalante and A. A. Garcia: Phys. Lett. B **517**, 243 (2001) [arXiv:astro-ph/0106020].
7. R. Brandenberger: Lectures on the Theory of Cosmological Perturbations, Lect. Notes Phys. 646, 127 (2004).
8. C. A. Terrero-Escalante and A. A. Garcia: Phys. Rev. D **65** 023515 (2002) [arXiv:astro-ph/0108188].
9. E. Ayon-Beato, A. Garcia, R. Mansilla and C. A. Terrero-Escalante: Phys. Rev. D **62** 103513 (2000) [arXiv:astro-ph/0007477].
10. K. Freese and M. Lewis, Phys. Lett. B **540**, 1 (2002) [arXiv:astro-ph/0201229].
11. A. Y. Kamenshchik, U. Moschella and V. Pasquier, Phys. Lett. B **511**, 265 (2001) [arXiv:gr-qc/0103004].
12. E. D. Stewart and D. H. Lyth: Phys. Lett. B **302** 171 (1993) [arXiv:gr-qc/9302019].
13. C. A. Terrero-Escalante: Is power-law inflation really attractive? In *Proceedings of IV DGFM-SMF Workshop on Gravitation and Mathematical-Physics*, ed by N. Bretón, J. Cervantes and M. Salgado (2001), arXiv:astro-ph/0204066.
14. R. Easther and W. H. Kinney, Phys. Rev. D **67**, 043511 (2003) [arXiv:astro-ph/0210345].
15. C. A. Terrero-Escalante, E. Ayón-Beato, and A. A. Garcia: Phys. Rev. D **64**, 023503 (2001).
16. F. Lucchin and S. Matarrese: Phys. Rev. D **32**, 1316 (1985).
17. C. A. Terrero-Escalante, J. E. Lidsey and A. A. Garcia: Phys. Rev. D **65** 083509 (2002) [arXiv:astro-ph/0111128].
18. C. A. Terrero-Escalante, Phys. Lett. B **563**, 15 (2003) arXiv:astro-ph/0209162.

Lectures on the Theory of Cosmological Perturbations

Robert H. Brandenberger

Brown University Physics Department, Providence, RI 02912, USA.
rhb@het.brown.edu

Abstract. The theory of cosmological perturbations has become a cornerstone of modern quantitative cosmology since it is the framework which provides the link between the models of the very early Universe such as the inflationary Universe scenario (which yield causal mechanisms for the generation of fluctuations) and the wealth of recent high-precision data on the spectrum of density fluctuations and cosmic microwave anisotropies. In these lectures, I provide an overview of the classical and quantum theory of cosmological fluctuations.

Crucial points in both the current inflationary paradigm [1, 2] of the early Universe and in proposed alternatives such as the Pre-Big-Bang [3] and Ekpyrotic [4] scenarios are that, first, the perturbations are generated on microscopic scales as quantum vacuum fluctuations, and, second, that via an accelerated expansion of the background geometry (or by a contraction of the background), the wavelengths of the fluctuations become much larger than the Hubble radius for a long period of cosmic evolution. Hence, both Quantum Mechanics and General Relativity are required in order to understand the generation and evolution of fluctuations.

As a guide to develop the physical intuition for the evolution of inhomogeneities, I begin with a discussion of the Newtonian theory of fluctuations. applicable at late times and on scales smaller than the Hubble radius. The analysis of super-Hubble fluctuations requires a general relativistic analysis. I first review the classical relativistic theory of fluctuations, and then discuss their quantization. I conclude with a brief overview of two applications of the theory of cosmological fluctuations: the trans-Planckian "problem" of inflationary cosmology and the current status of the study of the back-reaction of cosmological fluctuations on the background space-time geometry. Most of this article is based on the review [5] to which the reader is referred to for the details omitted in these lecture notes.

1 Motivation

As described in the lectures by Tegmark at this school [6], observational cosmology is currently in its golden years. Using a variety of observational techniques, physicists and astronomers are exploring the large-scale structure of the Universe. The Cosmic Microwave Background (CMB) is the observational window which in recent years has yielded the most information. The anisotropies in the CMB have now been detected on a wide range of angular scales, giving us a picture of the Universe at the time of recombination, the time that the cosmic photons last scattered. Large-scale galaxy redshift

R.H. Brandenberger, Lectures on the Theory of Cosmological Perturbations, Lect. Notes Phys. **646**, 127–167 (2004)
http://www.springerlink.com/

surveys are providing us with increasingly accurate power spectra of the distribution of objects in the Universe which emit light, which - modulo the question whether light in fact traces mass (this is the issue of the cosmic *bias*) - gives us the distribution of mass at the present time. Analyses of the spectra of quasar absorption line systems and weak gravitational lensing surveys are beginning to give us complementary information about the distribution of matter (independent of whether this matter in fact emits light, thus shedding light on the biasing issue). The analysis of weak gravitational lensing maps is in fact sensitive not only to the baryonic but also to the dark matter, and promises to give a technique which unambiguously reveals where the dark matter is concentrated. X-ray telescopes are providing additional information on the distribution of sources which emit X-rays.

The current data fits astonishingly well with the current paradigm of early Universe cosmology, the *inflationary Universe scenario* [1]. However, it is important to keep in mind that what is tested observationally is the paradigm that the primordial spectrum of inhomogeneities was scale-invariant and predominantly adiabatic (these terms will be explained in the following section), and that there might exist other scenarios of the very early Universe which do not yield inflation but predict a scale-invariant adiabatic spectrum. For example, within both the Pre-Big-Bang [3] and the Ekpyrotic scenarios [4] there may be models which yield such a spectrum [1] One should also not forget that topological defect models of structure formation (see e.g. [19, 20, 21] for reviews) naturally yield a scale-invariant spectrum, however of primordial isocurvature nature and thus no longer compatible with the latest CMB anisotropy results.

The theory of cosmological perturbations is what allows us to connect theories of the very early Universe with the data on the large-scale structure of the Universe at late times and is thus of central importance in modern cosmology. The techniques discussed below are applicable to most scenarios of the very early Universe. Most specific applications mentioned, however, will be within the context of the inflationary Universe scenario. To understand what the key requirements for a viable theory of cosmological perturbations are, recall the basic space-time diagram for inflationary cosmology (Fig. 1): Since, during the phase of standard cosmology $t_R < t < t_0$, where t_R corre-

[1] Note, however, that whereas the simplest inflationary models yield an almost scale-invariant $n = 1$ spectrum of fluctuations, as discussed in detail in these lectures, this is not the case for the simplest models of Pre-Big-Bang type nor for four dimensional descriptions of the Ekpyrotic scenario. In the case of single field realizations of Pre-Big-Bang cosmology, a spectrum with spectral index $n = 4$ emerges [7]. In Ekpyrotic cosmology, the value of the index of the final power spectrum is under active debate. Most studies conclude either that the spectral index is $n = 3$ [8, 9, 10, 11, 12], or that the result is ill-defined because of the singularities at the bounce [13, 14] (see, however, [15, 16, 17] for arguments in support of a final scale-invariant spectrum). See also [18] for criticisms of the basic setup of the Ekpyrotic scenario.

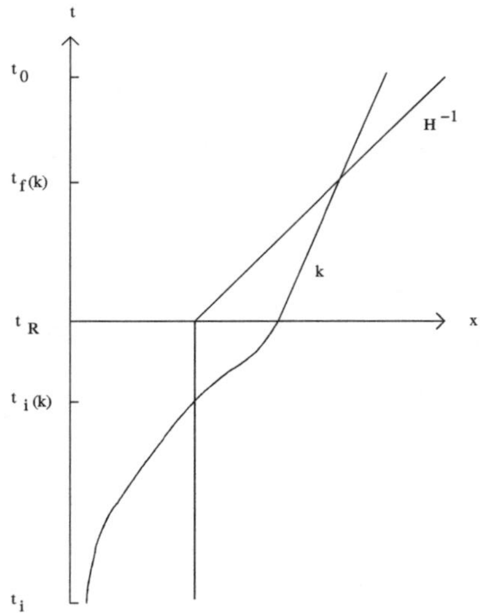

Fig. 1. Space-time diagram (sketch) showing the evolution of scales in inflationary cosmology. The vertical axis is time, and the period of inflation lasts between t_i and t_R, and is followed by the radiation-dominated phase of standard big bang cosmology. During exponential inflation, the Hubble radius H^{-1} is constant in physical spatial coordinates (the horizontal axis), whereas it increases linearly in time after t_R. The physical length corresponding to a fixed comoving length scale labelled by its wavenumber k increases exponentially during inflation but increases less fast than the Hubble radius (namely as $t^{1/2}$), after inflation.

sponds to the end of inflation, and t_0 denotes the present time, the Hubble radius $l_H(t) \equiv H^{-1}(t)$ expands faster that the physical wavelength associated with a fixed comoving scale, the wavelength becomes larger than the Hubble radius as we go backwards in time. However, during the phase of accelerated expansion (inflation), the physical wavelength increases much faster than the Hubble radius, and thus at early times the fluctuations emerged at micro-physical sub-Hubble scales. The idea is that micro-physical processes (as we shall see, quantum vacuum fluctuations) are responsible for the origin of the fluctuations. However, during the period when the wavelength is super-Hubble, it is essential to describe the fluctuations using General Relativity. Thus, both Quantum Mechanics and General Relativity are required to successfully describe the generation and evolution of cosmological fluctuations.

A similar conclusion can be reached when considering the space-time diagram in a model of Pre-Big-Bang or Ekpyrotic type, where the Universe starts out in a contracting phase during which the Hubble radius contracts faster than the physical length corresponding to a fixed comoving scale (see Fig. 2). The contracting phase ends at a cosmological bounce, after which the Universe is assumed to follow the same evolution history as it does in stan-

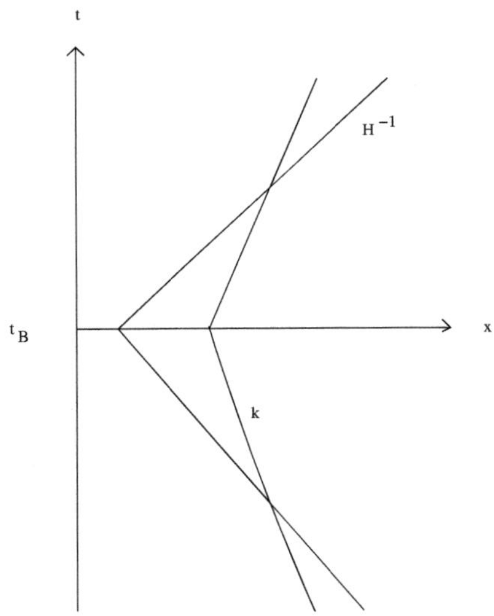

Fig. 2. Space-time diagram (sketch) showing the evolution of scales in a cosmology of PBB or Ekpyrotic type. The axes are as in Fig. 1. Times earlier than t_B correspond to the contracting phase, times after describe the post-bounce phase of expansion as described in standard cosmology. The Hubble radius decreases relative to a fixed comoving scale during the contracting phase, and increases faster in the expanding phase. Fluctuations of cosmological interest today are generated sub-Hubble but propagate super-Hubble for a long time interval.

dard Big Bang cosmology. As in inflationary cosmology, quantum vacuum fluctuation on sub-Hubble scales (in this case in the contracting phase) are assumed to be the seeds of the inhomogeneities observed today. For a long time period, the scale of the fluctuation is super-Hubble.

Thus, we see that in inflationary cosmology as well as in Pre-Big-Bang and Ekpyrotic-type models, both Quantum Mechanics and General Relativity are required to understand the generation and evolution of cosmological perturbations.

2 Newtonian Theory of Cosmological Perturbations

2.1 Introduction

The growth of density fluctuations is a consequence of the purely attractive nature of the gravitational force. Imagine (first in a non-expanding background) a density excess $\delta\rho$ localized about some point $\mathbf{x}$ in space. This fluctuation produces an attractive force which pulls the surrounding matter towards $\mathbf{x}$. The magnitude of this force is proportional to $\delta\rho$. Hence, by Newton's second law

$$\ddot{\delta\rho} \sim G\delta\rho\,, \tag{1}$$

where G is Newton's gravitational constant. Hence, there is an exponential instability of flat space-time to the development of fluctuations.

Obviously, in General Relativity it is inconsistent to consider density fluctuations in a non-expanding background. If we consider density fluctuations in an expanding background, then the expansion of space leads to a friction term in (1). Hence, instead of an exponential instability to the development of fluctuations, the growth rate of fluctuations in an expanding Universe will be as a power of time. It is crucial to determine what this power is and how it depends both on the background cosmological expansion rate and on the length scale of the fluctuations.

We will be taking the background space-time to be homogeneous and isotropic, with a metric given by

$$ds^2 = dt^2 - a(t)^2 d\mathbf{x}^2\,, \tag{2}$$

where t is physical time, $d\mathbf{x}^2$ is the Euclidean metric of the spatial hypersurfaces (here taken for simplicity to be spatially flat), and $a(t)$ denoting the scale factor, in terms of which the expansion rate is given by $H(t) = \dot{a}/a$. The coordinates $\mathbf{x}$ used above are "comoving" coordinates, coordinates painted onto the expanding spatial hypersurfaces. Note, however, that in the following two subsections $\mathbf{x}$ will denote the physical coordinates, and $\mathbf{q}$ the comoving ones.

The materials covered in this section are discussed in several excellent textbooks on cosmology, e.g. in [22, 23, 24, 25].

2.2 Perturbations About Minkowski Space-Time

To develop some physical intuition, we first consider the evolution of hydrodynamical matter fluctuations in a fixed non-expanding background. Note that in this case the background Einstein equations are **not** satisfied.

In this context, matter is described by a perfect fluid, and gravity by the Newtonian gravitational potential φ. The fluid variables are the energy density ρ, the pressure p, the fluid velocity $\mathbf{v}$, and the entropy density S. The basic hydrodynamical equations are

$$\begin{aligned}
\dot{\rho} + \nabla\cdot(\rho\mathbf{v}) &= 0 \\
\dot{\mathbf{v}} + (\mathbf{v}\cdot\nabla)\mathbf{v} + \frac{1}{\rho}\nabla p + \nabla\varphi &= 0 \\
\nabla^2\varphi &= 4\pi G\rho \\
\dot{S} + (\mathbf{v}\cdot\nabla)S &= 0 \\
p &= p(\rho, S)\,.
\end{aligned} \tag{3}$$

The first equation is the continuity equation, the second is the Euler (force) equation, the third is the Poisson equation of Newtonian gravity, the fourth expresses entropy conservation, and the last describes the equation of state of matter. The derivative with respect to time is denoted by an over-dot.

The background is given by the background energy density ρ_o, the background pressure p_0, vanishing velocity, constant gravitational potential φ_0 and constant entropy density S_0. As mentioned above, it does **not** satisfy the background Poisson equation.

The equations for cosmological perturbations are obtained by perturbing the fluid variables about the background,

$$\begin{aligned}
\rho &= \rho_0 + \delta\rho \\
\mathbf{v} &= \delta\mathbf{v} \\
p &= p_0 + \delta p \\
\varphi &= \varphi_0 + \delta\varphi \\
S &= S_0 + \delta S\,,
\end{aligned} \tag{4}$$

where the fluctuating fields $\delta\rho, \delta\mathbf{v}, \delta p, \delta\varphi$ and δS are functions of space and time, by inserting these expressions into the basic hydrodynamical equations (3), by linearizing, and by combining the resulting equations which are of first order in time to obtain the following second order differential equations for the energy density fluctuation $\delta\rho$ and the entropy perturbation δS

$$\begin{aligned}
\ddot{\delta\rho} - c_s^2 \nabla^2 \delta\rho - 4\pi G \rho_0 \delta\rho &= \sigma \nabla^2 \delta S \qquad (5)\\
\dot{\delta S} &= 0\,,
\end{aligned}$$

where the variables c_s^2 and σ describe the equation of state

$$\delta p \;=\; c_s^2 \delta\rho + \sigma \delta S \tag{6}$$

with

$$c_s^2 \;=\; \left(\frac{\delta p}{\delta\rho}\right)_{|S} \tag{7}$$

denoting the square of the speed of sound.

What can we learn from these equations? First of all, since the equations are linear, we can work in Fourier space. Each Fourier component $\delta\rho_k(t)$ of the fluctuation field $\delta\rho(\mathbf{x}, t)$

$$\delta\rho(\mathbf{x}, t) \;=\; \int e^{i\mathbf{k}\cdot\mathbf{x}} \delta\rho_k(t) \tag{8}$$

evolves independently.

There are various types of fluctuations. If the entropy fluctuation δS vanishes, we have **adiabatic** fluctuations. If the entropy fluctuation δS is non-vanishing but $\dot{\delta\rho} = 0$, we speak on an **entropy** fluctuation. The first conclusions we can draw from the basic perturbation equations (5) are that

1) entropy fluctuations do not grow,
2) adiabatic fluctuations are time-dependent, and
3) entropy fluctuations seed an adiabatic mode.

Taking a closer look at the equation of motion (5) for $\delta\rho$, we see that the third term on the left hand side represents the force due to gravity, a purely attractive force yielding an instability of flat space-time to the development of density fluctuations (as discussed earlier, see (1)). The second term on the left hand side of (5) represents a force due to the fluid pressure which tends to set up pressure waves. In the absence of entropy fluctuations, the evolution of $\delta\rho$ is governed by the combined action of both pressure and gravitational forces.

Restricting our attention to adiabatic fluctuations, we see from (5) that there is a critical wavelength, the Jeans length, whose wavenumber k_J is given by

$$k_J = \Big(\frac{4\pi G\rho_0}{c_s^2}\Big)^{1/2} . \tag{9}$$

Fluctuations with wavelength longer than the Jeans length ($k \ll k_J$) grow exponentially

$$\delta\rho_k(t) \sim e^{\omega_k t} \text{ with } \omega_k \sim 4(\pi G\rho_0)^{1/2} \tag{10}$$

whereas short wavelength modes ($k \gg k_J$) oscillate with frequency $\omega_k \sim c_s k$. Note that the value of the Jeans length depends on the equation of state of the background. For a background dominated by relativistic radiation, the Jeans length is large (of the order of the Hubble radius $H^{-1}(t)$), whereas for pressure-less matter the Jeans length goes to zero.

2.3 Perturbations About an Expanding Background

Let us now improve on the previous analysis and study Newtonian cosmological fluctuations about an expanding background. In this case, the background equations are consistent (the non-vanishing average energy density leads to cosmological expansion). However, we are still neglecting general relativistic effects (the fluctuations of the metric). Such effects turn out to be dominant on length scales larger than the Hubble radius $H^{-1}(t)$, and thus the analysis of this section is applicable only to scales smaller than the Hubble radius.

The background cosmological model is given by the energy density $\rho_0(t)$, the pressure $p_0(t)$, and the recessional velocity $\mathbf{v}_0 = H(t)\mathbf{x}$, where $\mathbf{x}$ is the Euclidean spatial coordinate vector ("physical coordinates"). The space- and time-dependent fluctuating fields are defined as in the previous section:

$$\begin{aligned} \rho(t,\mathbf{x}) &= \rho_0(t)\big(1+\delta_\epsilon(t,\mathbf{x})\big) \\ \mathbf{v}(t,\mathbf{x}) &= \mathbf{v}_0(t,\mathbf{x}) + \delta\mathbf{v}(t,\mathbf{x}) \\ p(t,\mathbf{x}) &= p_0(t) + \delta p(t,\mathbf{x}) , \end{aligned} \tag{11}$$

where δ_ϵ is the fractional energy density perturbation (we are interested in the fractional rather than in the absolute energy density fluctuation!), and the pressure perturbation δp is defined as in (6). In addition, there is the possibility of a non-vanishing entropy perturbation defined as in (4).

We now insert this ansatz into the basic hydrodynamical equations (3), linearize in the perturbation variables, and combine the first order differential equations for δ_ϵ and δp into a single second order differential equation for $\delta\rho_\epsilon$. The result simplifies if we work in "comoving coordinates" $\mathbf{q}$ which are the coordinates painted onto the expanding background, i.e.

$$\mathbf{x}(t) = a(t)\mathbf{q}(t) . \tag{12}$$

After a substantial amount of algebra, we obtain the following equation which describes the time evolution of density fluctuations:

$$\ddot{\delta_\epsilon} + 2H\dot{\delta_\epsilon} - \frac{c_s^2}{a^2}\nabla_q^2\delta_\epsilon - 4\pi G\rho_0\delta_\epsilon = \frac{\sigma}{\rho_0 a^2}\delta S , \tag{13}$$

where the subscript q on the ∇ operator indicates that derivatives with respect to comoving coordinates are used. In addition, we have the equation of entropy conservation

$$\dot{\delta S} = 0 . \tag{14}$$

Comparing with the equations (5) obtained in the absence of an expanding background, we see that the only difference is the presence of a Hubble damping term in the equation for δ_ϵ. This term will moderate the exponential instability of the background to long wavelength density fluctuations. In addition, it will lead to a damping of the oscillating solutions on short wavelengths. More specifically, for physical wavenumbers $k_p \ll k_J$ (where k_J is again given by (9)), and in a matter-dominated background cosmology, the general solution of (13) in the absence of any entropy fluctuations is given by

$$\delta_k(t) = c_1 t^{2/3} + c_2 t^{-1} , \tag{15}$$

where c_1 and c_2 are two constants determined by the initial conditions, and we have dropped the subscript ϵ in expressions involving δ_ϵ. There are two fundamental solutions, the first is a growing mode with $\delta_k(t) \sim a(t)$, the second a decaying mode with $\delta_k(t) \sim t^{-1}$. On short wavelength, one obtains damped oscillatory motion:

$$\delta_k(t) \sim a^{-1/2}(t) exp(\pm i c_s k \int dt' a^{-1}(t')) . \tag{16}$$

As a simple application of the Newtonian equations for cosmological perturbations derived above, let us compare the predicted cosmic microwave background (CMB) anisotropies in a spatially flat Universe with only baryonic matter - Model A - to the corresponding anisotropies in a flat Universe with mostly cold dark matter (pressure-less non-baryonic dark matter) - Model B. We start with the observationally known amplitude of the relative density fluctuations today (time t_0), and we use the fact that the amplitude of the CMB anisotropies on the angular scale $\theta(k)$ corresponding to the comoving wavenumber k is set by the value of the primordial gravitational potential ϕ - introduced in the following section - which in turn is related to the value of the primordial density fluctuations at Hubble radius crossing (and **not** to its value of the time t_{rec}). See e.g. Chapter 17 of [5]). In Model A, the dominant component of the pressure-less matter is coupled to radiation between t_{eq} and t_{rec}, the time of last scattering. Thus, the Jeans length is comparable to the Hubble radius. Therefore, for comoving galactic scales, $k \gg k_J$ in this time interval, and thus the fractional density contrast decreases as $a(t)^{-1/2}$. In contrast, in Model B, the dominant component of pressure-less matter couples only weakly to radiation, and hence the Jeans length is negligibly small. Thus, in Model B, the relative density contrast grows as $a(t)$ between t_{eq} and t_{rec}. In the time interval $t_{rec} < t < t_0$, the fluctuations scale identically in Models A and B. Summarizing, we conclude, working backwards in time from a fixed amplitude of δ_k today, that the amplitudes of $\delta_k(t_{eq})$ in Models A and B (and thus their primordial values) are related by

$$\delta_k(t_{eq})|_A \simeq \left(\frac{a(t_{rec})}{a(t_{eq})}\right)\delta_k(t_{eq})|_B \,. \tag{17}$$

Hence, in Model A (without non-baryonic dark matter) the CMB anisotropies are predicted to be a factor of about 30 larger [26] than in Model B, way in excess of the recent observational results. This is one of the strongest arguments for the existence of non-baryonic dark matter.

2.4 Characterizing Perturbations

Let us consider perturbations on a fixed comoving length scale given by a comoving wavenumber k. The corresponding physical length increases as $a(t)$. This is to be compared to the Hubble radius $H^{-1}(t)$ which scales as t provided $a(t)$ grows as a power of t. In the late time Universe, $a(t) \sim t^{1/2}$ in the radiation-dominated phase (i.e. for $t < t_{eq}$, and $a(t) \sim t^{2/3}$ in the matter-dominated period ($t_{eq} < t < t_0$). Thus, we see that at sufficiently early times, all comoving scales had a physical length larger than the Hubble radius. If we consider large cosmological scales (e.g. those corresponding to the observed CMB anisotropies or to galaxy clusters), the time $t_H(k)$ of "Hubble radius crossing" (when the physical length was equal to the Hubble radius) was in fact later than t_{eq}. As we will see in later sections, the time of Hubble radius crossing plays an important role in the evolution of cosmological perturbations.

Cosmological fluctuations can be described either in position space or in momentum space. In position space, we compute the root mean square mass fluctuation $\delta M/M(k,t)$ in a sphere of radius $l = 2\pi/k$ at time t. A scale-invariant spectrum of fluctuations is defined by the relation

$$\frac{\delta M}{M}(k, t_H(k)) = \text{const.}. \tag{18}$$

Such a spectrum was first suggested by Harrison [27] and Zeldovich [28] as a reasonable choice for the spectrum of cosmological fluctuations. We can introduce the "spectral index" n of cosmological fluctuations by the relation

$$(\frac{\delta M}{M})^2(k, t_H(k)) \sim k^{n-1}, \tag{19}$$

and thus a scale-invariant spectrum corresponds to $n = 1$.

To make the transition to the (more frequently used) momentum space representation, we Fourier decompose the fractional spatial density contrast

$$\delta_\epsilon(\mathbf{x}, t) = \int d^3k \tilde{\delta_\epsilon}(\mathbf{k}, t) e^{i\mathbf{k}\cdot\mathbf{x}}. \tag{20}$$

The **power spectrum** P_δ of density fluctuations is defined by

$$P_\delta(k) = k^3 |\tilde{\delta_\epsilon}(k)|^2, \tag{21}$$

where k is the magnitude of $\mathbf{k}$, and we have assumed for simplicity a Gaussian distribution of fluctuations in which the amplitude of the fluctuations only depends on k.

We can also define the power spectrum of the gravitational potential φ:

$$P_\varphi(k) = k^3 |\tilde{\delta\varphi}(k)|^2. \tag{22}$$

These two power spectra are related by the Poisson equation (3)

$$P_\varphi(k) \sim k^{-4} P_\delta(k). \tag{23}$$

In general, the condition of scale-invariance is expressed in momentum space in terms of the power spectrum evaluated at a fixed time. To obtain this condition, we first use the time dependence of the fractional density fluctuation from (15) to determine the mass fluctuations at a fixed time $t > t_H(k) > t_{eq}$ (the last inequality is a condition on the scales considered)

$$(\frac{\delta M}{M})^2(k, t) = (\frac{t}{t_H(k)})^{4/3} (\frac{\delta M}{M})^2(k, t_H(k)). \tag{24}$$

The time of Hubble radius crossing is given by

$$a(t_H(k)) k^{-1} = 2 t_H(k), \tag{25}$$

and thus

$$t_H(k)^{1/2} \sim k^{-1}. \tag{26}$$

Inserting this result into (24) making use of (19) we find

$$\left(\frac{\delta M}{M}\right)^2(k,t) \sim k^{n+3} . \tag{27}$$

Since, for reasonable values of the index of the power spectrum, $\delta M/M(k,t)$ is dominated by the Fourier modes with wavenumber k, we find that (27) implies

$$|\tilde{\delta_\epsilon}|^2 \sim k^n , \tag{28}$$

or, equivalently,

$$P_\varphi(k) \sim k^{n-1} . \tag{29}$$

2.5 Matter Fluctuations in the Radiation Era

Let us now briefly consider fluctuations in the radiation dominated epoch. We are interested in both the fluctuations in radiation and in matter (cold dark matter). In the Newtonian treatment, (13) is replaced by separate equations for each matter fluid component (these components are designated by the labels A or B):

$$\ddot{\delta_A} + 2H\dot{\delta_A} - v_A^2 a^{-2}\nabla^2\delta_A = 4\pi G \sum_B \rho_B \delta_B , \tag{30}$$

where ρ_B indicate the background densities, and δ_B the fractional density fluctuations. The velocities of the respective fluid components are denoted by v_B, with $v_r^2 = 1/3$ for radiation and $v_m = 0$ for cold dark matter.

In the radiation dominated epoch, the evolution of the fluctuations in radiation is to a first approximation (in the ratio of the background densities) independent of the cold matter content. Inserting the expansion rate for this epoch, we thus immediately obtain

$$\delta_r(t) \sim a(t)^2 \tag{31}$$

on scales much larger than the Hubble scale, i.e. $k \ll k_H$, whereas δ_r undergoes damped oscillatory motion on smaller scales.

The evolution of the matter fluctuation δ_m is more complicated. Its equation of motion is dominated by the source term coming from δ_r. What results is logarithmic growth of the amplitude of δ_m, instead of the growth proportional to $a(t)$ which would occur on these scales in the absence of radiation. This damping effect on matter fluctuations due to the presence of radiation is called the "Meszaros effect". It leads to a turnover in the spectrum of cosmological fluctuations at a scale k_{eq} which crosses the Hubble radius at the time of equal matter and radiation. On larger scales ($k < k_{eq}$), one has the primordial power spectrum with spectral index n, on smaller scales, to a first approximation, the spectral index changes to $n - 4$. The details of the power spectrum on small scales depend largely on the specifics of the matter content in the Universe. One can write

$$P_{final}(k,t) = T(k,t)P_0(k,t) \tag{32}$$

where P_0 is the primordial power spectrum extrapolated to late times with unchanged spectral index, and P_{final} denotes the actual power spectrum which depends on effects such as the ones mentioned above. For more details see e.g. [23, 24].

3 Relativistic Theory of Cosmological Fluctuations

3.1 Introduction

The Newtonian theory of cosmological fluctuations discussed in the previous section breaks down on scales larger than the Hubble radius because it neglects perturbations of the metric, and because on large scales the metric fluctuations dominate the dynamics.

Let us begin with a heuristic argument to show why metric fluctuations are important on scales larger than the Hubble radius. For such inhomogeneities, one should be able to approximately describe the evolution of the space-time by applying the first Friedmann-Lemâitre-Robertson-Walker (FLRW) equation of homogeneous and isotropic cosmology to the local Universe (this approximation is made more rigorous in [29]):

$$\left(\frac{\dot{a}}{a}\right)^2 = \frac{8\pi G}{3}\rho\,. \tag{33}$$

Based on this equation, a large-scale fluctuation of the energy density will lead to a fluctuation ("δa") of the scale factor a which grows in time. This is due to the fact that self gravity amplifies fluctuations even on length scales λ greater than the Hubble radius.

This argument is made rigorous in the following analysis of cosmological fluctuations in the context of general relativity, where both metric and matter inhomogeneities are taken into account. We will consider fluctuations about a homogeneous and isotropic background cosmology, given by the metric (2), which can be written in conformal time η (defined by $dt = a(t)d\eta$) as

$$ds^2 = a(\eta)^2\left(d\eta^2 - d\mathbf{x}^2\right). \tag{34}$$

The evolution of the scale factor is determined by the two FLRW equations, (33) and

$$\dot{\rho} = -3H(\rho + p)\,, \tag{35}$$

which determine the expansion rate and its time derivative in terms of the equation of state of the matter, whose background stress-energy tensor can be written as

$$T^\mu_\nu = \begin{pmatrix} \rho & 0 & 0 & 0 \\ 0 & -p & 0 & 0 \\ 0 & 0 & -p & 0 \\ 0 & 0 & 0 & -p \end{pmatrix}. \tag{36}$$

The theory of cosmological perturbations is based on expanding the Einstein equations to linear order about the background metric. The theory was initially developed in pioneering works by Lifshitz [30]. Significant progress in the understanding of the physics of cosmological fluctuations was achieved by Bardeen [31] who realized the importance of subtracting gauge artifacts (see below) from the analysis (see also [32]). The following discussion is based on Part I of the comprehensive review article [5]. Other reviews - in some cases emphasizing different approaches - are [33, 34, 35, 36].

3.2 Classifying Fluctuations

The first step in the analysis of metric fluctuations is to classify them according to their transformation properties under spatial rotations. There are scalar, vector and second rank tensor fluctuations. In linear theory, there is no coupling between the different fluctuation modes, and hence they evolve independently (for some subtleties in this classification, see [37]).

We begin by expanding the metric about the FLRW background metric $g_{\mu\nu}^{(0)}$ given by (34):

$$g_{\mu\nu} = g_{\mu\nu}^{(0)} + \delta g_{\mu\nu} \, . \tag{37}$$

The background metric depends only on time, whereas the metric fluctuations $\delta g_{\mu\nu}$ depend on both space and time. Since the metric is a symmetric tensor, there are at first sight 10 fluctuating degrees of freedom in $\delta g_{\mu\nu}$.

There are four degrees of freedom which correspond to scalar metric fluctuations (the only four ways of constructing a metric from scalar functions):

$$\delta g_{\mu\nu} = a^2 \begin{pmatrix} 2\phi & -B_{,i} \\ -B_{,i} & 2(\psi\delta_{ij} - E_{,ij}) \end{pmatrix} , \tag{38}$$

where the four fluctuating degrees of freedom are denoted (following the notation of [5]) ϕ, B, E, and ψ, a comma denotes the ordinary partial derivative (if we had included spatial curvature of the background metric, it would have been the covariant derivative with respect to the spatial metric), and δ_{ij} is the Kronecker symbol.

There are also four vector degrees of freedom of metric fluctuations, consisting of the four ways of constructing metric fluctuations from three vectors:

$$\delta g_{\mu\nu} = a^2 \begin{pmatrix} 0 & -S_i \\ -S_i & F_{i,j} + F_{j,i} \end{pmatrix} , \tag{39}$$

where S_i and F_i are two divergence-less vectors (for a vector with non-vanishing divergence, the divergence contributes to the scalar gravitational fluctuation modes).

Finally, there are two tensor modes which correspond to the two polarization states of gravitational waves:

$$\delta g_{\mu\nu} = -a^2 \begin{pmatrix} 0 & 0 \\ 0 & h_{ij} \end{pmatrix}, \tag{40}$$

where h_{ij} is trace-free and divergence-less

$$h_i^i = h_{ij}^j = 0 \,. \tag{41}$$

Gravitational waves do not couple at linear order to the matter fluctuations. Vector fluctuations decay in an expanding background cosmology and hence are not usually cosmologically important. The most important fluctuations, at least in inflationary cosmology, are the scalar metric fluctuations, the fluctuations which couple to matter inhomogeneities and which are the relativistic generalization of the Newtonian perturbations considered in the previous section.

3.3 Gauge Transformation

The theory of cosmological perturbations is at first sight complicated by the issue of gauge invariance (at the final stage, however, we will see that we can make use of the gauge freedom to substantially simplify the theory). The coordinates $t, \mathbf{x}$ of space-time carry no independent physical meaning. They are just labels to designate points in the space-time manifold. By performing a small-amplitude transformation of the space-time coordinates (called "gauge transformation" in the following), we can easily introduce "fictitious" fluctuations in a homogeneous and isotropic Universe. These modes are "gauge artifacts".

We will in the following take an "active" view of gauge transformation. Let us consider two space-time manifolds, one of them a homogeneous and isotropic Universe $\mathcal{M}_0$, the other a physical Universe $\mathcal{M}$ with inhomogeneities. A choice of coordinates can be considered to be a mapping $\mathcal{D}$ between the manifolds $\mathcal{M}_0$ and $\mathcal{M}$. Let us consider a second mapping $\tilde{\mathcal{D}}$ which will map the same point (e.g. the origin of a fixed coordinate system) in $\mathcal{M}_0$ into different points in $\mathcal{M}$. Using the inverse of these maps $\mathcal{D}$ and $\tilde{\mathcal{D}}$, we can assign two different sets of coordinates to points in $\mathcal{M}$.

Consider now a physical quantity Q (e.g. the Ricci scalar) on $\mathcal{M}$, and the corresponding physical quantity $Q^{(0)}$ on $\mathcal{M}_0$ Then, in the first coordinate system given by the mapping $\mathcal{D}$, the perturbation δQ of Q at the point $p \in \mathcal{M}$ is defined by

$$\delta Q(p) = Q(p) - Q^{(0)}\big(\mathcal{D}^{-1}(p)\big) \,. \tag{42}$$

Analogously, in the second coordinate system given by $\tilde{\mathcal{D}}$, the perturbation is defined by

$$\tilde{\delta Q}(p) = Q(p) - Q^{(0)}\big(\tilde{\mathcal{D}}^{-1}(p)\big) \,. \tag{43}$$

The difference

$$\Delta Q(p) = \tilde{\delta Q}(p) - \delta Q(p) \tag{44}$$

is obviously a gauge artifact and carries no physical significance.

Some of the metric perturbation degrees of freedom introduced in the first subsection are gauge artifacts. To isolate these, we must study how coordinate transformations act on the metric. There are four independent gauge degrees of freedom corresponding to the coordinate transformation

$$x^\mu \rightarrow \tilde{x}^\mu = x^\mu + \xi^\mu \,. \tag{45}$$

The zero (time) component ξ^0 of ξ^μ leads to a scalar metric fluctuation. The spatial three vector ξ^i can be decomposed

$$\xi^i = \xi^i_{tr} + \gamma^{ij}\xi_{,j} \tag{46}$$

(where γ^{ij} is the spatial background metric) into a transverse piece ξ^i_{tr} which has two degrees of freedom which yield vector perturbations, and the second term (given by the gradient of a scalar ξ) which gives a scalar fluctuation. To summarize this paragraph, there are two scalar gauge modes given by ξ^0 and ξ, and two vector modes given by the transverse three vector ξ^i_{tr}. Thus, there remain two physical scalar and two vector fluctuation modes. The gravitational waves are gauge-invariant.

Let us now focus on how the scalar gauge transformations (i.e. the transformations given by ξ^0 and ξ) act on the scalar metric fluctuation variables ϕ, B, E, and ψ. An immediate calculation yields:

$$\begin{aligned}
\tilde{\phi} &= \phi - \frac{a'}{a}\xi^0 - (\xi^0)' \\
\tilde{B} &= B + \xi^0 - \xi' \\
\tilde{E} &= E - \xi \\
\tilde{\psi} &= \psi + \frac{a'}{a}\xi^0 \,,
\end{aligned} \tag{47}$$

where a prime indicates the derivative with respect to conformal time η.

There are two approaches to deal with the gauge ambiguities. The first is to fix a gauge, i.e. to pick conditions on the coordinates which completely eliminate the gauge freedom, the second is to work with a basis of gauge-invariant variables.

If one wants to adopt the gauge-fixed approach, there are many different gauge choices. Note that the often used synchronous gauge determined by $\delta g^{0\mu} = 0$ does not totally fix the gauge. A convenient system which completely fixes the coordinates is the so-called **longitudinal** or **conformal Newtonian gauge** defined by $B = E = 0$.

If one prefers a gauge-invariant approach, there are many choices of gauge-invariant variables. A convenient basis first introduced by [31] is the basis Φ, Ψ given by

$$\Phi = \phi + \frac{1}{a}\left[(B - E')a\right]' \tag{48}$$

$$\Psi = \psi - \frac{a'}{a}(B - E') . \tag{49}$$

It is obvious from the above equations that the gauge-invariant variables Φ and Ψ coincide with the corresponding diagonal metric perturbations ϕ and ψ in longitudinal gauge.

Note that the variables defined above are gauge-invariant only under linear space-time coordinate transformations. Beyond linear order, the structure of perturbation theory becomes much more involved. In fact, one can show [38] that the only fluctuation variables which are invariant under all coordinate transformations are perturbations of variables which are constant in the background space-time.

3.4 Equation of Motion

We begin with the Einstein equations

$$G_{\mu\nu} = 8\pi G T_{\mu\nu} , \tag{50}$$

where $G_{\mu\nu}$ is the Einstein tensor associated with the space-time metric $g_{\mu\nu}$, and $T_{\mu\nu}$ is the energy-momentum tensor of matter, insert the ansatz for metric and matter perturbed about a FLRW background $\big(g^{(0)}_{\mu\nu}(\eta), \varphi^{(0)}(\eta)\big)$:

$$g_{\mu\nu}(\mathbf{x}, \eta) = g^{(0)}_{\mu\nu}(\eta) + \delta g_{\mu\nu}(\mathbf{x}, \eta) \tag{51}$$

$$\varphi(\mathbf{x}, \eta) = \varphi_0(\eta) + \delta\varphi(\mathbf{x}, \eta) , \tag{52}$$

(where we have for simplicity replaced general matter by a scalar matter field φ) and expand to linear order in the fluctuating fields, obtaining the following equations:

$$\delta G_{\mu\nu} = 8\pi G \delta T_{\mu\nu} . \tag{53}$$

In the above, $\delta g_{\mu\nu}$ is the perturbation in the metric and $\delta\varphi$ is the fluctuation of the matter field φ.

Note that the components δG^{μ}_{ν} and δT^{μ}_{ν} are not gauge invariant. If we want to use the gauge-invariant approach, we note [5] that it is possible to construct a gauge-invariant tensor $\delta G^{(gi)\,\mu}_{\nu}$ via

$$\begin{aligned}
\delta G^{(gi)\,0}_{0} &\equiv \delta G^{0}_{0} + ({}^{(0)}G'^{\,0}_{0})(B - E') \\
\delta G^{(gi)\,0}_{i} &\equiv \delta G^{0}_{i} + ({}^{(0)}G^{0}_{i} - \frac{1}{3}{}^{(0)}G^{k}_{k})(B - E')_{,i} \\
\delta G^{(gi)\,i}_{j} &\equiv \delta G^{i}_{j} + ({}^{(0)}G'^{\,i}_{j})(B - E') ,
\end{aligned} \tag{54}$$

where ${}^{(}0)G^{\mu}_{\nu}$ denote the background values of the Einstein tensor. Analogously, a gauge-invariant linearized stress-energy tensor $\delta T^{(gi)\,\mu}_{\nu}$ can be defined. In terms of these tensors, the gauge-invariant form of the equations of motion for linear fluctuations reads

$$\delta G^{(gi)}_{\mu\nu} = 8\pi G \delta T^{(gi)}_{\mu\nu} . \tag{55}$$

If we insert into this equation the ansatz for the general metric and matter fluctuations (which depend on the gauge), only gauge-invariant combinations of the fluctuation variables will appear.

In a gauge-fixed approach, one can start with the metric in longitudinal gauge

$$ds^2 = a^2\left[(1+2\phi)d\eta^2 - (1-2\psi)\gamma_{ij}dx^i dx^j\right] \tag{56}$$

and insert this ansatz into the general perturbation equations (53). The short-cut of inserting a restricted ansatz for the metric into the action and deriving the full set of variational equations is justified in this case.

Both approaches yield the following set of equations of motion:

$$\begin{aligned} -3\mathcal{H}\left(\mathcal{H}\Phi + \Psi'\right) + \nabla^2\Psi &= 4\pi G a^2 \delta T_0^{(gi)\,0} \\ \left(\mathcal{H}\Phi + \Psi'\right)_{,i} &= 4\pi G a^2 \delta T_i^{(gi)\,0} \\ \left[(2\mathcal{H}' + \mathcal{H}^2)\Phi + \mathcal{H}\Phi' + \Psi'' + 2\mathcal{H}\Psi'\right]\delta^i_j & \\ + \frac{1}{2}\nabla^2 D \delta^i_j - \frac{1}{2}\gamma^{ik} D_{,kj} &= -4\pi G a^2 \delta T_j^{(gi)\,i} , \end{aligned} \tag{57}$$

where $D \equiv \Phi - \Psi$ and $\mathcal{H} = a'/a$. If we work in longitudinal gauge, then $\delta T_j^{(gi)\,i} = \delta T^i_j$, $\Phi = \phi$ and $\Psi = \psi$.

The first conclusion we can draw is that if no anisotropic stress is present in the matter at linear order in fluctuating fields, i.e. $\delta T^i_j = 0$ for $i \neq j$, then the two metric fluctuation variables coincide:

$$\Phi = \Psi . \tag{58}$$

This will be the case in most simple cosmological models, e.g. in theories with matter described by a set of scalar fields with canonical form of the action, and in the case of a perfect fluid with no anisotropic stress.

Let us now restrict our attention to the case of matter described in terms of a single scalar field φ with action

$$S = \int d^4x\sqrt{-g}\left[\frac{1}{2}\varphi^{,\alpha}\varphi_{,\alpha} - V(\varphi)\right] \tag{59}$$

(where g denotes the determinant of the metric) and we expand the matter field as

$$\varphi(\mathbf{x},\eta) = \varphi_0(\eta) + \delta\varphi(\mathbf{x},\eta) \tag{60}$$

in terms of background matter φ_0 and matter fluctuation $\delta\varphi(\mathbf{x},\eta)$, then in longitudinal gauge (57) reduce to the following set of equations of motion (making use of (58))

$$\begin{aligned} \nabla^2\phi - 3\mathcal{H}\phi' - (\mathcal{H}' + 2\mathcal{H}^2)\phi &= 4\pi G\left(\varphi_0'\delta\varphi' + V' a^2 \delta\varphi\right) \\ \phi' + \mathcal{H}\phi &= 4\pi G \varphi_0' \delta\varphi \\ \phi'' + 3\mathcal{H}\phi' + (\mathcal{H}' + 2\mathcal{H}^2)\phi &= 4\pi G\left(\varphi_0'\delta\varphi' - V' a^2 \delta\varphi\right) , \end{aligned} \tag{61}$$

where $V^{'}$ denotes the derivative of V with respect to φ. These equations can be combined to give the following second order differential equation for the relativistic potential ϕ:

$$\phi^{''} + 2\left(\mathcal{H} - \frac{\varphi_0^{''}}{\varphi_0^{'}}\right)\phi^{'} - \nabla^2\phi + 2\left(\mathcal{H}^{'} - \mathcal{H}\frac{\varphi_0^{''}}{\varphi_0^{'}}\right)\phi = 0\,. \tag{62}$$

Let us now discuss this final result for the classical evolution of cosmological fluctuations. First of all, we note the similarities with the equation (13) obtained in the Newtonian theory. The final term in (62) is the force due to gravity leading to the instability, the second to last term is the pressure force leading to oscillations (relativistic since we are considering matter to be a relativistic field), and the second term is the Hubble friction term. For each wavenumber there are two fundamental solutions. On small scales ($k > H$), the solutions correspond to damped oscillations, on large scales ($k < H$) the oscillations freeze out and the dynamics is governed by the gravitational force competing with the Hubble friction term. Note, in particular, how the Hubble radius naturally emerges as the scale where the nature of the fluctuating modes changes from oscillatory to frozen.

Considering the equation in a bit more detail, observe that if the equation of state of the background is independent of time (which will be the case if $\mathcal{H}^{'} = \varphi_0^{''} = 0$), that then in an expanding background, the dominant mode of (62) is constant, and the sub-dominant mode decays. If the equation of state is not constant, then the dominant mode is not constant in time. Specifically, at the end of inflation $\mathcal{H}^{'} < 0$, and this leads to a growth of ϕ (see the following subsection).

To study the quantitative implications of the equation of motion (62), it is convenient to introduce [39, 40] the variable ζ (which, up to correction terms of the order $\nabla^2\phi$ which are unimportant for large-scale fluctuations is equal to the curvature perturbation $\mathcal{R}$ in comoving gauge [41]) by

$$\zeta \equiv \phi + \frac{2}{3}\frac{\left(H^{-1}\dot{\phi} + \phi\right)}{1+w}\,, \tag{63}$$

where

$$w = \frac{p}{\rho} \tag{64}$$

characterizes the equation of state of matter. In terms of ζ, the equation of motion (62) takes on the form

$$\frac{3}{2}\dot{\zeta}H(1+w) \;=\; 0 + \mathcal{O}(\nabla^2\phi)\,. \tag{65}$$

On large scales, the right hand side of the equation is negligible, which leads to the conclusion that large-scale cosmological fluctuations satisfy

$$\dot{\zeta}(1+w) \;=\; 0. \tag{66}$$

This implies that except possibly if $1 + w = 0$ at some points in time during cosmological evolution (which occurs during reheating in inflationary cosmology if the inflaton field undergoes oscillations - see [42] and [43, 44] for discussions of the consequences in single and double field inflationary models, respectively) ζ is constant. In single matter field models it is indeed possible to show that $\dot{\zeta} = 0$ on super-Hubble scales independent of assumptions on the equation of state [45, 46]. This "conservation law" makes it easy to relate initial fluctuations to final fluctuations in inflationary cosmology, as will be illustrated in the following subsection.

3.5 Application to Inflationary Cosmology

Let us now return to the space-time sketch of the evolution of fluctuations in inflationary cosmology (Fig. 1) and use the conservation law (66) - in the form $\zeta =$ const on large scales - to relate the amplitude of ϕ at initial Hubble radius crossing during the inflationary phase (at $t = t_i(k)$) with the amplitude at final Hubble radius crossing at late times (at $t = t_f(k)$). Since both at early times and at late times $\dot{\phi} = 0$ on super-Hubble scales as the equation of state is not changing, (66) leads to

$$\phi(t_f(k)) \simeq \frac{(1+w)(t_f(k))}{(1+w)(t_i(k))}\phi(t_i(k))\,. \tag{67}$$

This equation will allow us to evaluate the amplitude of the cosmological perturbations when they re-enter the Hubble radius at time $t_f(k)$, under the assumption (discussed in detail in the following section) that the origin of the primordial fluctuations is quantum vacuum oscillations.

The time-time perturbed Einstein equation (the first equation of (57)) relates the value of ϕ at initial Hubble radius crossing to the amplitude of the relative energy density fluctuations. This, together with the fact that the amplitude of the scalar matter field quantum vacuum fluctuations is of the order H, yields

$$\phi(t_i(k)) \sim H\frac{V'}{V}(t_i(k))\,. \tag{68}$$

In the late time radiation dominated phase, $w = 1/3$, whereas during slow-roll inflation

$$1 + w(t_i(k)) \simeq \frac{\dot{\varphi}_0^2}{V}(t_i(k))\,. \tag{69}$$

Making, in addition, use of the slow roll conditions satisfied during the inflationary period

$$\begin{aligned} H\dot{\varphi}_0 &\simeq -V' \\ H^2 &\simeq \frac{8\pi G}{3}V\,, \end{aligned} \tag{70}$$

we arrive at the final result

$$\phi(t_f(k)) \sim \frac{V^{3/2}}{V'}(t_i(k)), \tag{71}$$

which gives the position space amplitude of cosmological fluctuations on a scale labelled by the comoving wavenumber k at the time when the scale re-enters the Hubble radius at late times, a result first obtained in the case of the Starobinsky model [47] of inflation in [48], and later in the context of scalar field-driven inflation in [49, 50, 51, 39].

In the case of slow roll inflation, the right hand side of (71) is, to a first approximation, independent of k, and hence the resulting spectrum of fluctuations is scale-invariant.

4 Quantum Theory of Cosmological Fluctuations

4.1 Overview

As already mentioned in the last subsection of the previous section, in many models of the very early Universe, in particular in inflationary cosmology, but also in the Pre-Big-Bang and in the Ekpyrotic scenarios, primordial inhomogeneities emerge from quantum vacuum fluctuations on microscopic scales (wavelengths smaller than the Hubble radius). The wavelength is then stretched relative to the Hubble radius, becomes larger than the Hubble radius at some time and then propagates on super-Hubble scales until re-entering at late cosmological times. In the context of a Universe with a de Sitter phase, the quantum origin of cosmological fluctuations was first discussed in [48] - see [52] for a more general discussion of the quantum origin of fluctuations in cosmology, and also [53, 54] for earlier ideas. In particular, Mukhanov [48] and Press [53] realized that in an exponentially expanding background, the curvature fluctuations would be scale-invariant, and Mukhanov provided a quantitative calculation which also yielded the logarithmic deviation from exact scale-invariance.

To understand the role of the Hubble radius, consider the equation of a free scalar matter field φ on an unperturbed expanding background:

$$\ddot{\varphi} + 3H\dot{\varphi} - \frac{\nabla^2}{a^2}\varphi = 0. \tag{72}$$

The second term on the left hand side of this equation leads to damping of φ with a characteristic decay rate given by H. As a consequence, in the absence of the spatial gradient term, $\dot{\varphi}$ would be of the order of magnitude $H\varphi$. Thus, comparing the second and the third term on the left hand side, we immediately see that the microscopic (spatial gradient) term dominates on length scales smaller than the Hubble radius, leading to oscillatory motion, whereas this term is negligible on scales larger than the Hubble radius, and the evolution of φ is determined primarily by gravity.

To understand the generation and evolution of fluctuations in current models of the very early Universe, we thus need both Quantum Mechanics and General Relativity, i.e. quantum gravity. At first sight, we are thus faced with an intractable problem, since the theory of quantum gravity is not yet established. We are saved by the fact that today on large cosmological scales the fractional amplitude of the fluctuations is smaller than 1. Since gravity is a purely attractive force, the fluctuations had to have been - at least in the context of an eternally expanding background cosmology - very small in the early Universe. Thus, a linearized analysis of the fluctuations (about a classical cosmological background) is self-consistent.

From the classical theory of cosmological perturbations discussed in the previous section, we know that the analysis of scalar metric inhomogeneities can be reduced - after extracting gauge artifacts - to the study of the evolution of a single fluctuating variable. Thus, we conclude that the quantum theory of cosmological perturbations must be reducible to the quantum theory of a single free scalar field which we will denote by v. Since the background in which this scalar field evolves is time-dependent, the mass of v will be time-dependent. The time-dependence of the mass will lead to quantum particle production over time if we start the evolution in the vacuum state for v. As we will see, this quantum particle production corresponds to the development and growth of the cosmological fluctuations. Thus, the quantum theory of cosmological fluctuations provides a consistent framework to study both the generation and the evolution of metric perturbations. The following analysis is based on Part II of [5].

4.2 Outline of the Analysis

In order to obtain the action for linearized cosmological perturbations, we expand the action to quadratic order in the fluctuating degrees of freedom. The linear terms cancel because the background is taken to satisfy the background equations of motion.

We begin with the Einstein-Hilbert action for gravity and the action of a scalar matter field (for the more complicated case of general hydrodynamical fluctuations the reader is referred to [5])

$$S = \int d^4x \sqrt{-g} [- \frac{1}{16\pi G} R + \frac{1}{2} \partial_\mu \varphi \partial^\mu \varphi - V(\varphi)] \, , \tag{73}$$

where g is the determinant of the metric.

The simplest way to proceed is to work in a fixed gauge, longitudinal gauge, in which the metric and matter take the form

$$\begin{aligned} ds^2 &= a^2(\eta) \big[(1 + 2\phi(\eta, \mathbf{x})) d\eta^2 - (1 - 2\psi(t, \mathbf{x})) d\mathbf{x}^2 \big] \\ \varphi(\eta, \mathbf{x}) &= \varphi_0(\eta) + \delta\varphi(\eta, \mathbf{x}) \, . \end{aligned} \tag{74}$$

The next step is to reduce the number of degrees of freedom. First, as already mentioned in the previous section, the off-diagonal spatial Einstein equations force $\psi = \phi$ since $\delta T^i_j = 0$ for scalar field matter (no anisotropic stresses to linear order). The two remaining fluctuating variables ϕ and φ must be linked by the Einstein constraint equations since there cannot be matter fluctuations without induced metric fluctuations.

The two nontrivial tasks of the lengthy [5] computation of the quadratic piece of the action is to find out what combination of φ and ϕ gives the variable v in terms of which the action has canonical form, and what the form of the time-dependent mass is. This calculation involves inserting the ansatz (74) into the action (73), expanding the result to second order in the fluctuating fields, making use of the background and of the constraint equations, and dropping total derivative terms from the action. In the context of scalar field matter, the quantum theory of cosmological fluctuations was developed by Mukhanov [55, 56] (see also [57]). The result is the following contribution $S^{(2)}$ to the action quadratic in the perturbations:

$$S^{(2)} \, = \, \frac{1}{2} \int d^4x \big[v'^2 - v_{,i}v_{,i} + \frac{z''}{z} v^2 \big] \, , \tag{75}$$

where the canonical variable v (the "Mukhanov variable" introduced in [56] - see also [52]) is given by

$$v \, = \, a \big[\delta\varphi + \frac{\varphi_0'}{\mathcal{H}} \phi \big] \, , \tag{76}$$

with $\mathcal{H} = a'/a$, and where

$$z \, = \, \frac{a \varphi_0'}{\mathcal{H}} \, . \tag{77}$$

In both the cases of power law inflation and slow roll inflation, $\mathcal{H}$ and φ_0' are proportional and hence

$$z(\eta) \, \sim \, a(\eta) \, . \tag{78}$$

Note that the variable v is related to the curvature perturbation $\mathcal{R}$ in comoving coordinates introduced in [41] and closely related to the variable ζ used in [39, 40]:

$$v \, = \, z \mathcal{R} \, . \tag{79}$$

The equation of motion which follows from the action (75) is

$$v'' - \nabla^2 v - \frac{z''}{z} v \, = \, 0 \, , \tag{80}$$

or, in momentum space

$$v_k'' + k^2 v_k - \frac{z''}{z} v_k \, = \, 0 \, , \tag{81}$$

where v_k is the k'th Fourier mode of v. As a consequence of (78), the mass term in the above equation is given by the Hubble scale

$$k_H^2 \equiv \frac{z''}{z} \simeq H^2 . \tag{82}$$

Thus, it immediately follows from (81) that on small length scales, i.e. for $k > k_H$, the solutions for v_k are constant amplitude oscillations . These oscillations freeze out at Hubble radius crossing, i.e. when $k = k_H$. On longer scales ($k \ll k_H$), the solutions for v_k increase as z:

$$v_k \sim z , \quad k \ll k_H . \tag{83}$$

Given the action (75), the quantization of the cosmological perturbations can be performed by canonical quantization (in the same way that a scalar matter field on a fixed cosmological background is quantized [58]).

The final step in the quantum theory of cosmological perturbations is to specify an initial state. Since in inflationary cosmology, all pre-existing classical fluctuations are red-shifted by the accelerated expansion of space, one usually assumes (we will return to a criticism of this point when discussing the trans-Planckian problem of inflationary cosmology) that the field v starts out at the initial time t_i mode by mode in its vacuum state. Two questions immediately emerge: what is the initial time t_i, and which of the many possible vacuum states should be chosen. It is usually assumed that since the fluctuations only oscillate on sub-Hubble scales, that the choice of the initial time is not important, as long as it is earlier than the time when scales of cosmological interest today cross the Hubble radius during the inflationary phase. The state is usually taken to be the Bunch-Davies vacuum (see e.g. [58]), since this state is empty of particles at t_i in the coordinate frame determined by the FLRW coordinates (see e.g. [59] for a discussion of this point), and since the Bunch-Davies state is a local attractor in the space of initial states in an expanding background (see e.g. [60]). Thus, we choose the initial conditions

$$\begin{aligned} v_k(\eta_i) &= \frac{1}{\sqrt{2\omega_k}} \\ v_k'(\eta_i) &= \frac{\sqrt{\omega_k}}{\sqrt{2}} \end{aligned} \tag{84}$$

where here $\omega_k = k$, and η_i is the conformal time corresponding to the physical time t_i.

Let us briefly summarize the quantum theory of cosmological perturbations. In the linearized theory, fluctuations are set up at some initial time t_i mode by mode in their vacuum state. While the wavelength is smaller than the Hubble radius, the state undergoes quantum vacuum fluctuations. The accelerated expansion of the background redshifts the length scale beyond the Hubble radius. The fluctuations freeze out when the length scale is equal to the Hubble radius. On larger scales, the amplitude of v_k increases as the scale factor. This corresponds to the squeezing of the quantum state present at Hubble radius crossing (in terms of classical general relativity, it is self-gravity which leads to this growth of fluctuations). As discussed e.g. in [61], the squeezing of the quantum vacuum state leads to the emergence of the classical nature of the fluctuations.

4.3 Application to Inflationary Cosmology

In this subsection we will use the quantum theory of cosmological perturbations developed in this section to calculate the spectrum of curvature fluctuations in inflationary cosmology.

We need to compute the power spectrum $\mathcal{P}_{\mathcal{R}}(k)$ of the curvature fluctuation $\mathcal{R}$ defined in (79), namely

$$\mathcal{R} = z^{-1}v = \phi + \delta\varphi \frac{\mathcal{H}}{\varphi_0'} \tag{85}$$

The idea in calculating the power spectrum at a late time t is to first relate the power spectrum via the growth rate (83) of v on super-Hubble scales to the power spectrum at the time $t_H(k)$ of Hubble radius crossing, and to then use the constancy of the amplitude of v on sub-Hubble scales to relate it to the initial conditions (84). Thus

$$\begin{aligned} \mathcal{P}_{\mathcal{R}}(k,t) \equiv k^3\mathcal{R}_k^2(t) &= k^3 z^{-2}(t)|v_k(t)|^2 \\ &= k^3 z^{-2}(t)\big(\frac{z(t)}{z(t_H(k))}\big)^2|v_k(t_H(k))|^2 \\ &= k^3 z^{-2}(t_H(k))|v_k(t_H(k))|^2 \\ &\sim k^3 a^{-2}(t_H(k))|v_k(t_i)|^2 \,, \end{aligned} \tag{86}$$

where in the final step we have used (78) and the constancy of the amplitude of v on sub-Hubble scales. Making use of the condition

$$a^{-1}(t_H(k))k = H \tag{87}$$

for Hubble radius crossing, and of the initial conditions (84), we immediately see that

$$\mathcal{P}_{\mathcal{R}}(k,t) \sim k^3 k^{-2} k^{-1} H^2 \,, \tag{88}$$

and that thus a scale invariant power spectrum with amplitude proportional to H^2 results, in agreement with what was argued on heuristic grounds in Sect. 3.5.

4.4 Quantum Theory of Gravitational Waves

The quantization of gravitational waves parallels the quantization of scalar metric fluctuations, but is more simple because there are no gauge ambiguities. Note again that at the level of linear fluctuations, scalar metric fluctuations and gravitational waves are independent. Both can be quantized on the same cosmological background determined by background scale factor and background matter. However, in contrast to the case of scalar metric fluctuations, the tensor modes are also present in pure gravity (i.e. in the absence of matter).

Starting point is the action (73). Into this action we insert the metric which corresponds to a classical cosmological background plus tensor metric fluctuations:

$$ds^2 = a^2(\eta)\left[d\eta^2 - (\delta_{ij} + h_{ij})dx^i dx^j\right], \tag{89}$$

where the second rank tensor $h_{ij}(\eta, \mathbf{x})$ represents the gravitational waves, and in turn can be decomposed as

$$h_{ij}(\eta, \mathbf{x}) = h_+(\eta, \mathbf{x})e^+_{ij} + h_x(\eta, \mathbf{x})e^x_{ij} \tag{90}$$

into the two polarization states. Here, e^+_{ij} and e^x_{ij} are two fixed polarization tensors, and h_+ and h_x are the two coefficient functions.

To quadratic order in the fluctuating fields, the action separates into separate terms involving h_+ and h_x. Each term is of the form

$$S^{(2)} = \int d^4x \frac{a^2}{2}\left[h'^2 - (\nabla h)^2\right], \tag{91}$$

leading to the equation of motion

$$h_k^{''} + 2\frac{a'}{a}h_k^{'} + k^2 h_k = 0. \tag{92}$$

The variable in terms of which the action (91) has canonical kinetic term is

$$\mu_k \equiv a h_k, \tag{93}$$

and its equation of motion is

$$\mu_k^{''} + (k^2 - \frac{a''}{a})\mu_k = 0. \tag{94}$$

This equation is very similar to the corresponding equation (81) for scalar gravitational inhomogeneities, except that in the mass term the scale factor $a(\eta)$ is replaced by $z(\eta)$, which leads to a very different evolution of scalar and tensor modes during the reheating phase in inflationary cosmology during which the equation of state of the background matter changes dramatically.

Based on the above discussion we have the following theory for the generation and evolution of gravitational waves in an accelerating Universe (first developed by Grishchuk [62]): waves exit as quantum vacuum fluctuations at the initial time on all scales. They oscillate until the length scale crosses the Hubble radius. At that point, the oscillations freeze out and the quantum state of gravitational waves begins to be squeezed in the sense that

$$\mu_k(\eta) \sim a(\eta), \tag{95}$$

which, from (93) corresponds to constant amplitude of h_k. The squeezing of the vacuum state leads to the emergence of classical properties of this state, as in the case of scalar metric fluctuations.

5 The Trans-Planckian Window

Whereas the contents of the previous sections are well established, this and the following section deal with aspects of the theory of cosmological perturbations which are currently under investigation and are at the present time rather controversial. First, we consider the trans-Planckian issue (this section is adapted from [63]).

The same background dynamics which yields the causal generation mechanism for cosmological fluctuations, the most spectacular success of inflationary cosmology, bears in it the nucleus of the "trans-Planckian problem". This can be seen from Fig. 3. If inflation lasts only slightly longer than the minimal time it needs to last in order to solve the horizon problem and to provide a causal generation mechanism for CMB fluctuations, then the corresponding physical wavelength of these fluctuations is smaller than the Planck length at the beginning of the period of inflation. The theory of cosmological perturbations is based on classical general relativity coupled to a weakly coupled scalar field description of matter. Both the theories of gravity and of matter will break down on trans-Planckian scales, and this immediately leads to the trans-Planckian problem: are the predictions of standard inflationary cosmology robust against effects of trans-Planckian physics [64]?

The simplest way of modeling the possible effects of trans-Planckian physics, while keeping the mathematical analysis simple, is to replace the linear dispersion relation $\omega_{_{\rm phys}} = k_{\rm phys}$ of the usual equation for cosmological perturbations by a non standard dispersion relation $\omega_{_{\rm phys}} = \omega_{_{\bf phys}}(k)$ which differs from the standard one only for physical wavenumbers larger than the Planck scale. This method was introduced [65, 66] in the context of studying the dependence of the thermal spectrum of black hole radiation on trans-Planckian physics. In the context of cosmology, it has been shown [67, 68, 69] that this amounts to replacing k^2 appearing in (81) with $k_{\rm eff}^2(n,\eta)$ defined by

$$k^2 \;\rightarrow\; k_{\rm eff}^2(k,\eta) \equiv a^2(\eta)\omega_{_{\rm phys}}^2\left(\frac{k}{a(\eta)}\right). \tag{96}$$

For a fixed comoving mode, this implies that the dispersion relation becomes time-dependent. Therefore, the equation of motion of the quantity $v_k(\eta)$ takes the form (with $z(\eta) \propto a(\eta)$)

$$v_k'' + \left[k_{\rm eff}^2(k,\eta) - \frac{a''}{a}\right]v_k \;=\; 0\,. \tag{97}$$

A more rigorous derivation of this equation, based on a variational principle, has been provided [70] (see also [71]).

The evolution of modes thus must be considered separately in three phases, see Fig. 3. In Phase I the wavelength is smaller than the Planck scale, and trans-Planckian physics can play an important role. In Phase II, the wavelength is larger than the Planck scale but smaller than the Hubble radius.

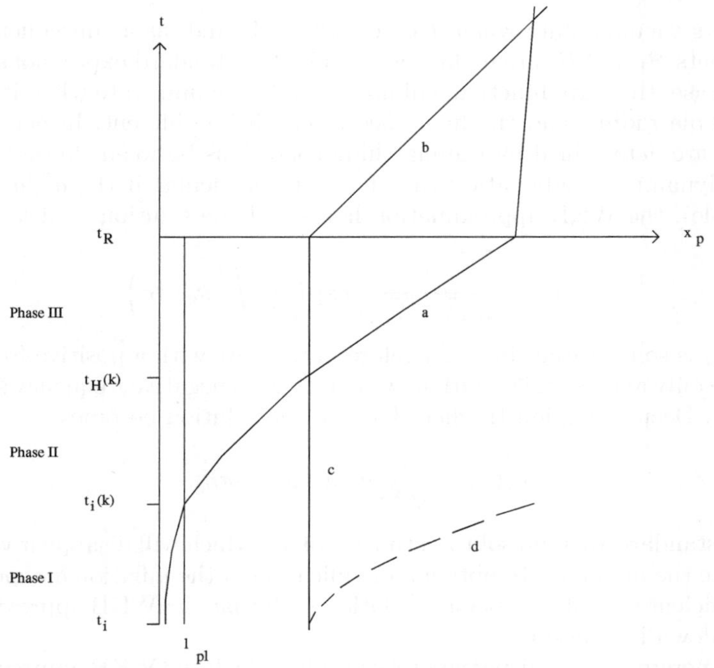

Fig. 3. Space-time diagram (physical distance vs. time) showing the origin of the trans-Planckian problem of inflationary cosmology: at very early times, the wavelength is smaller than the Planck scale ℓ_{Pl} (Phase I), at intermediate times it is larger than ℓ_{Pl} but smaller than the Hubble radius H^{-1} (Phase II), and at late times during inflation it is larger than the Hubble radius (Phase III). The line labeled a) is the physical wavelength associated with a fixed comoving scale k. The line b) is the Hubble radius or horizon in SBB cosmology. Curve c) shows the Hubble radius during inflation. The horizon in inflationary cosmology is shown in curve d).

In this phase, trans-Planckian physics will have a negligible effect (this statement can be quantified [72]). Hence, by the analysis of the previous section, the wave function of fluctuations is oscillating in this phase,

$$v_k = B_1 \exp(-ik\eta) + B_2 \exp(ik\eta) \tag{98}$$

with constant coefficients B_1 and B_2. In the standard approach, the initial conditions are fixed in this region and the usual choice of the vacuum state leads to $B_1 = 1/\sqrt{2k}$, $B_2 = 0$. Phase III starts at the time $t_H(k)$ when the mode crosses the Hubble radius. During this phase, the wave function is squeezed.

One source of trans-Planckian effects [67, 68] on observations is the possible non-adiabatic evolution of the wave function during Phase I. If this occurs, then it is possible that the wave function of the fluctuation mode is not in its vacuum state when it enters Phase II and, as a consequence, the coefficients B_1 and B_2 are no longer given by the standard expressions above. In this case, the wave function will not be in its vacuum state when it crosses the Hubble radius, and the final spectrum will be different. In general, B_1 and B_2 are determined by the matching conditions between Phase I and II. If the dynamics is adiabatic throughout (in particular if the a''/a term is negligible), the WKB approximation holds and the solution is always given by

$$v_k(\eta) \simeq \frac{1}{\sqrt{2k_{\rm eff}(k,\eta)}} \exp\left(-i\int_{\eta_{\rm i}}^{\eta} k_{\rm eff}\,{\rm d}\tau\right), \tag{99}$$

where η_i is some initial time. Therefore, if we start with a positive frequency solution only and use this solution, we find that no negative frequency solution appears. Deep in Region II where $k_{\rm eff} \simeq k$ the solution becomes

$$v_k(\eta) \simeq \frac{1}{\sqrt{2k}} \exp(-i\phi - ik\eta), \tag{100}$$

i.e. the standard vacuum solution times a phase which will disappear when we calculate the modulus. To obtain a modification of the inflationary spectrum, it is sufficient to find a dispersion relation such that the WKB approximation breaks down in Phase I.

A concrete class of dispersion relations for which the WKB approximation breaks down is

$$k_{\rm eff}^2(k,\eta) = k^2 - k^2|b_m|\left[\frac{\ell_{\rm pl}}{\lambda(\eta)}\right]^{2m}, \tag{101}$$

where $\lambda(\eta) = 2\pi a(\eta)/k$ is the wavelength of a mode. If we follow the evolution of the modes in Phases I, II and III, matching the mode functions and their derivatives at the junction times, the calculation [67, 68, 73] demonstrates that the final spectral index is modified and that superimposed oscillations appear.

However, the above example suffers from several problems. First, in inflationary models with a long period of inflationary expansion, the dispersion relation (101) leads to complex frequencies at the beginning of inflation for scales which are of current interest in cosmology. Furthermore, the initial conditions for the Fourier modes of the fluctuation field have to be set in a region where the evolution is non-adiabatic and the use of the usual vacuum prescription can be questioned. These problems can be avoided in a toy model in which we follow the evolution of fluctuations in a bouncing cosmological background which is asymptotically flat in the past and in the future. The analysis [74] shows that even in this case the final spectrum of fluctuations depends on the specific dispersion relation used.

An example of a dispersion relation which breaks the WKB approximation in the trans-Planckian regime but does not lead to the problems mentioned in the previous paragraph was investigated in [70]. It is a dispersion relation which is linear for both small and large wavenumbers, but has an intermediate interval during which the frequency decreases as the wavenumber increases, much like what happens in (101). The violation of the WKB condition occurs for wavenumbers near the local minimum of the $\omega(k)$ curve.

A justified criticism against the method summarized in the previous analysis is that the non-standard dispersion relations used are completely ad hoc, without a clear basis in trans-Planckian physics. There has been a lot of recent work [75, 76, 77, 78] on the implication of space-space uncertainty relations [79, 80] on the evolution of fluctuations. The application of the uncertainty relations on the fluctuations lead to two effects [81, 82]. Firstly, the equation of motion of the fluctuations in modified. Secondly, for fixed comoving length scale k, the uncertainty relation is saturated before a critical time $t_i(k)$. Thus, in addition to a modification of the evolution, trans-Planckian physics leads to a modification of the boundary condition for the fluctuation modes. The upshot of this work is that the spectrum of fluctuations is modified.

In [83], the implications of the stringy space-time uncertainty relation [84, 85]

$$\Delta x_{\rm phys} \Delta t \geq l_s^2 \tag{102}$$

on the spectrum of cosmological fluctuations was studied. Again, application of this uncertainty relation to the fluctuations leads to two effects. Firstly, the coupling between the background and the fluctuations is nonlocal in time, thus leading to a modified dynamical equation of motion (a similar modification also results [86] from quantum deformations, another example of a consequence of non-commutative basic physics). Secondly, the uncertainty relation is saturated at the time $t_i(k)$ when the physical wavelength equals the string scale l_s. Before that time it does not make sense to talk about fluctuations on that scale. By continuity, it makes sense to assume that fluctuations on scale k are created at time $t_i(k)$ in the local vacuum state (the instantaneous WKB vacuum state).

Let us for the moment neglect the nonlocal coupling between background and fluctuation, and thus consider the usual equation of motion for fluctuations in an accelerating background cosmology. We distinguish two ranges of scales. Ultraviolet modes are generated at late times when the Hubble radius is larger than l_s. On these scales, the spectrum of fluctuations does not differ from what is predicted by the standard theory, since at the time of Hubble radius crossing the fluctuation mode will be in its vacuum state. However, the evolution of infrared modes which are created when the Hubble radius is smaller than l_s is different. The fluctuations undergo *less* squeezing than they do in the absence of the uncertainty relation, and hence the final amplitude of fluctuations is lower. From the equation (86) for the power spectrum of fluctuations, and making use of the condition

$$a(t_i(k)) = k l_s \tag{103}$$

for the time $t_i(k)$ when the mode is generated, it follows immediately that the power spectrum is scale-invariant

$$\mathcal{P}_{\mathcal{R}}(k) \sim k^0 . \tag{104}$$

In the standard scenario of power-law inflation the spectrum is red ($\mathcal{P}_{\mathcal{R}}(k) \sim k^{n-1}$ with $n < 1$). Taking into account the effects of the nonlocal coupling between background and fluctuation mode leads [83] to a modification of this result: the spectrum of fluctuations in a power-law inflationary background is in fact blue ($n > 1$).

Note that, if we neglect the nonlocal coupling between background and fluctuation mode, the result of (104) also holds in a cosmological background which is NOT accelerating. Thus, we have a method of obtaining a scale-invariant spectrum of fluctuations without inflation. This result has also been obtained in [87], however without a micro-physical basis for the prescription for the initial conditions.

A key problem with the method of modified dispersion relations is the issue of back-reaction [88, 89]. If the mode occupation numbers of the fluctuations at Hubble radius crossing are significant, the danger arises that the back-reaction of the fluctuations will in fact prevent inflation. Another constraint arises from the observational limits on the flux of ultra-high-energy cosmic rays. Such cosmic rays would be produced [90] in the present Universe if Trans-Planckian effects of the type discussed in this section were present.

An approach to the trans-Planckian issue pioneered by Danielsson [91] which has recently received a lot of attention is to avoid the issue of the unknown trans-Planckian physics and to start the evolution of the fluctuation modes at the mode-dependent time when the wavelength equals the limiting scale. Obviously, the resulting spectrum will depend sensitively on which state is taken to be the initial state. The vacuum state is not unambiguous, and the choice of a state minimizing the energy density depends on the space-time splitting [92]. The signatures of this prescription are typically oscillations superimposed on the usual spectrum. The amplitude of this effect depends sensitively on the prescription of the initial state, and for a fixed prescription also on the background cosmology. For a discussion of these issues and a list of references on this approach the reader is referred to [93].

In summary, due to the exponential red-shifting of wavelengths, present cosmological scales originate at wavelengths smaller than the Planck length early on during the period of inflation. Thus, Planck physics may well encode information in these modes which can now be observed in the spectrum of microwave anisotropies. Two examples have been shown to demonstrate the existence of this "window of opportunity" to probe trans-Planckian physics in cosmological observations. The first method makes use of modified dispersion relations to probe the robustness of the predictions of inflationary cosmology, the second applies the stringy space-time uncertainty relation on the fluctuation modes. Both methods yield the result that trans-Planckian physics may lead to measurable effects in cosmological observables. An important issue which must be studied more carefully is the back-reaction of the cosmological fluctuations (see e.g. [94] for a possible formalism).

6 Back-Reaction of Cosmological Fluctuations

The presence of cosmological fluctuations influences the background cosmology in which the perturbations evolve. This back-reaction arises as a second order effect in the cosmological perturbation expansion. The effect is cumulative in the sense that all fluctuation modes contribute to the change in the background geometry, and as a consequence the back-reaction effect can be large even if the amplitude of the fluctuation spectrum is small. In this section (based on the review [95]) we discuss two approaches used to quantify back-reaction. In the first approach [96, 94], the effect of the fluctuations on the background is expressed in terms of an effective energy-momentum tensor. We show that in the context of an inflationary background cosmology, the long wavelength contributions to the effective energy-momentum tensor take the form of a negative cosmological constant, whose absolute value increases as a function of time since the phase space of infrared modes is increasing. This then leads to the speculation [97, 98] that gravitational back-reaction may lead to a dynamical cancellation mechanism for a bare cosmological constant, and yield a scaling fixed point in the asymptotic future in which the remnant cosmological constant satisfies $\Omega_\Lambda \sim 1$. We then discuss [99] how infrared modes effect local observables (as opposed to mathematical background quantities) and find that the leading infrared back-reaction contributions cancel in single field inflationary models. However, we expect non-trivial back-reaction of infrared modes in models with more than one matter field.

It is well known that gravitational waves propagating in some background space-time affect the dynamics of the background. This back-reaction can be described in terms of an effective energy-momentum tensor $\tau_{\mu\nu}$. In the short wave limit, when the typical wavelength of the waves is small compared with the curvature of the background space-time, $\tau_{\mu\nu}$ has the form of a radiative fluid with an equation of state $p = \rho/3$ (where p and ρ denote pressure and energy density, respectively). As we have seen in previous section, in inflationary cosmology it is the long wavelength scalar metric fluctuations which are more important. Like short wavelength gravitational waves, these cosmological fluctuations will contribute to the effective energy-momentum tensor $\tau_{\mu\nu}$. The work of [96, 94] is closely related to work by Woodard and Tsamis [100, 101] who considered the back-reaction of long wavelength gravitational waves in pure gravity with a bare cosmological constant. The recent paper [99] is related to the work of Abramo and Woodard [102, 103] who initiated the study of back-reaction of infrared modes on local observables. We first review the derivation of the effective energy-momentum tensor $\tau_{\mu\nu}$ which describes the back-reaction of linear cosmological fluctuations on the background cosmology, and summarize the evaluation of this tensor in an inflationary cosmological background. This gravitational back-reaction calculation is related to the early work on the back-reaction of gravitational waves by Brill, Hartle and Isaacson [104], among others. The idea is to expand the Einstein equations to second order in the perturbations, to assume that the first order terms satisfy the equations of motion for linearized cosmological perturbations discussed in previous section (hence these terms cancel), to take the spatial average of the remaining terms, and to regard the resulting equations as equations for a new homogeneous metric $g_{\mu\nu}^{(0,br)}$ which includes the effect of the perturbations to quadratic order:

$$G_{\mu\nu}(g_{\alpha\beta}^{(0,br)}) = 8\pi G \left[T_{\mu\nu}^{(0)} + \tau_{\mu\nu} \right] \tag{105}$$

where the effective energy-momentum tensor $\tau_{\mu\nu}$ of gravitational back-reaction contains the terms resulting from spatial averaging of the second order metric and matter perturbations:

$$\tau_{\mu\nu} = < T_{\mu\nu}^{(2)} - \frac{1}{8\pi G} G_{\mu\nu}^{(2)} > , \tag{106}$$

where pointed brackets stand for spatial averaging, and the superscripts indicate the order in perturbation theory.

As analyzed in detail in [96, 94], the back-reaction equation (105) is covariant under linear space-time coordinate transformations even though $\tau_{\mu\nu}$ is not invariant [2]. In the following, we will work in longitudinal gauge.

For simplicity, we shall take matter to be described in terms of a single scalar field. By expanding the Einstein and matter energy-momentum tensors to second order in the metric and matter fluctuations ϕ and $\delta\varphi$, respectively, it can be shown that the non-vanishing components of the effective back-reaction energy-momentum tensor $\tau_{\mu\nu}$ become

$$\begin{aligned}
\tau_{00} = &\frac{1}{8\pi G} \left[+12H\langle \phi\dot{\phi} \rangle - 3\langle (\dot{\phi})^2 \rangle + 9a^{-2}\langle (\nabla\phi)^2 \rangle \right] \\
&+ \frac{1}{2}\langle (\delta\dot{\varphi})^2 \rangle + \frac{1}{2}a^{-2}\langle (\nabla\delta\varphi)^2 \rangle \\
&+ \frac{1}{2}V''(\varphi_0)\langle \delta\varphi^2 \rangle + 2V'(\varphi_0)\langle \phi\delta\varphi \rangle \quad ,
\end{aligned} \tag{107}$$

and

[2] See [105], however, for important questions concerning the covariance of the analysis under higher order coordinate transformations.

$$\tau_{ij} = a^2\delta_{ij}\left\{\frac{1}{8\pi G}\left[(24H^2 + 16\dot{H})\langle\phi^2\rangle + 24H\langle\dot{\phi}\phi\rangle\right.\right.$$
$$+ \left.\langle(\dot{\phi})^2\rangle + 4\langle\phi\ddot{\phi}\rangle - \frac{4}{3}a^{-2}\langle(\nabla\phi)^2\rangle\right] + 4\dot{\varphi_0}^2\langle\phi^2\rangle$$
$$+ \frac{1}{2}\langle(\delta\dot{\varphi})^2\rangle - \frac{1}{6}a^{-2}\langle(\nabla\delta\varphi)^2\rangle - 4\dot{\varphi}_0\langle\delta\dot{\varphi}\phi\rangle$$
$$\left. - \frac{1}{2}V''(\varphi_0)\langle\delta\varphi^2\rangle + 2V'(\varphi_0)\langle\phi\delta\varphi\rangle\right\} \quad , \tag{108}$$

where H is the Hubble expansion rate.

The metric and matter fluctuation variables ϕ and $\delta\varphi$ are linked via the Einstein constraint equations, and hence all terms in the above formulas for the components of $\tau_{\mu\nu}$ can be expressed in terms of two point functions of ϕ and its derivatives. The two point functions, in turn, are obtained by integrating over all of the Fourier modes of ϕ, e.g.

$$\langle\phi^2\rangle \sim \int_{k_i}^{k_u} dk k^2 |\phi_k|^2 \,, \tag{109}$$

where ϕ_k denotes the amplitude of the k'th Fourier mode. The above expression is divergent both in the infrared and in the ultraviolet. The ultraviolet divergence is the usual divergence of a free quantum field theory and can be "cured" by introducing an ultraviolet cutoff k_u. In the infrared, we will discard all modes $k < k_i$ with wavelength larger than the Hubble radius at the beginning of inflation, since these modes are determined by the pre-inflationary physics. We take these modes to contribute to the background.

At any time t we can separate the integral in (109) into the contribution of infrared and ultraviolet modes, the separation being defined by setting the physical wavelength equal to the Hubble radius. Thus, in an inflationary Universe the infrared phase space is continually increasing since comoving modes are stretched beyond the Hubble radius, while the ultraviolet phase space is either constant (if the ultraviolet cutoff corresponds to a fixed physical wavelength), or decreasing (if the ultraviolet cutoff corresponds to fixed comoving wavelength). In either case, unless the spectrum of the initial fluctuations is extremely blue, two point functions such as (109) will at later stages of an inflationary Universe be completely dominated by the infrared sector. In the following, we will therefore restrict our attention to this sector, i.e. to wavelengths larger than the Hubble radius.

In order to evaluate the two point functions which enter into the expressions for $\tau_{\mu\nu}$, we make use of the known time evolution of the linear fluctuations ϕ_k discussed in previous section. On scales larger than the Hubble radius, and for a time-independent equation of state, ϕ_k is constant in time. From the Einstein constraint equations relating the metric and matter fluctuations, and making use of the inflationary slow roll approximation conditions we find

$$\delta\varphi = -\frac{2V}{V'}\phi \,. \tag{110}$$

Hence, in the expressions (107) and (108) for $\tau_{\mu\nu}$, all terms with space and time derivatives can be neglected, and we obtain

$$\rho_{br} \equiv \tau_0^0 \cong \left(2\frac{V''V^2}{V'^2} - 4V\right) < \phi^2 > \qquad (111)$$

and

$$p_{br} \equiv -\frac{1}{3}\tau_i^i \cong -\rho_{br}\,, \qquad (112)$$

The main result which emerges from this analysis is that the equation of state of the dominant infrared contribution to the energy-momentum tensor $\tau_{\mu\nu}$ which describes back-reaction takes the form of a *negative cosmological constant*

$$p_{br} = -\rho_{br} \text{ with } \rho_{br} < 0\,. \qquad (113)$$

The second crucial result is that the magnitude of ρ_{br} increases as a function of time. This is due in part to the fact that, in an inflationary Universe, as time increases more and more wavelengths become longer than the Hubble radius and begin to contribute to ρ_{br}.

How large is the magnitude of back-reaction? The basic point is that since the amplitude of each fluctuation mode is small, we need a very large phase space of infrared modes in order to induce any interesting effects. In models with a very short period of primordial inflation, the back-reaction of long-wavelength cosmological fluctuations hence will not be important. However, in many single field models of inflation, in particular in those of chaotic inflation type [106], inflation lasts so long that the infrared back-reaction effects can build up to become important for the cosmological background dynamics. To give an example, consider chaotic inflation with a potential

$$V(\varphi) = \frac{1}{2}m^2\varphi^2\,. \qquad (114)$$

In this case, the values of ϕ_k for long wavelength modes are well known (see e.g. [5]), and the integral in (109) can be easily performed, thus yielding explicit expressions for the dominant terms in the effective energy-momentum tensor. Comparing the resulting back-reaction energy density ρ_{br} with the background density ρ_0, we find

$$\frac{\rho_{br}(t)}{\rho_0} \simeq \frac{3}{4\pi}\frac{m^2\varphi_0^2(t_i)}{M_P^4}\left[\frac{\varphi_0(t_i)}{\varphi_0(t)}\right]^4, \qquad (115)$$

where M_P denotes the Planck mass. Without back-reaction, inflation would end [106] when $\varphi_0(t) \sim M_P$. Inserting this value into (115), we see that if $\varphi_0(t_i) > \varphi_{br} \sim m^{-1/3}M_P^{4/3}$, then back-reaction will become important before the end of inflation and may shorten the period of inflation. It is interesting to compare this value with the scale $\varphi_0(t_i) \sim \varphi_{sr} = m^{-1/2}M_P^{3/2}$ above which the stochastic terms in the scalar field equation of motion arising in the context of the stochastic approach to chaotic inflation [107, 108] are dominant. Notice that since $\varphi_{sr} \gg \varphi_{br}$ (recall that $m \ll M_P$), back-reaction effects can be very important in the entire range of field values relevant to stochastic inflation.

Since the back-reaction of cosmological fluctuations in an inflationary cosmology acts (see (113)) like a negative cosmological constant, and since the magnitude of the back-reaction effect increases in time, one may speculate [97] that back-reaction will lead to a dynamical relaxation of the cosmological constant (see Tsamis & Woodard [100] for similar speculations based on the back-reaction of long wavelength gravitational waves).

The background metric $g_{\mu\nu}^{(0,br)}$ including back-reaction evolves as if the cosmological constant at time t were

$$\Lambda_{\text{eff}}(t) = \Lambda_0 + 8\pi G \rho_{br}(t) \tag{116}$$

and not the bare cosmological constant Λ_0. Hence one might hope to identify (116) with a time dependent effective cosmological constant. Since $|\rho_{br}(t)|$ increases as t grows, the effective cosmological constant will decay. Note that even if the initial magnitude of the perturbations is small, eventually (if inflation lasts a sufficiently long time) the back-reaction effect will become large enough to cancel any bare cosmological constant.

Furthermore, one might speculate that this dynamical relaxation mechanism for Λ will be self-regulating. As long as $\Lambda_{\text{eff}}(t) > 8\pi G \rho_m(t)$, where $\rho_m(t)$ stands for the energy density in ordinary matter and radiation, the evolution of $g_{\mu\nu}^{(0,br)}$ is dominated by $\Lambda_{\text{eff}}(t)$. Hence, the Universe will be undergoing accelerated expansion, more scales will be leaving the Hubble radius and the magnitude of the back-reaction term will increase. However, once $\Lambda_{\text{eff}}(t)$ falls below $\rho_m(t)$, the background will start to decelerate, scales will enter the Hubble radius, and the number of modes contributing to the back-reaction will decrease, thus reducing the strength of back-reaction. Hence, it is likely that there will be a scaling solution to the effective equation of motion for $\Lambda_{\text{eff}}(t)$ of the form

$$\Lambda_{\text{eff}}(t) \sim 8\pi G \rho_m(t)\,. \tag{117}$$

Such a scaling solution would correspond to a contribution to the relative closure density of $\Omega_\Lambda \sim 1$.

There are important concerns about the above formalism, and even more so about the resulting speculations (many of these were first discussed in print in [105]). On a formal level, since our back-reaction effect is of second order in cosmological perturbation theory, it is necessary to demonstrate covariance of the proposed back-reaction equation (105) beyond linear order, and this has not been done. Next, it might be argued that by causality super-Hubble fluctuations cannot affect local observables. Thirdly, from an observational perspective one is not interested in the effect of fluctuations on the background metric (since what the background is cannot be determined precisely using local observations). Instead, one should compute the back-reaction of cosmological fluctuations on observables describing the local Hubble expansion rate. One might then argue that even if long-wavelength fluctuations have an effect on the background metric, they do not influence local observables. Finally, it is clear that the speculations in the previous section involve the extrapolation of perturbative physics deep into the non-perturbative regime.

These important issues have now begun to be addressed. Good physical arguments can be given [102, 103] supporting the idea that long-wavelength fluctuations can effect local physics. Consider, for example, a black hole of mass M absorbing a particle of mass m. Even after this particle has disappeared beyond the horizon, its gravitational effects (in terms of the increased mass of the black hole) remain measurable to an external observer. A similar argument can be given in inflationary cosmology: consider an initial localized mass fluctuation with a characteristic physical length scale λ in an exponentially expanding background. Even after the length scale of the fluctuation redshifts to be larger than the Hubble radius, the gravitational potential associated with this fluctuation remains measurable. On a more technical level, it has recently been shown that super-Hubble scale (but sub-horizon-scale) metric fluctuations can be parametrically amplified during inflationary reheating [109, 42, 43, 44]. This clearly demonstrates a coupling between local physics and super-Hubble-scale fluctuations.

These arguments, however, make it even more important to focus on back-reaction effects of cosmological fluctuations on local physical observables rather than on the mathematical background metric. In recent work [99], the leading infrared back-reaction effects on a local observable measuring the Hubble expansion rate were calculated.

Consider a perfect fluid with velocity four vector u^α in an inhomogeneous cosmological geometry, then the local expansion rate which generalizes the Hubble expansion rate $H(t)$ of homogeneous isotropic Friedmann-Robertson-Walker cosmology is given by $\frac{1}{3}\Theta$, where Θ is the four divergence of u^α:

$$\Theta = u^\alpha_{;\alpha} \, , \tag{118}$$

the semicolon indicating the covariant derivative. In [99], the effects of cosmological fluctuations on this variable were computed to second order in perturbation theory. To leading order in the infrared expansion, the result is

$$\Theta = 3\frac{a'}{a^2}(1 - \phi + \frac{3}{2}\phi^2) - 3\frac{\phi'}{a} \, , \tag{119}$$

where the prime denotes the derivative with respect to conformal time. If we now calculate the spatial average of Θ, the term linear in ϕ vanishes, and - as expected - we are left with a quadratic back-reaction contribution.

Superficially, it appears from (119) that there is a non-vanishing back-reaction effect at quadratic order which is not suppressed for super-Hubble modes. However, we must be careful and evaluate Θ not at a constant value of the background coordinates, but rather at a fixed value of some physical observable. For example, if we work out the value of Θ in the case of a matter-dominated Universe, and express the result as a function of the proper time τ given by $d\tau^2 = a(\eta)^2(1+2\phi)d\eta^2$ instead of as a function of conformal time η, then we find that the leading infrared terms proportional to ϕ^2 exactly cancel, and that thus there is no un-suppressed infrared back-reaction on the local measure of the Hubble expansion rate.

A more relevant example with respect to the discussion in earlier sections is a model in which matter is given by a single scalar field. In this case, the leading infrared back-reaction terms in Θ are again given by (119) which looks different from the background value $3H$. However, once again it is important to express Θ in terms of a physical background variable. If we choose the value of the matter field φ as this variable, we find after easy manipulations that, including only the leading infrared back-reaction terms,

$$\Theta(\varphi) = \sqrt{3}\sqrt{V(\varphi)}. \tag{120}$$

Hence, once again the leading infrared back-reaction contributions vanish, as already found in the work of [103] which considered the leading infrared back-reaction effects on a local observable different than the one we have used, and applied very different methods [3].

However, in a model with two matter fields, it is clear that if we e.g. use the second matter field as a physical clock, then the leading infrared back-reaction terms will not cancel in Θ, and that thus in such models infrared back-reaction will be physically observable. The situation will be very much analogous to what happens in the case of parametric resonance of gravitational fluctuations during inflationary reheating. This process is a gauge artifact in single field models of inflation [42] (see also [45, 110, 46]), but it is real and unsuppressed in certain two field models [43, 44]. In the case of two field models, work on the analysis of the back-reaction effects of infrared modes on the observable representing the local Hubble expansion rate is in progress.

Provided that it can indeed be shown that infrared modes have a nontrivial gravitational back-reaction effect in interesting models at second order in perturbation theory, it then becomes important to extend the analysis beyond perturbation theory. For initial attempts in this direction see [111, 112].

Acknowledgements

I wish to thank the organizers of the Vth Mexican School for inviting me to lecture at this wonderful place on the coast of the Yucatan peninsula. I am grateful to my collaborators Raul Abramo, Fabio Finelli, Ghazal Geshnizjani, Pei-Ming Ho, Sergio Joras, Jerôme Martin and in particular Slava Mukhanov

[3] For a different approach which also leads to the conclusion that there can be no back-reaction effects from infrared modes on local observables in models with a single matter component see [29].

for sharing their insights. This work has been supported in part by the U.S. Department of Energy under Contract DE-FG02-91ER40688, TASK A.

References

1. A. H. Guth, Phys. Rev. D **23**, 347 (1981).
2. A. Linde: *Particle Physics and Inflationary Cosmology*, (Harwood, Chur, 1990).
3. M. Gasperini and G. Veneziano, Astropart. Phys. **1**, 317 (1993) [arXiv:hep-th/9211021].
4. J. Khoury, B. A. Ovrut, P. J. Steinhardt and N. Turok, Phys. Rev. D **64**, 123522 (2001) [arXiv:hep-th/0103239].
5. V. F. Mukhanov, H. A. Feldman and R. H. Brandenberger, Phys. Rept. **215**, 203 (1992).
6. M. Tegmark, Measuring Spacetime: From Big Bang to Black Holes, Lect. Notes Phys. 646, 169 (2004).
7. R. Brustein, M. Gasperini, M. Giovannini, V. F. Mukhanov and G. Veneziano, Phys. Rev. D **51**, 6744 (1995) [arXiv:hep-th/9501066].
8. D. H. Lyth, Phys. Lett. B **524**, 1 (2002) [arXiv:hep-ph/0106153].
9. R. Brandenberger and F. Finelli, JHEP **0111**, 056 (2001) [arXiv: hep-th/0109004].
10. S. Tsujikawa, Phys. Lett. B **526**, 179 (2002) [arXiv:gr-qc/0110124].
11. J. c. Hwang, Phys. Rev. D **65**, 063514 (2002) [arXiv:astro-ph/0109045].
12. S. Tsujikawa, R. Brandenberger and F. Finelli, Phys. Rev. D **66**, 083513 (2002) [arXiv:hep-th/0207228].
13. D. H. Lyth, Phys. Lett. B **526**, 173 (2002) [arXiv:hep-ph/0110007].
14. J. Martin, P. Peter, N. Pinto Neto and D. J. Schwarz, Phys. Rev. D **65**, 123513 (2002) [arXiv:hep-th/0112128].
15. J. Khoury, B. A. Ovrut, P. J. Steinhardt and N. Turok, Phys. Rev. D **66**, 046005 (2002) [arXiv:hep-th/0109050].
16. R. Durrer and F. Vernizzi, Phys. Rev. D **66**, 083503 (2002) [arXiv:hep-ph/0203275].
17. C. Cartier, R. Durrer and E. J. Copeland, Phys. Rev. D **67**, 103517 (2003) [arXiv:hep-th/0301198].
18. R. Kallosh, L. Kofman and A. D. Linde, Phys. Rev. D **64**, 123523 (2001) [arXiv:hep-th/0104073].
19. A. Vilenkin and E.P.S. Shellard: *Cosmic Strings and Other Topological Defects*, (Cambridge Univ. Press, Cambridge, 1994).
20. M. B. Hindmarsh and T. W. Kibble, Rept. Prog. Phys. **58**, 477 (1995) [arXiv:hep-ph/9411342].
21. R. H. Brandenberger, Int. J. Mod. Phys. A **9**, 2117 (1994) [arXiv:astro-ph/9310041].
22. S. Weinberg: *Gravitation and Cosmology*, (Wiley, New York, 1972).
23. P.J.E. Peebles: *The Large-Scale Structure of the Universe*, (Princeton Univ. Press, Princeton, 1980).
24. T. Padmanabhan: *Structure Formation in the Universe*, (Cambridge Univ. Press, Cambridge, 1993).

25. J. Peacock: *Cosmological Physics*, (Cambridge Univ. Press, Cambridge, 1999).
26. R. K. Sachs and A. M. Wolfe, Astrophys. J. **147**, 73 (1967).
27. E. R. Harrison, Phys. Rev. D **1**, 2726 (1970).
28. Y. B. Zeldovich, Mon. Not. Roy. Astron. Soc. **160**, 1 (1972).
29. N. Afshordi and R. H. Brandenberger, Phys. Rev. D **63**, 123505 (2001) [arXiv:gr-qc/0011075].
30. E. Lifshitz, J. Phys. (USSR) **10**, 116 (1946);
E. M. Lifshitz and I. M. Khalatnikov, Adv. Phys. **12**, 185 (1963).
31. J. M. Bardeen, Phys. Rev. D **22**, 1882 (1980).
32. W. Press and E. Vishniac, Astrophys. J. **239**, 1 (1980).
33. H. Kodama and M. Sasaki, Prog. Theor. Phys. Suppl. **78**, 1 (1984).
34. M. Bruni, G. F. Ellis and P. K. Dunsby, Class. Quant. Grav. **9**, 921 (1992).
35. J. c. Hwang, Astrophys. J. **415**, 486 (1993).
36. R. Durrer, Helv. Phys. Acta **69**, 417 (1996).
37. J. Stewart, Class. Quant. Grav. **7**, 1169 (1990).
38. J. Stewart and M. Walker, Proc. R. Soc. London **A 341**, 49 (1974).
39. J. M. Bardeen, P. J. Steinhardt and M. S. Turner, Phys. Rev. D **28**, 679 (1983).
40. R. H. Brandenberger and R. Kahn, Phys. Rev. D **29**, 2172 (1984).
41. D. H. Lyth, Phys. Rev. D **31**, 1792 (1985).
42. F. Finelli and R. H. Brandenberger, Phys. Rev. Lett. **82**, 1362 (1999) [arXiv:hep-ph/9809490].
43. B. A. Bassett and F. Viniegra, Phys. Rev. D **62**, 043507 (2000) [arXiv:hep-ph/9909353].
44. F. Finelli and R. H. Brandenberger, Phys. Rev. D **62**, 083502 (2000) [arXiv:hep-ph/0003172].
45. S. Weinberg, Phys. Rev. D **67**, 123504 (2003) [arXiv:astro-ph/0302326].
46. W. B. Lin, X. H. Meng and X. M. Zhang, Phys. Rev. D **61**, 121301 (2000) [arXiv:hep-ph/9912510].
47. A. A. Starobinsky, Phys. Lett. B **91**, 99 (1980).
48. V. F. Mukhanov and G. V. Chibisov, JETP Lett. **33**, 532 (1981) [Pisma Zh. Eksp. Teor. Fiz. **33**, 549 (1981)].
49. A. H. Guth and S. Y. Pi, Phys. Rev. Lett. **49**, 1110 (1982).
50. A. A. Starobinsky, Phys. Lett. B **117**, 175 (1982).
51. S. W. Hawking, Phys. Lett. B **115**, 295 (1982).
52. V. N. Lukash, Pisma Zh. Eksp. Teor. Fiz. **31**, 631 (1980);
V. N. Lukash, Sov. Phys. JETP **52**, 807 (1980) [Zh. Eksp. Teor. Fiz. **79**, (1980)].
53. W. Press, Phys. Scr. **21**, 702 (1980).
54. K. Sato, Mon. Not. Roy. Astron. Soc. **195**, 467 (1981).
55. V. F. Mukhanov, JETP Lett. **41**, 493 (1985) [Pisma Zh. Eksp. Teor. Fiz. **41**, 402 (1985)].
56. V. F. Mukhanov, Sov. Phys. JETP **67**, 1297 (1988) [Zh. Eksp. Teor. Fiz. **94N7**, 1 (1988 ZETFA,94,1-11.1988)].
57. M. Sasaki, Prog. Theor. Phys. **76**, 1036 (1986).
58. N. Birrell and P.C.W. Davies: *Quantum Fields in Curved Space*, (Cambridge Univ. Press, Cambridge, 1982).
59. R. H. Brandenberger, Nucl. Phys. B **245**, 328 (1984).
60. R. H. Brandenberger and C. T. Hill, Phys. Lett. B **179**, 30 (1986).

61. D. Polarski and A. A. Starobinsky, Class. Quant. Grav. **13**, 377 (1996) [arXiv:gr-qc/9504030].
62. L. P. Grishchuk, Sov. Phys. JETP **40**, 409 (1975) [*Zh. Eksp. Teor. Fiz.* **67**, 825 (1974)].
63. R. H. Brandenberger: Trans-Planckian physics and inflationary cosmology. In: *2002 International Symposium on Cosmology and Particle Astrophysics, Taipei, Taiwan, 31 May - 2 Jun 2002*, arXiv:hep-th/0210186.
64. R. H. Brandenberger: Inflationary cosmology: Progress and problems. In: *1st Iranian International School on Cosmology: Large Scale Structure Formation, Kish Island, Iran, 22 Jan - 4 Feb 1999* Kluwer, Dordrecht, 2000. (Astrophys. Space Sci. Libr. ; 247), ed. by R. Mansouri and R. Brandenberger (Kluwer, Dordrecht, 2000). [arXiv:hep-ph/9910410].
65. W. G. Unruh, Phys. Rev. D **51**, 2827 (1995).
66. S. Corley and T. Jacobson, Phys. Rev. D **54**, 1568 (1996) [arXiv: hep-th/9601073].
67. J. Martin and R. H. Brandenberger, Phys. Rev. D **63**, 123501 (2001) [arXiv: hep-th/ 0005209].
68. R. H. Brandenberger and J. Martin, Mod. Phys. Lett. A **16**, 999 (2001) [arXiv:astro-ph/0005432].
69. J. C. Niemeyer, Phys. Rev. D **63**, 123502 (2001) [arXiv:astro-ph/0005533].
70. M. Lemoine, M. Lubo, J. Martin and J. P. Uzan, Phys. Rev.D **65**, 023510 (2002) [arXiv:hep-th/0109128].
71. T. Jacobson and D. Mattingly, Phys. Rev.D **63**, 041502 (2001) [arXiv:hep-th/0009052].
72. N. Kaloper, M. Kleban, A. E. Lawrence and S. Shenker, Phys. Rev. D **66**, 123510 (2002) [arXiv:hep-th/0201158].
73. J. Martin and R. H. Brandenberger, Phys. Rev. D **65**, 103514 (2002) [arXiv: hep-th/0201189].
74. R. H. Brandenberger, S. E. Joras and J. Martin, Phys. Rev. D **66**, 083514 (2002) [arXiv:hep-th/0112122].
75. R. Easther, B. R. Greene, W. H. Kinney and G. Shiu, Phys. Rev. D **64**, 103502 (2001) [arXiv:hep-th/0104102].
76. A. Kempf and J. C. Niemeyer, Phys. Rev. D **64**, 103501 (2001) [arXiv:astro-ph/0103225].
77. R. Easther, B. R. Greene, W. H. Kinney and G. Shiu, Phys. Rev. D **67**, 063508 (2003) [arXiv:hep-th/0110226].
78. F. Lizzi, G. Mangano, G. Miele and M. Peloso, JHEP **0206**, 049 (2002) [arXiv:hep-th/0203099].
79. D. Amati, M. Ciafaloni and G. Veneziano, Phys. Lett. B **197**, 81 (1987).
80. D. J. Gross and P. F. Mende, Nucl. Phys. B **303**, 407 (1988).
81. A. Kempf, Phys. Rev. D **63**, 083514 (2001) [arXiv:astro-ph/0009209].
82. S. F. Hassan and M. S. Sloth, Nucl. Phys. B **674**, 434 (2003). [arXiv:hep-th/0204110].
83. R. Brandenberger and P. M. Ho, Phys. Rev. D **66**, 023517 (2002) [arXiv:hep-th/0203119].
84. T. Yoneya, Mod. Phys. Lett. A **4**, 1587 (1989).
85. M. Li and T. Yoneya, arXiv:hep-th/9806240.
86. S. Cremonini, Phys. Rev. D **68**, 063514 (2003) [arXiv:hep-th/0305244].
87. S. Hollands and R. M. Wald, Gen. Rel. Grav. **34**, 2043 (2002) [arXiv:gr-qc/0205058].

88. T. Tanaka, arXiv:astro-ph/0012431.
89. A. A. Starobinsky, Pisma Zh. Eksp. Teor. Fiz. **73**, 415 (2001) [JETP Lett. **73**, 371 (2001)] [arXiv:astro-ph/0104043].
90. A. A. Starobinsky and I. I. Tkachev, JETP Lett. **76**, 235 (2002) [Pisma Zh. Eksp. Teor. Fiz. **76**, 291 (2002)] [arXiv:astro-ph/0207572].
91. U. H. Danielsson, Phys. Rev. D **66**, 023511 (2002) [arXiv:hep-th/0203198]; U. H. Danielsson, JHEP **0207**, 040 (2002) [arXiv:hep-th/0205227].
92. V. Bozza, M. Giovannini and G. Veneziano, JCAP **0305**, 001 (2003) [arXiv: hep-th/0302184].
93. J. Martin and R. Brandenberger, Phys. Rev. D **68**, 063513 (2003) [arXiv: hep-th/0305161].
94. L. R. Abramo, R. H. Brandenberger and V. F. Mukhanov, Phys. Rev. D **56**, 3248 (1997) [arXiv:gr-qc/9704037].
95. R. H. Brandenberger: Back reaction of cosmological perturbations and the cosmological constant problem. In *18th IAP Colloquium On The Nature Of Dark Energy: Observational And Theoretical Results On The Accelerating Universe*, arXiv:hep-th/0210165.
96. V. F. Mukhanov, L. R. Abramo and R. H. Brandenberger, Phys. Rev. Lett. **78**, 1624 (1997) [arXiv:gr-qc/9609026].
97. R. Brandenberger, Brown preprint BROWN-HET-1180, contribution to the 19th Texas Symposium on Relativistic Astrophysics, Paris, France, Dec. 14 - 18, 1998.
98. R. H. Brandenberger, in 'COSMO-99', ed. by U. Cotti et al, (World Scientific, Singapore, 2000), arXiv:hep-th/0004016.
99. G. Geshnizjani and R. Brandenberger, Phys. Rev. D **66**, 123507 (2002) [arXiv:gr-qc/0204074].
100. N. C. Tsamis and R. P. Woodard, Phys. Lett. B **301**, 351 (1993).
101. N. C. Tsamis and R. P. Woodard, Nucl. Phys. B **474**, 235 (1996) [arXiv:hep-ph/9602315].
102. L. R. Abramo and R. P. Woodard, Phys. Rev. D **65**, 043507 (2002) [arXiv:astro-ph/0109271].
103. L. R. Abramo and R. P. Woodard, Phys. Rev. D **65**, 063515 (2002) [arXiv:astro-ph/0109272].
104. D. Brill and J. Hartle, Phys. Rev. **135**, 1 (1964), B271-278; R. Isaacson, Phys. Rev. **166**, 2 (1968), 1263-1271 and 1272-1280.
105. W. Unruh, arXiv:astro-ph/9802323.
106. A. Linde, "Particle Physics and Inflationary Cosmology" (Harwood, Chur 1990); A. Linde, Phys. Scr. **T36**, 30 (1991).
107. A. Starobinsky, Current Topics in Field Theory, Quantum Gravity and Strings, Lect. Notes Phys. 246 (Springer Verlag, Berlin Heidelberg 1982).
108. A. S. Goncharov, A. D. Linde and V. F. Mukhanov, Int. J. Mod. Phys. A **2**, 561-591 (1987)
109. B. A. Bassett, D. I. Kaiser and R. Maartens, Phys. Lett. B **455**, 84 (1999) [arXiv:hep-ph/9808404].
110. M. Parry and R. Easther, Phys. Rev. D **59**, 061301 (1999) [arXiv:hep-ph/9809574].
111. N. C. Tsamis and R. P. Woodard, Annals Phys. **267**, 145 (1998) [arXiv:hep-ph/9712331].
112. L. R. Abramo, R. P. Woodard and N. C. Tsamis, Fortsch. Phys. **47**, 389 (1999) [arXiv:astro-ph/9803172].

Measuring Spacetime: From Big Bang to Black Holes

Max Tegmark

Dept. of Physics, Univ. of Pennsylvania, Philadelphia, PA 19104
max@physics.upenn.edu

Abstract. Space is not a boring static stage on which events unfold over time, but a dynamic entity with curvature, fluctuations and a rich life of its own which is a booming area of study. Spectacular new measurements of the cosmic microwave background, gravitational lensing, type Ia supernovae, large-scale structure, spectra of the Lyman α forest, stellar dynamics and x-ray binaries are probing the properties of spacetime over 22 orders of magnitude in scale. Current measurements are consistent with an infinite flat everlasting universe containing about 30% cold dark matter, 65% dark energy and at least two distinct populations of black holes.

1 Introduction

Traditionally, space was merely a three-dimensional static stage where the cosmic drama played out over time. Einstein's theory of general relativity [1, 2, 3] replaced this by four-dimensional spacetime, a dynamic geometric entity with a life of its own, capable of expanding, fluctuating and even curving into black holes. Now the focus of research is increasingly shifting from the cosmic actors to the stage itself. Triggered by progress in detector, space and computer technology, an avalanche of astronomical data is revolutionizing our ability to measure the spacetime we inhabit on scales ranging from the cosmic horizon down to the event horizons of suspected black holes, using photons and astronomical objects as test particles. The goal of this article is to review these measurements and future prospects, focusing on four key issues:

1. The global topology and curvature of space
2. The expansion history of spacetime and evidence for dark energy
3. The fluctuation history of spacetime and evidence for dark matter
4. Strongly curved spacetime and evidence for black holes

In the process, I will combine constraints from the cosmic microwave background [4], gravitational lensing, supernovae Ia, large-scale structure, the Lyman α forest[5], stellar dynamics and x-ray binaries. Although it is fashionable to use cosmological data to measure a small number of free "cosmological parameters", I will argue that improved data allow raising the ambition level

M. Tegmark, Measuring Spacetime: From Big Bang to Black Holes, Lect. Notes Phys. **646**, 169–189 (2004)
http://www.springerlink.com/

beyond this, testing rather than assuming the underlying physics. I will discuss how with a minimum of assumptions, one can measure key properties of spacetime itself in terms of a few cosmological functions: the expansion history of the universe, the spacetime fluctuation spectrum and its growth.

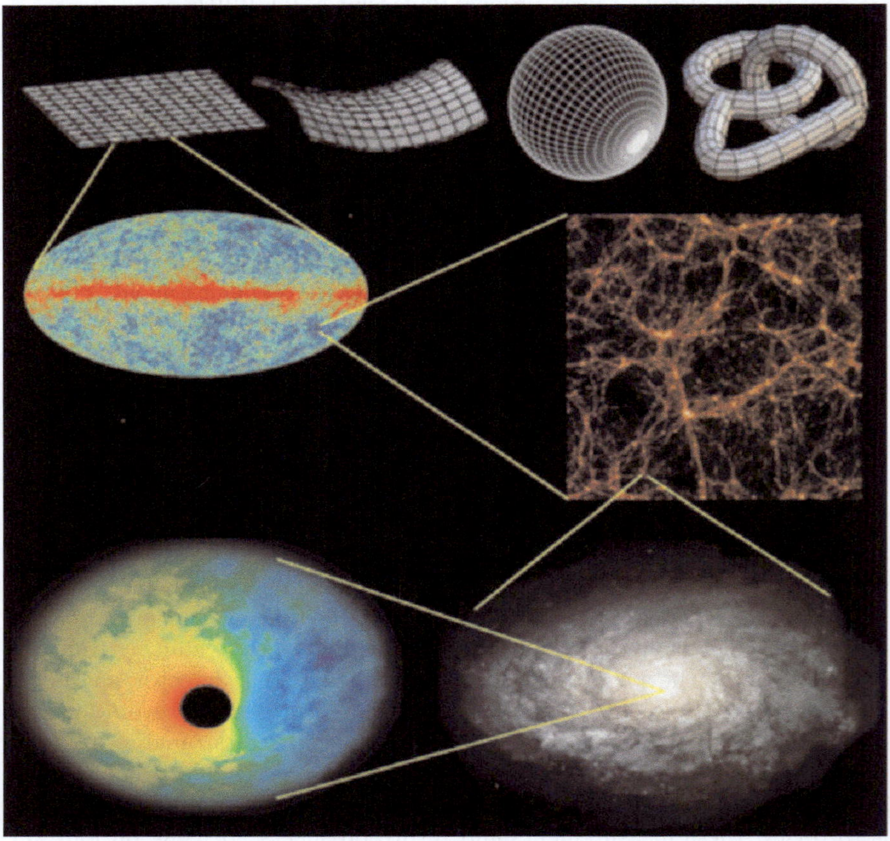

Fig. 1. Summary of the spacetime issues discussed in this article. One can use photons and astronomical objects as test particles to measure spacetime over 22 orders of magnitude in scale, ranging from the cosmic horizon (probing the global topology of and curvature of space — top) down to galaxies (giving evidence for dark matter), galactic nuclei and binary stellar systems (giving evidence for black holes). The figure illustrates how spacetime ripples at the 10^{-5} level will be imaged by the cosmic microwave background satellite MAP [6] and has grown via gravitational instability into cosmic large-scale structure [7], galaxies and, it seems, black holes [8].

1.1 Goals and Tools

Before embarking on our survey of spacetime, let us briefly review what it is we want to measure, the basic tools at our disposal [2, 3, 9] and the broad-brush picture of how our topics fit together. According to general relativity, spacetime is what mathematicians call a manifold, characterized by a topology and a metric. The topology gives the global structure (Fig. 1, top): is space infinite in all directions like in high-school geometry or multiply connected like say a hypersphere or doughnut so that traveling in a straight line could in principle bring you back home — from the other direction? The metric determines the local shape of spacetime, *i.e.*, the distances and time intervals we measure, and is mathematically specified by a 4×4 matrix at each point in spacetime.

General relativity theory (GR) consists of two parts, each providing a tool for measuring the metric. The first part of GR states that in the absence of non-gravitational forces, test particles (objects not heavy enough to have a noticeable effect on the metric) move along geodesics in spacetime, generalized straight lines, so the observed motions of photons and astronomical objects allow the metric to be reconstructed. I will refer to this as *geometric* measurements of the metric. The second part of GR states that the curvature of spacetime (expressions involving the metric and its first two derivatives) is related to its matter content — in most cosmological situations simply the density and pressure, but sometimes also bulk motions and stress energy. I will refer to such measurements of the metric as *indirect*, because they assume the validity of the Einstein field equations of GR.

1.2 The Broad Brush Picture

The current consensus in the cosmological community is that spacetime is extremely smooth, homogeneous and isotropic (translationally and rotationally invariant) on large ($\sim 10^{23}$m-10^{26}m) scales, with small fluctuations that have grown over time to form objects like galaxies and stars on smaller scales. Cosmic Microwave Background (CMB) observations have shown [10] that space is almost isotropic on the scale of our cosmic horizon ($\sim 10^{26}$m), with the metric fluctuating by only about one part in 10^5 from one direction to another, and combining this with the so-called cosmological principle, the assumption that there is nothing special about our vantage point, implies that space is homogeneous as well. Three-dimensional maps of the galaxy and quasar distribution give more direct evidence for large-scale homogeneity [11, 12, 13].

The fact that the CMB fluctuations are so small is useful, because it allows the intimidating nonlinear partial differential equations governing spacetime and its matter content to be accurately solved using a perturbation expansion. To zeroth order (ignoring the fluctuations), this fixes the global metric to be of the so-called Friedman-Robertson-Walker (FRW) form, which is completely specified except for a curvature parameter and a free function giving its expansion history. To first order, density perturbations grow due to gravitational instability and gravitational waves propagate through the FRW background spacetime, all governed by *linear* equations. Only on smaller scales ($\lesssim 10^{23}$m) do the fluctuations get large enough that nonlinear dynamics becomes important — in the realm of galaxies, stars and, perhaps, black holes. This review is organized analogously: Sections 2 and 3 discuss spacetime to 0th order (curvature, topology and expansion history), Section+4 describes spacetime to 1st order (fluctuations) and Sect. 5 focuses on nonlinear objects, mainly black holes.

2 Overall Shape of Spacetime

2.1 Curvature of Space

The question of whether space is infinite was answered last year with a resounding *maybe*. For an FRW metric, answering this question is equivalent to measuring the curvature of space as illustrated by the top left three cases in Fig. 1, specifically a single number R known as the *radius of curvature*. R is the radius of the hypersphere if space is finite, $R = \infty$ if space is flat, and R is an imaginary number($R^2 < 0$) for saddle-like curvature. Because the three angles of a triangle will add up to 180° in flat space, more if $R^2 > 0$ (like on a sphere) and less if $R^2 < 0$ (like on a saddle) cosmologists have measured R using the largest triangle available: one with us at one corner and the other two corners on the hot opaque surface of ionized hydrogen that delimits the visible universe and emits the CMB, merely 400,000 years after the Big Bang. Photographs of this surface reveal hot and cold spots of a characteristic angular size that can be predicted theoretically. This characteristic spot size (or, more rigorously, the first peak in the CMB power spectrum [14]) subtends about 0.5° — like the Moon — if space is flat. Sphere-like curvature would make all angles appear larger, so characteristic spots much larger than the Moon would indicate a finite universe curving back on itself, whereas smaller spots would indicate infinite space with negative curvature.

By 1994, evidence was mounting that there really was a peak in the CMB power spectrum [16], or at least a rise towards smaller scales. Data kept improving, and in 1998 the Toco experiment provided the first unambiguous detection and localization of a peak. The BOOMERanG experiment measured it with great precision in 2000, and by now the BOOMERanG, DASI and MAXIMA [17] teams have all seen both this peak and hints of additional smaller scale peaks matching theoretical predictions.

So is the universe infinite? The answer so far is still maybe, because the characteristic spot size has turned out to be so close to 0.5° that we still cannot tell whether space is perfectly flat or very slightly curved either way. The sharpest current limits on the curvature radius, obtained by combining all CMB experiments with galaxy clustering data [18, 19] to constrain other parameters affecting the spot sizes (mainly the cosmic matter budget), are $|R| > 20h^{-1}\text{Gpc} \approx 10^{27}\text{m}$. This is in sharp contrast to a few years ago, when the most popular models had negatively curved space with $|R| \approx 4h^{-1}$ Gpc. In other words, space now seems to be either infinite or much larger than the observable universe, whose radius is about $9h^{-1}$ Gpc.

In 1900, Karl Schwarzschild discussed the possibility that space was curved and published a paper with a lower limit $R > 2500$ light-years $\approx 2\times10^{19}$m [20]. A century later, we thus know that the universe is at least another 40 million times larger!

2.2 Topology of Space

Even if space turns out to be negatively curved or perfectly flat, it might be finite. General relativity does not prescribe the global topology, so various possibilities are possible (Fig. 1, top). The simplest non-trivial model has flat space and the topology of a three-dimensional torus, where opposing faces of a cube of size $L \times L \times L$ are identified to be one and the same. Living in such a universe would be indistinguishable from living in a perfectly periodic one: if $L = 10\,\text{m}$, you could see the back of your own head 10 meters away, and additional copies at 20 m, 30 m, and so on — searches for multiple images of cosmological objects have constrained such models [21]. Also, just as a finite guitar string has a fundamental tone and overtones, linear spacetime fluctuations in such a toroidal universe could have only certain discrete wavenumbers. As a result, its CMB power spectrum would differ on large scales, and the COBE [17] data was used to show that if the universe were such a torus, then L must be at least of the order of the cosmic horizon [22, 23]. Indeed, it was shown that all three dimensions of the torus must at least about this large to explain the absence of a type of approximate reflection symmetry in the COBE map [24]. This early work triggered dozens of papers in so-called cosmic crystallography, which turned out to be a rich mathematical subject — for an up-to-date review, see [25]. For instance, circles in the sky with near-identical temperature patterns were shown to be smoking-gun signals of compact topology. Unfortunately from an aesthetic point of view, many of the most mathematically elegant models, negatively curved yet compact spaces, have been abandoned after the recent evidence for spatial flatness. NASA's Microwave Anisotropy Probe(MAP) [6] will allow the cosmic topology to be probed with a new level of precision.

The interim conclusion about the overall shape of space is thus "back to basics" : although mathematicians have discovered a wealth of complicated manifolds to choose from and both positive and negative curvature would have been allowed *a priori*, all available data so far is consistent with the simplest possible space, the infinite flat Euclidean space that we learned about in high school. That is in regards to three-dimensional space. The global structure of our four-dimensional spacetime also depends on the beginning and end of time, to which we turn in the next section.

3 Spacetime Expansion History

One of the key quantities that cosmologists yearn to measure is the function $a(t)$, describing the expansion of the universe over time — if space is curved, a is simply the magnitude of the radius of curvature, $a = |R|$. A mathematically equivalent function more closely related to observations is the Hubble parameter as a function of redshift, $H(z)$, giving the cosmic expansion rate and defined by $H \equiv \frac{d}{dt}\ln a$, $1 + z \equiv a(t_{\text{now}})/a(t)$. Let us first discuss how this function encodes key information about the cosmic matter budget, the origin and the ultimate fate of the universe, and then turn to how it can be, has been and will be measured.

3.1 What $\rho(z)$ Tells Us About Dark Energy

As illustrated in Fig. 2, squaring our curve $H(z)$ gives us the cosmic matter density. If the Einstein Field equations of GR are correct, then the mean density of the universe is given by the Friedmann equation [2]

$$\rho(z) = \frac{3H(z)^2}{8\pi G}. \tag{1}$$

Here G is Newton's gravitational constant and, if space is curved, ρ is defined to include an optional curvature contribution $\rho_{\text{curv}} \equiv -\frac{3c^2}{8\pi G R^2}$, where c is the speed of light. Conveniently, all standard components of the cosmic matter budget contribute simple straight lines to this plot, because their densities drop as various power laws as the universe expands. For instance, the densities of both ordinary and cold dark matter particles are inversely proportional to the volume of space, scaling as $\rho \propto (1+z)^3$.

Figure 2 shows that although the cosmic density $\rho(z)$ measured from SN Ia and CMB was indeed higher in the past, the curve rises slower than this towards higher redshift, with a shallower slope than 3 at recent times. This is evidence for the existence of *dark energy*, a substance whose density does not rise rapidly with z. Adding a cosmological constant contribution $\rho_\Lambda \approx 4 \times 10^{-26}$ kg/m^3 (about 2/3 of the current matter budget) whose density is, by definition, constant, provides a good fit to the measurements (Fig. 2). This discovery, made independently by two teams in 1998 [27, 28], stunned the scientific community and triggered a worldwide effort to determine the nature of the dark energy. A model-independent approach will be to measure the curve $\rho(z)$ more accurately with a variety of different techniques as illustrated in the figure and described below, thereby answering two separate questions;

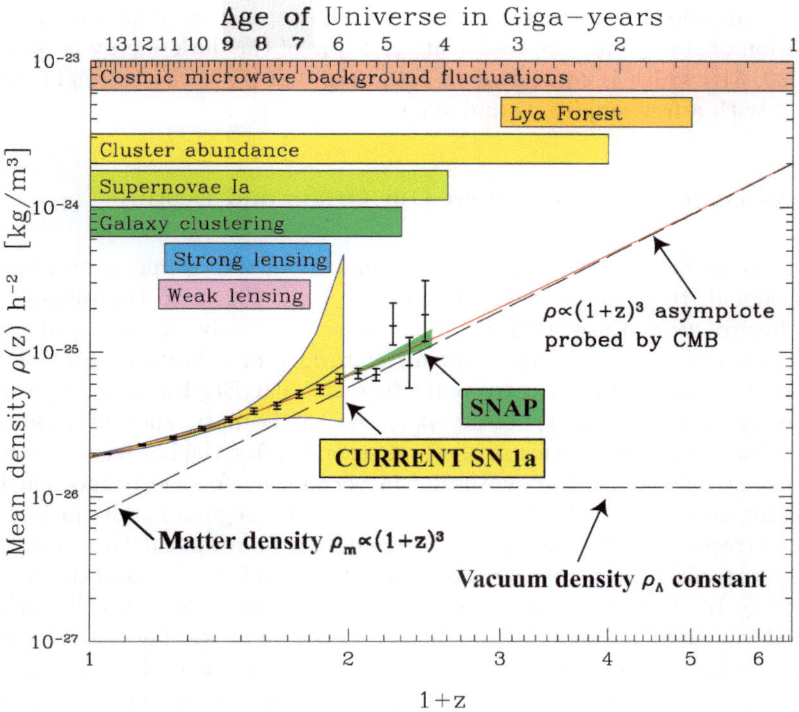

Fig. 2. Solid curve shows the concordance model [18] for the evolution of the cosmic mean density $\rho(z) \propto H(z)^2$. This curve uniquely characterizes the spacetime expansion history. The horizontal bars indicate the rough redshift ranges over which the various cosmological probes discussed are expected to constrain this function. Because the redshift scalings of all density contributions except that of dark energy are believed to be straight lines with known slopes in this plot (power laws), combining into a simple quartic polynomial, an estimate of the dark energy density $\rho_X(z)$ can be readily extracted from this curve. Specifically, $\rho \propto (1+z)^4$ for the cosmic microwave background (CMB), $\rho \propto (1+z)^3$ for baryons and cold dark matter, $\rho \propto (1+z)^2$ for spatial curvature, $\rho \propto (1+z)^0$ for a cosmological constant and $\rho \propto (1+z)^{3(1+w)}$ for dark energy with a constant equation of state w. Measurement errors are for current SN Ia constraints (yellow band) and a forecast for what the SNAP satellite [26] can do (green band), assuming flat space as favored by the CMB. Error bars are for a non-parametric reconstruction with SNAP.

1. Do independent measurements of $\rho(z)$ agree, so that we can rule out problems with observations and their interpretation?
2. Subtracting out the slope 3 line contributed by ordinary and dark matter, what is the time-dependence of dark energy density $\rho_X(z)$? If it is constant, we may have measured vacuum energy/Einstein's cosmological constant, and if not, we should learn interesting physics about a new scalar quintessence field, or whatever is responsible.

A less ambitious approach that is currently popular is assuming that the equation of state (pressure-to-density ratio) w of the dark energy is constant [29, 30, 31], which is equivalent to assuming that $\rho_X(z)$ is a straight line in Fig. 2 with a free amplitude and slope.

3.2 What $\rho(z)$ Tells Us about Our Origin and Destiny

If we can understand the different components of the cosmic matter budget well enough to extrapolate the curve $\rho(z)$ from Fig. 2 to the distant past and future, we can use the Friedmann equation to solve for $a(t)$ and obtain information about the origin and ultimate fate of spacetime. $a(t) = 0$ in the past or future would correspond to a singular Big Bang or big crunch, respectively, with infinite density $\rho(z)$. As to the past, such extrapolation seems justified at least back to the first seconds after the big bang, given the success of big bang nucleosynthesis in accounting for the primordial light element abundances [32, 33]. Regarding the very beginning, the jury is still out. Extrapolation back to the very beginning is more speculative. According to the currently most popular scenario, a large and nearly constant value of ρ at $t \lesssim 10^{-34}$ seconds caused exponential expansion $a(t) \propto e^{Ht}$ during a period known as inflation [34, 35, 36], successfully predicting both negligible spatial curvature and, as discussed in the next section, a nearly scale-invariant adiabatic scalar power spectrum[14] with subdominant gravitational waves. A rival "ekpyrotic" model inspired by string theory and a related eternally oscillating model have attracted recent attention[37, 38, 39]. If the density approaches the Planck density ($10^{97}\,\mathrm{kg/m^3}$) as $t \to 0$, quantum gravity effects for which we lack a fundamental theory should be important, and a host of speculative scenarios have been put forward for what happened at $t \sim 10^{-43}$ seconds. A very incomplete sample includes the Hawking-Hartle no-boundary condition [40], God creating the universe, the universe creating itself [41], and so-called pre-big-bang models [42]. Another possibility is that the Planck density was never attained and that there was no beginning, just an eternal fractal mess of replicating inflating bubbles, with our observed spacetime being merely one in an infinite ensemble of regions where inflation has stopped [35, 43].

As to the distant future, the expansion can clearly only stop ($H = 0$) if the effective density $\rho(z)$ drops to zero. The only two density contributions that can in principle be negative are those of curvature (which now seems to be negligible) and dark energy (which seems to be positive), suggesting that the universe will keep expanding forever. Indeed, if the dark energy density stays constant, we are now entering another inflationary phase of exponential expansion ($a(t) \propto e^{Ht}$), and in about 10^{11} years, our observable universe will be dark and lonely with almost all extragalactic objects having disappeared across our cosmic horizon [44]. However, such conclusions must clearly be taken with a grain of salt until the nature of dark energy is understood.

3.3 How to Measure $\rho(z)$

In conjunction with the curvature radius R, the curve $H(z)$ can be measured purely geometrically, using photons as test particles. Given objects of known luminosity ("standard candles") or known physical size ("standard yardsticks") at various redshifts, one simply compares their measured brightness or angular size with theoretical predictions. Because predictions, which follow from computing the trajectories of nearly parallel light rays, depend on only $H(z)$ and the (apparently negligible) curvature of space, objects at multiple redshifts can be used to reconstruct the curve $H(z)$ [45, 46].

The best standard candles to date are supernovae of type Ia, and the 92 SN Ia published by the two search teams [27, 28] were used [46] to measure $H(z)$ and thereby $\rho(z)$ in Fig. 2. These cosmic bombs all have the same mass, since they result when a white dwarf accretes enough gas from a companion star to exceed the Chandrashekar mass limit of 1.4 solar masses. They therefore have similar luminosity, and it has been shown that their actual luminosity can be accurately calibrated using their dimming rate and color [27, 28]. The best standard yardstick so far is the characteristic CMB spot size discussed above, suggesting that space is flat. As reviewed in [47, 46], numerous other candles and yardsticks have been discussed, especially in the Hubble parameter literature [48] focused on measuring $H(z)$ for $z \approx 0$, and although many have proven hard to standardize because of issues like galaxy evolution, it is far from clear that new multicolor surveys will not be able to measure $H(z)$ independently of SN Ia.

$H(z)$ can also be measured indirectly. As discussed in the next section, $H(z)$ affects the growth of density fluctuations and can therefore be probed by galaxy clustering and other techniques as indicated in Fig. 2. Such fluctuation measures have constrained matter to make up no more than about a third of the critical density needed to explain why space is flat. This Enron-like accounting situation provides supernova-independent evidence for dark energy [18, 19, 49].

4 Growth of Cosmic Structure

While SN Ia and CMB peak locations have recently revolutionized our knowledge of the metric to 0th order (curvature, topology and expansion history), other observations are probing its 1st order fluctuations with unprecedented accuracy. These perturbations come in two important types. The first are gravitational waves, hitherto undetected ripples in spacetime that propagate at the speed of light without growing in amplitude. The second are density fluctuations, which can get amplified by gravitational instability (Fig. 1) and are being measured by CMB, gravitational lensing and the clustering of extragalactic objects, notably galaxies and gas clouds absorbing quasar light (the so-called Lyman α forest, LyαF) over a range of scales and redshifts (Fig. 3).

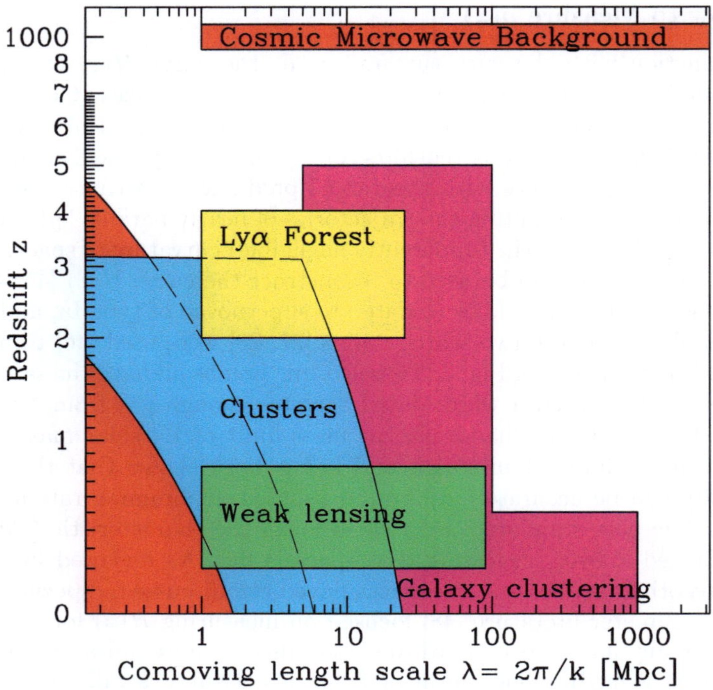

Fig. 3. Shaded regions show ranges of scale and redshift over which various observations are likely to measure spacetime fluctuations over the next few years. The lower left region, delimited by the dashed line, is the non-linear regime where rms density fluctuations exceed unity in the "concordance" model from [18].

Plane wave perturbations of different wavenumber evolve independently by linearity, and are so far consistent with having uncorrelated Gaussian-distributed amplitudes [50] as predicted by most inflation models [36]. The 1st order density perturbations are therefore characterized by a single function $P(k,z)$, the *power spectrum* [14], which gives the variance of the fluctuations as a function of wavenumber k and redshift z. $P(k,z)$ depends on (and can therefore teach us about) three things:

1. The cosmic matter budget
2. The seed fluctuations created in the Early Universe
3. Galaxy formation: reionization, "bias", *etc.*

A key challenge is to robustly disentangle the three. We are not there yet, but new data is making this increasingly feasible because each of the probes in Fig. 3 involve different physics and is affected by the three in different ways as outlined below.

Given the profusion of recent measurements of $H(z)$ and $P(k,z)$, it is striking that there is a fairly simple model that currently seems to fit everything

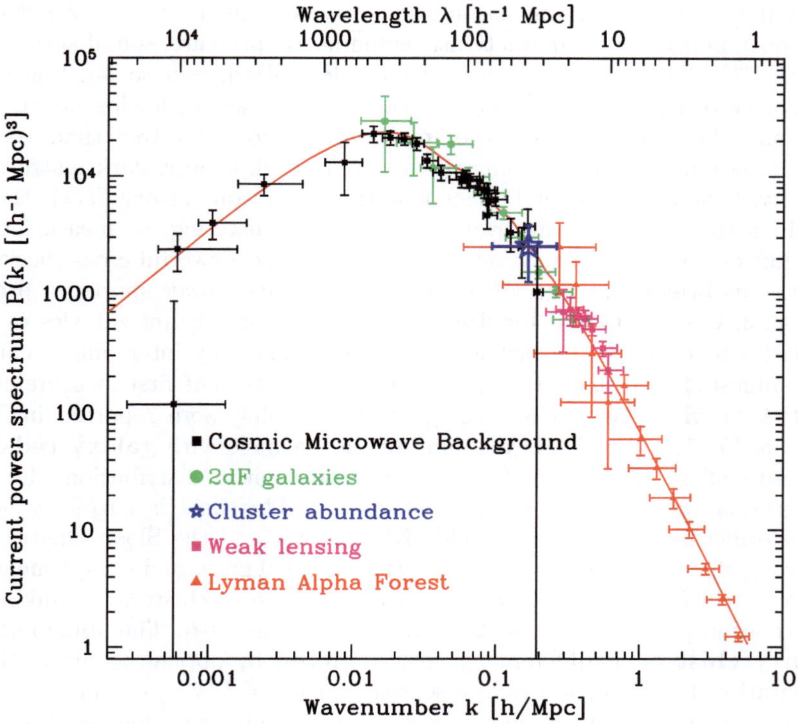

Fig. 4. Measurements of the current ($z = 0$) power spectrum of density fluctuations computed as described in [63], assuming the matter budget of [19] and reionization at $z = 8$. The CMB measurements combine the information from all experiments to date as in [63]. LSS points are from a recent analysis [64] of the 3D distribution of 2dF galaxies [11], and correcting them for bias shifts them vertically ($b = 1.3$ assumed here) and should perhaps blue-tilt them slightly. The cluster error bars reflect the spread in the literature. The lensing points are based on [65]. The LyαF points are from a reanalysis [66] of [52] and have an overall calibration uncertainty around 17%. The curve shows the concordance model of [19].

(Fig. 2 and Fig. 4). In this so-called concordance model [18, 19, 49, 51], the cosmic matter budget consists of about 5% ordinary matter (baryons), 30% cold dark matter, 0.1% hot dark matter (neutrinos) and 65% dark energy based on CMB and LSS observations, in good agreement with LyαF [18, 52, 53], lensing [54, 55, 56, 57, 58, 59, 60] and SN Ia [27, 28]. The seed fluctuations created in the early universe are consistent with the inflation prediction of a simple power law $P(k, z) \propto k^n$ early on, with $n = 0.9 \pm 0.1$ [18, 19]. Galaxy formation appears to have heated and reionized the universe not too long before redshift $z = 6$ based on the LyαF [61, 62].

Although the mere existence of a concordance model is a striking success, inferences about things like the expansion history, the matter budget and the early universe involve many assumptions — about the nature of dark energy and dark matter (*e.g.*, interactions, temperature, pressure, sound speed, viscosity [75]), about gravity, about galaxy formation, and so on. Since the avalanche of new cosmology data is showing no sign of slowing down, it is becoming feasible to to raise the ambition level to test rather than assume the underlying physics, probing the nature of dark energy, dark matter and gravity. Given the matter budget and the expansion history $H(z)$, theory predicts the complete time-evolution of linear clustering, so measuring its redshift dependence (Fig. 3) offers redundancy and powerful cross-checks.

Let us briefly summarize the status of our five power spectrum probes in Fig. 3. **Gravitational lensing** uses photons from distant galaxies as test particles to measure the metric fluctuations caused by intervening matter, as manifested by distorted images of distant objects. The first measurements of $P(k, z)$ with this "weak lensing" technique [54] were reported in 2000 [55, 56, 57, 58, 59, 60]. 3D mapping of the universe with **galaxy redshift surveys** offers another window on the cosmic matter distribution, through its gravitational effects on galaxy clustering. This field is currently being transformed by the 2 degree Field (2dF) survey and the Sloan Digital Sky Survey, which will jointly map more than 10^6 galaxies, and complementary surveys will map high redshifts and the evolution of clustering. Additional information can be extracted from galaxy velocities [67]. The abundance of **galaxy clusters** at different epochs, as probed by optical, x-ray, CMB or gravitational lensing surveys, is a sensitive probe of $P(k, z)$ on smaller scales [68, 69, 70] and the **LyαF** offers a new and exciting probe of matter clustering on still smaller scales when the universe was merely 10-20% of its present age [52, 71, 72, 66]. **CMB** experiments probe $P(k, z)$ through a variety of effects as far back as to redshifts $z > 10^3$ [73, 74]. The MAP satellite will publicly release CMB temperature measurements of unprecedented quality in December 2002 [6], and two new promising CMB fronts are opening up — CMB polarization (still undetected) and CMB fluctuations on tiny (arcminute) angular scales.

There is a rich literature on how all these complementary probes can be combined to break each others' degeneracies and independently measure the matter budget, the primordial power spectrum and galaxy formation details [45, 46, 75, 76, 77], so I will merely give a few examples here. The power spectra measured by CMB, LSS, lensing and LyαF are the product of the three terms: (i) the primordial power spectrum, (ii) a so-called transfer function quantifying the subsequent fluctuation growth, and (iii) (for LSS and LyαF only) a so called bias factor accounting for the fact that the measured galaxies/gas clouds may cluster differently than the underlying matter.

Disentangling Bias and Systematic Errors: Galaxy bias has now been directly measured from data and found to be of order unity for typical 2dF galaxies [51, 78], and LyαF bias may be computable with hydrodynamics simulations [52, 72, 66]. Although CMB, LSS, lensing and LyαF each comes with caveats of their own, their substantial overlap (Fig. 3) should allow disagreements between data sets to be distinguished from disagreements between data and theory.

Disentangling Primordial POWER from the Matter Budget: The transfer function can be disentangled from the primordial power because it depends on the matter budget, and conveniently in rather opposite ways for CMB than for low redshift $P(k)$ measurements (LSS, lensing, LyαF). For instance, increasing the cold dark matter density $h^2\Omega_c$ shifts the galaxy power spectrum up to the right and the CMB peaks down to the left if the primordial spectrum is held fixed. Adding more baryons boosts the odd-numbered CMB peaks but suppresses the galaxy power spectrum rightward of its peak and also makes it wigglier. Increasing the dark matter percentage that is hot (neutrinos) suppresses small-scale galaxy power while leaving the CMB almost unchanged. This means that combining CMB with other data allows unambiguous determination of the matter budget, and the primordial power spectrum can then be inferred. Combining CMB temperature and polarization measurements also helps in this regard, because the characteristic wiggles imprinted by the baryons and dark matter are out of phase for the two, whereas wiggles due to the primordial spectrum would of course line up for the two [63].

Although the best is still to come in this area, the basic conclusion that the universe is awash in nonbaryonic dark matter already appears quite solid, supported independently by CMB, LyαF, galaxy surveys, cluster counting and lensing — and by additional evidence in the next section. The agreement on the baryon density between fluctuation studies (CMB + galaxy surveys) and nucleosynthesis and on the dark energy density between fluctuation studies and SN Ia are both indications that spacetime fluctuation measurements are on the right track and will live up to their promise in this decade of precision cosmology.

5 Nonlinear Clustering and Black Holes

On small scales, the linear perturbation expansion eventually breaks down as density fluctuations grow to be of order unity, collapsing to form a variety of interesting astrophysical objects. Although the theoretical predictions are more difficult in this regime, the metric can still be accurately measured using photons and astrophysical objects as test particles. The gravitational potential well is probed by strong gravitational lensing of photons through its distorting effect on background objects [79] and also by the motions of massive objects like galaxies, stars or gas clouds. The orbital parameters in a binary system reveal the masses of the two objects, just as we once weighed the Sun by exploiting Earth's orbit around it. In more complicated systems, the central mass distribution can be inferred statistically from velocity dispersions observed in the vicinity. Below I review how these basic tools have revealed surprises on three vastly different scales: dark matter in galaxies and clusters ($\sim 10^{20-23}$m), supermassive black holes in galactic bulges ($\sim 10^{10} - 10^{13}$m) and stellar-mass black holes ($\sim 10^{4} - 10^{5}$m). Recent black hole reviews include [80, 81, 82, 83, 84].

5.1 Dark Matter in Galaxies and Clusters

As noted by Zwicky in 1933 [85], the amount of mass in galaxies and galaxy clusters inferred from rotation curves or velocity dispersions exceeds the mass of luminous matter by a large factor. Precision measurements with a variety of techniques have confirmed this finding, providing evidence that both galaxies and clusters are accompanied by roughly spherical halos of cold dark matter. This dark matter evidence is independent of that from linear perturbation theory described above, yet produces roughly consistent estimates of the total cosmic dark matter density [86, 87].

New measurements such as mapping tidal streamers, stripy remnants of galaxies cannibalized by the Milky Way in the past, are raising the ambition level towards a full 3D reconstruction of our own dark matter halo, and early results suggest that it may be elliptical rather than perfectly spherical [88]. Measurements of the shape and substructure of dark matter halos can probe the detailed nature of the dark matter. Indeed, computer simulations with cold dark matter composed of weakly interacting particles appear to predict overly dense cores in the centers of galaxies and clusters, and that there should be about 10^3 discrete dark matter halos in our Galactic neighborhood (the Local Group), in contrast to the less than 10^2 galaxies actually observed. These halo profile and substructure problems have triggered talk of a cold dark matter crisis and much recent interest in self-interacting dark matter [89], warm dark matter [90] and other more complicated dark matter models which suppress cores and substructure. It is not obvious that there really is a crisis, since baryonic feedback properties may be able to reconcile vanilla cold dark matter with observations and since substantial halo substructure has recently been detected with gravitational lensing [91], but this active research area should teach us more about dark matter properties whatever they turn out to be.

5.2 Supermassive Black Holes

Karl Schwarzschild was allegedly so distressed by his 1916 solution to the Einstein field equations that he hoped that such sinister objects, later christened black holes by Wheeler, did not exist in the real universe. The irony is that monstrous black holes are nowadays considered the *least* exotic explanation for the phenomena found in the centers of most — if not all — massive galaxies.

The spatial and velocity distribution of stars have unambiguously revealed compact objects weighing $10^6 - 10^{10}$ solar masses at the centers of over a dozen galaxies. The most accurate measurements are for our own Galaxy, giving a mass around $3 \times 10^6 M_\odot$ [92]. Here even individual stellar orbits have been measured and shown to revolve around a single point [92] that coincides with a strong source of radio and x-ray emission.

In many cases, gas disks have been found orbiting the mystery object. For instance, Hα emission from such a disk in the galaxy M87 has revealed a record mass of $3.2 \times 10^9 M_\odot$ in a region merely 10 light-years across, and 1.3 cm water maser emission from a disk in the galaxy NGC4258 has revealed $3.6 \times 10^7 M_\odot$ in a region merely 0.42 light-years across (1 light-year $\approx 10^{16}$ m). This is too compact to be a stable star cluster, so the only alternatives to the black hole explanation involve new physics — like a "fermion ball" made of postulated new particles [93].

Although impressive, all these spacetime measurements were still at $> 10^4$ Schwarzschild radii, and so probe no strong GR effects and give only indirect black hole evidence. X-ray spectroscopy provides another powerful probe, because x-rays can be produced closer to the event horizon, less than a light hour from the central engine where the material is hotter and the detailed shape of spacetime can imprint interesting signatures on the emitted radiation. For instance, a strong emission line from the Kα fluorescent transition of highly (photo-)ionized iron atoms has been observed by the ASCA and Beppo-SAX satellites [94] to have spectacular properties. Doppler shifts indicate a gas disk rotating with velocities up to 10% of the speed of light, and extremely broadened and asymmetric line profiles are best fit when including both Doppler and gravitational redshifts at 3-10 Schwarzschild radii.

In addition to all this geometric evidence for supermassive black holes, further support comes from the processes by which they eat and grow. Infalling gas is predicted to form a hot accretion disk around the hole that can radiate away as much as 10% of its rest energy. It was indeed this idea that led to the suggestions of supermassive black holes in the early 1960s, prompted by the discovery of quasars. About 50% of all galaxies are now known to have active galactic nuclei (AGN) at least at some low level — any black holes in the other half are presumed to have quieted down after consuming the gas in their vicinity. AGN's can produce luminosities exceeding that of 10^{12} suns in a region less than a light-year across, and no other mechanism is known for converting matter into radiation with the high efficiency required. In some cases, emission has been localized to a region $\lesssim$ a light-hour across (smaller than our solar system) by changing intensity in less than an hour.

Furthermore, magnetic phenomena in accretion discs can radiate beams of energetic particles, and such jets have been observed to up to 10^6 light-years long, perpendicular to the disk as predicted. This requires motions near the speed of light as well as a stable preferred axis over long ($\gg 10^6$ year) timescales, as naturally predicted for black holes [82, 95].

5.3 Stellar-Mass Black Holes

Numerous stars have been found to orbit a binary companion weighting too much to be a white dwarf or a neutron star ($\gtrsim 3M_\odot$), and being too faint (often invisible) to be a normal star. For example, after a transient outburst of soft x-rays in 1989, all orbital parameters of the binary system V404 Cygni were measured and the black hole candidate was found to weigh $12\pm2M_\odot$ [96]. Just as for supermassive BH's, x-ray variability has placed upper limits on the size of such objects that rule out all conventional black hole alternatives.

To counter such indirect arguments for black holes, unconventional compact objects such as "strange stars" and "Q-stars" have been proposed [97, 98]. However, the accretion disk model for soft x-ray transients such as V404 Cygni might require the object to have an event horizon that gas can disappear through — a hard surface could cause radiation to come back out. Indeed, the similarities between galactic and stellar accretion disk and jet observations are so striking that a single unified explanation seems natural, and black holes provide one.

There is thus strong evidence for existence of black holes in two separate mass ranges, each making up perhaps 10^{-6} or 10^{-5} of all mass in the universe. Still smaller classes of black holes have been speculated about without direct supporting evidence, both microscopic ones created in the early universe perhaps making up the dark matter [99] and transient ones constituting "spacetime foam" on the Planck scale [3].

5.4 Black Hole Prospects and Gravitational Waves

Whereas it is fairly well-understood how stellar-mass black holes can be formed by dying massive stars [100, 80], the origin and evolution of the apparently ubiquitous supermassive black holes are open questions, as is their relation to the formation of galaxies and galactic bulges. Another challenge involves measuring spacetime more accurately near the event horizon, particularly for evidence of black hole rotation [101]. Observations to look forward to include galactic center flashes as individual stars get devoured, multiwavelength accretion disk observations, and, in particular, detection of gravitational waves. These tiny ripples in spacetime should be produced whenever masses are accelerated, and binary pulsars have been measured to lose energy at precisely the rate gravitational wave emission predicts [102]. They should thus be copiously produced in inspiraling mergers involving black holes, both stellar-mass ones (measurable by ground-based detectors such as the Laser Interferometer Gravitational wave Observatory, LIGO) and and supermassive ones (measurable by space-based detectors such as the Laser Interferometer Space Antenna, LISA) [103]. At still longer wavelengths, the hunt for gravitational waves goes on using pulsar timing [104] and microwave background polarization that can constrain cosmological inflation [105, 106].

6 Outlook

I have surveyed recent measurements of spacetime over a factor of 10^{22} in scale, ranging from the cosmic horizon down to the event horizon of black holes. On the largest scales, evidence supports "back to basics" flat infinite space and eternal future time. The growth of spacetime fluctuations has suggested that about 30% of the cosmic matter budget is made up of (mostly cold) dark matter, about 5% ordinary matter and the remainder dark energy. There is further evidence for the same dark matter in the halos of galaxies and clusters. Finally, spacetime seems to be full of black holes, both supermassive ones in the centers of most galaxies and stellar mass ones wherever high mass stars have died. How much of this have we really measured and how much is based on assumptions? The above-mentioned geometric test particle observations have measured the spacetime metric, but all inferences about dark energy, dark matter and the inner parts of black holes assume that the Einstein Field Equations (EFEs) of GR are valid. Indeed, attempts have been made to explain away all three by modifying the EFEs. So-called scalar-tensor gravity has been found capable of giving accelerated cosmic expansion without dark energy [107]. Although not an ab initio theory, the approach known as Modified Newtonian Dynamics (MOND) attempts to explain galaxy rotation curves without dark matter [108, 109]. It is not inconceivable that the EFEs can be modified to avoid black hole singularities [110], even though the perhaps most publicized model with this property [111] has been argued to be flawed [112].

So could dark energy, dark matter and black holes be merely a modern form of epicycles, which just like those of Ptolemy can be eliminated by modifying the laws of gravity [46, 109, 113, 114]? The way to answer this question is clearly to test the EFEs observationally, by embedding them in a larger class of equations and quantifying the observational constraints. This program has been pioneered by Clifford Will and others [9, 115], showing that the true theory of gravity must be extremely close to GR in the regime probed by solar system dynamics and binary pulsars, and has also been pursued to close the MOND-loophole with some success [116, 117, 118]. However, this does *not* imply that the true theory of gravity must be indistinguishable from GR in all contexts, in particular for very compact objects [103] or for cosmology [9, 115], so testing gravity remains a fruitful area of research. Such tests continue even in the laboratory [119], testing the gravitational inverse square law down to millimeter scales to probe possible extra dimensions [120].

In conclusion, the coming decade will be exciting: an avalanche of astrophysical observations are measuring spacetime with unprecedented accuracy, allowing us to test whether it obeys Einstein's field equations, and consequently whether dark energy, dark matter and black holes are for real.

References

1. A. Einstein: *Relativity: the Special and the General Theory* (Random House; New York, 1920)
2. S. Weinberg: *Gravitation and Cosmology* (Wiley, New York, 1972)
3. C. W. Misner, K. S. Thorne and J. A. Wheeler: *Gravitation* (Freeman, San Francisco, 1973).
4. The cosmic microwave background (CMB) is the oldest light around, emanating from the hot opaque hydrogen plasma that filled the universe during its first 400,000 years. An up-to-date review is: M. White and J. Cohn, 2002;astro-ph/0203120
5. The Lyman α forest (LyαF) is the plethora of absorption lines in the spectra of distant quasars caused by neutral hydrogen in overdense intergalactic gas along the line of sight. It allows us to map the cosmic gas distribution out to great distances.
6. *http://map.gsfc.nasa.gov*
7. *http://www.mpa-garching.mpg.de/Virgo/*
8. B. C. Bromley, W. A. Miller and V. I. Pariev: Nature;**391**, 54 (1998).
9. C. M. Will: *Theory and Experiment in Gravitational Physics* (Cambridge U. P., Cambridge, 1993)
10. G. F. Smoot et al: Astrophys. J. ; **396**, L1 (1992)
11. M. Colless et al: Mon. Not. Roy. Astron. Soc., **328**,1039 (2001).
12. I. Zehavi et al: Astrophys. J., **571**, 172 (2002) [astro-ph/0106476].
13. F. Hoyle et al: Mon. Not. Roy. Astron. Soc. **329**,336 (2002).
14. The CMB power spectrum is the level of temperature fluctuations as a function of angular scale, defined as the variance of the spherical harmonic coefficients of a sky map. A matter power spectrum is the level of 3D density fluctuations as a function of spatial scale, defined as the variance of the Fourier coefficients of the density field.
15. The measured power spectrum of galaxies or LyAF clouds may differ from the power spectrum that we really care about — that for the underlying dark matter. Bias is defined as the square root of the ratio of the two power spectra.
16. D. Scott, J. Silk and M. White: Science, **268**, 829 (1995).
17. The CMB acronyms are the COsmic Background Explorer (COBE), Balloon Observations Of Millimetric Extragalactic Radiation and Geophysics (BOOMERanG), Degree Angular Scale Interferometer (DASI) and Millimeter Anisotropy eXperiment IMaging Array (MAXIMA).
18. X. Wang, M. Tegmark and M. Zaldarriaga: Phys. Rev. D, **65**, 123001 (2002) [astro-ph/0105091].
19. G. Efstathiou et al: Mon. Not. Roy. Astron. Soc.;**330** L29 (2002) [astro-ph/0109152].
20. K. Schwarzschild: Vier. d. Astr. Ges. **35**;337 (1900) English translation in Class. Quant. Grav. **15**, 2539 (1998)
21. B. Roukema:2002;astro-ph/0201092
22. D. Stevens, D. Scott and J. Silk: Phys. Rev. Lett. **71**, 20 (1993).
23. A. de Oliveira-Costa and G. F. Smoot: Astrophys. J. **448**; 477 (1995).
24. A. de Oliveira-Costa, G. F. Smoot and A. A. Starobinski: Astrophys. J.;**468**;457(1996).
25. J. Levin: Phys. Rep. **365**, 251 (2002) [gr-qc/0108043].

26. *http://snap.lbl.gov*
27. S. Perlmutter et al: Nature;**391**;51 (1998).
28. A. G. Riess et al: Astron. J.;**116**;1009 (1998).
29. P. M. Garnavich et al: Astrophys. J., **509**, 74 (1998).
30. I. Maor, R. Brustein and P. Steinhardt Phys. Rev. Lett. **86**, 6 (2001) [astro-ph/0007297].
31. D. Huterer and M. S. Turner: Phys. Rev. D **64**, 123527 (2001) [astro-ph/0012510].
32. S. Burles, K. M. Nollett and M. S. Turner: Astrophys. J., **552**, L1 (2001).
33. S. M. Carroll and M. Kaplinghat: Phys. Rev. D **65**, 063507 (2002).
34. A. Guth and P. Steinhardt: Scientific American, May (1984).
35. A. D. Linde: *Particle Physics and Inflationary Cosmology* (Harwood, Chur, Switzerland 1990).
36. A. R. Liddle and D. H. Lyth, *Cosmological Inflation and Large-Scale Structure* (Cambridge U. P. Cambridge, UK, 2000).
37. J. Khoury et al: Phys. Rev. D;**64**;123522 (2001).
38. R. Kallosh, L. Kofman and A. D. Linde: Phys. Rev. D;**64**;123523 (2001).
39. P. J. Steinhardt and N. Turok: Science (2002).
40. J. B. Hartle and S. W. Hawking, Phys. Rev. D **28**, 2960 (1983).
41. R. J. Gott and L. X. Li: Phys. Rev. D;**58**;023501 (1998).
42. G. Veneziano; 2000;hep-th/0002094
43. A. Vilenkin: Phys. Rev. Lett. **74**, 846 (1995).
44. A. Loeb: Phys. Rev. D **65**, 047301 (2002).
45. Y. Wang and P. M. Garnavich: Astrophys. J. **552**, 445 (2001). [astro-ph/0101040].
46. M. Tegmark: Phys. Rev. D **66**, 103507 (2002) [astro-ph/0101354].
47. M. Rowan-Robinson: *The Cosmological Distance Ladder* (Freeman, New York, 1985).
48. W. L. Freedman et al: Astrophys. J. **553**, 47 (2001).
49. N. Bahcall et al: Science **284**, 1481 (1999).
50. J. H. P. Wu: Phys. Rev. Lett. **87**, 251303 (2001) [astro-ph/0104248].
51. O. Lahav et al;2001;astro-ph/0112162
52. R. A. C. Croft et al: Astrophys. J. **581**, 20 (2002) [astro-ph/0012324].
53. S. Hannestad et al: Astropart. Phys. **17**, 375 (2002) [astro-ph/0103047].
54. M. Bartelmann and P. Schneider: Phys. Rep. **340**, 291 (2001).
55. D. M. Wittman et al: Nature, **405**, 143 (2000).
56. L. Van Waerbeke et al;2000;A&A;358;30
57. D. J. Bacon, A. R. Refregier and R. S. Ellis: Mon. Not. Roy. Astron. Soc. **318**, 625 (2000).
58. N. Kaiser, G. Wilson and G. A. Luppino ;2000;astro-ph/0003338
59. J. Rhodes, A. Refregier and E. J. Groth; 2001;astro-ph/0101213
60. L. Van Waerbeke et al;2001;astro-ph/0101511
61. R. H. Becker et al: Astron. J. **122**, 2850 (2001).
62. S. G. Djorgovski: 2001;astro-ph/010806.
63. M. Tegmark and M. Zaldarriaga: Phys. Rev. D **66**, 103508 (2002) [astro-ph/0207047].
64. M. Tegmark, A. J. S. Hamilton and Y. Xu: Mon. Not. Roy. Astron. Soc. **335**, 887 (2002) [astro-ph/0111575].
65. H. Hoekstra, H. Yee and M. Gladders M: Astrophys. J. **577**, 595 (2002) [astro-ph/0204295].

66. N. Y. Gnedin and A. J. S. Hamilton; 2001;astro-ph/0111194
67. I. Zehavi and A. Dekel: Nature, **401**, 252 (1999).
68. N. A. Bahcall and X. Fan: Astrophys. J. **504**, 1 (1998). [astro-ph/9803277].
69. E. Pierpaoli, D. Scott and M. White: Mon. Not. Roy. Astron. Soc. **325**, 77 (2001) [astro-ph/0010039].
70. G. Holder, Z. Haiman and J. Mohr: 2001;astro-ph/0105396
71. P. McDonald et al: Astrophys. J. **543**, 1 (2000).
72. M. Zaldarriaga, L. Hui, M. Tegmark: Astrophys. J. **557**, 519 (2001) [astro-ph/0011559].
73. W. Hu, N. Sugiyama, J. Silk: Nature, **386**, 37 (1997).
74. U. Seljak, M. Zaldarriaga: Astrophys. J. **469**, 437 (1996).
75. W. Hu: Astrophys. J. **506**, 485 (1998).
76. D. J. Eisenstein, W. Hu, M. Tegmark: Astrophys. J., **518**, 2 (1999).
77. W. Hu et al: Phys. Rev. D **59**, 023512 (1999).
78. L. Verde et al;2001;astro-ph/0112161
79. J. Wambsganss: 2000;astro-ph/0012423
80. A. Celotti, J. C. Miller, D. W. Sciama: Class. Quant. Grav. **16**, A3 (2000).
81. J. H. Krolik : *AGN : from the central black hole to the galactic environment* (Princeton U. P. Princeton 1999).
82. J. Kormendy and K. Gebhardt;2001;astro-ph/0105230
83. A. Marconi: 2002;astro-ph/0201504
84. J. Frank, A. King, D. Raine: *Accretion power in astrophysics*, 3rd ed. (Cambridge U. P. Cambridge, 2002).
85. F. Zwicky: Helv. Phys. Acta, **6**, 110 (1993).
86. H. Hoekstra et al: Astrophys. J. L. **548**, L5 (2001).
87. N. Bahcall et al: Astrophys. J. **541**, 1 (2000).
88. H. J. Newberg: Astrophys. J. **569**, 245 (2002).
89. D. N. Spergel, P. J. Steinhardt: Phys. Rev. Lett. **84**, 3760 (2000).
90. P. Bode, J. P. Ostriker, N. Turok: Astrophys. J. **556**, 93 (2001) [astro-ph/0010389].
91. N. Dalal, C. S. Kochanek: 2002; astro-ph/0111456
92. A. Eckhardt et al;2002;astro-ph/0201031
93. F. Munyaneza, R. D. Viollier: Astrophys. J. **564**, 274 (2002) [astro-ph/0103466].
94. K. Nandra: 2000;astro-ph/0007356
95. R. D. Blandford and R. L. Znajek: Mon. Not. Roy. Astron. Soc.,**179**, 433 (1997).
96. T. Shabhaz et al: Mon. Not. Roy. Astron. Soc., **271**, L10 (1994).
97. E. Witten: Phys. Rev. D;**30**, 272 (1984).
98. S. Bahcall: Astrophys. J., **362**, 251 (1990).
99. B. Carr: *The Renaissance of GR and Cosmology*, pp. 277. eds. G. F. R. Ellis et al (Cambridge U. P. Cambridge, 1993)
100. J. R. Oppenheimer and H. Snyder: Phys. Rev. **56**, 455 (1939).
101. J. M. Miller: Astrophys. J. L. , **570**, 69 (2002) [astro-ph/0202375].
102. I. H. Stairs et al: Astrophys. J. **505**, 352 (1998).
103. S. A. Hughes et al;2001;astro-ph/0110349.
104. A. N. Lommen and D. C. Backer: Astrophys. J. **562**, 297 (2001).[astro-ph/0107470].
105. M. Kamionkowski, A. Kosowsky, A. Stebbins: Phys. Rev. D, **55**, 7368 (1997).

106. M. Zaldarriaga and U. Seljak: Phys. Rev. D, **55**, 1830 (1997).
107. B. Boisseau et al: Phys. rev. Lett. **85**, 2236 (2000).
108. M. Milgrom: Astrophys. J., **270**, 365 (1983).
109. S. S. McGaugh: Astrophys. J.L **541**, 33 (2000).
110. P. O. Mazur and E. Mottol: 2001;astro-ph/0103466
111. H. Yilmaz: Nuovo Cim. B, **107**, 941 (1992).
112. C. W. Misner: Nuovo Cim. B, **114**, 1079 (1999).
113. P. J. E. Peebles: Comments on the State of our Subject. In *Clustering at High Redshift*, Marseilles, June 1999, eds. A. Mazure and O. Le Fevre, [astro-ph/9910234].
114. J. A. Sellwood and A. Kosowsky: Does Dark Matter Exist?. In *Gas & Galaxy Evolution* eds. Hibbard, Rupen & van Gorkom ;2000;astro-ph/0009074
115. C. M. Will: *The Confrontation between General Relativity and Experiment: A 1998 Update* (Washington University, St. Louis), [gr-qc/9811036].
116. L. M. Griffiths, A. Melchiorri, J. Silk: Astrophys. J. **553**, L5(2001)
117. M. White, D. Scott, E. Pierpaoli: Astrophys. J. **545**, 1 (2000) [astro-ph/0004385].
118. M. White, C. S. Kochanek: Astrophys. J. **560**, 539 (2001).
119. C. D. Hoyle et al: Phys. Rev. Lett. **86**, 1418 (2001).
120. L. Randall: Science, **296**, 1422 (2002).
121. The author wishes to thank Roger Blandford, Angélica de Oliveira-Costa and Harold Shapiro for helpful comments. Support for this work was provided by NSF grants AST-0071213 & AST-0134999, NASA grants NAG5-9194 & NAG5-11099, and the David and Lucile Packard Foundation.

The Accelerating Universe and Dark Energy: Evidence from Type Ia Supernovae

Alexei V. Filippenko

Department of Astronomy, University of California, Berkeley, CA 94720-3411, USA. alex@astro.berkeley.edu

Abstract. I discuss the use of Type Ia supernovae (SNe Ia) for cosmological distance determinations. Low-redshift SNe Ia ($z \lesssim 0.1$) demonstrate that the Hubble expansion is linear, that $H_0 = 72 \pm 8$ km s^{-1} Mpc^{-1}, and that the properties of dust in other galaxies are similar to those of dust in the Milky Way. The light curves of high-redshift ($z = 0.3$–1) SNe Ia are stretched in a manner consistent with the expansion of space; similarly, their spectra exhibit slower temporal evolution (by a factor of $1 + z$) than those of nearby SNe Ia. The measured luminosity distances of SNe Ia as a function of redshift have shown that the expansion of the Universe is currently accelerating, probably due to the presence of repulsive dark energy such as Einstein's cosmological constant (Λ). From about 200 SNe Ia, we find that $H_0 t_0 = 0.96 \pm 0.04$, and $\Omega_\Lambda - 1.4\Omega_M = 0.35 \pm 0.14$. Combining our data with the results of large-scale structure surveys, we find a best fit for Ω_M and Ω_Λ of 0.28 and 0.72, respectively — essentially identical to the recent *WMAP* results (and having comparable precision). The sum of the densities, ~ 1.0, agrees with extensive measurements of the cosmic microwave background radiation, including *WMAP*, and coincides with the value predicted by most inflationary models for the early Universe: the Universe is flat on large scales. A number of possible systematic effects (dust, supernova evolution) thus far do not seem to eliminate the need for $\Omega_\Lambda > 0$. However, during the past few years some very peculiar low-redshift SNe Ia have been discovered, and we must be mindful of possible systematic effects if such objects are more abundant at high redshifts. Recently, analyses of SNe Ia at $z = 1.0$–1.7 provide further support for current acceleration, and give tentative evidence for an early epoch of deceleration. The dynamical age of the Universe is estimated to be 13.1 ± 1.5 Gyr, consistent with the ages of globular star clusters and with the *WMAP* result of 13.7 ± 0.2 Gyr. Current projects include the search for additional SNe Ia at $z > 1$ to confirm the early deceleration, and the measurement of a few hundred SNe Ia at $z = 0.2$–0.8 to more accurately determine the equation of state of the dark energy, $w = P/(\rho c^2)$, whose value is now constrained by SNe Ia to be in the range $-1.48 \lesssim w \lesssim -0.72$ at 95% confidence.

1 Introduction

Supernovae (SNe) come in two main varieties (see Filippenko 1997b for a review). Those whose optical spectra exhibit hydrogen are classified as Type II, while hydrogen-deficient SNe are designated Type I. SNe I are further subdivided according to the appearance of the early-time spectrum: SNe Ia

A.V. Filippenko, The Accelerating Universe and Dark Energy: Evidence from Type Ia Supernovae, Lect. Notes Phys. **646**, 191–221 (2004)
http://www.springerlink.com/

are characterized by strong absorption near 6150 Å (now attributed to Si II), SNe Ib lack this feature but instead show prominent He I lines, and SNe Ic have neither the Si II nor the He I lines. SNe Ia are believed to result from the thermonuclear disruption of carbon-oxygen white dwarfs, while SNe II come from core collapse in massive supergiant stars. The latter mechanism probably produces most SNe Ib/Ic as well, but the progenitor stars previously lost their outer layers of hydrogen or even helium.

It has long been recognized that SNe Ia may be very useful distance indicators for a number of reasons; see Branch & Tammann (1992), Branch (1998), and references therein. (1) They are exceedingly luminous, with peak M_B averaging -19.0 mag if $H_0 = 72$ km s^{-1} Mpc^{-1}. (2) "Normal" SNe Ia have small dispersion among their peak absolute magnitudes ($\sigma \lesssim 0.3$ mag). (3) Our understanding of the progenitors and explosion mechanism of SNe Ia is on a reasonably firm physical basis. (4) Little cosmic evolution is expected in the peak luminosities of SNe Ia, and it can be modeled. This makes SNe Ia superior to galaxies as distance indicators. (5) One can perform *local* tests of various possible complications and evolutionary effects by comparing nearby SNe Ia in different environments.

Research on SNe Ia in the 1990s has demonstrated their enormous potential as cosmological distance indicators. Although there are subtle effects that must indeed be taken into account, it appears that SNe Ia provide among the most accurate values of H_0, q_0 (the deceleration parameter), Ω_M (the matter density), and Ω_Λ [the cosmological constant, $\Lambda c^2/(3H_0^2)$].

There have been two major teams involved in the systematic investigation of high-redshift SNe Ia for cosmological purposes. The "Supernova Cosmology Project" (SCP) is led by Saul Perlmutter of the Lawrence Berkeley Laboratory, while the "High-Z Supernova Search Team" (HZT) is led by Brian Schmidt of the Mt. Stromlo and Siding Springs Observatories. I have been privileged to work with both teams (see Filippenko 2001 for a personal account), but my primary allegiance is now with the HZT.

2 Homogeneity and Heterogeneity

Until the mid-1990s, the traditional way in which SNe Ia were used for cosmological distance determinations was to assume that they are perfect "standard candles" and to compare their observed peak brightness with those of SNe Ia in galaxies whose distances had been independently determined (e.g., with Cepheid variables). The rationale was that SNe Ia exhibit relatively little scatter in their peak blue luminosity ($\sigma_B \approx 0.4$–0.5 mag; Branch & Miller 1993), and even less if "peculiar" or highly reddened objects were eliminated from consideration by using a color cut. Moreover, the optical spectra of SNe Ia are usually rather homogeneous, if care is taken to compare objects at similar times relative to maximum brightness (Riess et al. 1997, and references therein). Over 80% of all SNe Ia discovered through the early 1990s were "normal" (Branch, Fisher, & Nugent 1993).

From a Hubble diagram constructed with unreddened, moderately distant SNe Ia ($z \lesssim 0.1$) for which peculiar motions are small and relative distances (given by ratios of redshifts) are accurate, Vaughan et al. (1995) find that

$$\langle M_B(\mathrm{max})\rangle = (-19.74 \pm 0.06) + 5 \log (H_0/50) \text{ mag.} \quad (1)$$

In a series of papers, Sandage et al. (1996) and Saha et al. (1997) combine similar relations with *Hubble Space Telescope (HST)* Cepheid distances to the host galaxies of seven SNe Ia to derive $H_0 = 57 \pm 4$ km s^{-1} Mpc^{-1}.

Over the past two decades it has become clear, however, that SNe Ia do *not* constitute a perfectly homogeneous subclass (e.g., Filippenko 1997a,b). In retrospect this should have been obvious: the Hubble diagram for SNe Ia exhibits scatter larger than the photometric errors, the dispersion actually *rises* when reddening corrections are applied (under the assumption that all SNe Ia have uniform, very blue intrinsic colors at maximum; van den Bergh & Pazder 1992; Sandage & Tammann 1993), and there are some significant outliers whose anomalous magnitudes cannot be explained by extinction alone.

Spectroscopic and photometric peculiarities have been noted with increasing frequency in well-observed SNe Ia. A striking case is SN 1991T; its pre-maximum spectrum did not exhibit Si II or Ca II absorption lines, yet two months past maximum brightness the spectrum was nearly indistinguishable from that of a classical SN Ia (Filippenko et al. 1992b; Phillips et al. 1993). The light curves of SN 1991T were slightly broader than the SN Ia template curves, and the object was probably somewhat more luminous than average at maximum. Another well-observed, peculiar SNe Ia is SN 1991bg (Filippenko et al. 1992a; Leibundgut et al. 1993; Turatto et al. 1996). At maximum brightness it was subluminous by 1.6 mag in V and 2.5 mag in B, its colors were intrinsically red, and its spectrum was peculiar (with a deep absorption trough due to Ti II). Moreover, the decline from maximum was very steep, the I-band light curve did not exhibit a secondary maximum like normal SNe Ia, and the velocity of the ejecta was unusually low. The photometric heterogeneity among SNe Ia is well demonstrated by Suntzeff (1996) with objects having excellent $BVRI$ light curves.

3 Cosmological Uses: Low Redshifts

Although SNe Ia can no longer be considered perfect "standard candles," they are still exceptionally useful for cosmological distance determinations. Excluding those of low luminosity (which are hard to find, especially at large distances), most of the nearby SNe Ia that had been discovered through the early 1990s were *nearly* standard (Branch et al. 1993; but see Li et al. 2001b for more recent evidence of a higher intrinsic peculiarity rate). Also, after many tenuous suggestions (e.g., Pskovskii 1977, 1984; Branch 1981), Phillips (1993) found convincing evidence for a correlation between light curve shape and the luminosity at maximum brightness by quantifying the photometric differences among a set of nine well-observed SNe Ia, using a parameter [$\Delta m_{15}(B)$] that measures the total drop (in B magnitudes) from B-band maximum to $t = 15$ days later. In all cases the host galaxies of his SNe Ia have accurate relative distances from surface brightness fluctuations or from the Tully-Fisher relation. The intrinsically bright SNe Ia clearly decline more slowly than dim ones, but the correlation is stronger in B than in V or I.

Using SNe Ia discovered during the Calán/Tololo survey ($z \lesssim 0.1$), Hamuy et al. (1995, 1996b) refine the Phillips (1993) correlation between peak luminosity and $\Delta m_{15}(B)$. Apparently the slope is steep only at low luminosities; thus, objects such as SN 1991bg skew the slope of the best-fitting single straight line. Hamuy et al. reduce the scatter in the Hubble diagram of normal, unreddened SNe Ia to only 0.17 mag in B and 0.14 mag in V; see also Tripp (1997). Yet another parameterization is the "stretch" method of Perlmutter et al. (1997) and Goldhaber et al. (2001): the B-band light curves of SNe Ia appear nearly identical when expanded or contracted temporally by a factor $(1 + s)$, where the value of s varies among objects. In a similar but distinct effort, Riess, Press, & Kirshner (1995) show that the luminosity of SNe Ia correlates with the detailed *shape* of the overall light curve.

By using light curve shapes measured through several different filters, Riess, Press, & Kirshner (1996a) extend their analysis and objectively eliminate the effects of interstellar extinction: a SN Ia that has an unusually red $B - V$ color at maximum brightness is assumed to be *intrinsically* subluminous if its light curves rise and decline quickly, or of normal luminosity but significantly *reddened* if its light curves rise and decline more slowly. With a set of 20 SNe Ia consisting of the Calán/Tololo sample and their own objects, they show that the dispersion decreases from 0.52 mag to 0.12 mag after application of this "multi-color light curve shape" (MLCS) method. The results from an expanded set of nearly 50 SNe Ia indicate that the dispersion decreases from 0.44 mag to 0.15 mag (Fig. 1). The resulting Hubble constant is 65 ± 2 (statistical) ± 7 (systematic) km s^{-1} Mpc^{-1}, with an additional systematic and zeropoint uncertainty of ± 5 km s^{-1} Mpc^{-1}. (Re-calibrations of the Cepheid distance scale, and other recent refinements, lead to a best estimate of $H_0 = 72 \pm 8$ km s^{-1} Mpc^{-1}, where the error bar includes both statistical and systematic uncertainties; Parodi et al. 2000; Freeman et al. 2001.) Riess et al. (1996a) also show that the Hubble flow is remarkably linear; indeed, SNe Ia now constitute the best evidence for linearity. Finally, they argue that the dust affecting SNe Ia is *not* of circumstellar origin, and show quantitatively that the extinction curve in external galaxies typically does not differ from that in the Milky Way (cf. Branch & Tammann 1992, but see Tripp 1998).

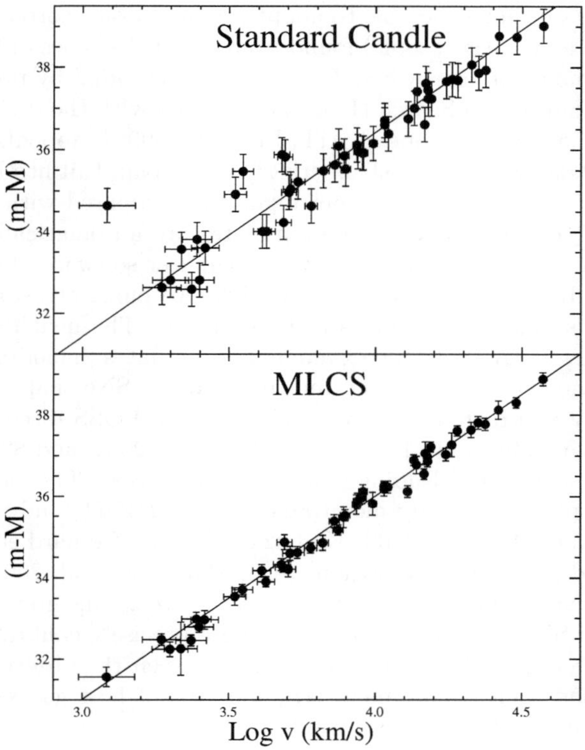

Fig. 1. Hubble diagrams for SNe Ia (A. G. Riess 2001, private communication) with velocities (km s^{-1}) in the *COBE* rest frame on the Cepheid distance scale. The ordinate shows distance modulus, $m - M$ (mag). *Top:* The objects are assumed to be *standard candles* and there is no correction for extinction; the result is $\sigma = 0.42$ mag and $H_0 = 58 \pm 8$ km s^{-1} Mpc^{-1}. *Bottom:* The same objects, after correction for reddening and intrinsic differences in luminosity. The result is $\sigma = 0.15$ mag and $H_0 = 65 \pm 2$ (statistical) km s^{-1} Mpc^{-1}, subject to changes in the zeropoint of the Cepheid distance scale. (Indeed, the latest results with SNe Ia favor $H_0 = 72$ km s^{-1} Mpc^{-1}.)

The advantage of systematically correcting the luminosities of SNe Ia at high redshifts rather than trying to isolate "normal" ones seems clear in view of evidence that the luminosity of SNe Ia may be a function of stellar population. If the most luminous SNe Ia occur in young stellar populations (e.g., Hamuy et al. 1996a, 2000; Branch, Baron, & Romanishin 1996; Ivanov, Hamuy, & Pinto 2000), then we might expect the mean peak luminosity of high-z SNe Ia to differ from that of a local sample. Alternatively, the use of Cepheids (Population I objects) to calibrate local SNe Ia can lead to a zeropoint that is too luminous. On the other hand, as long as the physics of SNe Ia is essentially the same in young stellar populations locally and at high redshift, we should be able to adopt the luminosity correction methods (photometric and spectroscopic) found from detailed studies of low-z SNe Ia.

Large numbers of nearby SNe Ia are now being found by my team's Lick Observatory Supernova Search (LOSS) conducted with the 0.76-m Katzman Automatic Imaging Telescope (KAIT; Li et al. 2000; Filippenko et al. 2001; Filippenko 2003; see http://astro.berkeley.edu/~bait/kait.html). CCD images are taken of $\sim$ 1000 galaxies per night and compared with KAIT "template images" obtained earlier; the templates are automatically subtracted from the new images and analyzed with computer software. The system reobserves the best candidates the same night, to eliminate star-like cosmic rays, asteroids, and other sources of false alarms. The next day, undergraduate students at UC Berkeley examine all candidates, including weak ones, and they glance at all subtracted images to locate SNe that might be near bright, poorly subtracted stars or galactic nuclei. LOSS discovered 20 SNe (of all types) in 1998, 40 in 1999, 38 in 2000, 68 in 2001, and 82 in 2002, making it by far the world's leading search for nearby SNe. The most important objects were photometrically monitored through $BVRI$ (and sometimes U) filters (e.g., Li et al. 2001a, 2003; Modjaz et al. 2001; Leonard et al. 2002a,b), and unfiltered follow-up observations (e.g., Matheson et al. 2001) were made of most of them during the course of the SN search. This growing sample of well-observed SNe Ia should allow us to more precisely calibrate the MLCS method, as well as to look for correlations between the observed properties of the SNe and their environment (Hubble type of host galaxy, metallicity, stellar population, etc.).

4 Cosmological Uses: High Redshifts

These same techniques can be applied to construct a Hubble diagram with high-redshift SNe Ia, from which the value of $q_0 = (\Omega_M/2) - \Omega_\Lambda$ can be determined. With enough objects spanning a range of redshifts, we can measure Ω_M and Ω_Λ independently (e.g., Goobar & Perlmutter 1995). Contours of peak apparent R-band magnitude for SNe Ia at two redshifts have different slopes in the Ω_M–Ω_Λ plane, and the regions of intersection provide the answers we seek.

4.1 The Search

Based on the pioneering work of Norgaard-Nielsen et al. (1989), whose goal was to find SNe in moderate-redshift clusters of galaxies, the SCP (Perlmutter et al. 1995a, 1997) and the HZT (Schmidt et al. 1998) devised a strategy that almost guarantees the discovery of many faint, distant SNe Ia "on demand," during a predetermined set of nights. This "batch" approach to studying distant SNe allows follow-up spectroscopy and photometry to be scheduled in

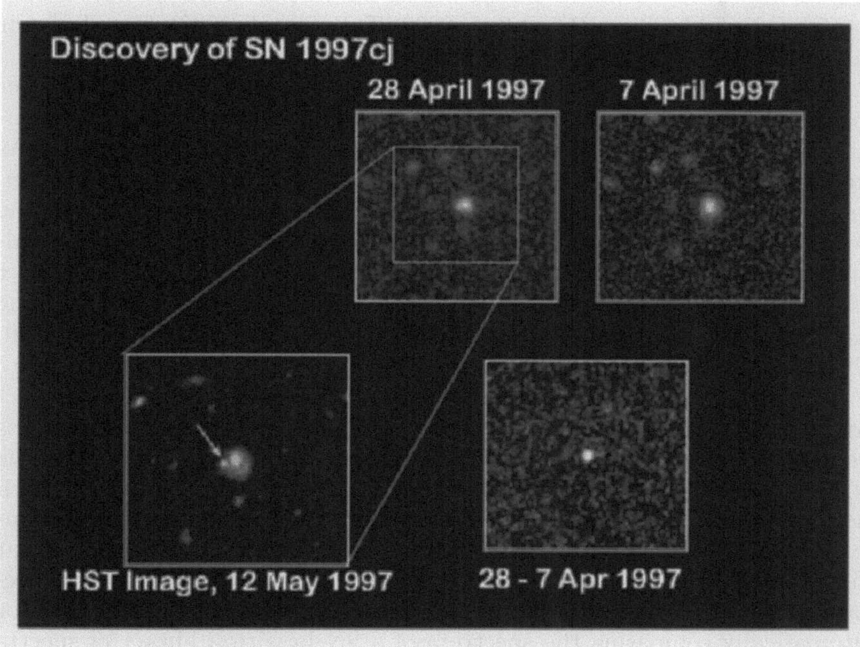

Fig. 2. Discovery image of SN 1997cj (28 April 1997), along with the template image and an *HST* image obtained subsequently. The net (subtracted) image is also shown.

advance, resulting in a systematic study not possible with random discoveries. Most of the searched fields are equatorial, permitting follow-up from both hemispheres. The SCP was the first group to convincingly demonstrate the ability to find SNe in batches.

Our approach is simple in principle; see Schmidt et al. (1998) for details, and for a description of our first high-redshift SN Ia (SN 1995K). Pairs of first-epoch images are obtained with wide-field cameras on large telescopes (e.g., the Big Throughput Camera on the CTIO 4-m Blanco telescope) during the nights around new moon, followed by second-epoch images 3–4 weeks later. (Pairs of images permit removal of cosmic rays, asteroids, and distant Kuiper-belt objects.) These are compared immediately using well-tested software, and new SN candidates are identified in the second-epoch images (Fig. 2). Spectra are obtained as soon as possible after discovery to verify that the objects are SNe Ia and determine their redshifts. Each team has already found over nearly 200 SNe in concentrated batches, as reported in numerous *IAU Circulars* (e.g., Perlmutter et al. 1995b, 11 SNe with $0.16 \lesssim z \lesssim 0.65$; Suntzeff et al. 1996, 17 SNe with $0.09 \lesssim z \lesssim 0.84$). The observed SN Ia rate at $z \approx 0.5$ is consistent with the low-z SN Ia rate together with plausible star-formation histories (Pain et al. 2002; Tonry et al. 2003), but the error bars on the high-z rate are still quite large.

Intensive photometry of the SNe Ia commences within a few days after procurement of the second-epoch images; it is continued throughout the ensuing and subsequent dark runs. In a few cases *HST* images are obtained. As expected, most of the discoveries are *on the rise or near maximum brightness*. When possible, the SNe are observed in filters that closely match the redshifted B and V bands; this way, the K-corrections become only a second-order effect (Kim, Goobar, & Perlmutter 1996; Nugent, Kim, & Perlmutter 2002). We try to obtain excellent multi-color light curves, so that reddening and luminosity corrections can be applied (Riess et al. 1996a; Hamuy et al. 1996a,b).

Although SNe in the magnitude range 22–22.5 can sometimes be spectroscopically confirmed with 4-m class telescopes, the signal-to-noise ratios are low, even after several hours of integration. Certainly Keck, Gemini, the VLT, or Magellan are required for the fainter objects (22.5–24.5 mag). With the largest telescopes, not only can we rapidly confirm a substantial number of candidate SNe, but we can search for peculiarities in the spectra that might indicate evolution of SNe Ia with redshift. Moreover, high-quality spectra allow us to measure the age of a SN: we have developed a method for automatically comparing the spectrum of a SN Ia with a library of spectra from many different epochs in the development of SNe Ia (Riess et al. 1997). Our technique also has great practical utility at the telescope: we can determine the age of a SN "on the fly," within half an hour after obtaining its spectrum. This allows us to decide rapidly which SNe are best for subsequent photometric follow-up, and we immediately alert our collaborators elsewhere.

4.2 Results

First, we note that the light curves of high-redshift SNe Ia are broader than those of nearby SNe Ia; the initial indications (Leibundgut et al. 1996; Goldhaber et al. 1997), based on small numbers of SNe Ia, are amply confirmed with the larger samples (Goldhaber et al. 2001). Quantitatively, the amount by which the light curves are "stretched" is consistent with a factor of $1+z$, as expected if redshifts are produced by the expansion of space rather than by "tired light" and other non-expansion hypotheses for the redshifts of objects at cosmological distances. [For non-standard cosmological interpretations of the SN Ia data, see Narlikar & Arp (1997) and Hoyle, Burbidge, & Narlikar (2000).] We also demonstrate this *spectroscopically* at the 2σ confidence level for a single object: the spectrum of SN 1996bj ($z = 0.57$) evolved more slowly than those of nearby SNe Ia, by a factor consistent with $1 + z$ (Riess et al. 1997). More recently, we have used observations of SN 1997ex ($z = 0.36$) at three epochs to conclusively verify the effects of time dilation: temporal changes in the spectra are slower than those of nearby SNe Ia by roughly the expected factor of 1.36. Although one might be able to argue that something other than universal expansion could be the cause of the apparent stretching of SN Ia light curves at high redshifts, it is much more difficult to attribute apparently slower evolution of spectral details to an unknown effect.

The formal value of Ω_M derived from SNe Ia has changed with time. The SCP published the first result (Perlmutter et al. 1995a), based on a single object, SN 1992bi at $z = 0.458$: $\Omega_M = 0.2 \pm 0.6 \pm 1.1$ (assuming that $\Omega_\Lambda = 0$). The SCP's analysis of their first seven objects (Perlmutter et al. 1997) suggested a much larger value of $\Omega_M = 0.88 \pm 0.6$ (if $\Omega_\Lambda = 0$) or $\Omega_M = 0.94 \pm 0.3$ (if $\Omega_{\text{total}} = 1$). Such a high-density universe seemed at odds with other, independent measurements of Ω_M. However, with the subsequent inclusion of just one more object, SN 1997ap at $z = 0.83$ (the highest known for a SN Ia at the time; Perlmutter et al. 1998), their estimates were revised back down to $\Omega_M = 0.2 \pm 0.4$ if $\Omega_\Lambda = 0$, and $\Omega_M = 0.6 \pm 0.2$ if $\Omega_{\text{total}} = 1$; the apparent brightness of SN 1997ap had been precisely measured with *HST*, so it substantially affected the best fits.

Meanwhile, the HZT published (Garnavich et al. 1998a) an analysis of four objects (three of them observed with *HST*), including SN 1997ck at $z = 0.97$, at that time a redshift record, although they cannot be absolutely certain that the object was a SN Ia because the spectrum is too poor. From these data, the HZT derived that $\Omega_M = -0.1 \pm 0.5$ (assuming $\Omega_\Lambda = 0$) and $\Omega_M = 0.35 \pm 0.3$ (assuming $\Omega_{\text{total}} = 1$), inconsistent with the high Ω_M initially found by Perlmutter et al. (1997) but consistent with the revised estimate in Perlmutter et al. (1998). An independent analysis of 10 SNe Ia using the "snapshot" distance method (with which conclusions are drawn from sparsely observed SNe Ia) gave quantitatively similar conclusions (Riess et al. 1998a). However, none of these early data sets carried the statistical discriminating power to detect cosmic acceleration.

The SCP's next results were announced at the 1998 January AAS meeting in Washington, DC. A press conference was scheduled, with the stated purpose of presenting and discussing the then-current evidence for a low-Ω_M universe as published by Perlmutter et al. (1998; SCP) and Garnavich et al. (1998a; HZT). When showing the SCP's Hubble diagram for SNe Ia, however, Perlmutter also pointed out tentative evidence for *acceleration*! He said that the conclusion was uncertain, and that the data were equally consistent with no acceleration; the systematic errors had not yet been adequately assessed. Essentially the same conclusion was given by the SCP in their talks at a conference on dark matter, near Los Angeles, in February 1998 (Goldhaber & Perlmutter 1998). Although it chose not to reveal them at the same 1998 January AAS meeting, the HZT already had similar, tentative evidence for acceleration in their own SN Ia data set. The HZT continued to perform numerous checks of their data analysis and interpretation, including fairly thorough consideration of various possible systematic effects. Unable to find any significant problems, even with the possible systematic effects, the HZT reported detection of a *nonzero* value for Ω_Λ (based on 16 high-z SNe Ia) at the Los Angeles dark matter conference in February 1998 (Filippenko &

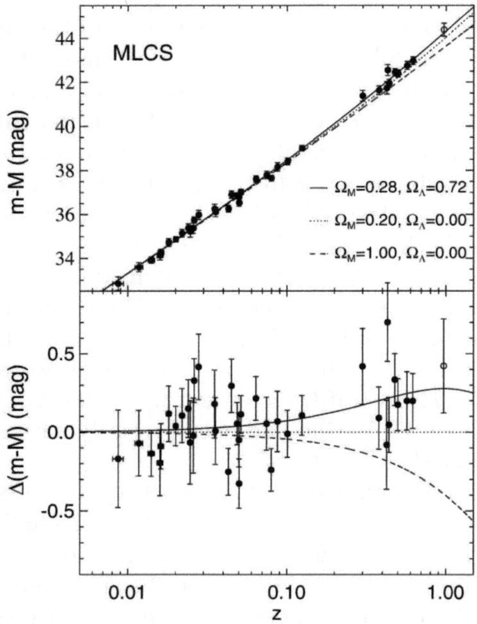

Fig. 3. The upper panel shows the Hubble diagram for the low-z and high-z HZT SN Ia sample with MLCS distances; see Riess et al. (1998b). Overplotted are three world models: "low" and "high" Ω_M with $\Omega_\Lambda = 0$, and the best fit for a flat universe ($\Omega_M = 0.28$, $\Omega_\Lambda = 0.72$). The bottom panel shows the difference between data and models from the $\Omega_M = 0.20$, $\Omega_\Lambda = 0$ prediction. Only the 10 best-observed high-z SNe Ia are shown. The average difference between the data and the $\Omega_M = 0.20$, $\Omega_\Lambda = 0$ prediction is ~ 0.25 mag.

Riess 1998), and soon thereafter submitted a formal paper that was published in September 1998 (Riess et al. 1998b). Their original Hubble diagram for the 10 best-observed high-z SNe Ia is given in Fig. 3. With the MLCS method applied to the full set of 16 SNe Ia, the HZT's formal results were $\Omega_M = 0.24 \pm 0.10$ if $\Omega_{\rm total} = 1$, or $\Omega_M = -0.35 \pm 0.18$ (unphysical) if $\Omega_\Lambda = 0$. If one demanded that $\Omega_M = 0.2$, then the best value for Ω_Λ was 0.66 ± 0.21. These conclusions did not change significantly when only the 10 best-observed SNe Ia were used (Fig. 3; $\Omega_M = 0.28 \pm 0.10$ if $\Omega_{\rm total} = 1$).

Another important constraint on the cosmological parameters could be obtained from measurements of the angular scale of the first acoustic peak of the CMB (e.g., Zaldarriaga, Spergel, & Seljak 1997; Eisenstein, Hu, & Tegmark 1998); the SN Ia and CMB techniques provide nearly complementary information. A stunning result was already available by mid-1998 from existing measurements (e.g., Hancock et al. 1998; Lineweaver & Barbosa 1998): the HZT's analysis of the SN Ia data in Riess et al. (1998b) demonstrated that $\Omega_M + \Omega_\Lambda = 0.94 \pm 0.26$ (Fig. 4), when the SN and CMB constraints

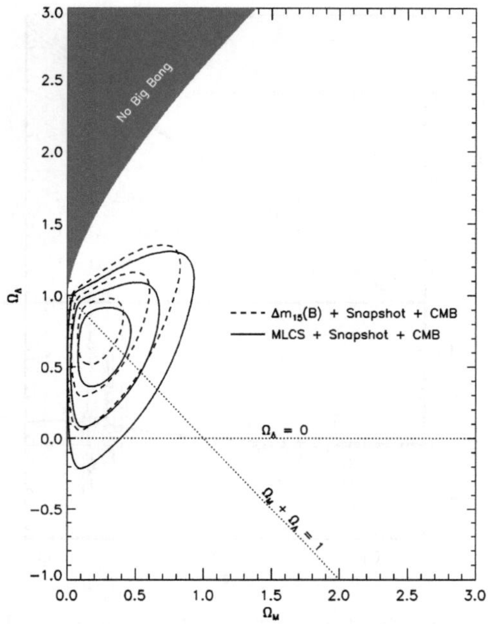

Fig. 4. The HZT's combined constraints from SNe Ia (left) and the position of the first acoustic peak of the cosmic microwave background (CMB) angular power spectrum, based on data available in mid-1998; see Garnavich et al. (1998b). The contours mark the 68.3%, 95.4%, and 99.7% enclosed probability regions. Solid curves correspond to results from the MLCS method; dotted ones are from the $\Delta m_{15}(B)$ method; all 16 SNe Ia in Riess et al. (1998b) were used.

were combined (Garnavich et al. 1998b; see also Lineweaver 1998, Efstathiou et al. 1999, and others).

Somewhat later (June 1999), the SCP published almost identical results, implying an accelerating expansion of the Universe, based on an essentially independent set of 42 high-z SNe Ia (Perlmutter et al. 1999). Their data, together with those of the HZT, are shown in Fig. 5, and the corresponding confidence contours in the Ω_Λ vs. Ω_M plane are given in Fig. 6. This incredible agreement suggested that neither group had made a large, simple blunder; if the result was wrong, the reason must be subtle. Had there been only one team working in this area, it is likely that far fewer astronomers and physicists throughout the world would have taken the result seriously.

Moreover, already in 1998–1999 there was tentative evidence that the "dark energy" driving the accelerated expansion was indeed consistent with the cosmological constant, Λ. If Λ dominates, then the equation of state of the dark energy should have an index $w = -1$, where the pressure (P) and density (ρ) are related according to $w = P/(\rho c^2)$. Garnavich et al. (1998b) and Perlmutter et al. (1999) were able to set an interesting limit, $w \lesssim -0.60$

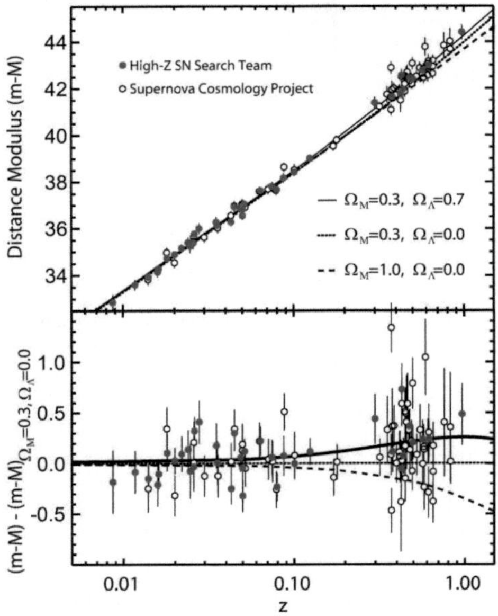

Fig. 5. As in Fig. 3, but this time including both the HZT (Riess et al. 1998b) and SCP (Perlmutter et al. 1999) samples of low-redshift and high-redshift SNe Ia. Overplotted are three world models: $\Omega_M = 0.3$ and 1.0 with $\Omega_\Lambda = 0$, and a flat universe ($\Omega_{\text{total}} = 1.0$) with $\Omega_\Lambda = 0.7$. The bottom panel shows the difference between data and models from the $\Omega_M = 0.3$, $\Omega_\Lambda = 0$ prediction.

at the 95% confidence level. However, more high-quality data at $z \approx 0.5$ are needed to narrow the allowed range, in order to test other proposed candidates for dark energy such as various forms of "quintessence" (e.g., Caldwell, Davé, & Steinhardt 1998).

Although the CMB results appeared reasonably persuasive in 1998–1999, one could argue that fluctuations on different scales had been measured with different instruments, and that suble systematic effects might lead to erroneous conclusions. These fears were dispelled only 1–2 years later, when the more accurate and precise results of the BOOMERANG collaboration were announced (de Bernardis et al. 2000, 2002). Shortly thereafter the MAXIMA collaboration distributed their very similar findings (Hanany et al. 2000; Balbi et al. 2000; Netterfield et al. 2002; see also the TOCO, DASI, and many other measurements). Figure 6 illustrates that the CMB measurements tightly constrain Ω_{total} to be close to unity; we appear to live in a flat universe, in agreement with most inflationary models for the early Universe! Combined with the SN Ia results, the evidence for nonzero Ω_Λ was fairly strong. Making the argument even more compelling was the fact that various studies of clusters of galaxies (see summary by Bahcall et al. 1999) showed that $\Omega_M \approx 0.3$, consistent with the results in Figs. 4 and 6. Thus, a

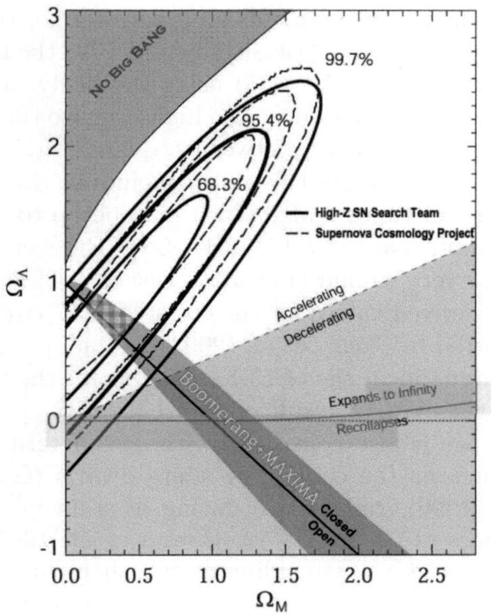

Fig. 6. The combined constraints from SNe Ia (see Fig. 5) and the position of the first acoustic peak of the CMB angular power spectrum, based on BOOMERANG and MAXIMA data. The contours mark the 68.3%, 95.4%, and 99.7% enclosed probability regions determined from the SNe Ia. According to the CMB, $\Omega_{\text{total}} \approx 1.0$.

"concordance cosmology" had emerged: $\Omega_M \approx 0.3$, $\Omega_\Lambda \approx 0.7$ — consistent with what had been suspected some years earlier by Ostriker & Steinhardt (1995; see also Carroll, Press, & Turner 1992).

Yet another piece of evidence for a nonzero value of Λ was provided by the Two-Degree Field Galaxy Redshift Survey (2dFGRS; Peacock et al. 2001; Percival et al. 2001; Efstathiou et al. 2002). Combined with the CMB maps, their results are inconsistent with a universe dominated by gravitating dark matter. Again, the implication is that about 70% of the mass-energy density of the Universe consists of some sort of dark energy whose gravitational effect is repulsive. Very recently, results from the *Wilkinson Microwave Anisotropy Probe (WMAP)* appeared; together with the 2dFGRS constraints, they confirm and refine the concordance cosmology ($\Omega_M = 0.27$, $\Omega_\Lambda = 0.73$, $\Omega_{\text{baryon}} = 0.044$, $H_0 = 71 \pm 4$ km s^{-1} Mpc^{-1}; Spergel et al. 2003).

The dynamical age of the Universe can be calculated from the cosmological parameters. In an empty Universe with no cosmological constant, the dynamical age is simply the "Hubble time" t_0 (i.e., the inverse of the Hubble constant); there is no deceleration. In the late-1990s, SNe Ia gave $H_0 = 65 \pm 7$ km s^{-1} Mpc^{-1}, and a Hubble time of 15.1 ± 1.6 Gyr. For a more complex cosmology, integrating the velocity of the expansion from the current epoch

($z = 0$) to the beginning ($z = \infty$) yields an expression for the dynamical age. As shown in detail by Riess et al. (1998b), by mid-1998 the HZT had obtained a value of $14.2^{+1.0}_{-0.8}$ Gyr (with $H_0 = 65$) using the likely range for $(\Omega_M, \Omega_\Lambda)$ that they measured. (The precision was so high because their experiment was sensitive to roughly the *difference* between Ω_M and Ω_Λ, and the dynamical age also varies in approximately this way.) Including the *systematic* uncertainty of the Cepheid distance scale, which may be up to 10%, a reasonable estimate of the dynamical age was 14.2 ± 1.7 Gyr (Riess et al. 1998b). Again, the SCP's result was very similar (Perlmutter et al. 1999), since it was based on nearly the same derived values for the cosmological parameters. The most recent results, reported by Tonry et al. (2003) and adopting $H_0 = 72 \pm 8$ km s^{-1} Mpc^{-1}, give a dynamical age of 13.1 ± 1.5 Gyr for the Universe — again, in agreement with the *WMAP* result of 13.7 ± 0.2 Gyr.

This expansion age is also consistent with ages determined from various other techniques such as the cooling of white dwarfs (Galactic disk > 9.5 Gyr; Oswalt et al. 1996), radioactive dating of stars via the thorium and europium abundances (15.2 ± 3.7 Gyr; Cowan et al. 1997), and studies of globular clusters (10–15 Gyr, depending on whether *Hipparcos* parallaxes of Cepheids are adopted; Gratton et al. 1997; Chaboyer et al. 1998). The ages of the oldest stars no longer seem to exceed the expansion age of the Universe; the long-standing "age crisis" has evidently been resolved.

5 Discussion

Although the convergence of different methods on the same answer is reassuring, and suggests that the concordance cosmology is correct, it is important to vigorously test each method to make sure it is not leading us astray. Moreover, only through such detailed studies will the accuracy and precision of the methods improve, allowing us to eventually set better constraints on the equation of state parameter, w. Here I discuss the systematic effects that could adversely affect the SN Ia results.

High-redshift SNe Ia are observed to be dimmer than expected in an empty Universe (i.e., $\Omega_M = 0$) with no cosmological constant. At $z \approx 0.5$, where the SN Ia observations have their greatest leverage on Λ, the difference in apparent magnitude between an $\Omega_M = 0.3$ ($\Omega_\Lambda = 0$) universe and a flat universe with $\Omega_\Lambda = 0.7$ is only about 0.25 mag. Thus, we need to find out if chemical abundances, stellar populations, selection bias, gravitational lensing, or grey dust can have an effect this large. Although both the HZT and SCP had considered many of these potential systematic effects in their original discovery papers (Riess et al. 1998b; Perlmutter et al. 1999), and had shown with reasonable confidence that obvious ones were not greatly affecting their conclusions, if was of course possible that they were wrong, and that the data were being misinterpreted.

5.1 Evolution

Perhaps the most obvious possible culprit is *evolution* of SNe Ia over cosmic time, due to changes in metallicity, progenitor mass, or some other factor. If the peak luminosity of SNe Ia were lower at high redshift, then the case for $\Omega_\Lambda > 0$ would weaken. Conversely, if the distant explosions are more powerful, then the case for acceleration strengthens. Theorists are not yet sure what the sign of the effect will be, if it is present at all; different assumptions lead to different conclusions (Höflich et al. 1998; Umeda et al. 1999; Nomoto et al. 2000; Yungelson & Livio 2000).

Of course, it is extremely difficult, if not effectively impossible, to obtain an accurate, independent measure of the peak luminosity of high-z SNe Ia, and hence to directly test for luminosity evolution. However, we can more easily determine whether *other* observable properties of low-z and high-z SNe Ia differ. If they are all the same, it is more probable that the peak luminosity is constant as well — but if they differ, then the peak luminosity might also be affected (e.g., Höflich et al. 1998). Drell, Loredo, & Wasserman (2000), for example, argue that there are reasons to suspect evolution, because the average properties of existing samples of high-z and low-z SNe Ia seem to differ (e.g., the high-z SNe Ia are more uniform).

The local sample of SNe Ia displays a weak correlation between light curve shape (or peak luminosity) and host galaxy type, in the sense that the most luminous SNe Ia with the broadest light curves only occur in late-type galaxies. Both early-type and late-type galaxies provide hosts for dimmer SNe Ia with narrower light curves (Hamuy et al. 1996a). The mean luminosity difference for SNe Ia in late-type and early-type galaxies is ~ 0.3 mag. In addition, the SN Ia rate per unit luminosity is almost twice as high in late-type galaxies as in early-type galaxies at the present epoch (Cappellaro et al. 1997). These results may indicate an evolution of SNe Ia with progenitor age. Possibly relevant physical parameters are the mass, metallicity, and C/O ratio of the progenitor (Höflich et al. 1998).

We expect that the relation between light curve shape and peak luminosity that applies to the range of stellar populations and progenitor ages encountered in the late-type and early-type hosts in our nearby sample should also be applicable to the range we encounter in our distant sample. In fact, the range of age for SN Ia progenitors in the nearby sample is likely to be *larger* than the change in mean progenitor age over the 4–6 Gyr lookback time to the high-z sample. Thus, to first order at least, our local sample should correct the distances for progenitor or age effects.

We can place empirical constraints on the effect that a change in the progenitor age would have on our SN Ia distances by comparing subsamples of low-redshift SNe Ia believed to arise from old and young progenitors. In the nearby sample, the mean difference between the distances for the early-type hosts (8 SNe Ia) and late-type hosts (19 SNe Ia), at a given redshift, is 0.04 $\pm$ 0.07 mag from the MLCS method. This difference is consistent with zero.

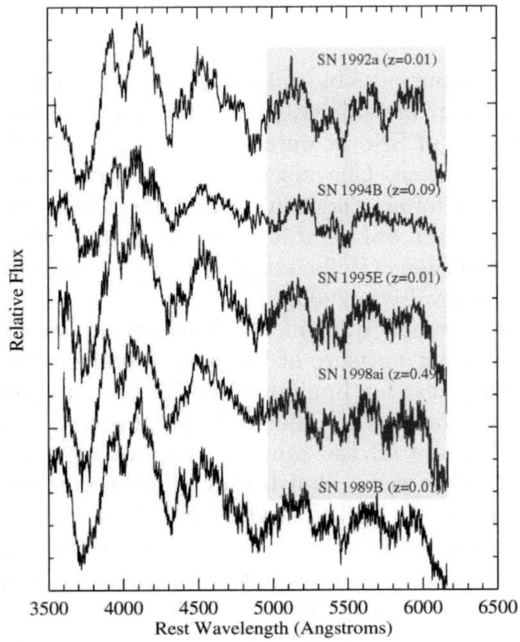

Fig. 7. Spectral comparison (in f_λ) of SN 1998ai ($z = 0.49$; Keck spectrum) with low-redshift ($z < 0.1$) SNe Ia at a similar age (~ 5 days before maximum brightness), from Riess et al. (1998b). The spectra of the low-redshift SNe Ia were resampled and convolved with Gaussian noise to match the quality of the spectrum of SN 1998ai. Overall, the agreement in the spectra is excellent, tentatively suggesting that distant SNe Ia are physically similar to nearby SNe Ia. SN 1994B ($z = 0.09$) differs the most from the others, and was included as a "decoy."

Even if the SN Ia progenitors evolved from one population at low redshift to the other at high redshift, we still would not explain the surplus in mean distance of 0.25 mag over the $\Omega_\Lambda = 0$ prediction. Moreover, in a major study of high-redshift SNe Ia as a function of galaxy morphology, the SCP found no clear differences (except for the amount of scatter; see Sect. 5.2) between the cosmological results obtained with SNe Ia in late-type and early-type galaxies (Sullivan et al. 2003).

It is also reassuring that initial comparisons of high-z SN Ia spectra appear remarkably similar to those observed at low redshift. For example, the spectral characteristics of SN 1998ai ($z = 0.49$) appear to be essentially indistinguishable from those of normal low-z SNe Ia; see Fig. 7. In fact, the most obviously discrepant spectrum in this figure is the second one from the top, that of SN 1994B ($z = 0.09$); it is intentionally included as a "decoy" that illustrates the degree to which even the spectra of nearby, relatively normal SNe Ia can vary. Nevertheless, it is important to note that a dispersion in lu-

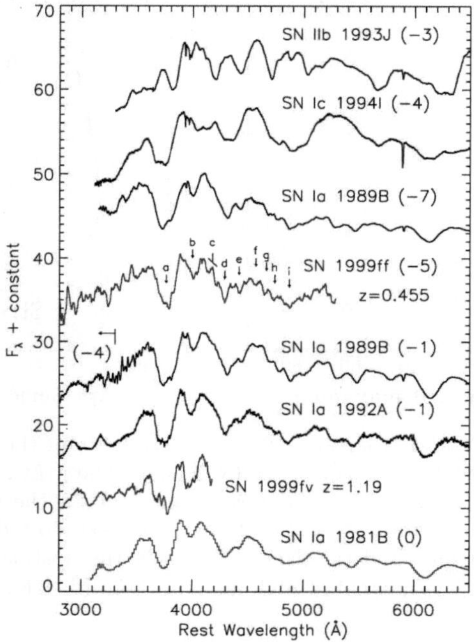

Fig. 8. Heavily smoothed spectra of two high-z SNe (SN 1999ff at $z = 0.455$ and SN 1999fv at $z = 1.19$; quite noisy below ~3500 Å) are presented along with several low-z SN Ia spectra (SNe 1989B, 1992A, and 1981B), a SN Ib spectrum (SN 1993J), and a SN Ic spectrum (SN 1994I); see Filippenko (1997) for a discussion of spectra of various types of SNe. The date of the spectra relative to B-band maximum is shown in parentheses after each object's name. Specific features seen in SN 1999ff and labeled with a letter are discussed by Coil et al. (2000). This comparison shows that the two high-z SNe are most likely SNe Ia.

minosity (perhaps 0.2 mag) exists even among the other, more normal SNe Ia shown in Fig. 7; thus, our spectra of SN 1998ai and other high-z SNe Ia are not yet sufficiently good for independent, *precise* determinations of peak luminosity from spectral features (Nugent et al. 1995). Many of them, however, are sufficient for ruling out other SN types (Fig. 8), or for identifying gross peculiarities such as those shown by SNe 1991T and 1991bg; see Coil et al. (2000).

We can help verify that the SNe at $z \approx 0.5$ being used for cosmology do not belong to a subluminous population of SNe Ia by examining restframe I-band light curves. Normal, nearby SNe Ia show a pronounced second maximum in the I band about a month after the first maximum and typically about 0.5 mag fainter (e.g., Ford et al. 1993; Suntzeff 1996). Subluminous SNe Ia, in contrast, do not show this second maximum, but rather follow a linear decline or show a muted second maximum (Filippenko et al. 1992a). As discussed by Riess et al. (2000), tentative evidence for the second maximum is seen from

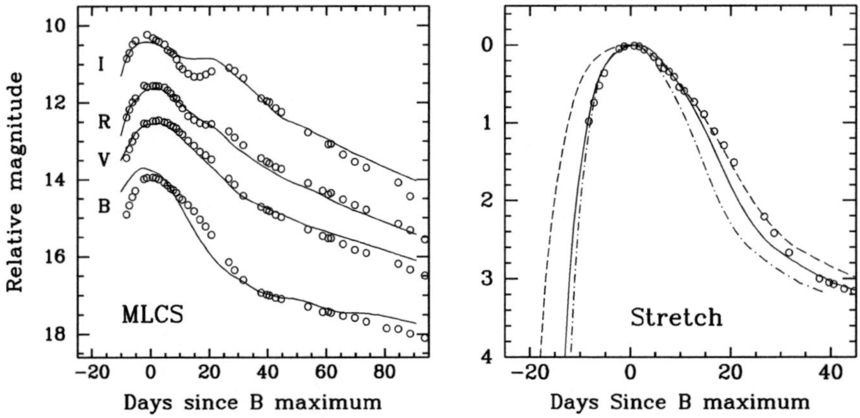

Fig. 9. The MLCS fit (Riess et al. 1998b; *left panel*) and the stretch method fit (Perlmutter et al. 1999; *right panel*) for SN 2000cx. The MLCS fit is the worst we had ever seen through 2000. For the stretch method fit, the solid line is the fit to all the data points from $t = -8$ to 32 days, the dash-dotted line uses only the premaximum datapoints, and the dashed line only the postmaximum datapoints. The three fits give very different stretch factors. From Li et al. (2001a).

the HZT's existing J-band (restframe I-band) data on SN 1999Q ($z = 0.46$); see Fig. 10. Additional tests with spectra and near-infrared light curves are currently being conducted.

Another way of using light curves to test for possible evolution of SNe Ia is to see whether the rise time (from explosion to maximum brightness) is the same for high-redshift and low-redshift SNe Ia; a difference might indicate that the peak luminosities are also different (Höflich et al. 1998). Riess et al. (1999c) measured the risetime of nearby SNe Ia, using data from KAIT, the Beijing Astronomical Observatory (BAO) SN search, and a few amateur astronomers. Though the exact value of the risetime is a function of peak luminosity, for typical low-redshift SNe Ia it is 20.0 ± 0.2 days. Riess et al. (1999b) pointed out that this differs by 5.8σ from the *preliminary* risetime of 17.5 ± 0.4 days reported in conferences by the SCP (Goldhaber et al. 1998a,b; Groom 1998). However, more thorough analyses of the SCP data (Aldering, Knop, & Nugent 2000; Goldhaber et al. 2001) show that the high-redshift uncertainty of ± 0.4 days that the SCP originally reported was much too small because it did not account for systematic effects. The revised discrepancy with the low-redshift risetime is about 2σ or less. Thus, the apparent difference in risetimes might be insignificant. Even if the difference is real, however, its relevance to the peak luminosity is unclear; the light curves may differ only in the first few days after the explosion, and this could be caused by small variations in conditions near the outer part of the exploding white dwarf that are inconsequential at the peak.

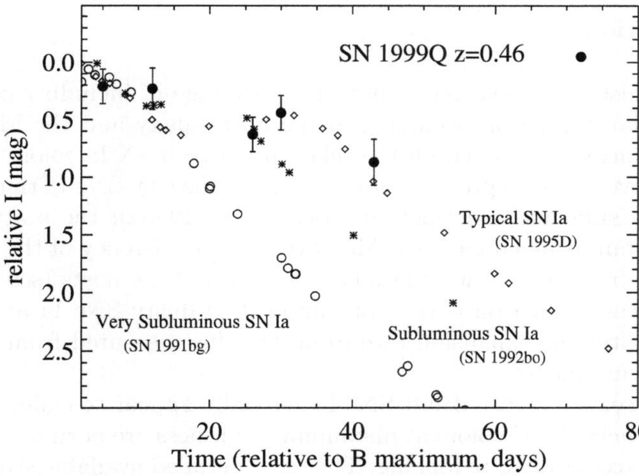

Fig. 10. Restframe I-band (observed J-band) light curve of SN 1999Q ($z = 0.46$, 5 solid points; Riess et al. 2000), and the I-band light curves of several nearby SNe Ia. Subluminous SNe Ia exhibit a less prominent second maximum than do normal SNe Ia.

Although there are no clear signs that cosmic evolution of SNe Ia seriously compromises our results, it is wise to remain vigilant for possible problems. At low redshifts, for example, we already know that *some* SNe Ia don't conform with the correlation between light curve shape and luminosity. SN 2000cx in the S0 galaxy NGC 524, for example, has light curves that cannot be fit well by any of the fitting techniques currently available (Li et al. 2001a; Filippenko 2003); see Fig. 9. Its late-time color is remarkably blue, inconsistent with the homogeneity described by Phillips et al. (1999). The spectral evolution of SN 2000cx is peculiar as well: the photosphere appears to have remained hot for a long time, and both iron-peak and intermediate-mass elements move at very high velocities. An even *more* peculiar object is SN 2002cx (Li et al. 2003; Filippenko 2003). It is spectroscopically bizarre, with extremely low expansion velocities and almost no evidence for intermediate-mass elements. The nebular phase was reached incredibly soon after maximum brightness, despite the low velocity of the ejecta, suggesting that the ejected mass is small. SN 2002cx was subluminous by ~ 2 mag at all optical wavelengths relative to normal SNe Ia, despite the early-time spectroscopic resemblance to the somewhat overluminous SN 1991T. The R-band and I-band light curves of SN 2002cx are completely unlike those of normal SNe Ia. No existing theoretical model successfully explains all observed aspects of SN 2002cx. If there are more strange beasts like SNe 2000cx and 2002cx at high redshifts than at low redshifts, systematic errors may creep into the analysis of high-z SNe Ia.

5.2 Extinction

Our SN Ia distances have the important advantage of including corrections for interstellar extinction occurring in the host galaxy and the Milky Way. Extinction corrections based on the relation between SN Ia colors and luminosity improve distance precision for a sample of nearby SNe Ia that includes objects with substantial extinction (Riess et al. 1996a); the scatter in the Hubble diagram is much reduced. Moreover, the consistency of the measured Hubble flow from SNe Ia with late-type and early-type hosts (see Sect. 5.1) shows that the extinction corrections applied to dusty SNe Ia at low redshift do not alter the expansion rate from its value measured from SNe Ia in low-dust environments.

In practice, the high-redshift SNe Ia generally appear to suffer very little extinction; their $B-V$ colors at maximum brightness are normal, suggesting little color excess due to reddening. The most detailed available study is that of the SCP (Sullivan et al. 2003): they found that the scatter in the Hubble diagram is minimal for SNe Ia in early-type host galaxies, but increases for SNe Ia in late-type galaxies. Moreover, on average the SNe in late-type galaxies are slightly fainter (by 0.14 ± 0.09 mag) than those in early-type galaxies. Finally, at peak brightness the colors of SNe Ia in late-type galaxies are marginally redder than those in early-type galaxies. Sullivan et al. (2003) conclude that extinction by dust in the host galaxies of SNe Ia is one of the major sources of scatter in the high-redshift Hubble diagram. By restricting their sample to SNe Ia in early-type host galaxies (presumably with minimal extinction), they obtain a very tight Hubble diagram that suggests a nonzero value for Ω_Λ at the 5σ confidence level, under the assumption that $\Omega_{\text{total}} = 1$. In the absence of this assumption, SNe Ia in early-type hosts still imply that $\Omega_\Lambda > 0$ at nearly the 98% confidence level. The results for Ω_Λ with SNe Ia in late-type galaxies are quantitatively similar, but statistically less secure because of the larger scatter.

Riess, Press, & Kirshner (1996b) found indications that the Galactic ratios between selective absorption and color excess are similar for host galaxies in the nearby ($z \leq 0.1$) Hubble flow. Yet, what if these ratios changed with lookback time (e.g., Aguirre 1999a)? Could an evolution in dust-grain size descending from ancestral interstellar "pebbles" at higher redshifts cause us to underestimate the extinction? Large dust grains would not imprint the reddening signature of typical interstellar extinction upon which our corrections necessarily rely.

However, viewing our SNe through such gray interstellar grains would also induce a *dispersion* in the derived distances. Using the results of Hatano, Branch, & Deaton (1998), Riess et al. (1998b) estimate that the expected dispersion would be 0.40 mag if the mean gray extinction were 0.25 mag (the value required to explain the measured MLCS distances without a cosmological constant). This is significantly larger than the 0.21 mag dispersion observed in the high-redshift MLCS distances. Furthermore, most of the observed scatter is already consistent with the estimated *statistical* errors, leaving little to be caused by gray extinction. Nevertheless, if we assumed that *all* of the observed scatter were due to gray extinction, the mean shift in the SN Ia distances would be only 0.05 mag. With the existing observations, it is difficult to rule out this modest amount of gray interstellar extinction.

Gray *intergalactic* extinction could dim the SNe without either telltale reddening or dispersion, if all lines of sight to a given redshift had a similar column density of absorbing material. The component of the intergalactic medium with such uniform coverage corresponds to the gas clouds producing Lyman-α forest absorption at low redshifts. These clouds have individual H I column densities less than about $10^{15}\,\mathrm{cm}^{-2}$ (Bahcall et al. 1996). However, they display low metallicities, typically less than 10% of solar. Gray extinction would require larger dust grains which would need a larger mass in heavy elements than typical interstellar grain size distributions to achieve a given extinction. It is possible that large dust grains are blown out of galaxies by radiation pressure, and are therefore not associated with Lyman-α clouds (Aguirre 1999b).

But even the dust postulated by Aguirre (1999a,b) and Aguirre & Haiman (1999) is not *completely* gray, having a size of about 0.1 μm. We can test for such nearly gray dust by observing high-redshift SNe Ia over a wide wavelength range to measure the color excess it would introduce. If $A_V = 0.25$ mag, then $E(U-I)$ and $E(B-I)$ should be 0.12–0.16 mag (Aguirre 1999a,b). If, on the other hand, the 0.25 mag faintness is due to Λ, then no such reddening should be seen. This effect is measurable using proven techniques; so far, with just one SN Ia (SN 1999Q; Fig. 11), our results favor the no-dust hypothesis to better than 2σ (Riess et al. 2000). More work along these lines is in progress.

5.3 The Smoking Gun

Suppose, however, that for some reason the dust is *very* gray, or our color measurements are not sufficiently precise to rule out Aguirre's (or other) dust. Or, perhaps some other astrophysical systematic effect is fooling us, such as possible evolution of the white dwarf progenitors (e.g., Höflich et al. 1998; Umeda et al. 1999), or gravitational lensing (Wambsganss, Cen, & Ostriker 1998). The most decisive test to distinguish between Λ and cumulative systematic effects is to examine the *deviation* of the observed peak magnitude of SNe Ia from the magnitude expected in the low-Ω_M, zero-Λ model. If Λ is positive, the deviation should actually begin to *decrease* at $z \approx 1$; we will be looking so far back in time that the Λ effect becomes small compared with Ω_M, and the Universe is decelerating at that epoch. If, on the other hand, a systematic bias such as gray dust or evolution of the white dwarf progenitors is the culprit, we expect that the deviation of the apparent magnitude will continue growing, unless the systematic bias is set up in such an unlikely way as to mimic the effects of Λ (Drell et al. 2000). A turnover, or decrease of

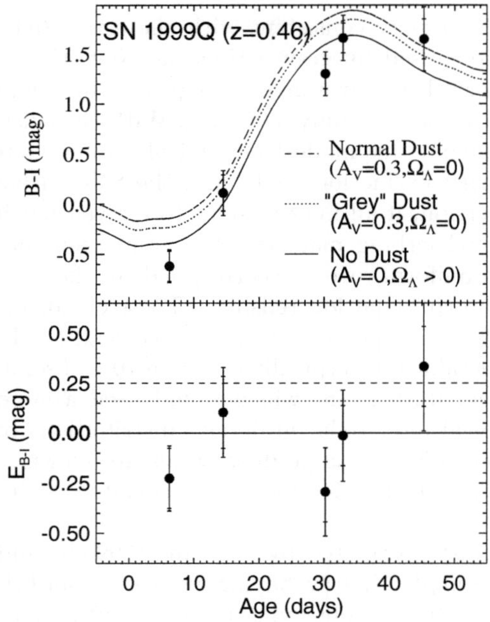

Fig. 11. Color excess, E_{B-I}, for SN 1999Q and different dust models (Riess et al. 2000). The data are most consistent with no dust and $\Omega_\Lambda > 0$.

the deviation of apparent magnitude at high redshift, can be considered the "smoking gun" of Λ.

In a wonderful demonstration of good luck and hard work, Riess et al. (2001) report on *HST* observations of a probable SN Ia at $z \approx 1.7$ (SN 1997ff, the most distant SN ever observed) that suggest the expected turnover is indeed present, providing a tantalizing glimpse of the epoch of deceleration. (See also Benítez et al. 2002, which corrects the observed magnitude of SN 1997ff for gravitational lensing.) SN 1997ff was discovered by Gilliland & Phillips (1998) in a repeat *HST* observation of the Hubble Deep Field–North, and serendipitously monitored in the infrared with *HST*/NICMOS. The peak apparent SN brightness is consistent with that expected in the decelerating phase of the concordance cosmological model, $\Omega_M \approx 0.3$, $\Omega_\Lambda \approx 0.7$ (Fig. 12). It is inconsistent with gray dust or simple luminosity evolution, when combined with the data for SNe Ia at $z \approx 0.5$. On the other hand, it is wise to remain cautious: the error bars are large, and it is always possible that we are being fooled by this single object. The HZT and SCP currently have programs to find and measure more SNe Ia at such high redshifts. For example, SN candidates at very high redshifts (e.g., Giavalisco et al. 2002) have been found by "piggybacking" on the Great Observatories Origins Deep Survey (GOODS) being conducted with the Advanced Camera for Surveys aboard *HST*; see Fig. 13

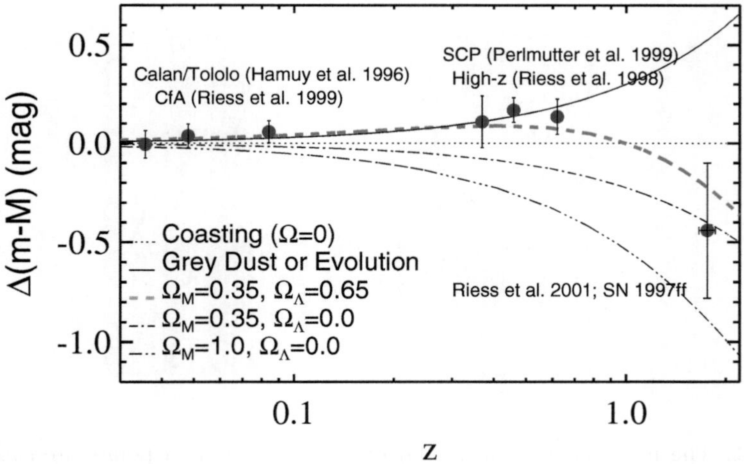

Fig. 12. Hubble diagram for SNe Ia relative to an empty universe ($\Omega = 0$) compared with cosmological and astrophysical models (Riess et al. 2001). Low-redshift SNe Ia are from Hamuy et al. (1996a) and Riess et al. (1999a). The magnitude of SN 1997ff at $z = 1.7$ has been corrected for gravitational lensing (Benítez et al. 2002). The measurements of SN 1997ff are inconsistent with astrophysical effects that could mimic previous evidence for an accelerating universe from SNe Ia at $z \approx 0.5$.

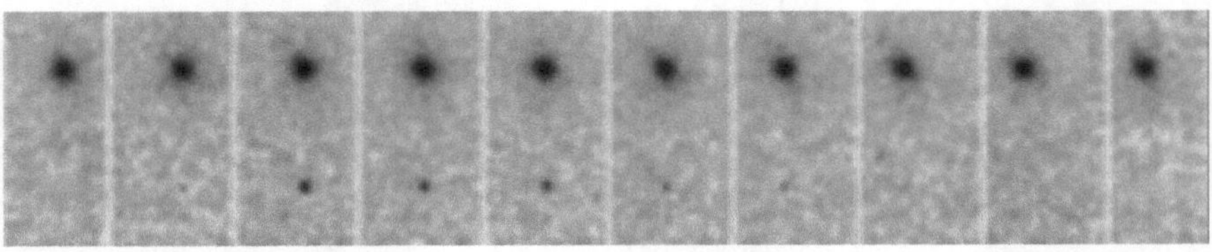

Fig. 13. SN 2002hp, a high-redshift supernova from the GOODS program using *HST*. One can see it brightening and subsequently fading with time. The assumed host galaxy is at the top of each frame.

Less ambitious programs, concentrating on SNe Ia at $z \gtrsim 0.8$, have already been completed (HZT: Tonry et al. 2003) or are nearing completion (SCP). Tonry et al. (2003) measured several SNe Ia at $z \approx 1$, and their deviation of apparent magnitude from the low-Ω_M, zero-Λ model is roughly the same as that at $z \approx 0.5$ (Fig. 14), in agreement with expectations based on the results of Riess et al. (2001). Moreover, the new sample of high-redshift SNe Ia presented by Tonry et al., analyzed with methods distinct from (but similar to) those used previously, confirm the result of Riess et al. (1998b) and Perlmutter et al. (1999) that the expansion of the Universe is accelerating.

By combining all of the available data sets, Tonry et al. (2003) are able to use about 200 SNe Ia, obtaining an incredibly firm detection of $\Omega_\Lambda > 0$ (Fig. 15). They place the following constraints on cosmological quantities:

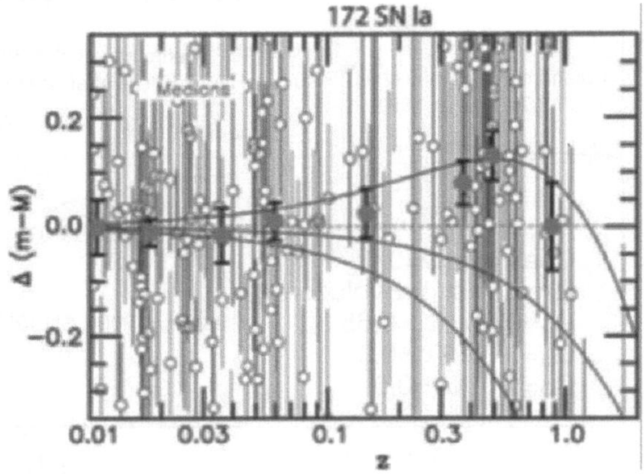

Fig. 14. The new Tonry et al. (2003) data and other data points are shown in a residual Hubble diagram: apparent magnitude difference between the expected magnitude in an empty universe and the observed magnitude of SNe Ia at each redshift. The highlighted points correspond to median values in eight redshift bins. From top to bottom the curves show $(\Omega_M, \Omega_\Lambda) = (0.3, 0.7)$, (0.3,0.0), and (1.0,0.0), respectively.

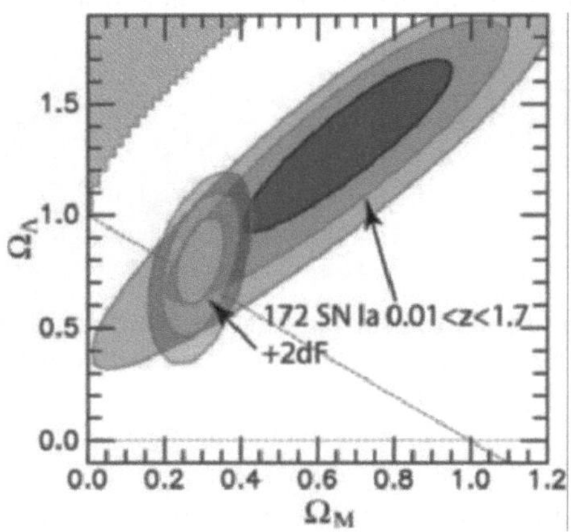

Fig. 15. From Tonry et al. (2003), the probability contours for Ω_Λ versus Ω_M are shown at 1σ, 2σ, and 3σ [assuming $w = -1$, where $w = P/(\rho c^2)$]. Also shown are the corresponding contours when a prior of $\Omega_M h = 0.20 \pm 0.03$ (where $H_0 = 100h$ km s^{-1} Mpc^{-1}) is adopted from the 2dFGRS (Percival et al. 2001). These constraints use the full sample of 172 SNe Ia with $z > 0.01$ and $A_V < 0.5$ mag.

(1) If the equation of state parameter of the dark energy is $w = -1$, then $H_0 t_0 = 0.96 \pm 0.04$, and $\Omega_\Lambda - 1.4\Omega_M = 0.35 \pm 0.14$. (2) Including the constraint of a flat universe, they find that $\Omega_M = 0.28 \pm 0.05$, independent of any large-scale structure measurements. (3) Adopting a prior based on the 2dF-

GRS constraint on Ω_M (Percival et al. 2001) and assuming a flat universe, they derive that $-1.48 < w < -0.72$ at 95% confidence. (4) Adopting the 2dFGRS results, they find $\Omega_M = 0.28$ and $\Omega_\Lambda = 0.72$, independent of any assumptions about $\Omega_{\rm total}$. These constraints are similar in precision and in value to very recent conclusions reported using *WMAP* (Spergel et al. 2003), also in combination with the 2dFGRS. Complete details on the SN Ia results can be found in Tonry et al. (2003).

5.4 Measuring the Dark Energy Equation of State

Every energy component in the Universe can be parameterized by the way its density varies as the Universe expands (scale factor a), with $\rho \propto a^{-3(1+w)}$, and w reflects the component's equation of state, $w = P/(\rho c^2)$, where P is the pressure exerted by the component. So for matter, $w = 0$, while an energy component that does not vary with scale factor has $w = -1$, as in the cosmological constant Λ. Some really strange energies may have $w < -1$: their density increases with time (Carroll, Hoffman, & Trodden 2003)! Clearly, a good estimate of w becomes the key to differentiating between models.

The CMB observations imply that the geometry of the universe is close to flat, so the energy density of the dark component is simply related to the matter density by $\Omega_x = 1 - \Omega_M$. This allows the luminosity distance as a function of redshift to be written as

$$D_L(z) = \frac{c(1+z)}{H_0} \int_0^z \frac{[1 + \Omega_x((1+\mathrm{z})^{3w} - 1)]^{-1/2}}{(1+\mathrm{z})^{3/2}} \,\mathrm{dz} \,,$$

showing that the dark energy density and equation of state directly influence the apparent brightness of standard candles. As demonstrated graphically in Fig. 16, SNe Ia observed over a wide range of redshifts can constrain the dark energy parameters to a cosmologically interesting accuracy.

But there are two major problems with using SNe Ia to measure w. First, systematic uncertainties in SN Ia peak luminosity limit how well $D_L(z)$ can be measured. While statistical uncertainty can be arbitrarily reduced by finding thousands of SNe Ia, intrinsic SN properties such as evolution and progenitor metallicity, and observational limits like photometric calibrations and K-corrections, create a systematic floor that cannot be decreased by sheer force of numbers. We expect that systematics can be controlled to at best 3%, with considerable effort.

Second, SNe at $z > 1.0$ are very hard to discover and study from the ground. As discussed above, both the HZT and the SCP have found a few SNe Ia at $z > 1.0$, but the numbers and quality of these light curves are insufficient for a w measurement. Large numbers of SNe Ia at $z > 1.0$ are best left to a wide-field optical/infrared imager in space, such as the proposed *Supernova/ Acceleration Probe* (*SNAP*; Nugent et al. 2000) satellite.

Fortunately, an interesting measurement of w can be made at present. The current values of Ω_M from many methods (most recently *WMAP*: 0.27;

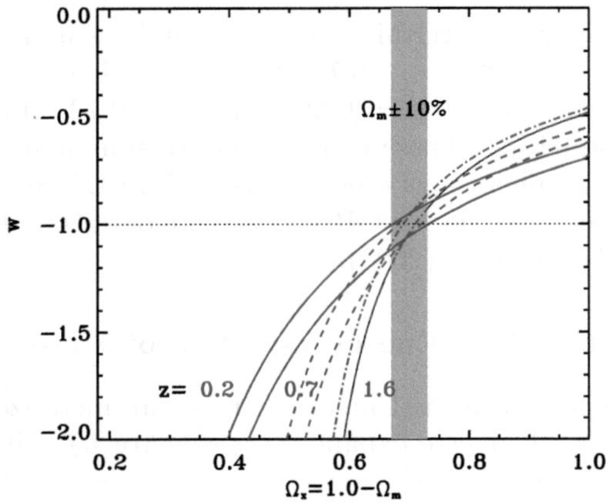

Fig. 16. Constraints on Ω_x and w from SN data sets collected at $z = 0.2$ (solid lines), $z = 0.7$ (dashed lines), and $z = 1.6$ (dash-dot lines). The shaded area indicates how an independent estimate of Ω_M with a 10% error can help constrain w.

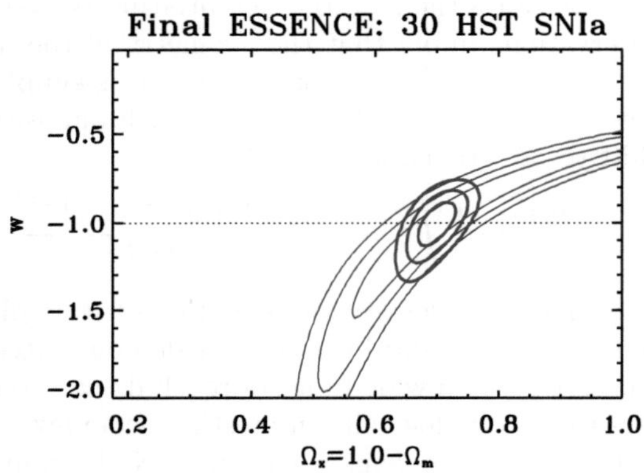

Fig. 17. Expected constraints on w with the desired final ESSENCE data set of 200 SNe Ia, 30 of which (in the redshift range $0.6 < z < 0.8$) are to be observed with *HST*. The thin lines are for SNe alone while the thick lines assume an uncertainty in Ω_M of 7%. The final ESSENCE data will constrain the value of w to ~10%.

Spergel et al. 2003) make an excellent substitute for those expensive SNe at $z > 1.0$. Figure 16 shows that a SN Ia sample with a maximum redshift of $z = 0.8$, combined with the current 10% error on Ω_M, will do as well as a SN Ia sample at much higher redshifts. Within a few years, the Sloan Digital Sky Survey and *WMAP* will solidify the estimate of Ω_M and sharpen w further.

Both the SCP and the HZT are involved in multi-year programs to discover and monitor hundreds of SNe Ia for the purpose of measuring w. For example, the HZT's project, ESSENCE (Equation of State: SupErNovae trace

Cosmic Expansion), is designed to discover 200 SNe Ia evenly distributed in the $0.2 < z < 0.8$ range. The CTIO 4-m telescope and mosaic camera are being used to find and follow the SNe by imaging on every other dark night for several consecutive months of the year. Keck and other large telescopes are being used to get the SN spectra and redshifts. Project ESSENCE will eventually provide an estimate of w to an accuracy of $\sim$10% (Fig. 17).

Farther in the future, large numbers of SNe Ia to be found by the *SNAP* satellite and the Large-area Synoptic Survey Telescope (the "Dark Matter Telescope"; Tyson & Angel 2001) could reveal whether the value of w depends on redshift, and hence should give additional constraints on the nature of the dark energy. High-redshift surveys of galaxies such as DEEP2 (Davis et al. 2001), as well as space-based missions to map the CMB (*Planck*), should provide additional evidence for (or against) Λ. Observational cosmology promises to remain exciting for quite some time!

Acknowledgements

I am grateful to N. Breton, J. Cervantes, and M. Salgado — the organizers of the Fifth Mexican School on Gravitation and Mathematical Physics, "The Early Universe and Observational Cosmology" — for the invitation to speak, for travel support to a beautiful setting, and for their patience while I wrote this review. I thank all of my HZT collaborators for their contributions to our team's research, and members of the SCP for their seminal complementary work on the accelerating Universe. My group's work at U.C. Berkeley has been supported by NSF grants AST–9987438 and AST–0206329, as well as by grants GO–7505, GO/DD–7588, GO–8177, GO–8641, GO–9118, and GO–9352 from the Space Telescope Science Institute, which is operated by the Association of Universities for Research in Astronomy, Inc., under NASA contract NAS 5–26555. Many spectra of high-redshift SNe were obtained at the W.M. Keck Observatory, which is operated as a scientific partnership among the California Institute of Technology, the University of California, and NASA; the observatory was made possible by the generous financial support of the W.M. Keck Foundation. KAIT has received donations from Sun Microsystems, Inc., the Hewlett-Packard Company, AutoScope Corporation, Lick Observatory, the National Science Foundation, the University of California, and the Sylvia and Jim Katzman Foundation.

References

1. Aguirre, A. N. 1999a, Astrophys. J., **512**, L19
2. ——. 1999b, Astrophys. J., **525**, 583
3. Aguirre, A. N., & Haiman, Z. 1999, Astrophys. J., **525**, 583
4. Aldering, G., Knop, R., & Nugent, P. 2000, Astron. J., **119**, 2110

5. Bahcall, J. N., et al. 1996, Astrophys. J., **457**, 19
6. Bahcall, N. A., Ostriker, J. P., Perlmutter, S., & Steinhardt, P. J. 1999, Science, **284**, 1481
7. Balbi, A., et al. 2000, Astrophys. J., **545**, L1
8. Benítez, N., Riess, A., Nugent, P., Dickinson, M., Chornock, R., & Filippenko, A. V. 2002, Astrophys. J., **577**, L1
9. Branch, D. 1981, Astrophys. J., **248**, 1076
10. ——. 1998, ARA&A, **36**, 17
11. Branch, D., Fisher, A., & Nugent, P. 1993, Astron. J., **106**, 2383
12. Branch, D., & Miller, D. L. 1993, Astrophys. J., **405**, L5
13. Branch, D., Romanishin, W., & Baron, E. 1996, Astrophys. J., **465**, 73 (erratum: 467, 473)
14. Branch, D., & Tammann, G. A. 1992, ARA&A, **30**, 359
15. Caldwell, R. R., Davé, R., & Steinhardt, P. J. 1998, Ap&SS, **261**, 303
16. Cappellaro, E., Turatto, M., Tsvetkov, D. Yu., Bartunov, O. S., Pollas, C., Evans, R., & Hamuy, M. 1997, A&A, **322**, 431
17. Carroll, S. M., Hoffman, M., & Trodden, M. Phys. Rev. D **68**, 023509 (2003), [astro-ph/0301273].
18. Carroll, S. M., Press, W. H., & Turner, E. L. 1992, ARA&A, **30**, 499
19. Chaboyer, B., Demarque, P., Kernan, P. J., & Krauss, L. M. 1998, Astrophys. J., **494**, 96
20. Coil, A. L., et al. 2000, Astrophys. J., **544**, L111
21. Cowan, J. J., McWilliam, A., Sneden, C., & Burris, D. L. 1997, Astrophys. J., **480**, 246
22. Davis, M., Newman, J. A., Faber, S. M., & Phillips, A. C. 2001, in Deep Fields, ed. S. Cristiani, A. Renzini, & R. E. Williams (Berlin: Springer), 241
23. de Bernardis, P., et al. 2000, Nature, **404**, 955
24. ——. 2002, Astrophys. J., **564**, 559
25. Drell, P. S., Loredo, T. J., & Wasserman, I. 2000, Astrophys. J., **530**,593
26. Efstathiou, G., et al. 1999, Mon. Not. Roy. Astron. Soc., **303**, L47
27. ——. 2002, Mon. Not. Roy. Astron. Soc., **330**, L29
28. Eisenstein, D. J., Hu, W., & Tegmark, M. 1998, Astrophys. J., **504**, L57
29. Filippenko, A. V. 1997a, in Thermonuclear Supernovae, ed. P. Ruiz-Lapuente et al. (Dordrecht: Kluwer), 1
30. ——. 1997b, ARA&A, **35**, 309
31. ——. 2001, PASP, **113**, 1441
32. ——. 2003, in From Twilight to Highlight: The Physics of Supernovae, ed. W. Hillebrandt & B. Leibundgut (Berlin: Springer-Verlag), 171.
33. Filippenko, A. V., et al. 1992a, Astron. J., **104**, 1543
34. ——. 1992b, Astrophys. J., **384**, L15
35. Filippenko, A. V., Li, W. D., Treffers, R. R., & Modjaz, M. 2001, in Small-Telescope Astronomy on Global Scales, ed. W. P. Chen, C. Lemme, & B. Paczyński (San Francisco: ASP), 121
36. Filippenko, A. V., & Riess, A. G. 1998, Phys. Rep., **307**, 31
37. Ford, C. H., et al. 1993, Astron. J., **106**, 1101
38. Freedman, W., et al. 2001, Astrophys. J., **553**, 47
39. Garnavich, P., et al. 1998a, Astrophys. J., **493**, L53
40. ——. 1998b, Astrophys. J., **509**, 74
41. Giavalisco, M., et al. 2002, IAUC 7981

42. Gilliland, R. L., & Phillips, M. M. 1998, IAUC 6810
43. Goldhaber, G., et al. 1997, in Thermonuclear Supernovae, ed. P. Ruiz-Lapuente et al. (Dordrecht: Kluwer), 777
44. ——. 1998a, BAAS, **30**, 1325
45. ——. 1998b, in Gravity: From the Hubble Length to the Planck Length, SLAC Summer Institute (Stanford, CA: SLAC)
46. ——. 2001, Astrophys. J., **558**, 359
47. Goldhaber, G., & Perlmutter, S. 1998, Phys. Rep., **307**, 325
48. Goobar, A., & Perlmutter, S. 1995, Astrophys. J., **450**, 14
49. Gratton, R. G., Fusi Pecci, F., Carretta, E., Clementini, G., Corsi, C. E., & Lattanzi, M. 1997, Astrophys. J., **491**, 749
50. Groom, D. E. 1998, BAAS, **30**, 1419
51. Hamuy, M., Phillips, M. M., Maza, J., Suntzeff, N. B., Schommer, R. A., & Aviles, R. 1995, Astron. J., **109**, 1
52. ——. 1996a, Astron. J., **112**, 2391
53. ——. 1996b, Astron. J., **112**, 2398
54. Hamuy, M., Trager, S. C., Pinto, P. A., Phillips, M. M., Schommer, R. A., Ivanov, V., & Suntzeff, N. B. 2000, Astron. J., **120**, 1479
55. Hanany, S., et al. 2000, Astrophys. J., **545**, L5
56. Hancock, S., Rocha, G., Lazenby, A. N., & Gutiérrez, C. M. 1998, Mon. Not. Roy. Astron. Soc., **294**, L1
57. Hatano, K., Branch, D., & Deaton, J. 1998, Astrophys. J., **502**, 177
58. Höflich, P., Wheeler, J. C., & Thielemann, F. K. 1998, Astrophys. J., **495**, 617
59. Hoyle, F., Burbidge, G., & Narlikar, J. V. 2000, A Different Approach to Cosmology (Cambridge: Cambridge Univ. Press)
60. Ivanov, V. D., Hamuy, M., & Pinto, P. A. 2000, Astrophys. J., **542**, 588
61. Kim, A., Goobar, A., & Perlmutter, S. 1996, PASP, **108**, 190
62. Leibundgut, B., et al. 1993, Astron. J., **105**, 301
63. ——. 1996, Astrophys. J., **466**, L21
64. Leonard, D. C., et al. 2002a, PASP, **114**, 35 (erratum: 114, 1291)
65. ——. 2002b, Astron. J., **124**, 2490
66. Li, W., et al. 2000, in Cosmic Explosions, ed. S. S. Holt & W. W. Zhang (New York: AIP), 103
67. ——. 2001a, PASP, **113**, 1178
68. ——. 2003, PASP, **115**, 453
69. Li, W., Filippenko, A. V., Treffers, R. R., Riess, A. G., Hu, J., & Qiu, Y. 2001b, Astrophys. J., **546**, 734
70. Lineweaver, C. H. 1998, Astrophys. J., **505**, L69
71. Lineweaver, C. H., & Barbosa, D. 1998, Astrophys. J., **496**, 624
72. Matheson, T., Filippenko, A. V., Li, W., Leonard, D. C., & Shields, J. C. 2001, Astron. J., **121**, 1648
73. Modjaz, M., Li, W., Filippenko, A. V., King, J. Y., Leonard, D. C., Matheson, T., Treffers, R. R., & Riess, A. G. 2001, PASP, **113**, 308
74. Narlikar, J. V., & Arp, H. C. 1997, Astrophys. J., **482**, L119
75. Netterfield, C. B., et al. 2002, Astrophys. J., **571**, 604
76. Nomoto, K., Umeda, H., Hachisu, I., Kato, M., Kobayashi, C., & Tsujimoto, T. 2000, in Type Ia Supernovae: Theory and Cosmology, ed. J. C. Niemeyer & J. W. Truran (Cambridge: Cambridge Univ. Press), 63

77. Norgaard-Nielsen, H., et al. 1989, Nature, **339**, 523
78. Nugent, P., 2000, in Particle Physics and Cosmology: Second Tropical Workshop, ed. J. F. Nieves (New York: AIP), 263
79. Nugent, P., Kim, A., & Perlmutter, S. 2002, PASP, **114**, 803
80. Nugent, P., Phillips, M., Baron, E., Branch, D., & Hauschildt, P. 1995, Astrophys. J., **455**, L147
81. Ostriker, J. P., & Steinhardt, P. J. 1995, Nature, **377**, 600
82. Oswalt, T. D., Smith, J. A., Wood, M. A., & Hintzen, P. 1996, Nature, **382**, 692
83. Parodi, B. R., et al. 2000, Astrophys. J., **540**, 634
84. Pain, R., et al. 2002, Astrophys. J., **577**, 120
85. Peacock, J. A., et al. 2001, Nature, **410**, 169
86. Percival, W., et al. 2001, Mon. Not. Roy. Astron. Soc., **327**, 1297
87. Perlmutter, S., et al. 1995a, Astrophys. J., **440**, L41
88. ——. 1995b, IAUC 6270
89. ——. 1997, Astrophys. J., **483**, 565
90. ——. 1998, Nature, **391**, 51
91. ——. 1999, Astrophys. J., **517**, 565
92. Phillips, M. M. 1993, Astrophys. J., **413**, L105
93. Phillips, M. M., et al. 1992, Astron. J., **103**, 1632
94. Phillips, M. M., et al. 1999, Astron. J., **118**, 1766
95. Pskovskii, Yu. P. 1977, Sov. Astron., **21**, 675
96. ——. 1984, Sov. Astron., **28**, 658
97. Riess, A. G., et al. 1997, Astron. J., **114**, 722
98. ——. 1998b, Astron. J., **116**, 1009
99. ——. 1999a, Astron. J., **117**, 707
100. ——. 1999c, Astron. J., **118**, 2675
101. ——. 2000, Astrophys. J., **536**, 62
102. ——. 2001, Astrophys. J., **560**, 49
103. Riess, A. G., Filippenko, A. V., Li, W. D., & Schmidt, B. P. 1999b, Astron. J., **118**, 2668
104. Riess, A. G., Nugent, P. E., Filippenko, A. V., Kirshner, R. P., & Perlmutter, S. 1998a, Astrophys. J., **504**, 935
105. Riess, A. G., Press, W. H., & Kirshner, R. P. 1995, Astrophys. J., **438**, L17
106. ——. 1996a, Astrophys. J., **473**, 88
107. ——. 1996b Astrophys. J., **473**, 588.
108. Saha, A., et al. 1997, Astrophys. J., **486**, 1
109. Sandage, A., et al. 1996, Astrophys. J., **460**, L15
110. Sandage, A., & Tammann, G. A. 1993, Astrophys. J., **415**, 1
111. Schmidt, B. P., et al. 1998, Astrophys. J., **507**, 46
112. Spergel, D. N., et al. , Astrophys. J. Suppl. **148**, 175 (2003) [astro-ph/0302209].
113. Sullivan, M., et al., Mon. Not. Roy. Astron. Soc., **340**, 1057 (2003) [astro-ph/0211444].
114. Suntzeff, N. 1996, in Supernovae and Supernova Remnants, ed. R. McCray & Z. Wang (Cambridge: Cambridge Univ. Press), 41
115. Suntzeff, N., et al. 1996, IAUC 6490
116. Tonry, J. L., et al., Astrophys. J., **594**, 1 (2003) [astro-ph/0305008].
117. Tripp, R. 1997, A&A, **325**, 871
118. ——. 1998, A&A, **331**, 815

119. Turatto, M., et al. 1996, Mon. Not. Roy. Astron. Soc., 283, 1
120. Tyson, J. A., & Angel, R. 2001, in The New Era of Wide Field Astronomy, ed. R. Clowes et al. (San Francisco: ASP), 347
121. Umeda, H., et al. 1999, Astrophys. J., **522**, L43
122. van den Bergh, S., & Pazder, J. 1992, Astrophys. J., **390**, 34
123. Vaughan, T. E., Branch, D., Miller, D. L., & Perlmutter, S. 1995, Astrophys. J., **439**, 558
124. Wambsganss, J., Cen, R., & Ostriker, J. P. 1998, Astrophys. J., **494**, 29
125. Yungelson, L. R., & Livio, M. 2000, Astrophys. J., **528**, 108
126. Zaldarriaga, M., Spergel, D. N., & Seljak, U. 1997, Astrophys. J., **488**, 1

Part II

Quintessence, Dark Energy, Dark Matter, and Other Topics

Quintessence and Dark Energy

Axel de la Macorra

Instituto de Física, UNAM; Apdo. Postal 20-364; 01000 México D.F., México.
macorra@fisica.unam.mx

Abstract. Dark energy accounts for about 70% of the content of our Universe. Perhaps the best candidate to parametrize the dark energy is a scalar field with only gravitational interactions called quintessence.

We first present a generic theoretical approach to quintessence. We show that if the minimum of the scalar potential is at $V|_{min} = 0$ (i.e. there is no arbitrary scale) then the behavior of scalar fields can be determined in a model independent way. Its equation of state parameter w_ϕ takes most of the time the values $w_\phi = 1, -1, w_{\phi o}$. The size of the different regions can also be calculated in a model independent way. The number of free parameters for quintessence models is therefore quite limited.

We show that late time phase transition models are good candidates for quintessence and they could explain why the acceleration of the Universe is at such a late time. We show how these models can be obtained from particle physics and in particular from non-abelian gauge dynamics. The only free parameter for these gauge models is its particle content as it is the case for the standard model of particle physics.

In the second part, we show how a phenomenological approach to the CMB can be implemented. We show that with only four parameters we cover a great number of theoretical models, including quintessence. By varying these parameters and comparing with the CMB we can, in principle, determine the relevant cosmological quantities such as the phase transition scale (when the quintessence field appears) and the present equation of state parameter $w_{\phi o}$.

1 Introduction

In recent time the cosmological observations on the cosmic microwave background radiation ("CMB") [1] and the supernova project SN1a [2] have lead to conclude that the universe is flat and it is expanding with an accelerating velocity. These conclusions show that the universe is now dominated by an energy density with negative pressure with $\Omega_{DE} = 0.7 \pm 0.1$ and $w_{DE} \equiv p_{DE}/\rho_{DE} < -0.78$ [1]. This energy is generically called the dark energy. Structure formation also favors a non-vanishing dark energy [3]. Besides dark energy we also have baryonic matter $\Omega_b \simeq 0.05$ and dark matter $\Omega_{DM} = 0.25 \pm 0.1$, necessary for structure formation, but we still do not know its origin. So, we have a universe which contains only 5% of particles of the well known Standard Model ("SM") of particle physics and 95% of matter unknown to us on earth.

A. de la Macorra, Quintessence and Dark Energy, Lect. Notes Phys. **646**, 225–257 (2004)
http://www.springerlink.com/

It is not clear yet what the dark energy is. It could be a true cosmological constant, quintessence (scalar field with gravitationally interaction) [9] or some other kind of exotic energy density. Perhaps the best way of determining the nature of dark energy is trough its equation of state parameter w_{DE}. The survey of redshifts of the different objects should in principle allow as to determine the value of w_{DEo} (o subscript referees to present day quantities) but only at small redshifts $z \leq 2$ with $z_o = 0$. The result from the SN1A project [2] sets an upper limit to $w_{\phi o} < -2/3$ but does not distinguish a true cosmological constant with $w_\Lambda \equiv -1$, quintessence or any other form of exotic energy with $w_{\phi o} < -2/3$. It would be very interesting if in the future the SN1a survey could constrain better the value of $w_{\phi o}$. On the other hand, the CMB could give us information not only on the value of w_{DEo} but also on its evolution during matter domination era, i.e. for a redshift $z \leq 8$.

Energy density of elementary particles of the Standard Model (e.g. quarks, leptons and bosons) have a non-negative pressure $p = w\rho$ with $w = 1/3$ for relativistic and $w = 0$ non-relativistic particles. Therefore, an energy density with negative pressure, the "dark energy", has to be explain from a particle physics point via particles that are not contained in the SM. The only particles that can give a negative pressure are particles with a non trivial self potential V and since fermions and bosons cannot have a vacuum expectation value (these particles transform non-trivially under the Lorentz transformation) the only possibility left are scalar fields. The scalar fields can be fundamental fields, as the Higgs field or the supersymmetric partners of the SM fermion particles, or they could be composite fields as meson fields in QCD.

In order to determine what the nature of the dark energy is we can proceed with two different approaches. On the one hand we can propose models derived from particle physics and see if these models give the correct observable data. On the other hand we could set a model independent analysis on the evolution of the equation of state parameter w_{DE} and determine its impact on the observed CMB spectrum and compare it with the data in order to infer the type of dark energy density.

In Sect. 2 we will discuss the theoretical approaches to obtain a dark energy from field theory. In Sect. 3 we introduce a particle physics model, based on gauge dynamics, that gives a dark energy field in a natural way. In Sect. 4 we study the possibility that the gauge group responsible for giving the dark energy gives at the same time the missing dark matter. In Sect. 5 we analyze a model independent phenomenological approach to dark energy and in Sect. 6 we present our conclusions.

2 Theoretical Approach

2.1 Cosmological Evolution of Quintessence

We will now determine the cosmological evolution of a scalar field ϕ with arbitrary potential $V(\phi)$ and with only gravitational interaction with all other fields. This field is called quintessence.

The cosmological evolution of ϕ with an arbitrary potential $V(\phi)$ can be determined from a system of differential equations describing a spatially flat Friedmann–Robertson–Walker universe in the presence of a barotropic fluid energy density ρ_b that can be either radiation or matter. The equations are

$$\begin{aligned} \dot{H} &= -\frac{1}{2}(\rho_b + p_b + \dot{\phi}^2), \\ \dot{\rho} &= -3H(\rho + p), \\ \ddot{\phi} &= -3H\dot{\phi} - \frac{dV(\phi)}{d\phi}, \end{aligned} \tag{1}$$

where H is the Hubble parameter, $\dot{\phi} = d\phi/dt$, ρ (p) is the total energy density (pressure). We will be working in a flat universe so that $H^2 = \rho/3$ and we use natural units $m_p^2 = G/8\pi \equiv 1$. If is useful to make a change of variables $x \equiv \frac{\dot{\phi}}{\sqrt{6}H}$ and $y \equiv \frac{\sqrt{V}}{\sqrt{3}H}$ and the equations (1) take the following form [27, 26]:

$$\begin{aligned} x_N &= -3x + \sqrt{\frac{3}{2}}\lambda\, y^2 + \frac{3}{2}x[2x^2 + \gamma_b(1 - x^2 - y^2)] \\ y_N &= -\sqrt{\frac{3}{2}}\lambda\, x\, y + \frac{3}{2}y[2x^2 + \gamma_b(1 - x^2 - y^2)] \\ H_N &= -\frac{3}{2}H[2x^2 + \gamma_b(1 - x^2 - y^2)] \end{aligned} \tag{2}$$

where N is the logarithm of the scale factor a, $N \equiv Log(a)$; $f_N \equiv df/dN$ for $f = x, y, H$; $\gamma_b = 1 + w_b$ and $\lambda(N) \equiv -V'/V$ with $V' = dV/d\phi$. In terms of x, y the energy density parameter is $\Omega_\phi = x^2 + y^2$ while the equation of state parameter is given by $\gamma_\phi - 1 = w_\phi \equiv p_\phi/\rho_\phi = \frac{x^2 - y^2}{x^2 + y^2}$. It is clear that $0 \leq x^2, y^2 \leq 1$.

The Friedmann or constraint equation for a flat universe $\Omega_b + \Omega_\phi = 1$ must supplement equations (2) which are valid for any scalar potential as long as the interaction between the scalar field and matter or radiation is gravitational only. This set of differential equations is non-linear and for most cases has no analytical solutions. A general analysis for arbitrary potentials is performed in [25, 26]. All model dependence falls on two quantities: $\lambda(N)$ and the constant parameter $\gamma_b = 1, 4/3$ for matter or radiation, respectively. We will be interested in studying scalar fields that lead to a late time accelerated universe, i.e. to quintessence, and in this case we will have a decreasing $\lambda(N)$

[26] and a late time behavior $\lambda(N) \to 0$. For constant $\lambda(N)$ (exponential potential) one can have an accelerating universe if $\lambda(N) < \sqrt{6}$ but its dynamics would lead to an accelerating universe too rapidly, i.e. not at a late time as ours, unless we fine tune the initial conditions.

It is also useful to have the evolution of $\Omega_\phi = \rho_\phi/3H^2 = x^2 + y^2$ and $\gamma_\phi = 1 + w_\phi$ $(0 \leq \gamma_\phi \leq 2)$ derived from (2), [21]

$$(\Omega_\phi)_N = 3(\gamma_b - \gamma_\phi)\Omega_\phi(1 - \Omega_\phi) \tag{3}$$

$$(\gamma_\phi)_N = 3\gamma_\phi(2 - \gamma_\phi)\left(\lambda\sqrt{\frac{\Omega_\phi}{3\gamma_\phi}} - 1\right). \tag{4}$$

2.2 Evolution of x, y, and H

We are interested in studying scalar potentials that lead to quintessence, i.e. a late time (present day) acceleration period of the universe. For this to happen one needs $\lambda = -m_{pl}V'/V \to 0$ in the asymptotic limit (or to a constant less then one). An accelerating universe (slow roll conditions) requires $|\lambda| < 1$ and we want this period to be at a late time. We will consider potentials with $V \geq 0$ and since the ϕ field evolves to its minimum $V' < 0$ and $\lambda \geq 0$ where we are assuming, without loss of generality, models with $\phi \geq 0$. We will define the phase transition scale Λ_c in terms of the potential by [35]

$$\Lambda_c = V(\phi_i)^{1/4} \tag{5}$$

where $V(\phi_i) \equiv V_i$ is the initial value of the potential and we will consider models that have an initial value

$$\lambda_i = -m_{pl}\frac{V'(\phi_i)}{V(\phi_i)} \gg 1. \tag{6}$$

From now on the subscript i stands for initial conditions, i.e. at the condensation Λ_c when V appears. From dimensional analysis we expect $\lambda_i = O(m_{pl}/\Lambda_c) \gg 1$. If we have a phase transition at a scale Λ_c which leads to the appearance of the ϕ field (e.g. composite field) then we would also expect $\phi_i \simeq \Lambda_c$ since Λ_c is the relevant scale of the process. We will be working with a late time phase transition but Λ_c could be as large as $10^{16} GeV$ and we would still have $\lambda_i \gg 1$. An interesting general property of these models is the presence of a many e-folds scaling period in which λ is practically constant and $\Omega_\phi \ll 1$.

A semi-analytic approach [18] is useful to study some properties of the differential equation system given by (2). To do this we initially consider only the terms that are proportional to λ, since $\lambda_i \gg 1$, then we follow the evolution of x, y and H so every period has a characteristic set of simplified differential equations. We see from (2) that the leading terms in x and y, for $\lambda \gg 1$, are $x_N = \sqrt{\frac{3}{2}}\lambda y^2$ and $y_N = -\sqrt{\frac{3}{2}}\lambda\, x\, y$. Combining these equations we have

$$x_N x = -y_N y \tag{7}$$

with a constant circular solution [18]

$$\Omega_\phi \equiv x^2 + y^2 = x_i^2(N_i) + y_i^2(N_i) \equiv \Omega_{\phi i}(N_i). \tag{8}$$

Since x_N is positive x will grow while y_N is negative giving a decreasing y. This initial period ends at a scale N_{min} with $x^2(N_{min}) \simeq \Omega_{\phi i}(N_i) \gg y^2_{min}$. Since $\lambda_i \gg 1$, the x and y derivatives are quite large and the amount of e-folds between the initial value y_i until y reaches its minimal value y_{min} is very short. An easy estimate can be derived from $y_N/y = -c\lambda \gg 1$, $c = \sqrt{3/2}x$ giving $1 \gg N_{min} - N_i = Log[y_{min}/y_i]/c\lambda_i = O(1/\lambda_i)$, in the assumption $c\lambda_i = cte$.

The minimal value of y, given at N_{min}, can be obtained from (2) with $y_N = 0$. At his point we have [35]

$$\lambda(N_{min}) = -\sqrt{\frac{2}{3}}\frac{H_N}{Hx} = \sqrt{\frac{3}{2}}\frac{[\gamma_b + \Omega_{\phi i}(2-\gamma_b)]}{\sqrt{\Omega_{\phi i}}} \simeq \frac{1}{\sqrt{\Omega_{\phi i}}} \tag{9}$$

where we have taken $x^2(N_{min}) \simeq \Omega_{\phi i}$ and $H_N/H = -3/2(\gamma_b + \Omega_{\phi i}(2-\gamma_b))$ since $y^2_{min} \ll 1$. We see that λ in (9) is of order $1/\sqrt{\Omega_{\phi i}}$ and we have $\lambda_i/\lambda(N_{min}) \gg 1$.

The value of y_{min} depends on the functional form of $V(\phi)$, which sets the functional form of $\lambda = -V'/V$. In general we have $y^2_{min} = V(\phi_{min})/(3H^2_{min})$ but without specifying $V(\phi)$ it is not possible to determine y_{min}. For an inverse power law potential with $V = \Lambda_c^{4+n}\phi^{-n} = 3y^2H^2$ one has

$$\begin{aligned} y_{min} &= \frac{\Lambda_c^{\frac{4+n}{2}}\phi_{min}^{-n/2}}{\sqrt{3}H_{min}} \\ &= y_i\left(\frac{\phi_i}{\phi_{min}}\right)^{\frac{n}{2}} = y_i\left(\frac{1}{\lambda_i\sqrt{\Omega_{\phi i}}}\right)^{\frac{n}{2}} \end{aligned} \tag{10}$$

where we have approximated $H^2_{min} \simeq H_i^2 = V_i/3y_i^2 = \Lambda_c^{4+n}\phi_i^{-n}/3y_i^2$ in (10) since $N_{min} - N_i \ll 1$ and we have taken from (9) $\phi_{min} = n/\lambda_{min} \simeq n\sqrt{\Omega_{\phi i}}$ and $\phi_i = n/\lambda_i$. Taking the initial value of $\phi_i = n/\lambda_i = n\Lambda_c$ then (10) gives

$$y_{min} = y_i\left(\frac{\Lambda_c}{\sqrt{\Omega_{\phi i}}}\right)^{n/2}. \tag{11}$$

We see that $y_{min} = O(\lambda_i^{-n/2}) \simeq O(\Lambda_c^{n/2}) \ll y_i$ if $\Omega_{\phi i}$ is not too small. At the end of the initial period we have $y^2 \ll 1$ and $\lambda x = O(1)$. Since x_N/x is now negative

$$x_N \simeq (-3 + 3/2\gamma_b)x \tag{12}$$

$|x|$ decreases as

$$x(N) = x(N_{min})e^{(-3+\frac{3}{2}\gamma_b)(N-N_{min})}. \tag{13}$$

leading to the scaling period. The transition between the initial period and the second (scaling) period is short because x decreases rapidly (for $x(N)/x(N_{min}) = 1/10$ one has $N - N_{min} \simeq 1.5$) and we get $1 \gg x \gg y$. The scaling period is defined by the validity of the equation

$$\frac{y_N}{y} = -\frac{H_N}{H}. \tag{14}$$

This period takes place when $\lambda x \ll 1$ as seen from (2). During the scaling period one has $yH = H_{min}y_{min} = cte$ which leads to a constant Hy and potential since $V = 3H^2y^2$. Therefore, λ and ϕ will also be constant during this scaling period [35], i.e.

$$\lambda(N_{min}) \simeq \lambda(N_2) \tag{15}$$

where we have defined the scale N_2 as the end of the scaling period. Neglecting the quadratic terms on x and y in the third equation of system (2) we get the expressions

$$\begin{aligned} H &= H_{min}e^{-\frac{3}{2}\gamma_b(N-N_{min})} \\ y &= y_{min}e^{\frac{3}{2}\gamma_b(N-N_{min})}. \end{aligned} \tag{16}$$

We can take in (16) $N_{min} \simeq N_i$ and $H_{min} \simeq H_i$ as discussed above, but $y_{min} \ll y_i$.

Since during the scaling period y increases as seen from (16) and λ is constant the term λy^2 in x_N will eventually dominate and lead to an increase of x. The end of the scaling period will happen when λx is again of order one and (14) is no longer valid. At this point we have $\lambda x \sim 1$ and $x \sim \lambda y^2$ which leads to an x of the same order of y, i.e. γ_ϕ will be significant larger then zero (say $\gamma_\phi \sim 0.1$). At the end of the scaling period we have $1/x_2 \sim \lambda(N_2) = \lambda(N_{min})$ and [35]

$$\Omega_\phi(N_2) = y^2(N_2) + x^2(N_2) \sim \lambda(N_{min})^{-2} \sim \Omega_{\phi i} \tag{17}$$

as seen from (9) and (15). The value of $\Omega_\phi(N_2)$ depends on the initial $\Omega_{\phi i}$ and can be much smaller than one. After the end of the scaling period $\Omega_\phi(N_2)$ grows to its present day value $\Omega_{\phi o} = 0.7 \pm 0.1$. If $\Omega_\phi(N_2) \ll 1$ then there is enough time for γ_ϕ to grow to its tracker value $\gamma_{\phi tr} = \lambda^2\Omega_\phi/n^2$. However, when $\Omega_\phi(N_2)$ is of the order 0.1 then there is not enough time to allow γ_ϕ to grow to its tracker value and one has at present day $0 < \gamma_{\phi o} \leq \gamma_{\phi tr}$. Finally, the late time behavior has $\lambda \to 0$ and $\Omega_\phi \sim y^2 \to 1$ with $\gamma_\phi \to 0$.

2.3 Parameters and Summary

There are only four independent parameters that fix the cosmological evolution of the models from its initial value to present day. These parameters are

$\Omega_{\phi i}, \Lambda_c, y_{min}$ and the value of $\gamma_{\phi o}$ today. All other quantities can be derived from them.

Let us summarize the evolution of x and y obtained in the previous section [35]
1) Regardless of the value of x_i, y_i we have a very short period ($N_{min} - N_i \ll 1$) with increasing x and decreasing y ending with $x(N_{min})^2 \simeq \Omega_{\phi i}$ and with $y_{min} \ll 1$ model dependent.
2) Shortly afterwards the scaling period starts with $x(N_{min})^2 \gg y_{min}^2$ and $\gamma_\phi \simeq 2$. During this period x decreases while y increases and it finishes when $x \sim y \ll 1$. The size of the period $\gamma_\phi = 2$ depends on how small y_{min} is.
3) After having $x \sim y \ll 1$, we still have a decreasing x and increasing y and the period with $\gamma_\phi = 0$ (with $1 \gg y \gg x$) starts. When λy^2 becomes of order of x, x_N becomes positive and x increases until $\lambda x \sim 1$ making (14) no longer valid and ending the period with $\gamma_\phi \simeq 0$ and the scaling period at N_2. The value of λ remains (almost) constant during all the scaling period which starts at N_{min} and finishes at N_2.
4) The tracking period starts with a increasing $\gamma_\phi \rightarrow \gamma_{\phi o}$ and Ω_ϕ.

3 Late Time Phase Transition as Dark Energy

The evolution of the scalar field ϕ depends on the functional form of its potential $V(\phi)$ and a late time accelerating universe constrains the form of the potential and when it appears [18, 30]. A late time appearance of a scalar field is a signal that a phase transition took place and that the scalar field is probably not a fundamental but a composite field. Here, we will present a model where quintessence field appears as a consequence of a phase transition due to a strong gauge coupling constant. This is a very physical assumption since it only requires to have an extra gauge group to the already known gauge groups of the SM. It is well known that the gauge coupling constant of a non-abelian asymptotically free gauge group increases with decreasing energy and the free elementary fields will eventually condense due to the strong interaction, e.g. mesons and baryons in QCD. The scale where the coupling constant becomes strong is called the condensation scale Λ_c and below it there are no more free elementary fields. These condensates, e.g. "mesons", develop a non trivial potential which can be calculated using Affleck's potential [22]. The potential is of the form $V = \Lambda_c^{4+n}\phi^{-n}$, where ϕ represents the "mesons", and depending on the value of n the potential V may lead to an acceptable phenomenology. The final value of $w_{\phi o}$ (from now on the subscript "o" refers to present day quantities) depends n and the initial condition $\Omega_{\phi i}$ [18]. A $w_{\phi o} < -2/3$, which is the upper limit of [6], requires $n < 2.74$ for $\Omega_{\phi i} \geq 0.25$ [18]. For smaller $\Omega_{\phi i}$ one obtains a larger $w_{\phi o}$ for a fixe n. The position of the third CMBR peak favors models with $n < 1$ [7] and for some class of models with $V = M^{4+n}\phi^{-n}e^{\phi^\beta/2}$, with $n \geq 1, \beta \geq 0$, the constraint on the equation of state is even stricter $-1 \leq w_{\phi o} \leq -0.93$

[8]. In this kind of inverse power potential models (i.e. $n < 2$) the tracker solution is not a good approximation to the numerical solution because the scalar field has not reached its tracker value by present day.

Here we focus on a non-abelian asymptotically free gauge group whose gauge coupling constant is unified with the couplings of the standard model ("SM") ones [17, 18]. We will call this group the dark group ("DG"). The cosmological picture in this case is very pleasing. We assume gauge coupling unification at the unification scale Λ_{gut} for all gauge groups (as predicted by string theory) and then let all fields evolve. At the beginning all fields, SM and DG model, are massless and red shift as radiation until we reach the condensation scale Λ_c of DG. Below this scale the fields of the quintessence gauge group will dynamically condense and we use Affleck's potential to study its cosmological evolution. The energy density of the quintessence field Ω_ϕ drops quickly, independently of its initial conditions, and it is close to zero for a long period of time, which includes nucleosynthesis (NS) if Λ_c is larger than the NS energy Λ_{NS} (or temperature $T_{NS} = 0.1 - 10 MeV$), and becomes relevant only until very recently. On the other hand, if $\Lambda_c < \Lambda_{NS}$ than the NS bounds on relativistic degrees of freedom must be imposed on the models. Finally, the energy density of ϕ grows and it dominates at present time the total energy density with the $\Omega_{\phi o} \simeq 0.7$ and a negative pressure $w_{\phi o} < -2/3$ leading to an accelerating universe [5].

The initial conditions at the unification scale and at the condensation scale are fixed by the number of degrees of freedom of the models given in terms of N_c, N_f. Imposing gauge coupling unification fixes N_c, N_f and we do not have any free parameters in the models (but for the susy breaking mechanism which we will comment in Sect. 2). It is surprising that such a simple model works fine. As we will see the restriction on N_c, N_f by gauge unification rules out models with a condensation energy scale between $2 \times 10^{-2} GeV < \Lambda_c < 6 \times 10^3 GeV$ or for models with $2 < n < 4.27$ (the scale Λ_c is given in terms of H_o and n by $\Lambda_c \simeq H_o^{2/(4+n)}$ [23],[18]). Since $w_{\phi o} < -2/3$ requires $n < 2.74$ all models must then have $\Lambda_c < 2 \times 10^{-2} GeV$. The number of models that satisfy gauge coupling unification with a $w_{\phi o} < -2/3$ is quite limited and in fact there are only three different models [18]. All acceptable models have $n \leq 2/3$ which implies that the condensation scale is smaller than the NS scale. The preferred model has $N_c = 3, N_f = 6$, $n = 2/3$ and it gives $w_{\phi o} = -0.90$ with an average value $w_{eff} = -0.93$ agreeing with recent CMBR analysis [6, 7].

3.1 Condensation Scale and Scalar Potential

We start be assuming that the universe has a matter content of the supersymmetric gauge groups $SU(1) \times SU(2) \times SU(3) \times SU(DG)$ where the first three are the SM gauge groups while the last one corresponds to the dark group and that the couplings are unified at Λ_{gut} with $g_1 = g_2 = g_3 = g_{DG} = g_{gut}$.

The condensation scale Λ_c of a gauge group $SU(N_c)$ with N_f (chiral + antichiral) matter fields has in $N = 1$ susy a one-loop renormalization group equation given by

$$\Lambda_c = \Lambda_{gut} e^{-\frac{8\pi^2}{b_o g_{gut}^2}} \tag{18}$$

where $b_o = 3N_c - N_f$ is the one-loop beta function and Λ_{gut}, g_{gut} are the unification energy scale and coupling constant, respectively. From gauge coupling unification we know that $\Lambda_{gut} \simeq 10^{16}$GeV and $g_{gut} \simeq \sqrt{4\pi/25.7}$ [33].

A phase transition takes place at the condensation scale Λ_c, since the elementary fields are free fields above Λ_c and condense at Λ_c. In order to study the cosmological evolution of these condensates, which we will call ϕ, we use Affleck's potential [22]. This potential is non-perturbative and exact [36].

The superpotential for a non-abelian $SU(N_c)$ gauge group with N_f (chiral + antichiral) massless matter fields is [22]

$$W = (N_c - N_f)\left(\frac{\Lambda_c^{b_o}}{det < Q\tilde{Q} >}\right)^{1/(N_c - N_f)} \tag{19}$$

where b_o is the one-loop beta function coefficient. Taking $det < Q\tilde{Q} >= \Pi_{j=1}^{N_f} \phi_j^2$ one has $W = (N_c - N_f)(\Lambda_c^{b_o} \phi^{-2N_f})^{1/(N_c - N_f)}$. The scalar potential in global supersymmetry is $V = |W_\phi|^2$, with $W_\phi = \partial W/\partial\phi$, giving [23, 24, 17, 30]

$$V = c^2 \Lambda_c^{4+n} \phi^{-n} \tag{20}$$

with $c = 2N_f$, $n = 2 + 4\frac{N_f}{N_c - N_f}$ and Λ_c is the condensation scale of the gauge group $SU(N_c)$. The natural initial value for the condensate is $\phi_i = \Lambda_c$ since it is precisely Λ_c the relevant scale of the physical process of the field binding.

In (20) we have taken ϕ canonically normalized, however the full Kahler potential K is not known and for $\phi \simeq 1$ other terms may become relevant [23] and could spoil the runaway and quintessence behavior of ϕ. Expanding the Kahler potential as a series power $K = |\phi|^2 + \Sigma_i a_i |\phi|^{2i}/2i$ the canonically normalized field ϕ' can be approximated[1] by $\phi' = (K_\phi^\phi)^{1/2}\phi$ and (20) would be given by $V = (K_\phi^\phi)^{-1}|W_\phi|^2 = (2N_f)^2 \Lambda_c^{4+n} \phi^{-n} (K_\phi^\phi)^{(n/2-1)}$. For $n < 2$ the exponent term of K_ϕ^ϕ is negative so it would not spoil the runaway behavior of ϕ [17, 18].

If we wish to study models with $0 < n < 2$, which are cosmologically favored [18] we need to consider the possibility that not all N_f condensates ϕ_i become dynamical but only a fraction ν are (with $N_f \geq \nu \geq 1$) and we also need $N_f > N_c$ [17, 18]. It is important to point out that even though it has been argued that for $N_f > N_c$ there is no non-perturbative superpotential W generated [22], because the determinant of $Q\tilde{Q}$ in (19) vanishes, this is not necessarily the case [29]. If we consider the elementary quarks $Q_i^\alpha, \tilde{Q}_i^\alpha$ ($i, j = 1, 2, ..., N_f,\ \alpha = 1, 2, ..., N_c$) to be the relevant degrees of freedom,

[1] The canonically normalized field ϕ' is defined as $\phi' = g(\phi, \bar{\phi})\phi$ with $K_\phi^\phi = (g + \phi g_\phi + \bar{\phi} g_{\bar{\phi}})^2$

then for $N_c < N_f$ the quantity $det(Q^i_\alpha \tilde{Q}^\alpha_j)$ vanishes since, being the sum of dyadics, always has zero eigenvalues. However, we are interested in studying the effective action for the "meson" fields $\phi^i_j =< Q^i_\alpha \tilde{Q}^\alpha_j >$, and the determinant of ϕ^i_j, i.e. $det < Q^i_\alpha \tilde{Q}^\alpha_j >$, being the product of expectation values does not need to vanish when $N_c < N_f$ (the expectation of a product of operators is not equal to the product of the expectations of each operator).

One can have $\nu \neq N_f$ with a gauge group with unmatching number of chiral and anti-chiral fields or if some of the chiral fields are also charged under another gauge group. In this case we have $c = 2\nu, n = 2+4\frac{\nu}{N_c-N_f}$ and $N_f - \nu$ condensates fixed at their v.e.v. $< Q\tilde{Q} >= \Lambda_c^2$ [17]. Another possibility is by giving a mass term to $N_f - \nu$ condensates $\varphi =< \bar{Q}_k Q_k >$, $(k = 1, ..., N_f - \nu)$ while leaving ν condensates $\phi^2 =< \bar{Q}_j Q_j >$, $(j = 1, ..., \nu)$ massless. Notice that we have chosen a different parameterization for φ and ϕ. The mass dimension for φ is 2 while for ϕ it is 1. The superpotential now reads [30]

$$W = (N_c - N_f)(\frac{\Lambda_c^{b_o}}{\phi^{2\nu}\varphi^{N_f-\nu}})^{1/(N_c-N_f)} + m\varphi \tag{21}$$

with m the mass of $\bar{Q}_k Q_k$. If we take the natural choice $\phi_i = \Lambda_c$, as discussed above, and $m = \Lambda_c$ [17] and we integrate out the condensates φ using

$$\frac{\partial W}{\partial \varphi} = \varphi^{-1}\left((\nu - N_f)\Lambda_c^{(b_o-2\nu)/(N_c-N_f)}\varphi^{-(N_f-\nu)/(N_c-N_f)} + m\varphi\right) = 0 \tag{22}$$

we obtain $\varphi = (N_f - \nu)^{(N_c-N_f)/(N_c-\nu)}\Lambda_c^2$. By integrating out the φ field the second terms in (21), which is proportional to the first term, can be eliminated. Substituting the solution of (22) into (21) one finds

$$W = (N_c - \nu)(N_f - \nu)^{(N_f-\nu)/(N_c-\nu)}\Lambda_c^{3+a}\phi^{-a} \tag{23}$$

with $a = 2\nu/(N_c - N_f)$.

The scalar potential $V = |\partial W|^2$ is now given by [30]

$$V = c'^2 \Lambda_c^{4+n'}\phi^{-n'} \tag{24}$$

with $c'^2 = 4\nu^2(\frac{N_c-\nu}{N_c-N_f})^2(N_f - \nu)^{(N_f-\nu)/(N_c-\nu)}$ and $n' = 2 + 4\nu/(N_c - N_f)$. Notice that for $\nu = N_f$ we recover (20). From now on we will work with (24) and we will drop the quotation on n'.

The radiative corrections to the scalar potential (24) are $V \sim \Lambda_c^{4+n}\phi^{-n}(1+O(\Lambda_c^2\phi^{-2}))$ [28]. They are not important because we have $\phi \geq \Lambda_c$ and are negligible at late times when $\phi \gg \Lambda_c$.

3.2 Gauge Unification Condition

In order to have a model with gauge coupling unification the scale Λ_c given in (20) or (24) must be identified with the energy scale in (18). However,

Table 1. We show the matter content for the three different models and we give the number of degrees of freedom for the susy and non susy Q group in the last two columns, respectively. Notice that the condensation scale and b_o is the same for all models.

Num	N_c	N_f	ν	n	b_o	$\Lambda_c(eV)$	g_{Qs}
I	3	6	1	2/3	3	42	97.5
II	6	15	3	2/3	3	42	468.5
III	7	18	4	6/11	3	42	652.5

not all values of Λ_c will give an acceptable cosmology. The correct values of Λ_c depend on the cosmological evolution of the scalar condensate ϕ which is determined by the power n in (24). The Λ_c scale can be expressed in terms of present day quantities from (20) by [18, 30]

$$\begin{aligned}\Lambda_c &= \left(3H_o^2 yo^2 \phi_o^n c^{-2}\right)^{\frac{1}{4+n}} \\ &\simeq \left(3H_o^2 \Omega_{\phi o}\right)^{\frac{1}{4+n}}\end{aligned} \tag{25}$$

where $y^2 \equiv V/3H^2 \simeq \Omega_\phi$, with $\Omega_{\phi o} = 0.7$. A rough estimate of (25) gives $\Lambda_c \simeq H_o^{2/(4+n)}$ since we also expect $\phi_o = O(1)$ [18] today (we are living at the beginning of an accelerating universe). The number of models that satisfy the unification and cosmological constraints of having $\Omega_{\phi o} = 0.7, h_o = 0.7$ (with the Hubble constant given by $H_o = 100 h_o$ km/Mpc sec) and $w_{\phi o} < -2/3$ [5] is quite limited [18]. In fact there are only three models given in Table 1. These models are obtained by equating Λ_c from (18), which is a function of N_c, N_f through b_o, and (24), which is also a function of N_c, N_f, ν through n. The exact value of y_o, ϕ_o must be determined by the cosmological evolution of ϕ (c.f. (1)) starting at Λ_c until present day. For an acceptable model the parameters N_c, N_f and ν must take integer values. We consider an acceptable model when Λ_c in (18) and (25) do not differ by more than 50%. With this assumption there are only 3 models, given in Table 1, that have (almost) integer values for N_f. In all these models one has $n \leq 2/3$ and the quantum corrections to the Kahler potential are, therefore, not dangerous. All other combinations of N_c, N_f, ν do not lead to an acceptable cosmological model. From (25) one has for $n \leq 4.27$ a scale $\Lambda_c \leq 6.5 \times 10^3 GeV$ and from (18) this implies that $b_o \leq 5.7$. Since $b_o = 3N_c - N_f = 2N_c + 4\nu/(n-2)$ and the minimum acceptable value for N_c is two one finds $b_o \geq 4 + 4\nu/(n-2)$. Taking $2 < n \leq 4.27$ gives a value of $b_o \geq 5.7$. The value of $n = 4.27$ gives the upper limiting value for which we can find a solution of (18) and (25). We see that it is not possible to have quintessence models with gauge coupling unification with $2 < n < 4.27$. In terms of the condensation scale the restriction for models with $2 \times 10^{-2} GeV < \Lambda_c < 6 \times 10^3 GeV$.

Using $n = 2 + 4\nu/(N_c - N_f)$ or equivalently $N_f = N_c + 4\nu/(n-2)$ with $b_o = 3N_c - N_f = 2N_c + 4\nu/(n-2)$ we can write from (18) as $b_o =$

$8\pi^2/g_{gut}^2(Log(\frac{\Lambda_{gut}}{\Lambda_c}))^{-1}$ and N_c [30]

$$
\begin{aligned}
N_c &= \frac{1}{2}b_o + \frac{2\nu}{2-n} \\
&= \frac{4\pi^2}{g_{gut}^2}(Log[\frac{\Lambda_{gut}}{\Lambda_c}])^{-1} + \frac{2\nu}{2-n}
\end{aligned}
\tag{26}
$$

From (25) we have Λ_c as a function of n (with the approximation of $y_o^2\phi_o^n = 1$) and N_c in (26) becomes a function of n and ν only. For $2\times 10^{-2}GeV < \Lambda_c < 6.5\times 10^3 GeV$ we have a $N_c < 2$ and therefore are ruled out. In terms of n the condition is that models with $2 < n < 4.27$ are not viable. In deriving these conditions, we have taken $\nu = 1$ which gives the smallest constraint to N_c as seen from (26).

The upper limit $\Lambda_c > 6.5\times 10^3 GeV$ has $n > 4.27$ (c.f. (25)). As mentioned in the introduction, the value of $w_{\phi o}$ depends on the initial condition $\Omega_{\phi i}$ and on n [18]. The larger n the larger $w_{\phi o}$ will be (same is true for the tracker value $w_{tr} = -2/(2+n)$). It has been shown that assuming an initial value of $\Omega_{\phi i}$ no smaller than 0.25 then the value of $w_{\phi o}$ will be less then $w_{\phi o} < -2/3$ only if $n < 2.74$ [18]. Therefore, the models with $n > 4.27$ are not phenomenological acceptable and since $4.27 > n > 2$ are also ruled out by the constraint on gauge coupling unification, we are left with models with [30]

$$
\Lambda_c < 2\times 10^{-2} GeV \qquad or \qquad n < 2. \tag{27}
$$

So, only models with a cosmological late time phase transition are allowed.

3.3 Thermodynamics, Nucleosynthesis Bounds, and Initial Conditions

Before determining the evolution of ϕ we must analyze the initial conditions for the $SU(DG)$ gauge group. The general picture is the following: The "DG" gauge group is by hypothesis, unified with the SM gauge groups at the unification energy Λ_{gut}. For energies scales between the unification and condensation scale, i.e. $\Lambda_c < \Lambda < \Lambda_{gut}$, the elementary fields of $SU(DG)$ are massless and weakly coupled and interact with the SM only gravitationally. The DG gauge interaction becomes strong at Λ_c and condense the elementary fields leading to the potential in (24).

Since for energies above Λ_{gut} we have a single gauge group it is naturally to assume that all fields (SM and DG) are in thermal equilibrium. However, at temperatures $T < T_{gut}$ the gauge group DG is decoupled since it interacts with the SM only via gravity.

The energy density at the unification scale is given by $\rho_{Tot} = \frac{\pi^2}{30} g_{Tot} T^4$, where $g_{Tot} = \Sigma Bosons + 7/8 \Sigma Fermions$ is the total number of degrees of freedom at the temperature T. The minimal models have $g_{Tot} = g_{smi} + g_{DGi}$, with $g_{smi} = 228.75$ and $g_{DGi} = (1+7/8)(2(N_c^2-1)+2N_f N_c)$ for the minimal supersymmetric standard model MSSM and for the $SU(DG)$ supersymmetric gauge group with N_c colors and N_f (chiral + antichiral) massless fields, respectively. The initial energy density at the unification scale for each group is simply given in terms of number of degrees of freedom, $\Omega = \rho/\rho_c$,

$$\Omega_{DGi}(\Lambda_{gut}) = \frac{g_{DGi}}{g_{Tot}}, \qquad \Omega_{smi}(\Lambda_{gut}) = \frac{g_{smi}}{g_{Tot}} \tag{28}$$

with $\Omega = \Omega_{DG} + \Omega_{sm} = 1$. Since the SM and DG gauge groups are decoupled below Λ_{gut}, their respective entropy, $S_k = g_k a^3 T^3$ with g_k the degrees of freedom of the k group and a the scale factor of the universe (see (2)), will be independently conserved. The total energy density ρ as a function of the photon's temperature T above Λ_c (i.e. $\Lambda_c < \Lambda < \Lambda_{gut}$), with the DG fields still massless and redshifting as radiation, is given by

$$\rho = \frac{\pi^2}{30} g_* T^4 \tag{29}$$

with

$$g_* = g_{smf} + g_{DGf} \left(\frac{T_{DG}}{T}\right)^4 = g_{smf} + g_{DGf} \left(\frac{g_{smf}\, g_{DGi}}{g_{smi}\, g_{DGf}}\right)^{4/3} \tag{30}$$

and $g_{smi}, g_{smf}, g_{DGi}, g_{DGf}$ are the initial (i.e. at decoupling) and final standard model and DG model relativistic degrees of freedom, respectively. From the entropy conservation, we know that the relative temperature between the standard model and the DG model is given by

$$\frac{T_{DG}}{T} = \left(\frac{g_{smf}\, g_{DGdec}}{g_{smdec}\, g_{DGf}}\right)^{1/3} \tag{31}$$

where g_{smdec} stands for the degrees of freedom when the DG-particle decouple from the SM. It is clear that the energy density for the DG model $\rho_{DG} = \pi^2/30 g_{DG} T_{DG}^4$ in terms of the photon's temperature T is fixed by the number of degrees of freedom,

$$\begin{aligned} \Omega_{DGf} &= \frac{g_{DGf} T_{DG}^4}{g_* T^4} \\ &= \frac{g_{DGf} (T_{DG}/T)^{4/3}}{g_{smf} + g_{DGf} (T_{DG}/T)^{4/3}} . \end{aligned} \tag{32}$$

Equation (32) permits us to determine the energy density of the DG group at any temperature above the condensation scale.

3.4 Energy Density at the Condensation Scale

We would like now to determine the energy density at the condensation scale which will set the initial energy density for the scalar composite field ϕ.

Just above the condensation scale Λ_c we take, for simplicity of argument, that all particles in the DG group are still massless and we can use (32) to determine the $\Omega_{DG}(\Lambda_c)$ with $g_{DGi} = g_{DGf}$ giving [30]

$$\Omega_{DGf} = \frac{g_{DGf}(T_{DG}/T)^{4/3}}{g_{smf} + g_{DGf}(T_{DG}/T)^{4/3}}. \tag{33}$$

If the decoupling of DG particles is above neutrino decoupling (around $1MeV$) then for temperatures below $1MeV$ one has $T_{DG}/T = (43/11/g_{smdec})^{1/3}$. At Λ_c we have a phase transition and we no longer have elementary free particles in the DG group. They are bind together through the strong gauge interaction and the acquire a non-perturbative potential and mass given by (24). In other words, below the condensation scale there are no free "quarks" DG and we have "meson" and "baryon" fields.

If we consider only the SM and the DG group, the energy density within the particles of the DG group must be conserved since they are decoupled from the SM (the interaction is by hypothesis only gravitational). All the energy density of the DG group is transmitted into dark energy (and possible dark matter, see Sect. 4) at the condensation scale Λ_c. This is a natural assumption from a particle point of view but is not crucial from a cosmological point of view, in the sense that any "reasonable" fraction of the energy density in the DG group would give a correct cosmological evolution of the ϕ field. We would like to stress out that the initial condition for ϕ is no longer a free parameter but it is given in terms of the degrees of freedom of the MSSM and the DG group.

3.5 Nucleosynthesis Constraint on the Energy Density

The big-bang nucleosynthesis (NS) bound on the energy density from non SM fields, relativistic or non-relativistic, is quite stringent $\Omega_{DG} < 0.1 - 0.2$ [31, 32].

If the DG gauge group condense at temperatures much higher than NS then, the evolution of the condensates will be given by (2) with the potential of (20) and we must check that Ω_{DG} at NS is no larger than 0.1-0.2. This will be, in general, no problem since it was shown that even for a large initial Ω_{DG} at the condensation scale the evolution of ϕ is such that Ω_{DG} decreases quite rapidly and remains small for a long period of time (see figure 2) [17, 18].

On the other hand, if the gauge group condenses after NS we must determine if the DG energy density is smaller than $\Omega_{DG} < 0.1 - 0.2$ at NS. Since the condensations scale Λ_c is smaller than the NS scale, all fields in the DG group are still massless and the energy density is given in terms of the relativistic degrees of freedom and from (32) to set a limit on g_{DGf} and g_{DGi},

$$\Delta g_{DG} \equiv g_{DGf}^{-1/3} g_{DGi}^{4/3} = \frac{\Omega_{DG}}{1 - \Omega_{DG}} g_{smf}^{-1/3} g_{smi}^{4/3} \tag{34}$$

and for $g_{DGf} = g_{DGi} = g_{DG}$

$$\Delta g_{DG} = g_{DG} = \frac{\Omega_{DG}}{1 - \Omega_{DG}} g_{smf}^{-1/3} g_{smi}^{4/3} \tag{35}$$

where we should take $g_{smf} = 10.75$ at the final stage (i.e. NS scale) and $g_{smi} = 228.75$ at the initial stage (i.e. at unification) for the minimal supersymmetric standard model MSSM. For $\Omega_{DG} \leq 0.1, 0.2$ (34) gives un upper limit on the number of relativistic degrees of freedom $\Delta g_{DG} \leq 70, 158$ respectively (or $g_{DG} \leq 70, 158$ if $g_{DGf} = g_{DGi} = g_{DG}$).

The l.h.s. of (34) depends on the initial (i.e. at unification) and final (at NS) number of degrees of freedom of the gauge group DG. The smaller (larger) the initial (final) degrees of freedom of DG the smaller Δg_{DG} and Ω_{DG} will be.

4 Dark Matter

We would like to see now if the dark group responsible for giving the dark energy through its scalar condensates can at the same time account for the missing dark matter [34].

The restrictions on DM is that it must have $\Omega_{DM} = 0.25 \pm 0.1$ and it should allow for structure formation at scales larger than Mpc. As we will see later our models have a warm DM with a mass of the order of keV. There are still problems with cold and warm DM models. Cold DM have an overproduction of substructure of galactic halos which a warm DM model does not have [11]. On the other hand, recent observations on the reionization redshift [1] seem to indicate that warm dark matter is not a good candidate. However, the value of the parameters used are still not well established which makes the conclusion not definite [10]. So, we believe that further studies need to be done in order to fully set the nature of dark matter.

If we have a dark gauge group with $N_c < N_f$ then on top of the gauge singlet meson fields we can have gauge singlet dark baryons $B^{i,\ldots,N_c} = \prod_i^{N_c} Q^i$ and anti–baryons. These particles get a non-vanishing mass due to non-perturbative effects (like protons and neutrons in QCD). These baryons could be degenerated in mass or there could be a lightest massive stable baryon. The order of magnitude of the mass of the DM particle can be estimated by dimensional analysis and it must be given by the condensation scale

$$m = k\,\Lambda_c \tag{36}$$

with $k = O(1)$ a constant. In this picture we have at high energies $E > \Lambda_c$ a DG with massless particles. At Λ_c non-perturbative effects, due to a strong coupling, generate a mass for dark baryons and a scalar potential for dark meson. The DM is the massive stable particle with mass given by (36) while the quintessence with potential (20) gives the DE.

Before studying the dynamics of the DG let us determine the constraint on the temperature and mass for DM in order to agree with structure formation. The relative temperature of the decoupled particle compared to the photon temperature T is given by (31). For a neutrino one has at decoupling $g_{smdec} = 11/2, g_{smf} = 2$ and with $g_{DGf} = g_{DGdec}$ one has $T_\nu = T(4/11)^{1/3} = (1/1.76)T$. However, if the decoupling is at a high energy scale, say $T \gg 10^3 GeV$, then all particles of the standard model are still relativistic and $g_{dec} = 106.75, 228.75$ for the non-susy and susy standard model respectively giving a temperature $T_D = T(43/11/g_{smdec})^{1/3}$ (below 1 MeV with $g_{smf} = 43/11$ that takes into account neutrino decoupling) with $T_D \simeq (1/3)T$ for non-susy SM and $T_D \simeq (1/3.88)T$ for the susy-SM. The temperature of DG is in these cases $3-4$ times smaller then the photon temperature and $2-3$ times smaller then T_ν. If there are more relativistic degrees of freedom coupled to the susy-SM (could be Kaluza-Klein states or other gauge groups, e.g. gauge group responsible for susy breaking [30]) at decoupling then T_D would be even smaller.

We can set an upper and lower limit to Ω_{DG}. The smallest number of degrees of freedom would be for a gauge group with $N_c = 2, N_f = 1$ giving $g_{DG} = 18.75$. While the upper limit on g_{DG} comes from Nucleosynthesis "NS" bounds which requires an upper limit to any extra energy density. This limit is $\Omega_{DG}(NS) \leq 0.1 - 0.2$ [31]. Since from (34) $g_{DG}/g_{dec}^{4/3} = (10.75)^{-1/3}\Omega_{DG}(NS)/(1 - \Omega_{DG}(NS))$ the NS bound sets un upper limit $g_{DG} \leq 0.05 g_{dec}^{4/3}, 0.113 g_{dec}^{4/3}$ for $\Omega_{DG}(NS) \leq 0.1, 0.2$, respectively. Taking $g_{DG} \leq 0.113 g_{dec}^{4/3} \sim 158(g_{dec}/228.75)^{4/3}$ we obtain un upper limit $\Omega_{DGc} \leq 0.17$ at any condensation scale below $1MeV$.

The free streaming scale λ_{fs} gives the minimum size at which perturbations survive. For scales smaller than the λ_{fs} the perturbations are wiped out. For structure formation it is required that $\lambda_{fs} \leq O(1)Mpc$. One has [12]

$$\begin{aligned}\lambda_{fs} &\simeq 0.2(\Omega_{DM}h^2)^{1/3}(1.5/g'_{DM})^{1/3}(keV/m)^{4/3} \\ &= 0.079(\Omega_{DM}h^2)^{-1}(g'_{DM}/1.5)(228.75/g_{dec})^{4/3}\end{aligned} \tag{37}$$

where $g'_{DM} = g_{bDM} + 3/4 g_{fDM}$ with g_{bDM} the bosonic, g_{bfDM} the fermion degrees of freedom of DM and we used (38) in the second equality of (37).

The energy density of the DG will be divided in DE (quintessence) and DM. For DM the entropy conservation gives $n_{DM}/n_\gamma = (g'_{DM}/2)(T_D/T)^3$ where $n_{DM}, n_\gamma = 2(\zeta(3)/\pi^2)T^3$ are the number density for DM and photon respectively. Since the energy density for matter is $\rho_m = nm$ and using $\rho_\gamma = n_\gamma(\pi^4/30\zeta(3))T$ we can write $\Omega_{DMo} = \Omega_{\gamma o}(\zeta(3)30/\pi^4)(n_{DM}/n_\gamma)(m/T_{\gamma o}) = \Omega_{\gamma o}(\zeta(3)30/\pi^4)(g'_{DM}/2)(m/T_{\gamma o})(T_D/T)^3$ giving [34]

$$\Omega_{DMo} = 0.25\left(\frac{0.71}{h_o}\right)^2\left(\frac{g'_{DM}m}{g_{dec}1.66\,eV}\right) \tag{38}$$

where we have used in the last equation the present day quantities $h_o^2\Omega_\gamma = 2.47 \times 10^{-5}$, $T_{\gamma o} = 2.37 \times 10^{-13} GeV$. Equation (38) is valid for all DM that decouples at temperature $T_i \gg 10^3 GeV$ from the susy-SM. Taking the central values of wmap [1] $\Omega_{DMo}h_o^2 = 0.135 - 0.0224 = 0.1126$ (where $\Omega_b h^2 = 0.0224$) one gets a neutrino mass $m = 12\, eV$ and $\lambda_{fs} = 36\, Mpc$ giving the usual hot DM problem. It cannot form structure at small scales. For a model that decouples from the susy-SM at $T \gg 10^3 GeV$ one has $T_D/T \leq 1/3.88$ with $g_{dec} \geq 228.75$, a mass $m \geq 381(254)\, eV$ for $g'_{DM} = 1(1.5)$, respectively, and (37) gives $\lambda_{fs} \simeq 0.62(0.41)\, Mpc$. Allowing for a more conservative variation of $\Omega_{DMo} = 0.25 \pm 0.1$ and $h_o = 0.7 \pm 0.05$ the constraint on $g'_{DM}\, m/g_{dec}$ from (38) is $0.83 g_{dec} eV \leq g'_{DM}\, m \leq 2.59 g_{dec} eV$. The number of degrees of freedom g'_{DM} is not arbitrary since $0.113 g_{dec}^{4/3} \geq g_{DG} > g'_{DM} \geq 1$, as discussed above. This bound implies that the mass of the DM particle must be [34]

$$1.2(228.75/g_{dec})^{1/3}\, eV \leq m \leq 593(g_{dec}/228.75)\, eV. \tag{39}$$

For $g_{dec} \leq 228.75$ one has $m \leq 593\, eV$ while for $m \geq 750\, eV, 1 keV$ one requires $g_{dec} \geq 450(676), 600(901)$ for $g'_{DM} = 1(1.5)$, respectively.

If we do not want to relay on having the same initial temperature between the SM and DG we can estimate the amount of DM by the backward evolution of DM from present day to the phase transition scale Λ_c where the particles acquire a mass. The evolution of the DM is $\rho_{DMo} = \rho_{DM}(a/a_o)^3$ where $a(t)$ is the scale factor. In terms of $\Omega_{DM} = 3H^2 \rho_{DM}$ (we have taken $8\pi G = 1/m_{pl}^2 = 1$) we can write the DM energy density as

$$\Omega_{DMo} = \Omega_{DMc}(\Omega_{ro}/\Omega_{rc})^{\frac{3}{4}}(H_c^2/H_o^2)^{\frac{1}{4}} \tag{40}$$

where we have expressed $a_c/a_o = (\Omega_{ro}H_o^2/\Omega_{rc}H_c^2)^{1/4}$. The evolution of the DE depends on the specific potential. However, the non-abelian gauge dynamics leads to an inverse power potential of the form [23, 17, 30]

$$V = \Lambda_c^{4+n}\phi^{-n} \tag{41}$$

where $\phi =< \bar{Q}Q >$ is the condensate of the elementary fields. Here we will treat n as a free parameter but it can be related to N_c, N_f by $n = 2 + 4\nu/(N_c - N_f)$ and ν counts the number of light condensates [17, 30]. When the kinetic term is much smaller than the potential energy one has $\Omega_{DE} \simeq \Lambda_c^{4+n}\phi^{-n}/3H^2$. This is certainly valid for present day since we require ρ_{DE} to accelerate the universe and the slow roll condition $E_k \ll V$ must be satisfied. Since the beginning of an accelerated epoch is very recently one has $\phi_o \simeq 1$ [23]. Of course, a numerical analysis must be performed [17, 30] in order to obtain the precise values of $\phi_o, w_{\phi o}$ but the analytic solution is a reasonable approximation. At the condensation scale Λ_c the initial value of the condensate ϕ_c must be giving by Λ_c and taking $\phi_c = \Lambda_c$ [17] we have

$$\Omega_{DEc} = \frac{\Lambda_c^4}{3H_c^2}, \qquad \Omega_{DEo} = \frac{\Lambda_c^{n+4}}{3H_o^2}. \tag{42}$$

Using (40) and (25) we can write [34]

$$\Omega_{DMo} = \Omega_{DMc}(\Omega_{ro}/\Omega_{rc})^{\frac{3}{4}}(\Omega_{DEo}/\Omega_{DEc})^{\frac{1}{4}}\Lambda_c^{-\frac{n}{4}} \tag{43}$$

where we have used $H_o^2/H_c^2 = (\Omega_{DEc}/\Omega_{DEo})\Lambda_c^n$. Using $\Lambda_c = (3\Omega_{DEo}H_o^2)^{\frac{1}{4+n}}$ in (43) we can determine the allowed range of values of n and Λ_c. The allowed range is quite limited. Taking the central wmap values $h_o = 0.71$, $\Omega_{DMo} = 0.25$ [1] with $g'_{DM} = 1.5$ (i.e. $g_{DM} = 7/4$) and taking as examples $g_{DG} = 97.5(160)$ we get for $g_{dec} = 228.75, 676, 901$ an inverse power $n = 0.78(0.79), 0.87(0.88), 0.90(0.91)$ and $\Lambda_c = 189(214), 559(634), 745(845)\, eV$, respectively. If we allow for a conservative variation $\Omega_{DMo} = 0.25 \pm 0.1$ and $h_o = 0.7 \pm 0.05$ and taking $g_{DGc} = 97.5$ and the upper value $g_{DGc} = 0.113 g_{dec}^{4/3}$ (results in parenthesis) then the range for n and Λ_c for $g_{dec} = 228.75, 676, 901$ is 0.34, 0.42, 0.44 (0.31, 0.29, 0.28) $\leq n \leq$ 0.87,0.96,0.98 (0.88, 1.0, 1.04) and 0.55, 1.63, 2.17 (0.34, 0.23, 0.21) $eV \leq \Lambda_c \leq$518, 1530, 2040 (585, 2484, 3645) eV, respectively, where the lower limit has $g_{DMc} = g_{DGc} - 1, h_o = 0.65, \Omega_{DMo} = 0.15$ and the upper limit has $g_{DMc} = 1, h_o = 0.75, \Omega_{DMo} = 0.35$. We see that the allowed range is [34]

$$0.28 \leq n \leq 1.04 \quad \Leftrightarrow \quad 0.21\, eV \leq \Lambda_c \leq 3645\, eV \tag{44}$$

given for $g_{dec} \leq 901$. Increasing g_{dec} would enlarge the range of n, Λ_c but not significantly. In Fig. 1 we show the behavior of Ω_{DMo} as a function of n for different values of $g_{DGc} = 0.113 g_{dec}^{4/3}$ with $g_{dec} = 228.75, 676, 901$ (solid,dashed and dotted lines, respectively) for the extreme values of g_{DMc} given by $g_{DG} - 1 \geq g_{DMc} \geq 1$. The allowed region is in between the horizontal lines $\Omega_{DMo} = 0.15 - 0.35$. From (44) we see that there is only a limited range of condensation energy scales and IPL parameter n that allows for a gauge group to give the correct DM and DE densities. It is also interesting to note that the lower limit on Λ_c is very similar to the one obtain by CMB analysis [30] where the minimum scale was $\Lambda_c = 0.2\, eV$. On the other hand, the evolution of quintessence requires for $\Omega_{DEc} < 0.17$ an IPL parameter n to be smaller than $n \leq 1.6$ for $w_{DEo} \leq -0.78$ which is the upper value of wmap. For smaller Ω_{DEc} we will need a smaller n, e.g. $\Omega_{DEc} = 0.05$ requires $n \leq 1.05$. So, once again there is a consistency within the acceptable values of n coming form different considerations (amount of DM and observable w_{DEo}). The constraint on Λ_c is very similar to the constraint obtained for the DM particle mass m obtained in (39). The similarity $m \sim \Lambda_c$ is required by non-abelian gauge dynamics and it is indeed satisfied as can be verified using (38), (25) and (43) [34]

$$k \equiv \frac{m}{\Lambda_c} = \frac{\pi^4}{\zeta(3)30} \frac{g_{DMc}}{g_{DEc}^{1/4} g'_{DM}} \tag{45}$$

with $\pi^4/(\zeta(3)30) \simeq 2.7$. Equation (36) should be compared with (45). There is a subtle point on the values of $g_{DMc}, g_{DEc}, g'_{DM}$. The "true" degrees of freedom of the dark matter particles (i.e. the lightest field of the dark gauge

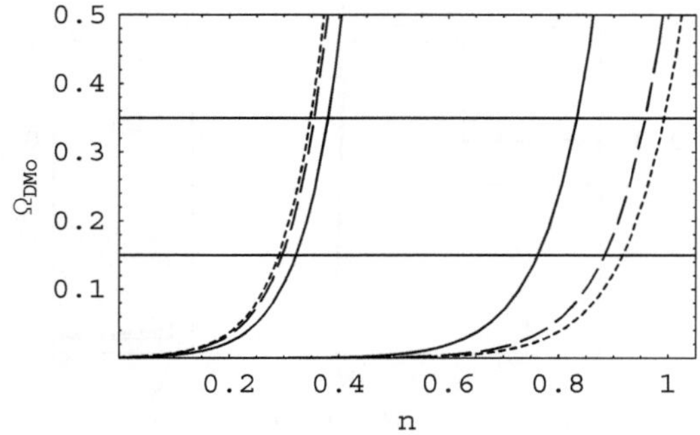

Fig. 1. We plot Ω_{DMo} as a function of the IPL parameter n. The allowed region is the one between the horizontal lines $\Omega_{DMo} = 0.15 - 0.35$ and the curves with the limiting values of $g_{DM} = 1.5$ and $g_{DG} - 1$ for $g_{dec} = 228.75, 676, 901$ (solid,dashed and dotted lines, respectively).

group) are given by g'_{DM} while g_{DMc} and g_{DEc} represent the proportion of energy density that goes into Ω_{DMc} and Ω_{DEc}. It is reasonable to assume that the particles of the dark group will decay into the lightest state. Therefore we expect $g_{DMc} > g'_{DM}$ and $m > \Lambda_c$.

5 Phenomenological Approach

The best way to determine what kind of energy is the dark energy is trough the equation of state parameter w_{DE} we and through its imprint on the CMB. This presentation is based on [35]. We will analyze the contribution to the CMB from a dark energy with a $\gamma_{DE} = w_{DE} + 1$ that takes four different values [35]. It will have a $w_{DE} = 1/3$ for energies above a certain scale Λ_c, which we will call the phase transition scale. Starting at Λ_c we will have a region with $w_{DE} = 1$ and duration ΔN_1, where N is the logarithm of the scale factor a ($N = Log[a]$). Thirdly we will have $w_{DE} = -1$ for almost the same amount of time as in the previous period, $\Delta N_2 \simeq \Delta N_1$, and finally we will end up in a region with $-1 \leq w_{DEo} = cte \leq -2/3$ for a duration of ΔN_o. The cosmological evolution and the resulting CMB will have only four new parameters $\Delta N_1, \Delta N_2, \Delta N_o$ and w_{DEo}. By varying these parameters we will cover a wide range of models. In particular we will cover all quintessence models.

The analysis of the CMB with this kind of dark energy does not depend on its nature, it could be a scalar field (quintessence) or any other form of dark energy that gives the four sectors described above. In Fig 2 we show an example with an IPL potential with $n = 1$ and $\Omega_{DEi} = 0.05$ and we see that the $w = 1/3, 1, -1, w_{tr} = cte$ approximation fits well with the numerical w_{DE}. The strategy is to analyze the spectra of CMB, using a modified version

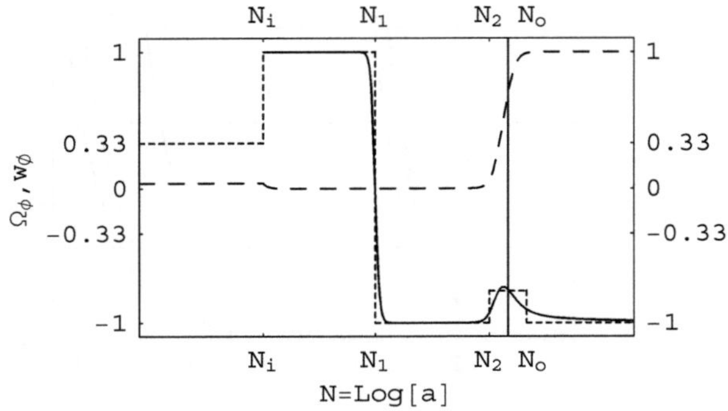

Fig. 2. We show the evolution of w_{DE} and Ω_{DE}, solid and dashed lines respectively for an IPL potential with $n = 1$ and $\Omega_{DEi} = 0.05$ as a function of $N = Log[a]$, where a is the scale factor. The dotted line represents the theoretical approximation w_{DE} and we see that it makes a good fit to the numerical solution. N_i is given at the initial scale Λ_c and N_1, N_2 give the end of the regions with $w_{DE} = 1, -1$ respectively while the solid vertical line at N_o denotes present day. Notice that for $N < N_i$ we are assuming that the energy density ρ_{DE} redshifts as radiation with $w_{DE} = 1/3$.

[13, 14] of CMBFAST [15] to include an energy density with a varying w, in a model independent way and see from its result if we can distinguish between different quintessence models, tracker, cosmological constant or other kinds of exotic energy densities.

In the case of a scalar field, we will assume that the scalar field appears at a scale Λ_c with an energy density $\Omega_{DE}(\Lambda_c)$. The late time appearance of the ϕ field suggests that a phase transition takes place creating the scalar field. We are not concern with the precise mechanism of its appearance (see [17, 30]). However, energy conservation would suggest that the energy density of the ϕ field after the phase transition would be given in terms of the energy density of the system before the phase transition and we will take them to be equal. It is natural to assume that all the energy density before the phase transition, in this sector, was relativistic. If the phase transition takes place after nucleosynthesis "NS" then the primordial creation of nuclei puts an upper limit to the relativistic energy density to be less than 0.1-0.2 of the critical energy density [31, 32]. If Λ_c is larger than the NS scale then we do not need to worry about the NS bound since independent of its initial value, Ω_{DE} will drop rapidly and remain small for a long period of time (covering NS).

In a chronological order, we would start with a universe filled with the SM particles and a DG sector (could be another gauge group) and with gravitational interaction between the two sectors only. In both sectors all fields start massless, i.e. they redshift as radiation. The evolution of the SM is the standard one and we have nothing new to say. However, the DG sector will have a phase transition at Λ_c leading to the appearance of a scalar field

ϕ with a potential $V(\phi)$, the quintessence field. Above Λ_c the fields in this sector behave as radiation. The evolution of ϕ for energies below Λ_c is that of a scalar field with given potential V. However, the precise form of V is unknown. In Table 1 we show the different model independent regions that we consider. The model dependence lies only on the size of these different periods and on the value of γ_{DE} in the last region.

From a cosmological point of view we have only 4 free parameters $\Delta N_1 = N_i - N_1, \Delta N_2 = N_1 - N_2$, $\Delta N_o = N_2 - N_o$ and γ_{DEo} (the value of γ_{DE} during the third period), where $N \equiv log[a]$ with a the scale factor. With these parameters we cover all models. The cosmological parameters can be expressed in terms of the field theoretical parameters $\Omega_{DEi}, \Lambda_c, y_{min}$ and γ_{DEo}.

Table 2. We show the different regions, its duration and the value of γ_{DE} in each region with $N = Log[a]$ and N_o is at present day.

Sector	Energy	Duration	$\gamma_{DE} = w_{DE} + 1$
Radiation	$E > \Lambda_c$	$N_i < N$	$4/3$
First	$E_1 < \Lambda_c$	$N_i < N < N_1$	2
Second	$E_2 < E_1$	$N_1 < N < N_2$	0
Third	$E_3 < E_2$	$N_o < N < N_2$	$\lambda^2 \Omega_{DE}/3$

5.1 Evolution of w

We have seen the evolution of x, y, H in the preceding subsection and we would like now to show how w_{DE} evolves in a general framework.

The evolution of the equation of state parameter, $\gamma_{DE} = 1+w_{DE}$, as given by (4) has a generic behavior for all scalar fields independent of its potential. We see that $(\gamma_{DE})_N = 0$ has three solutions, $\gamma_{DE} = 2, 0$ and $\lambda^2 \Omega_{DE}/3$ [35] (or $w_{DE} = 1, -1$ and $\lambda^2 \Omega_{DE}/3 - 1$). The parameter γ_{DE} will be most of the time in either of the three critical points. Independent of its initial value it will go quite rapidly to $\gamma_{DE} = 2$ and remain there for a long period of time. The fast increase in γ_{DE} is because $\lambda_i \gg 1$. This stage represents a scalar field whose kinetic energy density dominates ($E_k \gg V$), it is called the kinetic region, and the energy density redshifts as $\rho_{DE} \sim a^{-6} = e^{-6N}$. Afterwards it will sharply go to $\gamma_{DE} = 0$ and stay there during almost the same amount of time as in the first stage. This second period is valid when the potential energy dominates ($E_k \ll V$) and the energy density redshifts as a cosmological constant with $\rho_{DE} \sim a^0 \sim cte$. Finally it will go to its last period given by the tracker value $\gamma_{tr} = \lambda^2 \Omega_{DE}/3$ where it will remain. This last critical value $\gamma_{tr} = \lambda^2 \Omega_{DE}/3$ is not necessarily constant (λ, Ω_{DE} evolve in time). The energy density redshifts as a tracker field $\rho_{DE} \sim a^{-3\gamma_{tr}} = e^{-3N\gamma_{tr}}$.

Let us define the quantity

$$A \equiv \lambda\sqrt{\frac{\Omega_{DE}}{3\gamma_{DE}}}. \tag{46}$$

We see from (4) that the sign of $\gamma_{\phi N}$ depends if $A > 1$ or $A < 1$. For $A > 1$ we have $\gamma_{\phi N} \geq 0$ and the value $\gamma_{\phi max} = 2$ or $w_{DE} = 1$, which is the maximum value for γ_{DE}, is a stable point. For $A < 1$ we have $\gamma_{\phi N} \leq 0$ and the value $\gamma_{DE} = w_{DE} + 1 = 0$ will be a stable point also.

We will denote the beginning of the kinetic term by N_i and the end by N_1. The second period ($\gamma_{DE} = 0$) starts at N_1 and finishes at N_2 while the last period starts at N_2 and lasts until present day N_o. We have then, $\Delta N_1 \equiv N_1 - N_i$ and $\Delta N_2 \equiv N_2 - N_1$, the amount of e-folds during the $\gamma_{DE} = 2$ and $\gamma_{DE} = 0$ periods, respectively, and $\Delta N_o \equiv N_o - N_2$ the size of the tracking period.

First Period, w=1

At the initial time since $\lambda_i \gg 1$ we have $A > \lambda\sqrt{\Omega_{DEi}/6} \gg 1$ since $\gamma_{DE} \leq 2$. From (4) we see that the derivative $(\gamma_{DE})_N \gg 1$ and γ_{DE} will rapidly go to its maximum value 2. The period of $\gamma_{DE} \simeq 2$ remains valid for a long period of time since for $x(N_{min})^2 = \Omega_{DEi} \gg y^2_{min}$ one has $\gamma_{DE}(N_{min}) = 2x(N_{min})^2/(x(N_{min})^2 + y^2_{min}) \simeq 2(1 - y^2_{min}/x(N_{min})^2) \ll 1$. So we expect γ_{DE} to be close to two until $y \sim x$. We will have at the end of the period $N = N_1$, $\gamma_{DE} \sim 2$ and $\Omega_{DE}(N_1) = r_1/(1 + r_1) \ll 1$ with

$$r_1 \equiv \frac{\rho_{DE}(N_1)}{\rho_b(N_1)} = \frac{\rho_{DE}(N_i)}{\rho_b(N_i)}\ e^{-3(N_1-N_i)(2-\gamma_b)}. \tag{47}$$

Second Period, w=-1

The second stage starts when $1 \gg x \sim y$ and $\gamma_{DE} \sim 0$. The transition time between $\gamma_{DE} = 1.9$ and $\gamma_{DE} = 0.1$ is quite short, about $\Delta N = 1 - 1.5$, so we do not take it into account. In this second region we are still in the scaling regime with $yH = cte$ and since we have $\Omega_{DE} \ll 1$ we have $A \ll 1$ and $(\gamma_{DE})_N < 0$. The quantity γ_{DE} has been decreasing and it will arrive at its minimum possible value $\gamma_{DE} \simeq 0$ or $w_{DE} \simeq -1$. As long as $A < 1$ the value of $\gamma_{DE} \sim 0$ will remain constant.

During the second period we have, $\gamma_{DE} \sim 0, \phi \sim cte, \lambda = \lambda_{min}$ and the evolution of $\Omega_{DE}(N_2) = r_2/(1 + r_2)$ is given by

$$r_2 \equiv \frac{\rho_{DE}(N_2)}{\rho_b(N_2)} = \frac{\rho_{DE}(N_1)}{\rho_b(N_1)}\ e^{3(N_2-N_1)\gamma_b}. \tag{48}$$

Since in this period ρ_{DE} redshifts much slower than radiation or matter, Ω_{DE} will increase and A will eventually become larger than one again. The second period ends (as the scaling period) when (14) is no longer valid and the first term in the equation y_N of (2) cannot be neglected. This happens for $x(N_2) \sim \lambda(N_2)^{-1}$ (c.f. discussion below (16)).

Third Period, w=wtr

The third period starts when γ_{DE} is not too small (i.e. x is comparable to y and $\gamma_{DE} = O(1/10)$). During this region we will have $A > 1$ and a growing γ_{DE}. However, in this case γ_{DE} will not arrive at the maximum value $\gamma_{DE} = 2$ since λ is not very large and γ_{DE} will be stabilized at the critical point $\gamma_{\phi N} \simeq 0$ with

$$\gamma_{tr} = \lambda^2 \Omega_{DE}/3. \tag{49}$$

We will have $\Omega_{DE}(N_o) = r_3/(1 + r_3)$ with

$$r_3 \equiv \frac{\rho_{DE}(N_o)}{\rho_b(N_o)} = \frac{\rho_{DE}(N_2)}{\rho_b(N_2)}\; e^{-3(N_o - N_2)(\gamma_{tr} - \gamma_b)} \tag{50}$$

If $\gamma_{tr} < \gamma_b$ then $\Omega_{DE} \to 1$. While $\lambda^2 \Omega_{DE}$ remains constant we have the constant tracker value for γ_{DE} or w_{DE}. A constant γ_{DE} is possible when $\Omega_{DE} \ll 1$. However, at late times the attractor value will be $\gamma_{tr} \to 0$ and $\Omega_{DE} \to 1$ since Ω_{DE} is constrained to be smaller than one and $\lambda \to 0$. But, even for γ_{tr} not constant the evolution of γ_{tr} in (49) is valid and the value generalizes the tracker behavior. For an inverse power law potential $V = V_o \phi^{-n}$ we have $\lambda = n/\phi$ and $\gamma_{tr} = n^2 \Omega_{DE}/3\phi^2$ which is the valued obtained by [9],[18].

5.2 Duration of the Periods and Relation to the Field Parameters

In order to know the relative size of the different periods we can use (47) and (48). Combining both (47) and (48) we have

$$\frac{r_2}{r_i} = \frac{\rho_{DE}(N_2)\rho_b(N_i)}{\rho_b(N_2)\rho_{DE}(N_i)} = e^{-3\Delta N_1 (2-\gamma_b) + 3\Delta N_2 \gamma_b} \tag{51}$$

Solving for ΔN_2 in (51) we obtain [35]

$$\Delta N_2 = \Delta N_1 \left(\frac{2}{\gamma_b} - 1\right) + \frac{1}{3\gamma_b} Log\left[\frac{r_2}{r_i}\right] \tag{52}$$

If we use the result of quintessence evolution at the beginning and end of the scaling period $\Omega_{DE}(N_2) = \Omega_{DEi}(N_i)$ given in (17) we have $r_2 = r_i$. For matter, $\gamma_b = 1$, and (52) gives $\Delta N_2 = \Delta N_1$ while for radiation, $\gamma_b = 4/3$, and $\Delta N_2 = \Delta N_1/2$. The universe has been dominated by matter for a period of $N_o - N_{rm} \simeq 8$, where N_o stands for present day value and N_{rm} for the scale at radiation-matter equivalence.

Including the third period we have from (47), (48) and (50) [35]

$$\begin{aligned}\frac{r_3}{r_i} &= \frac{\rho_{DE}(N_o)\rho_b(N_i)}{\rho_b(N_o)\rho_{DE}(N_i)} = e^{-3\Delta N_1 (2-\gamma_b(N_1)) + 3\Delta N_2 \gamma_b(N_2) - 3\Delta N_o(\gamma_{tr} - 1)} \\ &= \frac{r_2}{r_i}\; e^{-3\Delta N_o (\gamma_{tr} - 1)},\end{aligned} \tag{53}$$

$$\Delta N_o = \frac{3}{1-\gamma_{tr}} Log\left[\frac{r_3}{r_2}\right] = \frac{3}{1-\gamma_{tr}} Log\left[\frac{\Omega_{DEo}}{1-\Omega_{DEo}}\frac{1-\Omega_{DE}(N_2)}{\Omega_{DE}(N_2)}\right] \tag{54}$$

where we have assumed that the third period is already at the matter dominated epoch, $\gamma_b(N_o) = 1$. If we take in (54) the equality $\Omega_{DEi}(N_i) = \Omega_{DE}(N_2)$ the size ΔN_o and the value of γ_{tr} will set the initial energy density Ω_{DEi}. Of course, on the other hand, if we know Ω_{DEi} then we can determine $\Delta N_o \gamma_{tr}$.

Let us now relate the four field parameters $\Lambda_c, \Omega_{DEi}, y_{min}, \gamma_{DEo}$ to the size of the different periods. The amount of e-folds between the initial time N_i at Λ_c and N_1, the scale where w goes from $w = 1$ to $w = -1$ is set by the condition $x \sim y \ll 1$. We use the evolution of x from (16) and (13) to get

$$\Delta N_1 \equiv N_1 - N_i = \frac{1}{3} Log\left[\frac{x(N_{min})}{y_{min}}\right] \tag{55}$$

were we have assumed $N_i \simeq N_{min}$. Equation (55) is independent of γ_b. We can take $x(N_{min}) = \sqrt{\Omega_{DEi}}$, $y_i \simeq \sqrt{\Omega_{DEi}}$ and for an IPL model we have $y_{min} \simeq y_i(\Lambda_c/\sqrt{\Omega_{DEi}})^{n/2}$ and (55) gives

$$\Delta N_1 = \frac{n}{6} Log\left[\frac{\sqrt{\Omega_{DEi}}}{\Lambda_c}\right]. \tag{56}$$

The amount of e-folds between the initial time N_i at Λ_c and the end of the scaling period N_2 is given by (16), (9) and (17) with $y \sim x$ but this time we have with $x = 1/\lambda(N_{min}) \sim \sqrt{\Omega_{DEi}}, y_i \sim \sqrt{\Omega_{DEi}}$ giving

$$\Delta N_1 + \Delta N_2 = N_2 - N_i = \frac{2}{3\gamma_b} Log\left[\frac{y_2}{y_{min}}\right] = \frac{n}{3\gamma_b} Log[\frac{\sqrt{\Omega_{DEi}}}{\Lambda_c}]. \tag{57}$$

Notice that (57) minus (56) gives ΔN_2 in (52). Summing (57) and (54) we have [35]

$$\Delta N_T \equiv N_o - N_i = \frac{n}{3\gamma_b} Log[\frac{\sqrt{\Omega_{DEi}}}{\Lambda_c}] + \frac{3}{1-\gamma_{tr}} Log\left[\frac{\Omega_{DEo}}{1-\Omega_{DEo}}\frac{1-\Omega_{DEi}}{\Omega_{DEi}}\right] \tag{58}$$

which gives the total scale ΔN_T between the initial time at Λ_c and present day and we taken $\Omega_{DEi} = \Omega_{DE}(N_2)$ in (58). Alternatively we can estimate the magnitude of the phase transition scale Λ_c. From $\Lambda_c \equiv V_i^{1/4} = (3\Omega_{DEi}H_i^2)^{1/4}$ and using the approximation that $\Omega_{DE} \ll 1$ during almost all the time between present day and initial time (at Λ_c) we have

$$H_i = H_o\, e^{3\gamma_b \Delta N_T/2} \tag{59}$$

giving a scale

$$\Lambda_c = (3\Omega_{DEi}H_o^2)^{1/4}\, e^{3\gamma_b \Delta N_T/4}. \tag{60}$$

The scale Λ_c increases with larger ΔN_T. From (58) and (60) we can derive the order of magnitude for Λ_c in terms of n and H_o giving $\Lambda_c \simeq H_o^{2/(4+n)}$

which is the well known result for IPL potentials [18]. If we know ΔN_T then we can determine Λ_c and the power n for IPL models. We see that the size of the different regions can be determine by the four parameters $\Lambda_c, \Omega_{DEi}, y_{min}$ and γ_{DEo}.

How long do the periods last depends on the models and by varying the size of $\Delta N_2, \Delta N_o$ and γ_{tr} we cover all models. If $\Delta N_o = 0$ and $\Delta N_2 > \Delta N_{rm} = N_o - N_{rm}$ then the model would be undistinguishable from a true cosmological constant $\gamma_{DE} = w_{DE} + 1 = 0$ since during all the matter domination era the equation of state would be $\gamma_{DE} = 0$. If we have $\Delta N_o > \Delta N_{rm}$ then the model reduces to tracker models with a constant γ_{DEo} during all the matter domination era. So, our model contains the tracker and cosmological constant as limiting cases.

More interesting is to see if we can determine the nature and scale of the dark energy. For this to happen a late time phase transition must take place such that Λ_c is at $\Delta N_T = O(\Delta N_{rm})$.

5.3 Analysis of CMB spectra

We will now analyze the generic behavior of a fluid with equation of state divided in four different regions with $w_{DE} = 1/3, 1, -1, w_{DEo}$ [35]. We will vary the sizes of the regions and we will determine the effect of having regions with $w_{DE} = 1/3$ or $w_{DE} = 1$ in contrast to a cosmological constant or a tracker field (with $-1 < w_{DEo} = cte < -2/3$). We compare the CMB spectra with the data given in RADPACK [16]. This analysis is valid for all kinds of fluids with the specific equation of state and it is also the generic behavior of scalar fields. We will compare to the model $w_{tr} = -0.82$ which was found to be a better fit to CMB than a true cosmological constant [6].

5.4 Effect of Radiation Period, w=1/3

The first section we have $w_{DE} = 1/3$ and the fluid (scalar field) redshifts as radiation. As long as the fluid has $w_{DE} = 1/3$ its energy density will remain the same compared to radiation. If during nucleosynthesis the fluid has $w_{DE} = 1/3$ then the BBNS bound requires the $\Omega_{DE}(NS) < 0.1 - 0.2$ [31, 32].

In Fig. 3 we show the different CMB for $w_{DE} = 1/3, 0, -1$ for $\Delta N_T = N_o - N_i = 9, \Delta N_2 = N_2 - N_1 = 4.5$ and $\Delta N_o = N_o - N_2 = 0$. We have chosen $\Delta N_T = 9$ because it is the smallest value satisfying the condition $\Delta N_1 = \Delta N_2, \Delta N_o = 0$ and giving the correct CMB spectrum. We have taken $w_{DE} = 1, -1$ for $N_i < N < N_1$ and $N_1 < N < N_2 = N_o$, respectively.

We see that the first and second peaks are suppressed for $w_{DE} = 1/3$ compared to $w_{DE} = -1$ while the third peak is enhanced. The positions of the first two peaks is basically the same and the position of the third peak is moved from 868 to 864 (0.4%), for $w = 1/3, -1$ respectively. For smaller N_i, i.e. more distant from present day, the effect is suppressed. It is not surprising

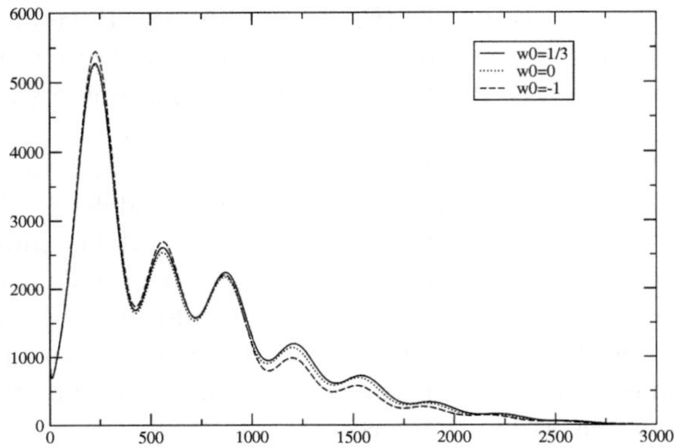

Fig. 3. We show the effect of the radiation on the CMB era for $N < N_i$ by changing $w_{DE} = 1/3, 0, -1$ with $\Delta N_T = \Delta N_1/2\Delta N_2/2 = 9$. The vertical axes is $l(l+1)c_l/2\pi(\mu K^2)$.

since the N_i would be further way from energy-matter equality and its effect on CMB would be less important. The total χ^2 obtained by comparing the CMB spectrum with the data [16] gives $\chi^2 = 75, 74, 80$ for $w_{DE} = 1/3, 0, -1$ respectively.

5.5 Effect of First Period, w=1

In Fig. 4 we show the CMB for different values of $w_{DE} = 1, 0, -1$ during $N_i < N < N_1$ and take $w = 1/3$ for $N < N_i$ while $w_{DE} = -1$ for $N_1 < N_2 = N_o$. The effect of having a kinetic period enhances the first three peaks and shifts the spectrum to higher modes, i.e. higher l. The curve for $w_{DE} = 0$ is indi-

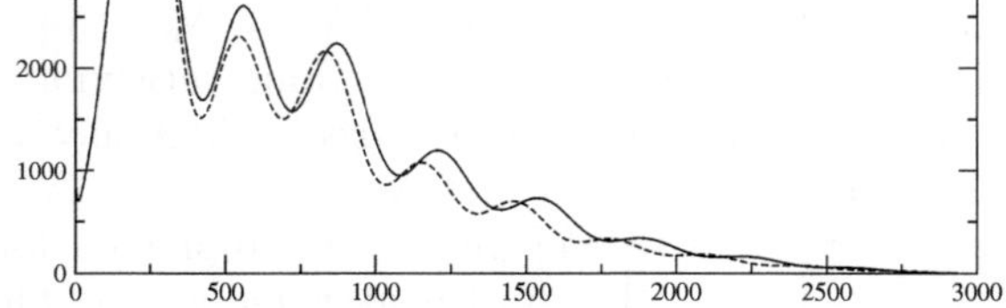

Fig. 4. We show the effect of the kinetic era with $N_i < N < N_1$ by varying $w_{DE} = 1, 0, -1$ with $\Delta N_T = \Delta N_1/2 = \Delta N_2/2 = 9$. The curves with $w_{DE} = 0$ and $w_{DE} = -1$ cannot be distinguished. The vertical axes is $l(l+1)c_l/2\pi(\mu K^2)$.

stinguishable from the $w_{DE} = -1$ one. The position and height of the peaks are $p_1 = (227, 5275), p_2 = (559, 2605), p_3 = (868, 2240)$ for $w_{DE} = 1$ while for $w_{DE} = -1$ we have $p_1 = (224, 5138), p_2 = (545, 2310), p_3 = (832, 2165)$ giving a percentage difference $p_1 = (1.3\%, 2.6\%), p_2 = (2.5\%, 12.7\%), p_3 = (4.3\%, 3.4\%)$. We see that the largest discrepancy is the altitude of the second peak. The total χ^2 gives 75, 78, 78 for $w = 1, 0, -1$ respectively.

The difference in height and positions may in principle distinguish between a cosmological constant and a scalar field, or any fluid with the specific equation of state behavior.

5.6 Equal Length Periods

We have studied the case with $\Delta N_1 = \Delta N_2$, $\Delta N_o = 0$. In Fig. 5 we show the behavior for different values of $\Delta N_T = 2\Delta N_2 = 6, 8, 9, 12, 16$ giving a total χ^2 of 1685, 465, 75, 75, 78, respectively.

There is a lower limit of ΔN_T that gives an acceptable CMB spectrum. The lower limit is $\Delta N_T \geq 9$. For smaller ΔN_T the peaks move to the right of the spectrum and the height increases giving a spectrum not consistent with the CMB data.

For larger $\Delta N_T > 9$ the curves tend to the cosmological constant. It is not surprising since for large $\Delta N_T = 2\Delta N_2$ it means that we have a larger time with $w = -1$ and in the case that $\Delta N_2 > N_o - N_{rm}$ the universe content, after matter radiation equality, would have been given by matter and a fluid with $w_{DE} = -1$, i.e. a cosmological constant.

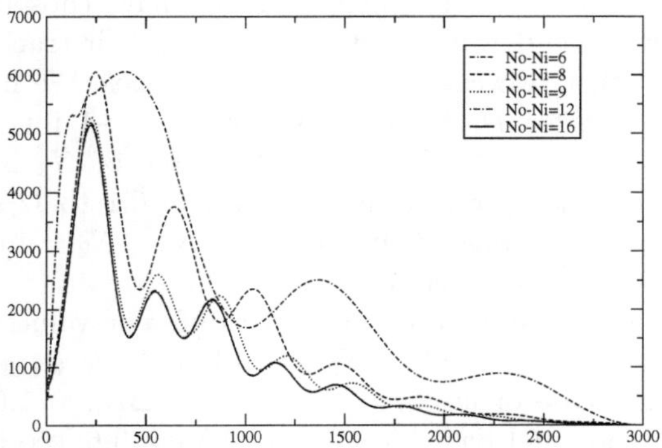

Fig. 5. We show the effect on the CMB by varying the $\Delta N_T = \Delta N_1/2 = \Delta N_2/2 = 6, 8, 9, 12, 16$ with $\Delta N_o = 0$. The curves with $\Delta N_T = 12, 16$ cannot be easily distinguished. The vertical axes is $l(l+1)c_l/2\pi(\mu K^2)$.

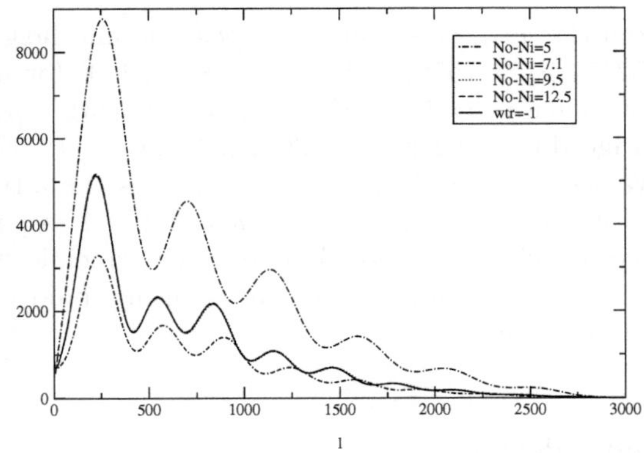

Fig. 6. We show the effect on the CMB by varying the $\Delta N_T = N_o - N_i = 5, 7.1, 9.5, 12.5$, with the constraint $\Omega_{DEi}(N_i) = \Omega_{DE}(N_2) = 0.1$ and $w_{DEo} = -1, \Delta N_2 = 1.03$ and we include the cosmological constant $w_{tr} \equiv -1$, for comparison. The curves with $\Delta N_T = 9.5, 12.5$ cannot be easily distinguished from the cosmological constant. The vertical axes is $l(l+1)c_l/2\pi(\mu K^2)$.

5.7 Scaling Condition

Following the discussion in Sect. 2.2, we now that a scalar field will end up its scaling period with a Ω_{DE} equal to its starting value, i.e. $\Omega_{DE}(N_i) = \Omega_{DE}(N_2) = 0.1$. We have taken this value of Ω_{DE} since for $N > N_i$ the energy density behaved as radiation and we have to impose the nucleosynthesis bound on relativistic degrees of freedom $\Omega_{DE}(NS) \leq 0.1 - 0.2$. Imposing this condition we have determined the evolution of the CMB for three different values of $w_{DEo} = -1, -0.82, -0.7$. We have chosen to analyze the $w_{DEo} = -0.82$ because it was found to be the best fit tracker model by [6] and we take the other two cases as the extreme ones. We have $w_{DE} = 1/3$ for $N \leq N_i$, $w = 1$ for $N_i \leq N \leq N_1$, $w_{DE} = -1$ for $N_1 \leq N \leq N_2$ and $w_{DEo} = w_{tr}$ for $N_2 \leq N \leq N_o$. The value of N_2 is determined so that the energy density grows from $\Omega_{DE}(N_2) = 0.1$ to $\Omega_{DE} = 0.7$ today. This conditions sets the range of the period to $N_o - N_2 = 1.03, 1.25, 1.47$ for $w_{DEo} = -1, -0.82, -0.7$ respectively.

In Figs. 6 and 7 we show the curves for different values of N_i with the restriction that $\Omega_{DE}(N_i) = \Omega_{DE}(N_2) = 0.1$ and for $w_{DEo} = -1, -0.82$, respectively. In the case of $w_{DEo} = -1$ we have $\Delta N_o = 1.03$ and that the smallest acceptable model has $\Delta N_T = 8.5$, $\Delta N_2 = 3.6$, see Fig. 6. The best model has $\Delta N_T = 8.88, N_1 - N_o = 3.7$ and peaks $p_1 = (224, 5133), p_2 = (549, 2363), p_3 = (840, 2178)$ with $\chi^2 = 75$. The total χ^2 obtained gives $697, 597, 77.3, 75.3$ for $\Delta N_T = 5, 7.1, 9.5, 12.5$, respectively, and $\chi^2 = 78$ for a cosmological constant.

For $w_{DEo} = -0.82$, Fig. 7 we have $\Delta N_o = N_o - N_2 = 1.25$ and the minimum acceptable distance is $\Delta N_T = N_i - N_o = 7.19, \Delta N_2 = 3.9$. Smal-

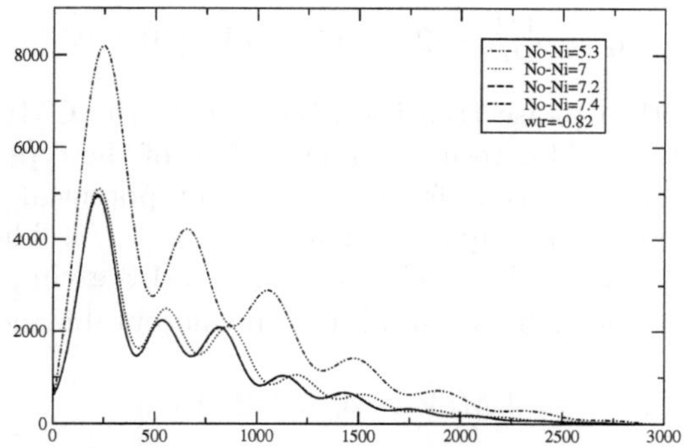

Fig. 7. We show the effect on the CMB by varying the $\Delta N_T = N_o - N_i = 5.3, 7, 7.2, 7.4$, with the constraint $\Omega_{DEi}(N_i) = \Omega_{DE}(N_2) = 0.1$ and $w_{DEo} = -0.82, \Delta N_2 = 1.25$. We also include the tracker with constant $w_{tr} \equiv -0.82$. The curves with $\Delta N_1 = 7.2, 7.4$ cannot be distinguished from the tracker one. The vertical axes is $l(l+1)c_l/2\pi(\mu K^2)$.

ler values of ΔN_T give a spectrum with peaks too large and second and third peaks moved to the right (high l modes). For large ΔN_T the spectrum tends to the tracker spectrum $w_{tr} = -0.82$. The total χ^2 obtained gives $1004, 1285, 78, 75$ for $\Delta N_T = 5.3, 7, 7.2, 7.4$, respectively, and $\chi^2 = 98$ for the tracker $w_{tr} = -0.82$.

The best model has $\Delta N_T = N_1 - N_o = 7.4, \Delta N_2 = 3$ with peaks and position $p_1 = (225, 5101), p_2 = (555, 2493), p_3 = (860, 2110)$ and it has a better fit than the tracker model with constant $w_{tr} = -0.82$ which was found to be the best tracker fit [6]. We see that having a dynamical w_{DE} is not only more reasonable from a theoretical point of view but it fits the data better.

Finally, we consider $w_{DEo} = -0.7$ for $N > N_2$. In this case we have $\Delta N_o = N_o - N_2 = 1.47$ and the minimum acceptable model has $\Delta N_T = 6.8, \Delta N_2 = 3.6$, while the best model has $\Delta N_T = 7.3, \Delta N_2 = 3.8$ with peaks $p_1 = (222, 4954), p_2 = (550, 2422), p_3 = (853, 2035)$. The total χ^2 obtained gives $711, 218, 82$ for $\Delta N_T = 5.5, 6.8, 7.3$, respectively, and $\chi^2 = 144$ for the tracker $w_{tr} = -0.7$

We see that in all three cases $w_{DEo} = -1, -0.82, -0.7$, with condition $\Omega_{DE}(N_i) = \Omega_{DE}(N_2) = 0.1$ we have a minimum acceptable value of ΔN_T and for smaller ΔN_T the peaks move to the right of the spectrum and the height of the peaks increases considerably. This conclusion is generic and sets a lower limit to ΔN_T, the distance to the phase transition scale Λ_c, or equivalently it sets a lower limit to Λ_c.

The smallest ΔN_T is set by the largest acceptable w_{DEo} (here we have taken it to be $w_{DEo} = -0.7$) giving in our case a $\Delta N_T = 6.8$ for $\Omega_{DEi} = 0.1$. This result puts a constraint on how late the phase transition can take place. In terms of the energy $\Lambda_c = \rho_{DEi}^{1/4} = [\Omega_{DEi} 3H_i^2]^{1/4}$ we can set a lower value for the transition scale. Using (60) with $\Omega_{DEi} = 0.1$ and$\Delta N_T = 6.82$ we get

$$\Lambda_c = \rho_{DEi}^{1/4} = 2 \times 10^{-10}\, GeV = 0.2\ eV \tag{61}$$

i.e. for models with a phase transition below (61) the CMB will not agree with the observations. This result is independent of the type of potential.

Furthermore, we now that for inverse power potential there is an upper limit to Λ_c coming by requiring that $w_{DEo} < -2/3$. The limiting value assuming $\Omega_{DEi} \leq 0.1$, for $V = \Lambda^{4+n}\phi^{-n}$, is $n < 1.8$ giving $\Lambda_c = 4\, MeV \simeq H_o^{2/(4+n)}$. Therefore, for IPL potentials the only acceptable models have phase transition scale

$$4\ MeV > \Lambda_c > 0.2\ eV. \tag{62}$$

6 Conclusions

We have studied the dark energy necessary for explaining the positive acceleration and flatness of the universe and structure formation. We have derived the model independent evolution of quintessence,i.e. a scalar field with only gravitational coupling with the SM particles.

We proposed a quintessence model based on a non-abelian asymptotically free gauge group. This group forms dynamically gauge invariant particles below the condensation scale (as mesons and baryons in QCD) and it is these scalar condensates that acquire a non trivial potential V and parameterize the quintessence field. We have shown that an unification scheme, where all coupling constants are unified, as predicted by string theory, leads to an acceptable dark energy parameterized in terms of the condensates of a non-abelian gauge group. Above the unification scale we have all fields in thermal equilibrium and the number of degrees of freedom for the SM and DG model determines the initial conditions for each group. Below Λ_{gut} the DG group decouples, since it interacts with the SM only through gravity. For temperatures above the condensation scale of the DG group its fields are relativistic and redshift as radiation. Below Λ_c we have the gauge invariant condensates.

We have also studied the possibility that the dark gauge group contains the dark matter and energy. The allowed values of the different parameters are severely restricted by different considerations. However, the constraints on the dark energy and dark matter overlap allowing for the possibility of having a gauge group containing both dark energy and dark matter. The NS constraint on g_{DG} sets a limit to the dark energy density at Λ_c of $\Omega_{DGc} \leq 0.17$. The evolution and acceptable values of DM and DE leads to a constraint of Λ_c and n giving $0.21\, eV \leq \Lambda_c \leq 3645\, eV$ and $0.24 \leq n \leq 0.104$ for $g_{dec} \leq 901$. The evolution of the quintessence field requires also a small n in order to have a small w_{DEo}. For $\Omega_{DE} \leq 0.17$ and $w_{DEo} \leq -0.78$ one needs $n < 1.6$. On the other hand, the analysis of the CMB spectrum sets also a lower scale for the condensation scale $\Lambda_c > 0.2\ eV$ with $n > 0.27$. So, from three different analysis (quintessence, dark matter and CMB spectrum) we are led to conclude that the most acceptable models have a low condensation scale Λ_c of the order of $1 - 10^3\,eV$. The fact that the condensation is low explains why the acceleration of the universe is at such a late time.

Finally, we have analyzed the CMB spectra for a fluid with an equation of state that takes different values. The values are $w_{DE} = 1/3, 1, -1, w_{DEo}$ for N in the regions $N_i - N_{Planck}, N_1 - N_i, N_2 - N_1, N_o - N_2$, respectively. The results are independent of the type of fluid we have. The cosmological constant and the tracker models are special cases of our general set up.

We have shown that the evolution of a scalar field, for any potential that leads to an accelerating universe at late times, has exactly the kind of behavior described above. It starts at the condensation scale Λ_c and enters a period with $w_{DE} = 1$, then it undergoes a period with $w_{DE} = -1$ and finally ends up in a region with $-1 \leq w_{DEo} \leq -2/3$. We have shown that the energy density at the end of the scaling period (end of $w_{DE} = -1$ region) has the same energy ratio as in the beginning, i.e. $\Omega_{DE}(N_i) = \Omega_{DE}(N_2)$. The time it spends on the last region depends on the value of $\Omega_{DE}(N_2)$ and on w_{DEo} during this time. Before the phase transition scale Λ_c we are assuming that all particles were at thermal equilibrium and massless in the quintessence sector. At the phase transition scale Λ_c the particles acquire a mass and a non trivial potential.

We have shown that models with $w_{DE} = 1/3, 1, -1, w_{DEo}$ have a better fit to the data then tracker or cosmological constant. Furthermore, we have determined the effect on the CMB of the first two periods $w_{DE} = 1/3$ and $w_{DE} = 1$ compared to a cosmological constant and even though the effect is small it is nonetheless observable.

In general, the CMB spectrum sets a lower limit to ΔN_T, which implies a lower limit to the phase transition scale Λ_c. For smaller ΔN_T the CMB peaks are moved to the right of the spectrum and the height increases considerably. For any Ω_{DEi} the CMB sets a lower limit to the phase transition scale. In the case of $\Omega_{DEi}(N_i) = 0.1$ the limit is $\Lambda_c = 0.2eV$ for any scalar potential. We do not take Ω_{DEi} much larger because we should comply with the NS bound on relativistic degrees of freedom $\Omega_{DEi} \leq 0.1 - 0.2$. If we take $\Omega_{DEi} \ll 0.1$ then the constraint on the phase transition scale will be less stringent since the effect of the scalar field is only relevant recently ($\Omega_{DE} \ll 1$ during all the time before present time). For inverse power law potentials we can also set an upper limit to Λ_c and for $\Omega_{DEi} \leq 0.1$ it gives an inverse power $n \leq 1.8$ and $\Lambda_c \leq 4 \times\ MeV$. In this class of potentials only models with $4\,MeV > \Lambda_c > 0.2\,eV$ would give the correct w_{DEo} and CMB spectrum.

This work allows for the possibility of distinguishing the kind of physical process that gives rise to the dark energy.

Acknowledgments

We would like to thank useful discussions with C. Terrero. This work was supported in part by CONACYT project 32415-E and DGAPA, UNAM project IN-110200.

References

1. D.N. Spergel et al. astro-ph/0302209
2. A.G. Riess *et al.*, Astron. J. 116 (1998) 1009; S. Perlmutter *et al*, ApJ 517 (1999) 565; P.M. Garnavich *et al*, Ap.J 509 (1998) 74.
3. G. Efstathiou, S. Maddox and W. Sutherland, Nature 348 (1990) 705. J. Primack and A. Klypin, Nucl. Phys. Proc. Suppl. 51 B, (1996), 30
4. P.S. Corasaniti astro-ph/0210257;P.S. Corasaniti, E.J. Copeland astro-ph/0205544,S. Hannestad, E. Mortsell Phys.Rev.D66:063508,2002; J. P. Kneller, G. Steigman astro-ph/0210500
5. S. Perlmutter, M. Turner and M. J. White, Phys.Rev.Lett.83:670-673, 1999; T. Saini, S. Raychaudhury, V. Sahni and A.A. Starobinsky, Phys.Rev.Lett.85:1162-1165,2000
6. Carlo Baccigalupi, Amedeo Balbi, Sabino Matarrese, Francesca Perrotta, Nicola Vittorio, Phys.Rev. D65 (2002) 063520
7. Michael Doran, Matthew J. Lilley, Jan Schwindt, Christof Wetterich, astro-ph/0012139; Michael Doran, Matthew Lilley, Christof Wetterich astro-ph/0105457
8. P.S. Corasaniti, E.J. Copeland, Phys.Rev.D65:043004,2002
9. I. Zlatev, L. Wang and P.J. Steinhardt, Phys. Rev. Lett.82 (1999) 8960; Phys. Rev. D59 (1999)123504
10. A. Knebe, J. Devriendt, B. Gibson, J. Silk astro-ph/0302443
11. V. Avila-Reese, P. Colin, O. Valenzuela, E. D'Onghia, C. Firmani, ApJ, 559 (2001) 516; V. Avila-Reese, C. Firmani. astro-ph/0202195
12. J.M. Bardeen, J.R. Bond, N. Kaiser, A.S. Szalay, ApJ,304(1986)15
13. see www.camb.info by A. Lewis and A. Challinor
14. C. Terrero and A. de la Macorra
15. M. Tegmark, M. Zaldarriaga, A. J. S. Hamilton Phys.Rev. D63 (2001) 043007
16. L.Knox, http://www.cita.utoronto.ca/ knox/radical.html
17. A. de la Macorra and C. Stephan-Otto, Phys.Rev.Lett.87:271301,2001
18. A. de la Macorra and C. Stephan-Otto, Phys.Rev.D65:083520,2002
19. P.J.E. Peebles and B. Ratra, ApJ 325 (1988) L17; Phys. Rev. D37 (1988) 3406
20. J.P. Uzan, Phys.Rev.D59:123510,1999
21. A. de la Macorra, Int.J.Mod.Phys.D9 (2000) 661
22. I. Affleck, M. Dine and N. Seiberg, Nucl. Phys.B256 (1985) 557
23. P. Binetruy, Phys.Rev. D60 (1999) 063502, Int. J.Theor. Phys.39 (2000) 1859
24. A. Masiero, M. Pietroni and F. Rosati, Phys. Rev. D61 (2000) 023509
25. A.R. Liddle and R.J. Scherrer, Phys.Rev. D59, (1999)023509
26. A. de la Macorra and G. Piccinelli, Phys. Rev.D61 (2000) 123503
27. E.J. Copeland, A. Liddle and D. Wands, Phys. Rev. D57 (1998) 4686
28. P. Brax, J. Martin, Phys.Rev.D61:103502,2000

29. C.P. Burgess, A. de la Macorra, I. Maksymyk and F. Quevedo Phys.Lett.B410 (1997) 181
30. A. de la Macorra JHEP01(2003)033 (hep-ph/0111292)
31. K. Freese, F.C. Adams, J.A. Frieman and E. Mottola, Nucl. Phys. B 287 (1987) 797; M. Birkel and S. Sarkar, Astropart. Phys. 6 (1997) 197.
32. C. Wetterich, Nucl. Phys. B302 (1988) 302, R.H. Cyburt, B.D. Fields, K. A. Olive, Astropart.Phys.17:87-100,2002
33. U. Amaldi, W. de Boer and H. Furstenau, Phys. Lett.B260 (1991) 447, P.Langacker and M. Luo, Phys. Rev.D44 (1991) 817
34. A. de la Macorra astro-ph/0212275
35. A. de la Macorra astro-ph/0211519 (to appear in PRD)
36. K. Intriligator and N. Seiberg, Nucl.Phys.Proc.Suppl.45BC:1-28,1996
37. M. Dine, R. Rhom, N. Seiberg and E. Witten, Phys. Lett. B156 (1985) 55, G. Veneziano and S. Yankielowicz, Phys. Lett. B113 (1984)231; S. Ferrara, L. Girardello and H.P. Nilles, Phys. Lett. B125 (1983) 457, D. Amati, K. Konishi, Y. Meurice, G. Rossi and G. Veneziano, Phys. Rep. 162 (1988) 169

Quintessential Inflation at the Maxima of the Potential

Gabriel Germán[1] and Axel de la Macorra[2]

[1] Centro de Ciencias Físicas, Universidad Nacional Autónoma de México, Apartado Postal 48-3, 62251 Cuernavaca, Morelos, México. gabriel@fis.unam.mx

[2] Instituto de Física, Universidad Nacional Autónoma de México, Apartado Postal 20-364, 01000 México D.F., México. macorra@fisica.unam.mx

Abstract. There is great interest in understanding a possible late accelerated expansion of the universe. Data suggest that the universe was still decelerating around redshift 1 and started to accelerate more recently at redshift 0.5. In models where the expansion is driven by a cosmological constant the acceleration should become increasingly greater with time, thus inflation never ends. This could also be the case with most models of quintessence or quintessential inflation where the late accelerated expansion is produced by a monotonically decreasing scalar potential. Here we would like to explore the possibility that a recent inflation has already or is about to end. This possibility is not ruled out by existing data and could be testable with far more, higher accuracy, supernovae on the Hubble diagram. We construct a two-stage inflationary model which can accommodate early inflation as well as a second stage of inflation (quintessence) with a single scalar field ϕ. Using an analogy from a mechanical problem we propose an inflaton field solution to the equations of motion which can account for two inflationary epochs. Inflation occurs close to the maxima of the potential. As a consequence both inflations are necesarilly finite. A first inflation is produced when fluctuations displace the inflaton field from its higher maximum rolling down the potential as in new inflation. Instead of rolling towards a global minimum the inflaton approaches a lower maximum where a second inflation takes place. The model is not realistic, however, because matter has not been taken into account at the end of the first inflation where particle production should occur as in non-oscillatory models. This is a delicate problem which will be treated elsewhere.

1 Introduction

The idea that the universe underwent an early inflationary expansion is now widely accepted [1]. This era of inflation makes plausible certain initial conditions for standard cosmology and provides a mechanism for structure formation. More speculatively the idea that the universe is at present undergoing inflation (usually denoted by the term quintessence) is the subject of much current interest [2]. Several models have been proposed where typically the potential energy of a scalar field, in general different from the one producing early inflation, is dominating the dynamics of the universe. Usually the potential is an inverse power of the field decreasing monotonically towards

G. Germán and A. de la Macorra, Quintessential Inflation at the Maxima of the Potential, Lect. Notes Phys. **646**, 259–271 (2004)
http://www.springerlink.com/

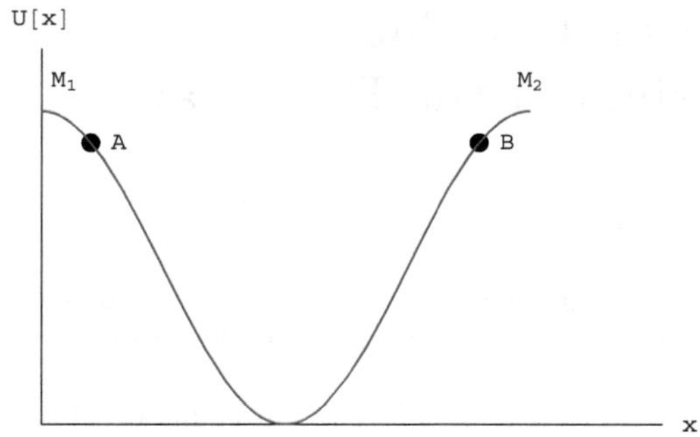

Fig. 1. The potential energy $U(x)$ of a particle in classical mechanics. In the absence of friction if we leave the particle at the point A with vanishing velocity it will eventually reach B, with $U(A) = U(B)$, also with vanishing velocity. In the limit when $A \to M_1$ it will take an infinite amount of time for the particle to reach $B \to M_2$.

zero. In the present work we are interested in studying a model which accommodates two stages of inflation by the evolution of a single scalar field [3]. Here, however, we look at the possibility that both inflations are produced when the inflaton is close to the maxima of the potential. The fact that both inflations occur at the maxima implies that they are necesarilly finite. This opens the interesting possibility where the second inflation has already or is about to end. This possibility is not ruled out by existing data and could be testable with far more, higher accuracy, supernovae on the Hubble diagram [4]. In what follows we construct a two-stage inflationary model by using an analogy with a problem from classical mechanics. The resulting potential could be obtained from supergravity (see Appendix).

Let us consider a potential $U(x)$ as shown in Fig. 1. When there is no friction the equation of motion for a particle of mass $m = 1$ is given by

$$\ddot{x} + U'(x) = 0. \tag{1}$$

We study the problem of a particle that leaves with vanishing velocity somewhere from the left of the minimum, let us say A and reaches B some time later. If we fix the origin of time at the minimum of $U(x)$ then the particle leaves A in the past reaching B sometime in the future. As A becomes close to the maximum at M_1 the particle spends longer close to the maxima. In the limit when $A \to M_1$ it takes an infinite amount of time for the particle to reach M_2. The particle would spend most of the time leaving M_1 and trying to reach M_2. As a result the kinetic energy is negligible close to the maxima; the potential energy dominates. We call this the limiting solution. The maximum at M_1 is located at $x = 0$ thus we require $x(t = -\infty) = 0$ and $x(t = +\infty)$ locates the maximum at M_2.

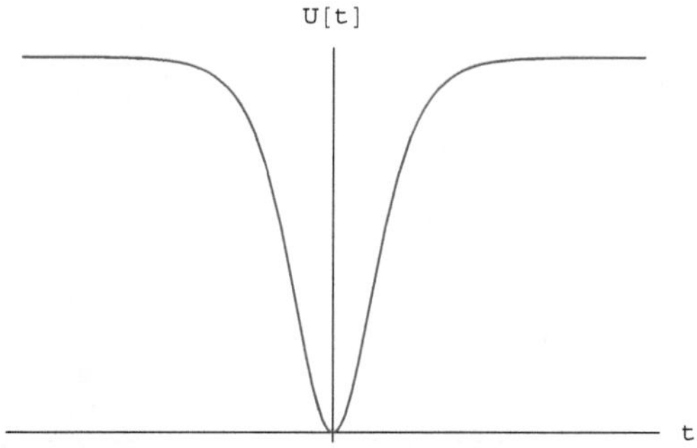

Fig. 2. The potential energy of the particle of Fig. 1 now as a function of time. The particle spends most of its time close to the maxima M_1, M_2 where the kinetic energy is negligible.

As a concrete example let us consider the potential

$$U(x) = \cos^2(x). \tag{2}$$

It is easy to check that the limiting solution is

$$x(t) = 2\arctan[\tanh(\frac{t}{\sqrt{2}})] + \frac{\pi}{2}, \tag{3}$$

where $x(t=-\infty)=0$, $x(t=+\infty)=\pi$ and $\dot{x}(t=-\infty)=\dot{x}(t=+\infty)=0$. The potential $U(x)$ is already illustrated in Fig. 1. As a function of time the potential is shown in Fig. 2. If we could lower the r.h.s. branch of this potential we could use this mechanical problem as an analogy to construct a model with two stages of inflation. Actually this can be done as follows. Instead of the potential $U(x)$ let us consider a new potential $\bar{U}(x)$ illustrated in Fig. 3. Now the maximum at M_2 is much smaller than the maximum at M_1. If we impose a solution of the type given by (3) it is clear that we need a friction term in the corresponding (1) to stop the particle precisely at M_2. Thus imposing a limiting solution to the potential $\bar{U}(x)$ determines the friction term and, as before the particle will spend most of the time close to M_1 and close to M_2 with negligible kinetic energy. As a function of time the potential of Fig. 4 shows the two plateaus at $t \to -\infty$ and $t \to +\infty$ corresponding to the maxima at M_1 and M_2 respectively.

In inflationary models of the "new" type one typically starts with a very flat potential and inflation occurs close to the maximum at $\phi = 0$, where ϕ is the inflaton field. There could be a previous "primordial" stage of inflation probably of the chaotic type setting the initial conditions for new inflation. For simplicity in what follows we will call this new inflationary epoch a first stage or simply first inflation (although probably there was inflation before) characterized by a scale Λ_1. This scenario is illustrated in Fig. 5. Here we study the possibility of a second stage of inflation at a scale Λ_2, where $\Lambda_2 \ll$

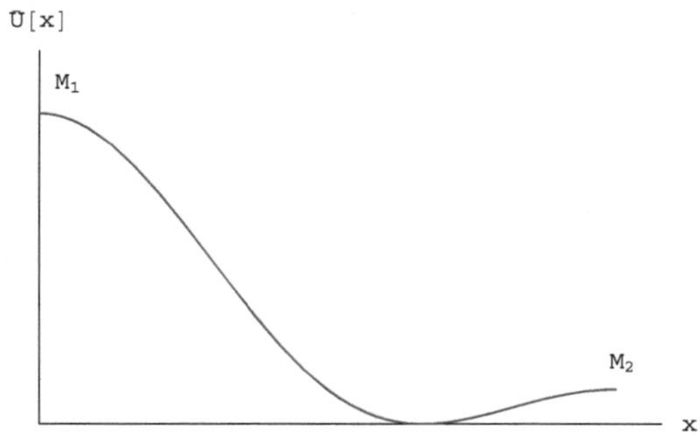

Fig. 3. A particle leaves the maximum at M_1 with vanishing velocity. It will just reach M_2 also with vanishing velocity if there is a friction term which stops the particle precisely at M_2.

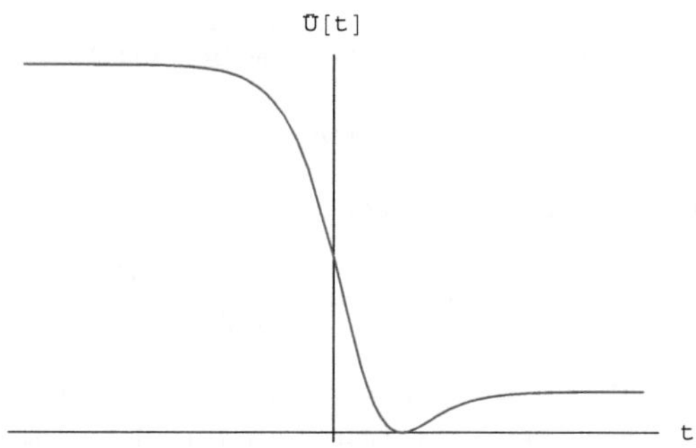

Fig. 4. The potential $\bar{U}(x)$ of Fig. 3 as a function of time. With an appropriate friction term the particle which leaves M_1 with vanishing velocity will just manage to reach M_2 in an infinite time. The particle spends most of its time close to the maxima with vanishing kinetic energy. Two plateaus appear at different energy scales.

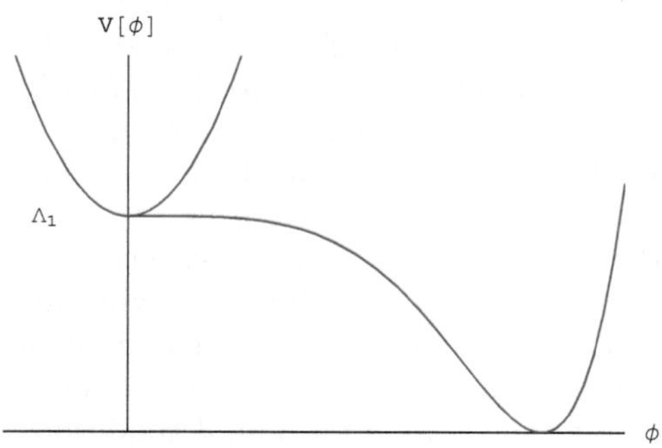

Fig. 5. A typical new inflationary potential gives rise to an early epoch of inflation (which in this work we call first inflation). Note, however, that there could have been a previous "primordial" stage of inflation providing initial conditions for new inflation to occur.

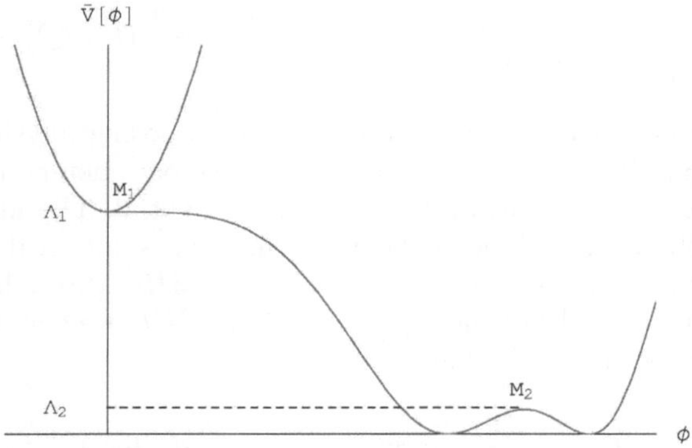

Fig. 6. The scenario studied here requires the presence of a second maximum at a scale Λ_2. There is a solution to the field equations wich makes the inflaton (due to the expansion of the universe) to evolve very slowly close to both maxima thus providing an inflationary model with two stages of inflation. In the limiting solution it takes an infinite amount of time for the scalar field to reach M_2 starting to roll from M_1. In a more realistic situation fluctuations displace ϕ from the maximum at M_1 thus allowing for the possibility of a finite second inflation. The scalar field ending in oscillations around one of its minima (see Fig. 9).

Λ_1. The mechanical analogy indicates that the second inflation will occur also close to a maximum, we then expect something like Fig. 6.

In Sect. 2 we construct a two-stage inflationary model. Section 3 deals with the problem of initial conditions and conclude in Sect. 4 with a discussion of the main results.

2 The Model

The inflaton field and Friedmann's equations are, as usual, given by

$$\ddot{\phi} + 3H\dot{\phi} + V'(\phi) = 0, \tag{4}$$

$$3H^2 = V + \frac{1}{2}\dot{\phi}^2, \tag{5}$$

where we have set the reduced Planck mass $M = 2.44 \times 10^{18} GeV$ to unity. The equations above can be rewritten as

$$3H^2 + \dot{H} = V, \tag{6}$$

$$\dot{H} = -\frac{1}{2}\dot{\phi}^2, \tag{7}$$

where the dot means derivative w.r.t. cosmic time and a prime denotes a field derivative. The limiting solution is conveniently written as

$$\phi(t) = \frac{2b}{\sqrt{6(1-b^2)}}\left(2\arctan[\tanh[\frac{\sqrt{3}a(1-b)}{2b}(t+t_o)]] + \frac{\pi}{2}\right), \tag{8}$$

where the parameters a, b, t_o are determined by imposing physical conditions on the potential. The peculiar way in which these parameters are introduced above simplifies the analysis on the resulting potential. The main difference w.r.t. the mechanical problem is that the limiting solution determines (up to an integration constant) the "friction" term $3H\dot{\phi}$ through (7) *and* the potential through (6). Equivalently by providing $H(t)$ we could get $\phi(t)$ and V. The derivative of the Hubble function is

$$\dot{H} = -a^2\frac{1-b}{1+b}sech^2[\frac{\sqrt{3}a(1-b)}{b}(t+t_o)]. \tag{9}$$

Integrating this expression we get the Hubble function

$$H(t) = \frac{a}{\sqrt{3}(1+b)}\left(1 - b\tanh[\frac{\sqrt{3}a(1-b)}{b}(t+t_o)]\right), \tag{10}$$

the arbitray integration constant $a/(\sqrt{3}(1+b))$ has been chosen this way to get an overall scale for $H(t)$ and to guarantee a positive Hubble function for $b < 1$. This choice also makes the potential very simple when $t \to -\infty$. From (6) the potential follows

$$V(t) = \frac{a^2}{(1+b)^2}\left(b - \tanh[\frac{\sqrt{3}a(1-b)}{b}(t+t_o)]\right)^2. \tag{11}$$

We can invert (8) and from (11) obtain the potential as a function of ϕ, this is given by

$$V(\phi) = \frac{a^2}{(1+b)^2}\left(b + \cos[\sqrt{\frac{3(1-b^2)}{2b^2}}\phi]\right)^2. \tag{12}$$

A potential of this type could follow from supergravity (see Appendix). The parameters are determined from the following conditions. The potential at $t \to -\infty$ should reach the scale Λ_1 of the first inflation, this fixes a

$$V(t \to -\infty) = a^2 \equiv \Lambda_1^4, \qquad \Rightarrow a = \Lambda_1^2. \tag{13}$$

At $t \to +\infty$ the potential reaches the second scale Λ_2 of inflation, this fixes b

$$V(t \to +\infty) = a^2\frac{(1-b)^2}{(1+b)^2} \equiv \Lambda_2^4, \qquad \Rightarrow b = \frac{1-d}{1+d}; \qquad d \equiv (\frac{\Lambda_2}{\Lambda_1})^2. \tag{14}$$

Actually, (14) has two solutions for b but the second solution leads eventually to a negative Hubble function and thus to a contracting universe. This case is not studied here. Inflation ends (or starts) at points where the acceleration of the scale factor $a(t)$ vanishes

$$\frac{\ddot{a}(t)}{a(t)} = H^2 + \dot{H} = 0. \tag{15}$$

This equation has two solutions. The end of the first inflation is taken at $t = 0$ thus fixing t_o

$$t_o = -\frac{b}{\sqrt{3}a(1-b)} arctanh[\frac{\sqrt{6}(1-b^2)-b}{3-2b^2}]. \tag{16}$$

The beginning of the second inflation is given by

$$t_{s2} = -\frac{\ln(49-20\sqrt{6})}{2\sqrt{3}} \frac{b}{a(1-b)}, \tag{17}$$

which, by virtue of (14), can be written as

$$t_{s2} = -\frac{\ln(49-20\sqrt{6})}{4\sqrt{3}\Lambda_2^2}(1-(\frac{\Lambda_2}{\Lambda_1})^2) \approx \frac{0.662}{\Lambda_2^2}, \tag{18}$$

the last result follows because $\Lambda_2/\Lambda_1 \ll 1$.

Corresponding to $\rho_\phi \approx 0.7\rho_c$ with $\rho_\phi \approx V(\phi)$, we take $\Lambda_2 = 2.744 \times 10^{-12}\sqrt{h}GeV$, where h is somewhere between 0.68 and 0.75. Thus, (18) gives the time when the second inflation starts with respect to the end of the first inflation at $t = 0$, which for any practical purpose could be taken as the Big-Bang. Using the reduced Planck time $T = 2.7 \times 10^{-43} sec$, (18) gives

$$t_{s2} \approx \frac{4.5 \times 10^9}{h} years. \tag{19}$$

This is not, however, a realistic model because particle production at the end of the first inflation t_{end1} has not been considered. Shortly after t_{end1} the radiation, followed by matter energy density should dominate the inflaton energy density. Because we want a second stage of inflation to occur at a second maximum then reheating after the first inflation should be produced as in non-oscillatory models [5], a delicate problem which will be dealt with elsewhere.

In Fig. 7 we show the total, potential and kinetic energies, as well as the acceleration of the scale factor of the universe as functions of time for the limiting solution given by (8). Finally Fig. 8 shows the equation of state parameter $\omega = p/\rho$. The example above was developed using the limiting solution. This is only an approximation to a more general situation where the scalar field is initially displaced from its maximum at M_1.

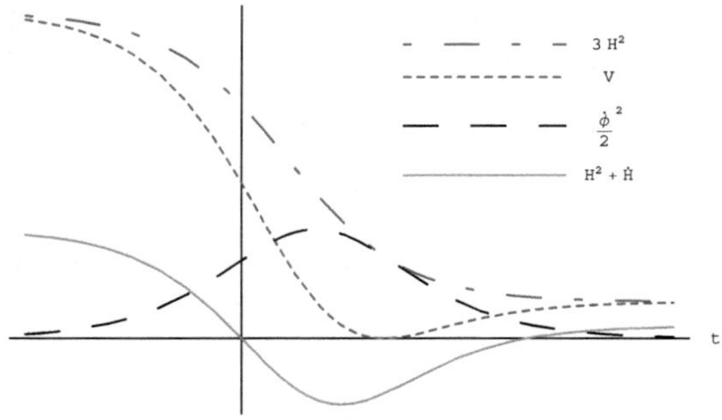

Fig. 7. We show the total, potential and kinetic energies, as well as the acceleration of the scale factor of the universe, $\ddot{a}(t)/a(t)$, as functions of time for the limiting solution given by (8). The origin of time has been chosen so that the end of the first inflation occurs at $t = 0$. Thus we find that the start of the second inflation denoted by t_{s2} is given by $t_{s2} \approx 4.5 \times 10^9/h$ years.

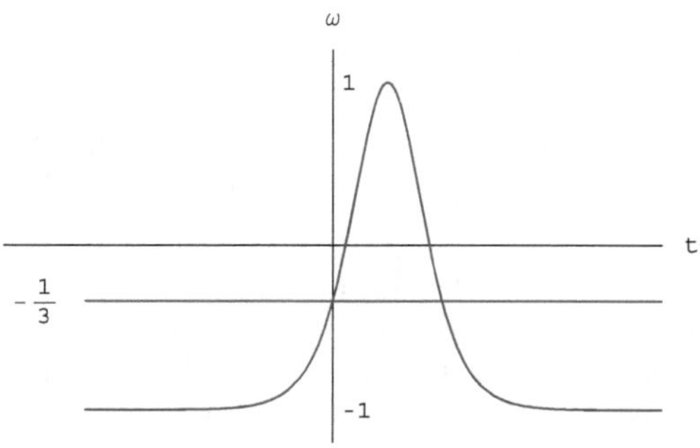

Fig. 8. The equation of state parameter $\omega = p/\rho$ as a function of time. For $t \to \pm\infty$ ω takes the cosmological constant value of -1. Inflation occurs for $\omega \leq -1/3$.

3 Initial Conditions

In a more realistic situation the inflaton leaves not from the maximum at M_1 but from a slightly displaced position. The potential is shown in Fig. 6. A mechanism setting the field away from M_1 is provided by its fluctuations. We have that

$$\delta\phi \approx \frac{H(t \to -\infty)}{2\pi} \approx \frac{\Lambda_1^2}{2\pi\sqrt{3}}. \tag{20}$$

Depending on the initial conditions the scalar field approaches M_2 ending in oscillations around one of the minima. The time evolution of ϕ is illustrated in Fig. 9. Figure 9a corresponds to a field which is unable to reach the maximum at M_2 ending in oscillations around the first minimum while Fig. 9b shows the time evolution of the field when this is able to overcome the maximum M_2 ending at the second minimum. In both cases the flat part of the figure

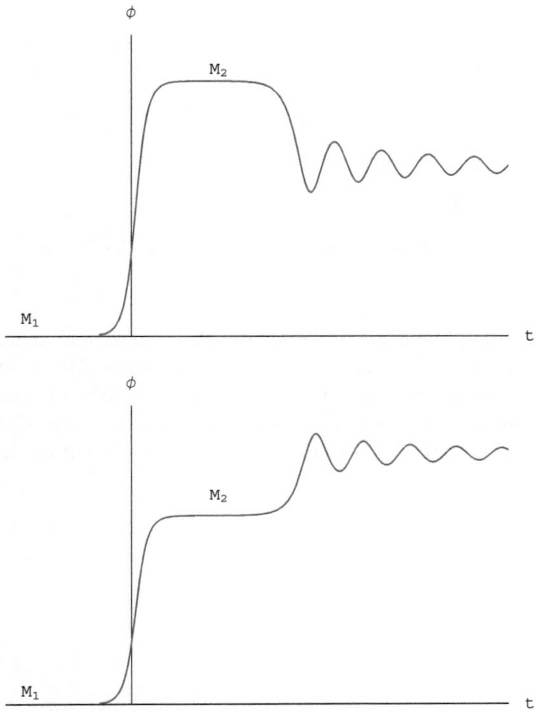

Fig. 9. The inflaton leaves from close to M_1 where it has been displaced due to its fluctuations $\delta\phi \approx H(t \to -\infty)/2\pi \approx \Lambda_1^2/2\pi\sqrt{3}$. After some time it approaches the second maximum at M_2 (see Fig. 6) ending in oscillations around the first minimum Fig. 9a or the second Fig. 9b.

is where the second inflation occurs and its duration clearly depends on the initial conditions with which the universe was prepared.

The equivalent to Fig. 7 for this case is shown in Fig. 10 where we plot the total, potential and kinetic energies as well as the acceleration of the scale factor of the universe. Finally Fig. 11 shows the behaviour of the equation of state parameter $\omega = p/\rho$. This should be compared with Fig. 8 for the case of the limiting solution.

4 Conclusions

We have studied a model of inflation which can accommodate two inflationary eras. Both stages of inflation are drived by the potential energy of a single scalar field. The new feature is that inflation occurs close to the maxima of the potential where the kinetic energy is negligible. As a consequence both inflations are of finite duration. It is then possible that the second inflation

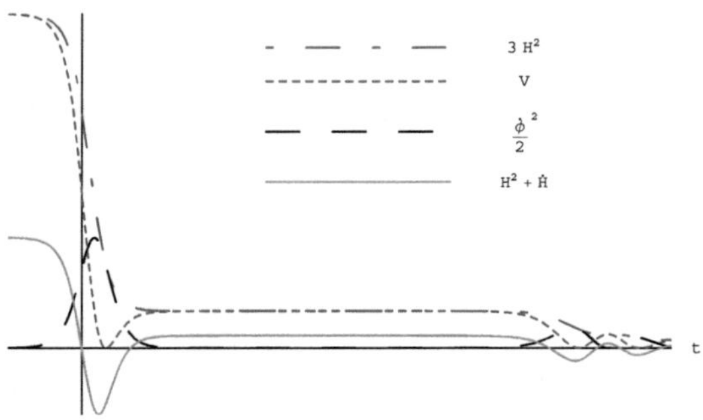

Fig. 10. Corresponding to Fig. 7 where now the scalar field starts its rolling displaced from the maximum at M_1 due to its fluctuations. All the curves are flat to the left of the figure with the inflaton close to M_1. It finally approaches M_2 (second plateau). After some time close to the second maximum at M_2 the inflaton rolls to a minimum (see Fig. 6) with the oscillatory behaviour shown at the far right of the figure.

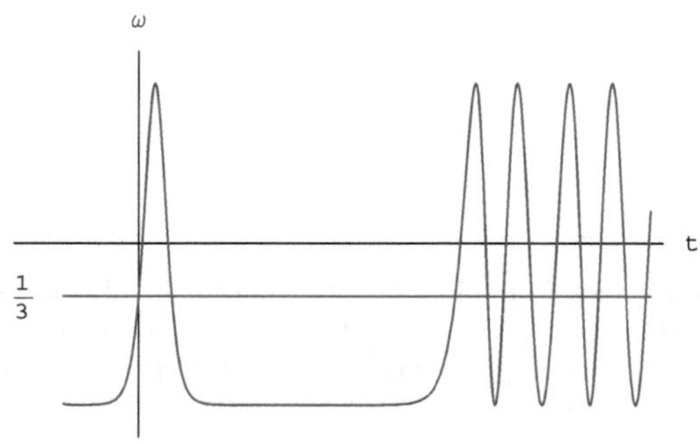

Fig. 11. The equation of state parameter $\omega = p/\rho$ as a function of time when the scalar field has been displaced from the maximum at M_1. Note how during the oscillations of the inflaton there are short periods of inflation.

has already or is about to end which should be testable by substantially increasing the number and accuracy of supernovae on the Hubble diagram. In the ideal case, which we call the limiting solution, the scalar field takes an infinite amount of time to reach the second, smaller, maximum. In a more realistic case the scalar field is displaced from the higher maximum by its fluctuations ending in oscillations in one of the minima of the potential. The origin of time is fixed by the requirement that $H^2 + \dot{H}$ vanishes at $t = 0$. Thus the end of the first inflation defines the origin of time which for any practical purpose could be taken as the Big-Bang. A realistic model should incorporate an era of radiation followed by matter domination after the end of the first inflation. Potentially problematic is that the initial conditions are fine-tuned to avoid the scalar field undershoot or overshoot the second maximum of the potential. On the other hand we have been able to show that a potential of the type (12) could be derived from supergravity. In supergravity the only natural scale is the Planck scale and we can find arguments to explain the possible origin of the first scale of inflation [6] while the second could be understood in terms of the first one by considering friction terms due to the expansion of the universe and possible interactions of the inflaton with matter fields.

Appendix

Let us consider the supergravity potential for one chiral superfield with scalar component z and without D-terms [7]

$$V = e^{K}\left[F^{*}(K_{zz^{*}})^{-1}F - 3|W|^{2}\right], \tag{21}$$

where

$$F \equiv \frac{\partial W}{\partial z} + \left(\frac{\partial K}{\partial z}\right)W, \qquad K_{zz^{*}} \equiv \frac{\partial^{2} K}{\partial z \partial z^{*}} \; . \tag{22}$$

The reduced Planck mass $M \sim 2.4 \times 10^{18}$ GeV has been set equal to one. The superpotential and Kähler potential denoted W and K respectively. Here we are interested in models where W and K are given by polynomial expressions such as

$$W = \sum_{n=0}^{\infty} a_n z^n, \tag{23}$$

and

$$K = \sum_{n=1}^{\infty} b_n (zz^{*})^n, \tag{24}$$

where a_n and b_n are real coefficients. In general this structure leads to expressions that contain cos-form potentials for the angular field ϕ which is a real field defined from z in the following way

$$z = \chi e^{i\phi} \; . \tag{25}$$

By using the superpotential and Kähler potential as given by (23) and (24), it is straightforward to show that the supergravity potential can be written in the form

$$V = e^{K} \sum_{n=0}^{\infty} \sum_{m=0}^{\infty} \left[\frac{(n+K_1)(m+K_1)}{K_2} - 3\right] a_n a_m z^n z^{*m}, \tag{26}$$

where K_i denote the sums

$$K_1 = \sum_{n=1}^{\infty} n b_n (zz^{*})^n \; , \tag{27}$$

$$K_2 = \sum_{n=1}^{\infty} n^2 b_n (zz^*)^n \; . \tag{28}$$

Notice that for superpotentials and Kähler potentials of the form (23) and (24), respectively, (26) is entirely equivalent to the supergravity potential given by (21). Let us now insert the radial and angular fields by writing z in the way expressed by (25), $z = \chi e^{i\phi}$. The potential is then given by [8]

$$V = e^K \sum_{n=0}^{\infty}\sum_{m=0}^{\infty} \left[\frac{(n+K_1)(m+K_1)}{K_2} - 3\right] a_n a_m \chi^{n+m} \cos[(n-m)\phi], \tag{29}$$

It is easy to show that (29) can give rise to potentials of the type (12). Let us write the superpotential and Käler potential in the form

$$W = a_0 + a_1 z + a_2 z^2, \tag{30}$$

and

$$K = zz^* = \chi^2, \tag{31}$$

Assuming that the χ field has relaxed to its v.e.v., χ_0 and eliminating e.g., a_1 we get

$$V(\phi) = c_1(c_2 + \cos[\phi])^2, \tag{32}$$

where

$$c_1 = 2e^{\chi_0^2}\chi_0\sqrt{(\chi_0^2-1)a_0a_2}, \tag{33}$$

and

$$c_2 = \frac{((\chi_0^2-2)a_0 + (\chi_0^4+2)a_2)\sqrt{(\chi_0^2-3)a_0^2 - 2\chi_0^2(\chi_0^2-1)a_0a_2 + \chi_0^2(\chi_0^4+\chi_0^2+4)a_2^2}}{\sqrt{(\chi_0^2-2)^2a_0^2 - 2(\chi_0^6-2\chi_0^4+2\chi_0^2+2)a_0a_2 + (\chi_0^4+2)^2a_2^2}}. \tag{34}$$

It is then not untinkable that a model of the type (12) could arise from a particle physics model.

Acknowledgements

This work was supported in part by CONACYT project 32415-E and DGAPA, UNAM project IN-110200.

References

1. For a review and extensive references, see, D. H. Lyth, A. Riotto: Phys. Rep. **314**, 1 (1999); A.R. Liddle, D. Lyth: *Cosmological Inflation and Large Scale Structure*, (Cambridge U.P., 2000).
2. P.J. Peebles and B. Ratra: Astrophys. J. **325**, L17 (1988). B. Ratra, and P.J.E. Peebles: Phys. Rev. D **37**, 3406 (1988). C. Wetterich: Nucl. Phys. B **302**, 668 (1988). I. Zlatev, L. Wang and P.J. Steinhardt: Phys. Rev. Lett. **82**, 896 (1999). R.R. Caldwell, R. Dave and P.J. Steinhardt: Phys. Rev. Lett. **80**,1582 (1998). A. de la Macorra and C. Stephan-Otto: Phys. Rev. Lett. **87**, 271301 (2001); Phys. Rev. D **65**, 083520 (2002); A. de la Macorra: JHEP **0301**, 033 (2003). For a recent reviews and further references see, V. Sahni: Class. Quant. Grav. **19**, 3435 (2002); P.J.E. Peebles, B. Ratra: Rev. Mod. Phys. **75**, 559 (2003) [astro-ph/0207347].
3. P.J.E. Peebles, A. Vilenkin: Phys. Rev. D **59**, 063505 (1999); W.H. Kinney, A. Riotto: Astropart. Phys. **10**,387 (1999); M. Peloso, F. Rosati: JHEP **9912**, 026 (1999). N.J. Nunes, E.J.Copeland: Phys. Rev. D **66**, 043524 (2002); K. Dimopoulos, J.W.F. Valle: Astropart. Phys. **18**, 287 (2002). For a recent review and further references see, P.J.E. Peebles, B. Ratra, Rev. Mod. Phys. **75**, 559 (2003) [astro-ph/0207347].
4. A. Filippenko, private communication.
5. G. Felder, L. Kofman, A. Linde: Phys. Rev. D **60**, 103505 (1999). A. H. Campos, H. C. Reis, R. Rosenfeld: Phys. Lett. B **575**, 151 (2003) [hep-ph/0210152].
6. G. Germán, G.G Ross, S. Sarkar: Phys. Lett. B **469**, 46 (1999). G. Germán, G.G Ross, S. Sarkar: Nucl. Phys. B **608**, 423 (2001).
7. D. Bailin, A. Love: *Supersymmetric Gauge Field Theory and String Theory*, (Hilger, 1994).
8. G. Germán, A. Mazumdar, A. Perez-Lorenzana: Mod. Phys. Lett. A **17**, 1627 (2002).

Quantum Corrections to Scalar Quintessence Potentials

Michael Doran[1] and Joerg Jaeckel[2]

[1] Dept. of Physics and Astronomy, Dartmouth College, 6127 Wilder Laboratory, Hanover, NH 03755. Michael.Doran@dartmouth.edu
[2] Institut für Theoretische Physik der Universität Heidelberg, Philosophenweg 16 69120 Heidelberg, Germany. jaeckel@thphys.uni-heidelberg.de

Abstract. We give a brief introduction into the formalism of the effective action. We use the effective action to investigate the stability of scalar quintessence potentials under quantum fluctuations both for uncoupled models and models with a coupling to fermions. We find that uncoupled models are usually stable in the late universe. However, a coupling to fermions is severely restricted. We check whether a graviton induced fermion-quintessence coupling is compatible with this restriction.

1 Introduction

Observations indicate that dark energy constitutes a substantial fraction of our Universe [1, 2, 3, 4, 5]. The range of possible candidates includes a cosmological constant and – more flexibly – some form of dark energy with a time dependent equation of state, called quintessence [6]. Commonly, realizations of quintessence scenarios feature a light scalar field [7, 8, 9].

The evolution of the scalar field is usually treated at the classical level. However, quantum fluctuations may alter the classical quintessence potential. In this contribution which is mainly based on [10, 11], we will investigate one-loop contributions to the effective potential from both quintessence and fermion fluctuations. We will show that in the late universe, quintessence fluctuations are harmless for most of the potentials used in the literature. For inverse power laws and SUGRA inspired models, this has already been demonstrated in [12]. That the smallness of the quintessence mass needs to be protected by some symmetry has been pointed out in [13, 14].

In contrast with the rather harmless quintessence field fluctuations, fermion fluctuations severely restrict the magnitude of a possible coupling of quintessence to fermionic dark matter, as we will show.

In Euclidean conventions, the action we use for the quintessence field Φ and a fermionic species Ψ to which it may couple [15, 16, 17] is

$$S = \int d^4x \, \sqrt{g} \Bigg[\mathrm{M}_{\mathrm{P}}^2 R + \frac{1}{2}\, \partial_\mu \Phi(x) \partial^\mu \Phi(x) + V(\Phi(x)) \\ + \overline{\Psi}(x) \left[i \nabla\!\!\!\!/ + \gamma^5 m_{\mathrm{f}}(\Phi) \right] \Psi(x) \Bigg], \quad (1)$$

M. Doran and J. Jaeckel, Quantum Corrections to Scalar Quintessence Potentials, Lect. Notes Phys. **646**, 273–291 (2004)
http://www.springerlink.com/

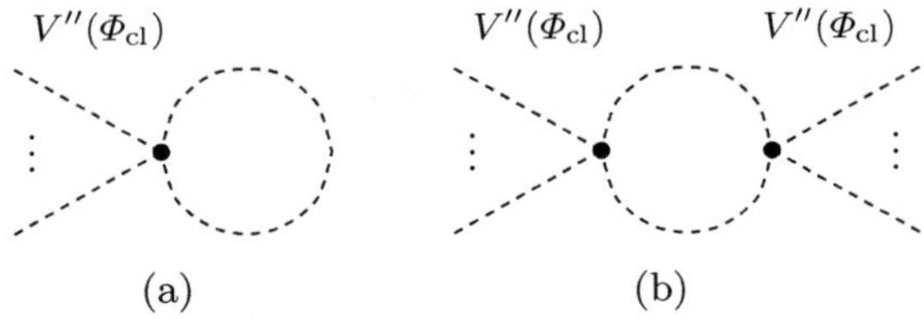

Fig. 1. Pure quintessence fluctuations (depicted as dashed lines). The loop of the fluctuating quintessence field modifies the potential. Since the potential involves in principle arbitrary powers of Φ, we depict V'' as multiple external lines.

with $m_{\text{f}}(\Phi)$ as a Φ dependent fermion mass. This Φ dependence (if existent in a model) determines the coupling of the quintessence field to the fermions. As long as one is not interested in quantum gravitational effects, one may set $\sqrt{g} = 1$, $R = 0$ and replace $\nabla\!\!\!/ \to \partial\!\!\!/$ in the action (1).

By means of a saddle point expansion, we arrive at the effective action $\Gamma[\Phi_{\text{cl}}]$ to one-loop order of the quintessence field. The equation governing the dynamics of the quintessence field is then determined by $\delta\Gamma[\Phi_{\text{cl}}]_{|\Phi_{\text{cl}}=\Phi^{\star}_{\text{cl}}} = 0$. When estimating the magnitude of the loop corrections, we will assume that $\Phi^{\star}_{\text{cl}}$ is close to the solution of the classical field equations: $\delta S = 0$. Evaluating Γ for constant fields, we can factor out the space-time volume U from $\Gamma = UV$. This gives the effective potential

$$V_{\text{1-loop}}(\Phi_{\text{cl}}) = V(\Phi_{\text{cl}}) + \frac{\Lambda^2}{32\pi^2} V''(\Phi_{\text{cl}}) - \frac{\Lambda^2_{\text{ferm}}}{8\pi^2} \left[m_{\text{f}}(\Phi_{\text{cl}})\right]^2 . \tag{2}$$

Here, primes denote derivatives with respect to Φ; Φ_{cl} is the classical field value and Λ and Λ_{ferm} are the ultra violet cutoffs of scalar and fermion fluctuations. The last term in (2) accounts for the fermionic loop corrections as shown in Fig. 5. The second term in (2), is the leading order scalar loop, depicted in Fig. 1(a). We neglect graphs of the order $(V''_{|\text{cl}})^2$ and higher like the one in Fig. 1(b), because V and its derivatives are of the order 10^{-120} (see Sect. 4). We have also ignored Φ-independent contributions, as these will not influence the quintessence dynamics.

However, the Φ-independent contributions add up to a cosmological constant of the order $\Lambda^4 \approx \mathcal{O}(\text{M}_{\text{P}}^4)$. This is the old cosmological constant problem, common to most field theories. We hope that some symmetry or a more fundamental theory will force it to vanish. The same symmetries or theories could equally well remove the loop contribution by some cancelling mechanism. After all, this mechanism must be there, for the observed cosmological constant is far less than the naively calculated $\mathcal{O}(\text{M}_{\text{P}}^4)$. Unfortunately, SUSY is broken too badly to be this symmetry [13].

In addition, none of the potentials under investigation can be renormalized in the strict sense. However, as we will see, terms preventing renormalization may in some cases be absent to leading order in $V''_{|\text{cl}}$. As the mass of the quintessence field is extremely small, one may for all practical purposes view these specific potentials (such as the exponential potential) as renormalizable.

There is also a loophole for all models that will be ruled out in the following: The potential used in a given model could be the full effective potential including all quantum fluctuations, down to macroscopic scales. For coupled quintessence models, this elegant argument is rather problematic and the loophole shrinks to a point (see Sect. 4).

After a brief review of some basic features of the effective action and its calculation we derive (2) and apply it to various quintessence models in order to check their stability against one-loop corrections. We do this separately for coupled and uncoupled models. We use units in which $M_P = 1$. For clarity, we restore it when appropriate.

2 Effective Action

The effective action Γ [18, 19, 20] is a very useful tool in quantum field theory (QFT). It allows us to calculate interesting quantities like vacuum expectation values, propagators and correlation functions more or less by simply taking (functional) derivatives. Indeed, we can promote a classical equation to full quantum status by replacing the action by the effective action $S \to \Gamma$ and the fields by their expectation values $\phi \to \langle\phi\rangle$. E.g. the equation of motion becomes

$$\frac{\delta\Gamma[\langle\phi\rangle]}{\delta\langle\phi\rangle} = 0. \tag{3}$$

Knowledge of the effective action is equivalent to knowledge of the full quantum theory. From this one can already deduce that calculating the effective action is a quite difficult task and can usually be done only approximately.

Before going into more detail let us briefly review the definition of the (1PI) effective action[3], and some of its basic properties. In the following we will write $\tilde{\phi}$ for the fluctuating quantum field and $\phi = \langle\tilde{\phi}\rangle$. We suppress all indices. Indeed, ϕ might also contain fermionic degrees of freedom. A typical ϕ might therefore look like $(\sigma, \sigma^\star, V^\mu, A^\mu, \ldots, \psi_i, \overline{\psi}_j \ldots)$ with several bosonic and fermionic species. If we keep track of the order of fields and differential operators, no problems arise from this notation.

Moreover, we work in Euclidean space. That is why we have a minus sign in the path integrals instead of an i in front of the action. The transition to Euclidean time is usually done via a Wick rotation. We do not want to go into detail, here, however, for fermions there are some slight difficulties because the action is no longer necessarily Hermitian [21, 22].

The generating functional (or partition function if one prefers the statistical mechanics language) of a quantum field theory is defined by the following functional integral

[3] 1PI abbreviates one particle irreducible (cf. Fig. 2).

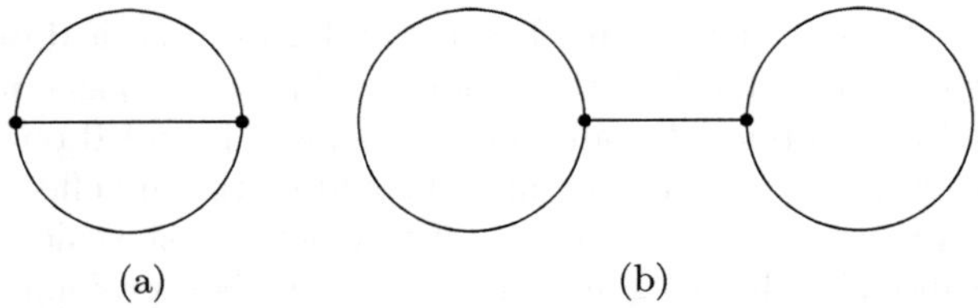

Fig. 2. An example of a diagram which is 1PI 2(a) and one which is not 2(b). The latter one can be split into two by cutting the line between the two bubbles.

$$Z[j] = \int \mathcal{D}\tilde{\phi} \exp(-S[\tilde{\phi}] + j\tilde{\phi}). \tag{4}$$

Here, $S[\tilde{\phi}]$ is the classical action and j is an external source. We recall that in our matrix notation $j\tilde{\phi} = \int d^d x j(x)\tilde{\phi}(x)$.

Using the generating functional, expectation values of fields (and products of fields) can be calculated by taking derivatives with respect to j

$$\phi[j](q) = \langle\tilde{\phi}(q)\rangle = \frac{\int \mathcal{D}\tilde{\phi}\, \tilde{\phi}(q) \exp(-S[\tilde{\phi}] + j\tilde{\phi})}{\int \mathcal{D}\tilde{\phi} \exp(-S[\tilde{\phi}] + j\tilde{\phi})} = \frac{1}{Z[j]} \frac{\delta Z[j]}{\delta j(q)} = \frac{\delta W[j]}{\delta j(q)}, \tag{5}$$

with

$$W[j] = \ln(Z[j]). \tag{6}$$

Physical values are obtained at vanishing external sources e.g. $\phi[0]$.

The (1PI) effective action is now the Legendre transform of W,

$$\Gamma[\phi] = -W[j[\phi]] + j[\phi]\phi \tag{7}$$

and depends on the expectation value of the field. Combining (4), (6), (7) and shifting the integration variable to $\hat{\phi} = \tilde{\phi} - \phi$ we obtain the following very useful formula

$$\Gamma[\phi] = -\ln \int \mathcal{D}\hat{\phi} \exp\left(-S[\phi + \hat{\phi}] + \frac{\delta\Gamma[\phi]}{\delta\phi}\hat{\phi}\right), \tag{8}$$

since

$$\frac{\delta\Gamma[\phi]}{\delta\phi} = j. \tag{9}$$

We note that due to the shift of the integration variable $\langle\hat{\phi}\rangle = 0$. Finally, we would like to comment on the notion of one particle irreducibility (1PI). This is explained most easily in terms of Feynman diagrams. A diagram is one particle irreducible if it is impossible to split it into two parts by cutting an internal line (cf. Fig. 2). The effective action is now the generating functional of the 1PI diagrams. For a proof of this statement and further details about the 1PI effective action we refer to textbooks as e.g. [22, 23, 24, 25].

2.1 Calculating the Effective Action – Loop Expansion

The loop expansion is a perturbative technique to calculate the effective action. It can be shown that the effective action is nothing but the sum of all 1PI vacuum diagrams in a background field ϕ as depicted in Fig. 3. An easy way to obtain this expansion is to make a saddle-point approximation of (8) about the solution of the field equation. In lowest non-trivial order one obtains,

$$\Gamma[\phi] = S[\phi] + \frac{1}{2}\,\mathrm{STr}\ln(S^{(2)}[\phi]) + \cdots, \tag{10}$$

where

$$S^{(2)}[\phi] = \frac{\overrightarrow{\delta}}{\delta\phi^{\mathrm{T}}}\,S[\phi]\,\frac{\overleftarrow{\delta}}{\delta\phi}\,. \tag{11}$$

In the diagrammatic language (Fig. 3) the second term of (10) corresponds to the one-loop diagram, and we have omitted the higher loop diagrams in (10). The supertrace (STr) comes around due to our notation where bosonic as well as fermionic degrees of freedom are contained in ϕ. However, its effect is very simple as it only provides a minus sign in the fermionic sector of the matrix.

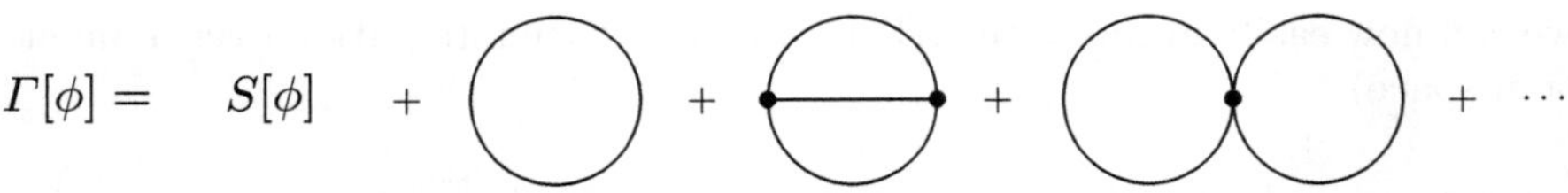

Fig. 3. Perturbative expansion of the effective action. To be explicit we choose a theory which has a quartic interaction. The propagators are propagators in a background field ϕ. For the example of a theory with a $\frac{\lambda}{12}\tilde{\phi}^4$ interaction the propagator in the background field would be $(p^2 + m^2 + \lambda\phi^2)^{-1}$.

An advantage of perturbation theory is that it can usually be constructed to preserve symmetries order by order in the expansion. However, it is usually not suited for calculations in the non-perturbative domain (hence the name) where the coupling is not small. Nevertheless, for our modest aim of getting an overview over the size of the quantum contributions to the effective action one-loop perturbation theory seems reasonable. Moreover, in the late universe the couplings of the quintessence field are usually reasonably small.

2.2 Details of the One-Loop Calculation

Let us now demonstrate how the somewhat abstract formula given in (10) works in our case and leads to (2). In our case of a one component scalar quintessence field Φ and one species of Dirac fermions Ψ we have $\phi(p) =$

$(\Phi(p), \Psi(p), \overline{\Psi}^{\mathrm{T}}(-p))$. For the moment we are only interested in the correction to the quintessence potential, therefore we can restrict ourselves to a background field constant in space, i.e. $\Phi(p) = \Phi\delta(p)$ and $\overline{\Psi} = \Psi = 0$. Moreover, we neglect graviton fluctuations and use a flat background metric, $\sqrt{g} = 1$, $R = 0$ and $\nabla\!\!\!/ = \partial\!\!\!/$. With this we find,

$$S^{(2)}(p_1, p_2) = \frac{\overrightarrow{\delta}}{\delta\phi^{\mathrm{T}}(-p_1)} S[\phi] \frac{\overleftarrow{\delta}}{\delta\phi(p_2)} \tag{12}$$
$$= \begin{pmatrix} p_1^2 + V''(\Phi) & 0 & 0 \\ 0 & 0 & -p\!\!\!/_1^{\mathrm{T}} - m_{\mathrm{f}}(\Phi)\gamma^5 \\ 0 & -p\!\!\!/_1 + m_{\mathrm{f}}(\Phi)\gamma^5 & 0 \end{pmatrix} \delta(p_1 - p_2).$$

We could now insert directly into (10). However, the following trick simplifies the calculation,

$$\Delta\Gamma^{\text{1-loop}} = \frac{1}{2}\,\mathrm{STr}(S^{(2)}) = \frac{1}{4}\,\mathrm{STr}((S^{(2)}S^{(2)\star})), \tag{13}$$

where the last equality holds because the eigenvalues of $S^{(2)}$ are real. With

$$(S^{(2)}S^{(2)\star})(p_1, p_2) = \begin{pmatrix} (p_1^2 + V'')^2 & 0 & 0 \\ 0 & p_1^2 + m_{\mathrm{f}}^2(\Phi) & 0 \\ 0 & 0 & p_1^2 + m_{\mathrm{f}}^2(\Phi) \end{pmatrix} \delta(p_1 - p_2) \tag{14}$$

we can now easily evaluate the STr ln (including the integration over momentum space),

$$\Delta\Gamma^{\text{1-loop}} = V_4 \left(\frac{1}{16\pi^2} \int_0^{\Lambda} dp\, p^3 \ln(p^2 + V''(\Phi)) - \frac{1}{4\pi^2} \int_0^{\Lambda_{\mathrm{ferm}}} dp\, p^3 \ln(p^2 + m_{\mathrm{f}}^2(\Phi)) \right). \tag{15}$$

We note that the STr includes integration over momentum space. Hence, we have to set $p_1 = p_2 = p$ and integrate over p. The four volume factor V_4 originates from the $\delta(0)$ appearing under the integral. The minus sign is from the STr. Dividing by the volume factor V_4 and expanding to lowest order in powers of V'' and m_{f}^2, respectively, yields the correction equation (2) to the effective potential.

Finally, let us mention the connection to the diagrams depicted in Figs. 1, 5, 6. The trick (13), does not always diagonalize the matrix. In particular, this is true when we are interested in the corrections to the fermionic couplings and evaluate at non-vanishing $\overline{\Psi}$ and Ψ. In this case we can split $S^{(2)} = \mathcal{P} + \mathcal{F}$ into a field-independent part $\mathcal{P}$, which is the inverse propagator, and a field-dependent part $\mathcal{F}$. Expanding the logarithm,

$$\Delta\Gamma^{\text{1-loop}} = \frac{1}{2}\,\mathrm{STr}\{\left(\frac{1}{\mathcal{P}}\mathcal{F}\right)\} - \frac{1}{4}\,\mathrm{STr}\{\left(\frac{1}{\mathcal{P}}\mathcal{F}\right)^2\} \tag{16}$$
$$+ \frac{1}{6}\,\mathrm{STr}\{\left(\frac{1}{\mathcal{P}}\mathcal{F}\right)^3\} - \frac{1}{8}\,\mathrm{STr}\{\left(\frac{1}{\mathcal{P}}\mathcal{F}\right)^4\} + \cdots,$$

effectively corresponds to an expansion in the number of vertices. In this language Fig. 1(a) corresponds to a first order term, while Fig. 1(b) is second order and Fig. 4.2(a) is of third order. In this way the evaluation of Feynman diagrams reduces to calculating powers of matrices.

3 Uncoupled Quintessence

Here, we are going to discuss inverse power law, pure and modified exponential, and cosine-type potentials.

3.1 Inverse Power Law and Exponential Potentials

Inverse power laws [7, 8], exponential potentials [9, 26] and mixtures of both [27] can be treated by considering the potential $V = A\Phi^{-\alpha}\exp(-\lambda\Phi^{\gamma})$ [28]. Limiting cases include inverse power laws, exponentials, and SUGRA inspired models. Deriving twice with respect to Φ, we find

$$V'' = A\Phi^{-\alpha}\exp(-\lambda\Phi^{\gamma})\Big\{\alpha(\alpha+1)\Phi^{-2} + 2\alpha\lambda\gamma\Phi^{\gamma-2} + \lambda^2\gamma^2\Phi^{2\gamma-2} - \lambda\gamma(\gamma-1)\Phi^{\gamma-2}\Big\}. \quad (17)$$

Inverse Power Laws

For inverse power laws, we set $\gamma = \lambda = 0$. This gives the classical potential $V_{\mathrm{cl}}^{\mathrm{IPL}} = A\Phi_{\mathrm{cl}}^{-\alpha}$ and by means of (2) the loop corrected potential

$$V_{\text{1-loop}}^{\mathrm{IPL}} = V_{\mathrm{cl}}^{\mathrm{IPL}}\left(1 + \frac{1}{32\pi^2}\Lambda^2\alpha(\alpha+1)\Phi_{\mathrm{cl}}^{-2}\right). \quad (18)$$

The potential is form stable if $\frac{1}{32\pi^2}\Lambda^2\alpha(\alpha+1)\Phi^{-2} \ll 1$, which today is satisfied, as $\Phi \approx \mathrm{M_P}$ [27].

However, if the field is on its attractor today, then $\Phi \propto (1+z)^{-3/(\alpha+2)}$, where z is the redshift [27]. Using this, we have for $z \gg 1$

$$V_{\text{1-loop}}^{\mathrm{IPL}} \approx V_{\mathrm{cl}}^{\mathrm{IPL}}\left(1 + \frac{1}{32\pi^2}\Lambda^2\alpha(\alpha+1)z^{6/(\alpha+2)}\right). \quad (19)$$

Thus, the cutoff needs to satisfy $\Lambda^2 \ll \frac{32\pi^2}{\alpha(\alpha+1)} \times z^{-6/(\alpha+2)}$. Cosmologically viable inverse power law potentials seem to be restricted to $\alpha < 2$ [29, 30]. Using $\alpha = 1$ and $z \approx 10^4$ for definiteness, the bound becomes $\Lambda^2 \ll 10^{-6}$.

So, at equality (and even worse before that epoch), the cutoff needs to be well below 10^{12} Gev, if classical calculations are meant to be valid. In [12] it is argued that for inverse power laws, the quintessence content in the early universe is negligible and hence the fluctuation corrections are important only at an epoch where quintessence is subdominant. As the loop corrections introduce only higher negative powers in the field, it is hoped that, even though one does not know the detailed dynamics, the field will nevertheless roll down its potential (which at that time is supposed to be much steeper) and by the time it is is cosmologically relevant, the classical treatment is once again valid. Having no means of calculating the true effective potential for the inverse power law in the early universe, this view is certainly appealing.

Pure Exponential Potentials

The pure exponential potential is special because its derivatives are multiples of itself. The classical potential (with $\alpha = 0,\ \gamma = 1$) is $V_{\rm cl}^{\rm EP} = A\exp(-\lambda\Phi_{\rm cl})$ and to one-loop order

$$V_{\text{1-loop}}^{\rm EP} = V_{\rm cl}^{\rm EP}\left\{1 + \frac{1}{32\pi^2}\,\Lambda^2\lambda^2\right\}. \tag{20}$$

It is easy to see that a rescaling of $A \to A/\left(1 + \frac{1}{32\pi^2}\,\Lambda^2\lambda^2\right)$ absorbs the loop correction, leading to a stable potential up to order $V_{\rm cl}''$. Working to next to leading order, i.e. restoring terms of order $(V_{\rm cl}'')^2$, we get

$$V_{\text{1-loop, n.l.}}^{\rm EP} = \frac{1}{32\pi^2}\left(V_{\rm cl}''\right)^2\ln\left(\frac{V_{\rm cl}''}{\Lambda^2}\right).$$

It is this term which in four dimensions spoils strict renormalizability.

3.2 Nambu-Goldstone Cosine Potentials

Cosine type potentials resulting from a quintessence axion were introduced in [31, 32] and their implications for the CMB have been studied in [33]. They take on the classical potential $V_{\rm cl}^{\rm NG} = A\left[1 - \cos\left(\Phi_{\rm cl}/f_Q\right)\right]$ and including loop corrections

$$V_{\text{1-loop}}^{\rm AS} = A\left[1 - \left\{1 - \frac{1}{32\pi^2}\frac{\Lambda^2}{f_Q^2}\right\}\cos\left(\frac{\Phi_{\rm cl}}{f_Q}\right)\right].$$

Upon a redefinition $A \to A/\left\{1 - \frac{1}{32\pi^2}\frac{\Lambda^2}{f_Q^2}\right\}$ and, recalling that the loop correction is only defined up to a constant, one arrives at the same functional form as the classical potential.

3.3 Modified Exponentials

In the model proposed by Albrecht and Skordis [34], the classical potential is $V_{\rm cl}^{\rm AS} = V_p\exp(-\lambda\Phi_{\rm cl})$, where V_p is a polynomial in the field. To one-loop order, this leads to

$$V_{\text{1-loop}}^{\text{AS}} = V_{\text{cl}}^{\text{AS}} \left\{ 1 + \frac{1}{32\pi^2} \Lambda^2 \left(\frac{V_p''}{V_p} - 2\lambda \frac{V_p'}{V_p} + \lambda^2 \right) \right\}. \tag{21}$$

Let us for definiteness discuss the example given in [34], where the authors chose $V_p(\Phi) = (\Phi - B)^2 + C$. With this choice, we have

$$V_{\text{1-loop}}^{\text{AS, EXMPL}} = V_{\text{cl}}^{\text{AS, EXMPL}} \left\{ 1 + \frac{1}{32\pi^2} \Lambda^2 \left(\frac{1}{V_p} [2 - 4\lambda(\Phi_{\text{cl}} - B)] + \lambda^2 \right) \right\}. \tag{22}$$

Now consider field values close to the minimum of V_p, i.e., let the absolute value of $\xi \equiv \Phi_{\text{cl}} - B$ be small compared to $\sqrt{C}$. Then

$$V_{\text{1-loop}}^{\text{AS, EXMPL}} = V_{\text{cl}}^{\text{AS, EXMPL}} \left\{ 1 + \frac{\Lambda^2}{32\pi^2} \left(\frac{2 - 4\lambda\xi}{C + \xi^2} + \lambda^2 \right) \right\}, \tag{23}$$

and to leading order in ξ

$$V_{\text{1-loop}}^{\text{AS, EXMPL}} \approx V_{\text{cl}}^{\text{AS, EXMPL}} \left\{ 1 + \frac{\Lambda^2}{32\pi^2} \left(\frac{1}{C} [2 - 4\lambda\xi] + \lambda^2 \right) \right\}. \tag{24}$$

Now consider, as was the case in the example given in [34], $C = 0.01$ for definiteness. If we assume a cutoff Λ and a Plank mass of approximately the same order, we get

$$V_{\text{1-loop}}^{\text{AS, EXMPL}} \approx V_{\text{cl}}^{\text{AS, EXMPL}} \left\{ 1 + \frac{1}{32\pi^2} \left(100 [2 - 4\lambda\xi] + \lambda^2 \right) \right\}. \tag{25}$$

The ξ (and hence Φ_{cl}) dependent contribution in the curly brackets of (25) is $-25/(2\pi^2)\lambda\xi$ which for the value $\lambda = 8$ chosen in the example gives $-200/(2\pi^2)\xi \approx -10\xi$.

If we now look at the behaviour of the loop correction as a function of Φ_{cl} and hence ξ in the vicinity of the minimum of this example polynomial, we see that for, e.g., $\xi = 0.01$, the one-loop contribution dominates the classical potential giving rise to a linear term in Φ_{cl} unaccounted for in the classical treatment. For many values of the parameters B and C, this just changes the form and location of the bump in the potential. In principle, however the loop correction can remove the local minimum altogether (see Fig. 4).

Needless to say that this finding depends crucially on the cutoff. If it is chosen small enough, the conclusion is circumvented. In addition, only the specific choice of V_p above has been shown to be potentially unstable. The space of polynomials is certainly large enough to provide numerous stable potentials of the Albrecht and Skordis form.

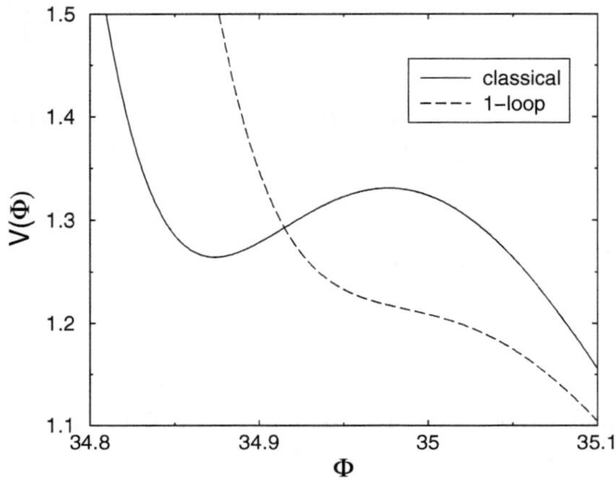

Fig. 4. Classical and 1-loop corrected potential [in $10^{-123}\mathrm{M}_\mathrm{P}^4$] for $V_{\mathrm{cl}}^{\mathrm{AS}} = \left[(\Phi - B)^2 + C\right] \exp(-\lambda\Phi_{\mathrm{cl}})$ with $B = 34.8$, $C = 0.013$, $\Lambda = 1.2$. The classical potential has a local minimum, which is absent for the loop corrected one. This is a hand-picked example and in most cases, the bump will not vanish but move and change its form.

4 Coupled Quintessence

Various models featuring a coupling of quintessence to some form of dark matter have been proposed [35, 16, 36, 37, 17]. From the action equation (1), we see that the mass of the fermions could be Φ dependent: $m_\mathrm{f} = m_\mathrm{f}(\Phi_{\mathrm{cl}})$. Two possible realization of this mass dependence are, for instance, $m_\mathrm{f} = m_\mathrm{f}^0 \exp(-\beta\Phi_{\mathrm{cl}})$ and $m_\mathrm{f} = m_\mathrm{f}^0 + c(\Phi_{\mathrm{cl}})$, where in the second case, we may have a large field independent part together with small couplings to quintessence.[4] For the model discussed in [16], the coupling is of the first form, whereas in [17], the coupling is realized by multiplying the cold dark matter Lagrangian by a factor $f(\Phi)$. This factor is usually taken of the form $f(\Phi) = 1 + \alpha(\Phi - \Phi_0)^\beta$. Hence, the coupling is $m_\mathrm{f}(\Phi) = f(\Phi)\, m_\mathrm{f}^0$, if we assume that dark matter is fermionic. If it were bosonic, the following arguments would be similar.

We will first discuss general bounds on the coupling and in a second step check whether these bounds are broken via an effective gravitational coupling.

4.1 General Bounds on a Coupling

We will discuss only the new effects coming from the coupling and set

$$V_{\text{1-loop}} = V_{\mathrm{cl}} - \Delta V, \tag{26}$$

[4] The constant m_f^0 is *not* the fermion mass today, which would rather be $m_{\mathrm{today}} = m_\mathrm{f}(\Phi_{\mathrm{cl}}(\mathrm{today}))$.

where $\Delta V = \Lambda_{\mathrm{ferm}}^2 \left[m_{\mathrm{f}}(\Phi_{\mathrm{cl}})\right]^2 / \left(8\pi^2\right)$. If we assume that the potential energy of the quintessence field constitutes a considerable part of the energy density of the universe today, i.e. $\rho_{\mathrm{q}} \sim \rho_{\mathrm{critical}}$, we see from the Friedmann equation

$$3H^2 = \rho_{\mathrm{critical}}, \tag{27}$$

that $V_{\mathrm{cl}} \sim H^2$. With today's Hubble parameter $H = 8.9 \times 10^{-61}\, h$ ($h = 0.5 \ldots 0.9$), we have

$$V_{\mathrm{cl}} \sim 7.9 \times 10^{-121}\, h^2. \tag{28}$$

The ratio of the 'correction' to the classical potential is

$$\frac{\Delta V}{V_{\mathrm{cl}}} = \frac{1}{8\pi^2} \frac{\Lambda_{\mathrm{ferm}}^2 \left[m_{\mathrm{f}}(\Phi_{\mathrm{cl}})\right]^2}{V_{\mathrm{cl}}}. \tag{29}$$

Let us first consider the case that all of the fermion mass is field dependent, i.e., we consider cases like $m_{\mathrm{f}} = m_{\mathrm{f}}^0 \exp(-\beta\Phi_{\mathrm{cl}})$. As an example, we choose a fermion cutoff at the GUT scale $\Lambda_{\mathrm{ferm}} = 10^{-3}$, and a fermion mass, $m_{\mathrm{f}}(\Phi_{\mathrm{cl}})$ of the order of 100 Gev $= 10^{-16}\mathrm{M_P}$. Then (29) gives the overwhelmingly large ratio

$$\frac{\Delta V}{V_{\mathrm{cl}}} \approx 10^{80}. \tag{30}$$

Thus, the classical potential is negligible relative to the correction induced by the fermion fluctuations.

Having made this estimate, it is clear that the fermion loop corrections are harmless only, if the square of the coupling takes on *exactly* the same form as the classical potential itself. If, for example, we have an exponential potential $V_{\mathrm{cl}} = A\exp(-\lambda\Phi_{\mathrm{cl}})$ together with a coupling $m_{\mathrm{f}}(\Phi_{\mathrm{cl}}) = m_{\mathrm{f}}^0 \exp(-\beta\Phi_{\mathrm{cl}})$, then this coupling can only be tolerated, if $2\beta = \lambda$.[5] Taken at face value, this finding restricts models with these types of coupling. It is however interesting to note that for exponential coupling, the case $2\beta = \lambda$ is not ruled out by cosmological observations [37].

Turning to the possibility of a fermion mass that consists of a field independent part and a coupling, i.e., $m_{\mathrm{f}} = m_{\mathrm{f}}^0 + c(\Phi_{\mathrm{cl}})$, (29) becomes

$$\frac{\Delta V}{V_{\mathrm{cl}}} = \frac{1}{8\pi^2} \frac{\Lambda_{\mathrm{ferm}}^2 \left[2m_{\mathrm{f}}^0 c(\Phi_{\mathrm{cl}}) + c(\Phi_{\mathrm{cl}})^2\right]}{V_{\mathrm{cl}}}, \tag{31}$$

where we have ignored a quintessence field independent contribution proportional to $(m_{\mathrm{f}}^0)^2$. Assuming $c(\Phi_{\mathrm{cl}}) \ll m_{\mathrm{f}}^0$, and demanding that the loop corrections should be small compared to the classical potential, (31) yields the bound

$$c(\Phi_{\mathrm{cl}}) \ll \frac{4\pi^2\, V_{\mathrm{cl}}}{\Lambda_{\mathrm{ferm}}^2\, m_{\mathrm{f}}^0}. \tag{32}$$

[5] Of course, a sufficiently small β will lead to a more or less constant contribution, where $m_{\mathrm{f}}(\Phi_{\mathrm{cl}}) \approx m_{\mathrm{f}}^0 - \beta\Phi_{\mathrm{cl}}$.

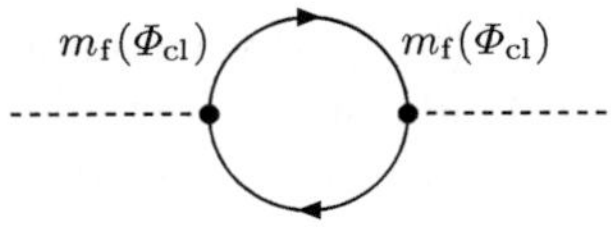

Fig. 5. Correction to the quintessence potential due to fermion fluctuations. Fermion lines are solid, quintessence lines dashed. Shown is the case where $m_f(\Phi)$ gives a Yukawa coupling, i.e. $c(\Phi) = \beta\Phi$, corresponding to one quintessence line. Of course, for more complicated $m_f(\Phi)$ such as $m_f(\Phi_{cl}) = m_f^0 \exp(-\beta\Phi_{cl})$, several external lines as in Fig. 1 would appear.

If, as above, we assume $\Lambda_{\text{ferm}} = 10^{-3}\text{M}_\text{P}$, $m_f^0 = 10^{-16}\text{M}_\text{P}$ and V_{cl} from (28), this gives

$$c(\Phi_{cl}) \ll 3 \times 10^{-97}, \tag{33}$$

in units of the Planck mass. Once again, the bound from (32) applies only if the functional form of the loop correction differs from the classical potential. Assuming a Yukawa-type coupling $c(\Phi_{cl}) = \beta\Phi_{cl}$ and field values of at least the order of the Plank mass, we get $\beta \ll 10^{-97}$.

For the coupling $c(\Phi) = m_f^0\alpha(\Phi - \Phi_0)^\beta$ with the values $\alpha = 50$, $\beta = 8$, $\Phi_0 = 32.5$ given in [17], $c(\Phi)$ is usually larger than m_f^0. Therefore we take $m_f(\Phi_{cl}) \approx c(\Phi_{cl})$. With $m_f(\Phi_{cl}) = 10^{-16}$ as before, we get the same result as in (30).

The coupled models share one property: the loop contribution from the coupling is by far larger than the classical potential. At first sight, the golden way out of this seems to be to view the potential as already effective: all fluctuations would be included from the start. However, there is no particular reason, why *any* coupling of quintessence to dark matter should produce just exactly *the* effective potential used in a particular model: there is a relation between the coupling and the effective potential generated. Put another way, *if* the effective potential is of an elegant form and we have a given coupling, then it seems unlikely that the *classical* potential could itself be elegant or natural.

4.2 Effective Gravitational Fermion Quintessence Coupling

The bound in (32) is so severe that the question arises whether gravitational coupling between fermions and the quintessence field violates it. To give an estimate, we calculate[6] two simple processes depicted in Fig. 6. We evaluate

[6] Unfortunately, the field-dependent propagator matrix is non-diagonal ($\Phi_{cl} \neq 0$ usually). This is a subtle point. We split the full propagator into a field-independent part P and a field-dependent part F. The logarithm in $\text{STr}\log(P + F)$ is then expanded in powers of F. For the Weyl-Frame calculation in Section 5 this is not longer possible, as the graviton-graviton propagator involves the field χ^2 and thus the field-independent part P is non-invertible. For simplicity, we

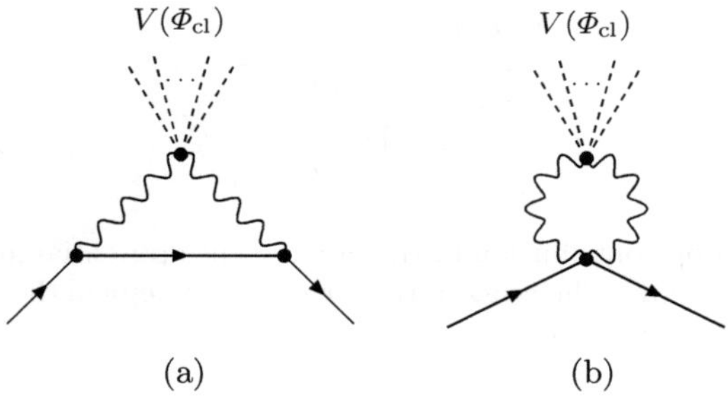

Fig. 6. Effective fermion-quintessence coupling via graviton exchange. The fermions (solid lines) emit gravitons (wiggled lines) which are caught by the quintessence field (dashed lines). As the graphs involve couplings of the gravitons to the classical quintessence potential, the generated coupling is proportional to the classical potential. Since the potential involves arbitrary powers of Φ, we depict it as several Φ-lines. A Yukawa type coupling, corresponding to just one line, is then generated by power expanding $V(\Phi) = V(\Phi_{cl}) + V'_{|cl}\,(\Phi - \Phi_{cl})$ in the fluctuating field.

the diagrams for vanishing external momenta. This is consistent with our derivation of the fermion loop correction equation (2), in which we have assumed momentum independent couplings. The effective coupling due to the graviton exchange contributes to the fermion mass, which becomes Φ_{cl} dependent. We assume that this coupling is small compared to the fermion mass and write $m_f(\Phi_{cl}) = m_f^0 + c(\Phi_{cl})$.

From the first diagram, Fig. 4.2(a) we get (see the Appendix 6):

$$c(\Phi_{cl}) = \frac{1}{8\pi^2}\, m_f^0 V(\Phi_{cl}) \times \left[\ln\left(\frac{\Lambda^2}{\Lambda^2 + [m_f^0]^2}\right) - \ln\left(\frac{\lambda^2}{\lambda^2 + [m_f^0]^2}\right)\right], \quad (34)$$

whereas 4.2(b) gives

$$c(\Phi_{cl}) = \frac{5}{8\pi^2}\, m_f^0 V(\Phi_{cl}) \ln\left(\frac{\Lambda}{\lambda}\right). \quad (35)$$

Here, we have introduced infrared and ultraviolet cutoffs λ and Λ for the graviton momenta. We assume Λ to be of the order M_P and λ about the inverse size of the horizon. Since the results depend only logarithmically on the cutoffs, this choice is not critical, and in addition $\ln(M_P/H) \approx 140$, which is small. From (31, 35, 34), we see that, in leading order, the change in the quintessence potential due to this effective fermion coupling would

ignored the gravity part in the Weyl-Frame calculation (including the coupling of gravitons to χ).

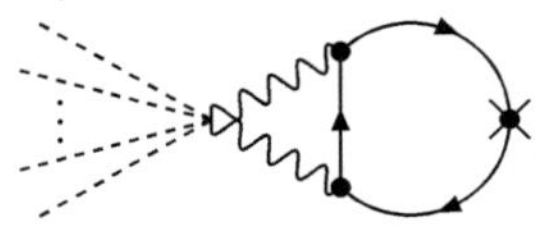

Fig. 7. Fermion loop contribution to the quintessence potential involving the effective coupling Fig. 4.2(a). The cross in the fermion line depicts the field independent fermion mass $m_{\rm f}^0$.

be proportional to $V(\Phi_{\rm cl})$ and could hence be absorbed upon redefining the pre-factor of the potential (see also Fig. 7). In next to leading order, the contribution is proportional to $V(\Phi_{\rm cl})^2$, which is negligible.

From the Appendix 6, in which we present the calculation in more detail, it is clear that there are processes where the vertices are more complicated. However, to this order all diagrams are proportional to $V(\Phi_{\rm cl})$. Thus, they can be absorbed just like the two processes presented above.

5 Weyl Transformed Fields

So far, we have assumed a constant Planck mass together with a field independent cutoff. We could, however, assume that the Planck mass is not constant, but rather given by the expectation value of a scalar field χ. We will call the frame resulting from this Weyl scaling the Weyl frame, as opposed to the frame with a constant Plank mass which we will call the Einstein frame. From the classical point of view, both frames are equivalent. On calculating quantum corrections, we have to evaluate a functional integral. Usually, the functional measure in the Einstein frame is set to unity. In principle, the variable change associated with the Weyl scaling leads to a non-trivial Jacobian and therefore a different functional measure. Taking the position that the Weyl frame is fundamental, this measure could equally well be set to unity in the Weyl frame. Therefore, it is a priori unclear whether the loop corrected potential in the Weyl frame, when transformed back into the Einstein frame, will be the same as the one from (2).

As the cutoff in the Einstein frame is a constant mass scale and hence proportional to the Plank mass, it seems natural to assume that the cutoff in the Weyl frame is proportional to χ. We restrict our discussion to this case. For other choices of the χ-dependence of the cutoff, the results may differ.

The Weyl transformation is achieved by scaling the metric, the curvature scalar, all fields, and the tetrad by appropriate powers of $\chi/\mathrm{M_P}$ (see Table 1) [9, 35]:

$$\tilde{S} = \int d^4x \sqrt{\tilde{g}} \Bigg[\chi^2 \tilde{R} + \frac{z}{2}\, \partial_\mu \chi \partial^\mu \chi + W(\chi) + \bar{\tilde{\Psi}} \left(i\, \tilde{\gamma}^\mu(x) \nabla_\mu + \chi \, \frac{m_{\mathrm{f}}(\Phi_{\mathrm{cl}})}{\mathrm{M_P}}\, \gamma^5 - \frac{3}{2}\, i \tilde{\gamma}^\mu(x) \ln \chi_{,\mu} \right) \tilde{\Psi} \Bigg], \quad (36)$$

where $\Phi = (12+z)^{1/2} \mathrm{M_P} \ln(\chi/\mathrm{M_P})$ and

$$W(\chi) \equiv \left(\frac{\chi}{\mathrm{M_P}} \right)^4 V(\Phi(\chi)). \quad (37)$$

Table 1. Weyl scaling of various quantities. The transformation of the curvature scalar R follows from the scaling of the metric. This scaling, in turn, originates from the condition that instead of the Plank mass squared multiplying R in the action in the Einstein frame, a factor χ^2 should appear. Here, we have set $\sigma = \ln(\chi/\mathrm{M_P})$.

$$\begin{aligned} g_{\mu\nu} &\to (\chi/\mathrm{M_P})^2\, \tilde{g}_{\mu\nu} \\ g^{\mu\nu} &\to (\chi/\mathrm{M_P})^{-2}\, \tilde{g}^{\mu\nu} \\ \sqrt{g} &\to (\chi/\mathrm{M_P})^4 \sqrt{\tilde{g}} \\ R &\to (\chi/\mathrm{M_P})^{-2} \left(\tilde{R} - 6 \tilde{g}^{\mu\nu} \sigma_{;\mu\nu} - 6 \tilde{g}^{\mu\nu} \sigma_{,\mu} \sigma_{,\nu} \right) \\ e^\mu_a(x) &\to (\chi/\mathrm{M_P})^{-1}\, \tilde{e}^\mu_a(x) \\ \Psi &\to (\chi/\mathrm{M_P})^{-3/2}\, \tilde{\Psi} \end{aligned}$$

The term proportional to $\ln \chi_{,\mu}$ in (36) is somewhat inconvenient. Adopting the position that the Weyl frame is fundamental, this term is unnatural. Instead, one could formulate the theory with canonical couplings for the fermions. Dropping this term,

$$\tilde{S}_{\mathrm{can.}} = \int d^4x \sqrt{\tilde{g}} \Bigg[\chi^2 \tilde{R} + \frac{z}{2}\, \partial_\mu \chi \partial^\mu \chi + W(\chi) + \bar{\tilde{\Psi}} \left(i\, \tilde{\gamma}^\mu(x) \nabla_\mu + \frac{\chi}{\mathrm{M_P}}\, m_{\mathrm{f}}(\Phi(\chi))\, \gamma^5 \right) \tilde{\Psi} \Bigg], \quad (38)$$

we observe by going back to the usual action $\tilde{S}_{\mathrm{can.}} \to S$,

$$S = \int d^4x\, \sqrt{g} \Bigg[\frac{1}{2}\, \partial_\mu \Phi(x) \partial^\mu \Phi(x) + V(\Phi(x)) + \overline{\Psi}(x) \left(i \not{\nabla} + \gamma^5 m_{\mathrm{f}}(\Phi) + \frac{3}{2\mathrm{M_P}}\, i\, \gamma^\mu(x) \phi_{,\mu} \right) \Psi(x) \Bigg], \quad (39)$$

that the canonical form of the action in the Weyl frame gives rise to a derivative coupling of the quintessence field to the fermions in the Einstein frame, which we can safely ignore.[7]

Working with (38), we get the loop correction in the Weyl frame by replacing $V \to W$ and $\Phi \to \chi$ in (2). In addition, the constant cutoffs Λ and Λ_{ferm} are replaced by $const \cdot \chi$:

$$W_{\text{1-loop}} = W(\chi) + \frac{(\mathcal{C}\chi)^2}{32\pi^2 z^2} W''(\chi) - \frac{(\mathcal{C}_{\text{f}}\chi)^2}{8\pi^2} \left[\frac{\chi}{\text{M}_\text{P}} m_\text{f}(\chi)\right]^2 . \tag{40}$$

Transforming $W_{\text{1-loop}}$ back into the Einstein frame, the potential V is modified by

$$V_{\text{1-loop}} = V(\Phi_{\text{cl}}) + \frac{(\mathcal{C}_\text{f}\text{M}_\text{P})^2}{8\pi^2} [m_\text{f}(\Phi_{\text{cl}})]^2 + \frac{(\mathcal{C}\text{M}_\text{P})^2}{32\pi^2 z^2} \times \left[12 \frac{V(\Phi_{\text{cl}})}{\text{M}_\text{P}^2} + 7\sqrt{12+z}\, \frac{V'(\Phi_{\text{cl}})}{\text{M}_\text{P}} + (12+z)V''(\Phi_{\text{cl}})\right]. \tag{41}$$

As an example, lets calculate the correction to the pure exponential potential $V_{\text{cl}}^{\text{EP}} = A \exp(-\lambda\Phi_{\text{cl}})$, once again setting $\text{M}_\text{P} = 1$. The Weyl frame potential is

$$W(\chi) = A\chi^4 \exp(-\lambda\Phi_{\text{cl}}(\chi)) = A\chi^{(4-\lambda\sqrt{12+z})}. \tag{42}$$

Neglecting fermion fluctuations and choosing $z = 1$,

$$W_{1-\text{loop}} = \left[1 + \frac{\mathcal{C}^2}{32\pi^2 z^2} (4 - \lambda\sqrt{13})(3 - \lambda\sqrt{13})\right] W(\chi). \tag{43}$$

Again (and not surprisingly) we can absorb the terms in the square brackets in a redefinition of the pre-factor A. In the case of an inverse power law, the term proportional to V' in (41) leads to a slightly different contribution compared to (18) (a term $\propto \Phi_{\text{cl}}^{-\alpha-1}$ arises). For the modified exponential potentials the expressions corresponding to V' in (41) make no structural difference.

6 Conclusions

We have calculated quantum corrections to the classical potentials of various quintessence models. In the late universe, most potentials are stable with

[7] Actually, this coupling is non-renormalizable in the strict sense. Since the theory is non-renormalizable anyway, this is not of great concern. In addition, if one believes that the Weyl frame is fundamental, there is no need to go back to the Einstein frame and hence no need to face this nuisance.

respect to the scalar quintessence fluctuations. The pure exponential and Nambu-Goldstone type potentials are form invariant up to order V'', yet terms of order $(V'')^2$ prevent them from being renormalizable in the strict sense.

For the modified exponential potential introduced by Albrecht and Skordis, stability depends on the specific form of the polynomial factor V_p in the potential. In some cases the local minimum in the potential can even be removed by the loop. An explicit coupling of the quintessence field to fermions (or similarly to dark matter bosons) seems to be severely restricted. The effective potential to one-loop level would be completely dominated by the contribution from the fermion fluctuations. All models in the literature share this fate. One way around this conclusion could be to view these potentials as already effective. They must, however, not only be effective in the sense of an effective quantum field theory originating as a low–energy limit of an underlying theory, but also include all fluctuations from this effective QFT. In this case, there is a strong connection between coupling and potential and it is rather unlikely that the *correct* pair can be guessed.

The bound on the coupling is so severe that for consistency, we have calculated an effective coupling due to graviton exchange. To lowest order in $V(\Phi)$, this coupling leads to a fermion contribution which can be absorbed by redefining the pre-factor of the potential.

To check that the results are not artefacts from the Einstein frame, we switched to the Weyl frame. As the transition from $\Phi \to \chi$ involves a non-trivial Jacobian, the details of the results differ. However, the basic results stay the same.

Surely, the one-loop calculation does not give the true effective potential. Symmetries or more fundamental theories that make the cosmological constant as small as it is, could force loop contributions to cancel. In addition, the back reaction of the changing effective potential on the fluctuations remains unclear in the one-loop calculation. A renormalization group treatment would therefore be of great value. We leave this to future work.

Acknowledgments

We would like to thank Gert Aarts, Luca Amendola, Jürgen Berges, Matthew Lilley, Volker Schatz, and Christof Wetterich for helpful discussions. Joerg Jaeckel would like to thank the organizers of the “Fifth Mexican School: The Early Universe and Observational Cosmology” for the nice conference.

Appendix: Coupling to Gravitons

Fermions in general relativity are usually treated within the tetrad formalism. The γ matrices become space-time dependent: $\gamma^\mu(x) \equiv \gamma^a e_a^\mu(x)$. Together with the spin connection ω, one uses (see, e.g., [22, 38]):

$$\not{\nabla} = e^{\mu}_{a}(x)\gamma^{a}\left(\partial_{\mu} + \frac{i}{4}\sigma_{bc}\omega^{bc}_{\mu}\right). \tag{44}$$

The action (1) can then be expanded in small fluctuations around flat space: $g_{\mu\nu} = \delta_{\mu\nu} + h_{\mu\nu}/\mathrm{M_P}$.

Using the gauge fixing term $-\frac{1}{2}(\partial^{\nu}h_{\mu\nu} - \frac{1}{2}\partial_{\mu}h^{\nu}_{\nu})^2$ and expanding the action to second order in h, we find the propagator [38]:

$$P^{-1}_{\mathrm{grav}}(k) = \frac{\delta_{\mu\alpha}\delta_{\nu\beta} + \delta_{\mu\beta}\delta_{\nu\alpha} - \delta_{\mu\nu}\delta_{\alpha\beta}}{k^2}. \tag{45}$$

The diagrams in Fig. 6 are generated by the expansion of $\sqrt{g} = 1 + \frac{1}{2}h^{\mu\mu} - \frac{1}{4}(h^{\mu\nu})^2 + \frac{1}{8}(h^{\mu\mu})^2$ multiplying the matter Lagrangian. Additional (and more complicated) vertices originate from the spin connection and the tetrad.

However, we do not consider external graviton lines, which would only give corrections to the couplings and wave function renormalization of the gravitons. Therefore only internal gravitons appear. In order to contribute a quintessence dependent part to the fermion mass, the gravitons starting from the fermion-graviton vertices (complicated as they may be) have to touch quintessence-graviton vertices. As these quintessence vertices are proportional to $V(\Phi_{\mathrm{cl}})$, all diagrams to lowest order in $V(\Phi_{\mathrm{cl}})$ will only produce mass contributions proportional to $V(\Phi_{\mathrm{cl}})$.

Evaluating the diagrams in Fig. 6 for vanishing external momenta we get (34) and (35).

References

1. A. G. Riess *et al.*, Astron. J. **116**, 1009 (1998) [astro-ph/9805201].
2. S. Perlmutter *et al.*, Astrophys. J. **517**, 565 (1999) [astro-ph/9812133].
3. C. B. Netterfield *et al.*, Astrophys. J. **575**, 604 (2002) [astro-ph/0104460].
4. A. T. Lee *et al.*, Astrophys. J. **561**, L1 (2001) [astro-ph/0104459].
5. W. J. Percival et. al., Mon. Not. Roy. Astron. Soc. **327**, 1297 (2001) [astro-ph/0105252].
6. R. R. Caldwell, R. Dave and P. J. Steinhardt, Phys. Rev. Lett. **80**, 1582 (1998) [astro-ph/9708069].
7. P. J. Peebles and B. Ratra, Astrophys. J. **325**, L17 (1988).
8. B. Ratra and P. J. Peebles, Phys. Rev. D **37**, 3406 (1988).
9. C. Wetterich, Nucl. Phys. B **302**, 668 (1988).
10. M. Doran and J. Jaeckel, Phys. Rev. D **66**, 043519 (2002) [astro-ph/0203018].
11. M. Doran and J. Jaeckel: "Loop corrections to scalar quintessence potentials," astro-ph/0205206.
12. P. Brax and J. Martin, Phys. Rev. D **61**, 103502 (2000) [astro-ph/9912046].
13. C. Kolda and D. H. Lyth, Phys. Lett. B **458**, 197 (1999) [hep-ph/9811375].
14. R. D. Peccei: "Light scalars in cosmology," hep-ph/0009030.

15. C. Wetterich, Nucl. Phys. B **302**, 645 (1988).
16. L. Amendola, Phys. Rev. D **62**, 043511 (2000) [astro-ph/9908023].
17. R. Bean and J. Magueijo, Phys. Lett. B **517**, 177 (2001) [astro-ph/0007199].
18. J. Goldstone, A. Salam and S. Weinberg, Phys. Rev. **127**, 965 (1962).
19. G. Jona-Lasinio, Nuovo Cim. **34**, 1790 (1964).
20. B. De Witt: Relativity, Groups and Topology. In: *Lectures delivered at Les Houches during the 1963 Session of the Summer School of Theoretical Physics*, ed. by C. De Witt and B. De Witt (Gordon and Breach, New York, 1964).
21. K. Osterwalder and R. Schrader, Commun. Math. Phys. **31**, 83 (1973).
22. J.Zinn-Justin: *Quantum Field Theory and Critical Phenomena*, (Oxford Science Publications, USA, 1995)
23. M. E. Peskin and D. V. Schroeder: *Quantum Field Theory*, (Perseus Books, Cambridge MA, 1995).
24. Steven Weinberg: *The Quantum Theory of Fields* (Cambridge University Press, Cambridge UK, 1996).
25. Brian Hatfield: *Quantum Field Theory of Point Particles and Strings* (Addison-Wesley, Reading MA, 1992).
26. P. G. Ferreira and M. Joyce, Phys. Rev. D **58**, 023503 (1998) [astro-ph/9711102].
27. P. Brax and J. Martin, Phys. Lett. B **468**, 40 (1999) [astro-ph/9905040].
28. P. S. Corasaniti and E. J. Copeland, Phys. Rev. D **65**, 043004 (2002) [astro-ph/0107378].
29. A. Balbi *et al.* Astrophys. J. **547**, L89 (2001) [astro-ph/0009432].
30. M. Doran, M. Lilley and C. Wetterich, Phys. Lett. B **528**, 175 (2002) [astro-ph/0105457].
31. J. E. Kim, JHEP **9905**, 022 (1999) [hep-ph/9811509].
32. Y. Nomura, T. Watari and T. Yanagida, Phys. Lett. B **484**, 103 (2000) [hep-ph/0004182].
33. M. Kawasaki, T. Moroi and T. Takahashi, Phys. Rev. D **64**, 083009 (2001) [astro-ph/0105161].
34. A. Albrecht and C. Skordis, Phys. Rev. Lett. **84**, 2076 (2000) [astro-ph/9908085].
35. C. Wetterich, Astron. Astrophys. **301**, 321 (1995) [hep-th/9408025].
36. O. Bertolami and P. J. Martins, Phys. Rev. D **61**, 064007 (2000) [gr-qc/9910056].
37. D. Tocchini-Valentini and L. Amendola, Phys. Rev. D **65**, 063508 (2002) [astro-ph/0108143].
38. M.J.G. Veltman: In: *Méthodes en théories des champs*, Les Houches Summer School of Theoretical Physics, 1975, edited by R. Baliau (North-Holland, Amsterdam, 1975).

Electroweak Baryogenesis and Primordial Hypermagnetic Fields

Gabriella Piccinelli[1] and Alejandro Ayala[2]

[1] Centro Tecnológico Aragón, Universidad Nacional Autónoma de México, Av. Rancho Seco S/N, Bosques de Aragón, Nezahualcóyotl, Estado de México 57130, México. gabriela@astroscu.unam.mx

[2] Instituto de Ciencias Nucleares, Universidad Nacional Autónoma de México, Apartado Postal 70-543, México Distrito Federal 04510, México. ayala@nuclecu.unam.mx

Abstract. The origin of the matter-antimatter asymmetry of the universe remains one of the outstanding questions yet to be answered by modern cosmology, and also one of only a handful of problems where the need of a larger number of degrees of freedom than those contained in the standard model (SM) is better illustrated. An appealing scenario for the generation of baryon number is the electroweak phase transition that took place when the temperature of the universe was about 100 GeV. Though in the minimal version of the SM, and without considering the interaction of the SM particles with additional degrees of freedom, this scenario has been ruled out given the current bounds for the Higgs mass, this still remains an open possibility in supersymmetric extensions of the SM. In recent years it has also been realized that large scale magnetic fields could be of primordial origin. A natural question is what effect, if any, these fields could have played during the electroweak phase transition in connection to the generation of baryon number. Prior to the electroweak symmetry breaking, the magnetic modes able to propagate for large distances belonged to the $U(1)$ group of hypercharge and hence receive the name of *hypermagnetic* fields. In this contribution, we summarize recent work aimed to explore the effects that these fields could have introduced during a first order electroweak phase transition. In particular, we show how these fields induce a CP asymmetric scattering of fermions off the true vacuum bubbles nucleated during the phase transition. The segregated axial charge acts as a seed for the generation of baryon number. We conclude by mentioning possible research venues to further explore the effects of large scale magnetic fields for the generation of the baryon asymmetry.

1 Introduction

The standard model (SM) of electroweak interactions meets all the requirements –known as Sakharov conditions [1] – to generate a baryon asymmetry during the electroweak phase transition (EWPT), provided that this last be of first order. However, it is also well known that neither the amount of CP violation within the minimal SM nor the strength of the EWPT are enough to generate a sizable baryon number [2, 3]. Supersymmetric extensions of the SM, with a richer particle content, contain new sources of CP violation [4].

G. Piccinelli and A. Ayala, Electroweak Baryogenesis and Primordial Hypermagnetic Fields, Lect. Notes Phys. **646**, 293–308 (2004)
http://www.springerlink.com/

They also allow a stronger first order phase transition [5]. In spite of these improvements with respect to the SM, the minimal supersymmetric model (MSSM) is severely constrained from experimental bounds on the chargino properties [6] leaving only a small corner of parameter space for the MSSM as a viable candidate for baryogenesis. Further possibilities to accommodate an explanation for the generation of baryon number during the EWPT include non-minimal supersymmetric models which, nonetheless, all share the unappealing feature of containing an even larger set of parameters than the already extensive number contained in the MSSM.

Though it might appear tempting to abandon the idea of electroweak baryogenesis (EWB) given the above difficulties, in recent years this possibility has been revisited due to the observation that magnetic fields are able to generate a stronger first order EWPT [7, 8, 9]. The situation is similar to what happens to a type I superconductor where an external magnetic field modifies the nature of the superconducting phase transition, changing it from second to first order due to the Meissner effect.

In spite of this observation, it has also been realized that the sphaleron bound becomes more restrictive due to the interaction between the sphaleron's magnetic dipole moment and the external field [10]. Nevertheless, these arguments are either classical or resort to perturbation theory to lowest order. However, in the absence of magnetic fields, it is well known that the phase transition picture is influenced by non-perturbative effects cast in terms of the resummation of certain classes of diagrams [11] and it might also be expected that the same is true in the presence of magnetic fields. The situation with regards to the strength of the phase transition in the presence of magnetic fields is thus far from being settled and requires further research.

However, the influence of the magnetic fields on the enhancement of CP violation has received much less attention. In a series of recent papers [12, 13, 14], it has been shown that the external field is able to produce an axially asymmetric scattering of fermions off first order phase transition bubbles during the EWPT. This CP violating reflection is due to the chiral nature of the couplings of right- and left-handed modes with the external field in the symmetric phase (where right- and left-handed modes can be thought of as the spin projection with respect to the movement direction). This mechanism produces an axial charge segregation which can then be transported in the symmetric phase where sphaleron induced transitions can convert it into baryon number [15]. The main purpose of our work is the description of the mechanism for the generation of this axial charge segregation.

As was briefly presented in this introduction, the process of baryogenesis involves several physical ingredients which all deserve to be addressed in order to cover the entire topic; here we will concentrate on those aspects related to hypermagnetic fields, which are also one of the most recent and less explored parts of this field. For other aspects on the subject, we refer the reader to recent reviews [16, 17, 18].

The work is organized as follows: in Sect. 2 we present the basic framework for electroweak baryogenesis. Section 3 is devoted to generalities of hypermagnetic fields and phase transitions. In Sect. 4 we summarize the current ideas for the origin of large scale magnetic fields as well as the experimental bounds on their strength set by different observations. In Sect. 5 we describe the mechanism whereby the asymmetric reflection of fermions off first order EWPT bubbles in the presence of external magnetic fields leads to an axial charge segregation that can then be converted into baryon number in the symmetric phase. Finally, in Sect. 6 we summarize and give a brief account of some possible research venues to further explore the influence of primordial magnetic fields in the generation of the baryon asymmetry of the universe.

2 Electroweak Baryogenesis

2.1 Baryogenesis

The theory of baryogenesis is an intent to explain the existence of matter in the Universe. As Cohen, Kaplan and Nelson [4] put it: why is there something rather than nothing?

From the point of view of elementary particle physics, there is a symmetry between particles and antiparticles which suggests that there should be an overall balance between the amount of matter and antimatter in the universe. However the observed universe is composed almost entirely of matter, with no traces of present or primordial antimatter (see. e.g. [19] pp 158–159, or [20] for a recent review). On the other hand, from the cosmological approach, there is also some evidence that some ingredients are missing. In the hot early epoch of the universe evolution one expects to have particles and antiparticles in thermal equilibrium with radiation; particle/antiparticle pairs would then annihilate each other until their annihilation rate becomes smaller than the rate of expansion of the universe. The remaining density of all the species can thus be estimated and it comes to be only a very small fraction of the closure density of the universe. In this way, if we do not wish to postulate that the universe was just born with ad hoc asymmetric initial conditions, we must find a mechanism to generate a net baryon number ($B = n_b - n_{\bar{b}}$).

In 1966, Sakharov [1] laid out the conditions for the development of a net excess of baryons over antibaryons: (1) Existence of interactions that violate baryon number; (2) C and CP violation and (3) departure from thermal equilibrium. (The implications of each one of these criteria is discussed in many places, see e.g. the review [16]).

It must be stressed however that some scenarios (possibly exotic) have been recently proposed where one of these conditions is not achieved (see e.g. [21] and references therein).

It is important to notice that a successful model for baryon generation has to put together two ingredients:

1) the generation of a baryon asymmetry

2) its preservation In the following subsection, we will review the necessary conditions for both situations in the framework of EWB.

2.2 Electroweak Baryogenesis

Sakharov Conditions in the Standard Model

The sphaleron (the name is based on the classical greek adjective meaning "ready to fall") [22] is an static and unstable solution of the field equations of the EW model, corresponding to the top of the energy barrier between two topologically distinct vacua. Transitions between these vacua are associated with the violation of baryon (B) and lepton number (L) [23], in the combination B + L, with leptons and baryons produced at the same rate (i.e. B - L conservation). For this reason, they can either induce baryogenesis, or be a mechanism for washing out the previously created baryon asymmetry. It is therefore important to define the epoch at which the sphaleron transitions fall out of thermal equilibrium.

As we mentioned, C and CP violation are present in the SM but are too tiny to be at the origin of the present baryon asymmetry [2, 24]. The generation of a sizable CP violation in the SM is the central part of this work and we will return to it later.

For the out of equilibrium requirement, we rely on the phase transition (PT). This is the only possible source of departure from thermal equilibrium, since, at electroweak scale, the rate of expansion of the universe is small compared to the rate of baryon number violating processes. But the PT is efficient in producing out-of-equilibrium conditions only if it is strongly first order [25], i.e. if the Higgs field -which is the order parameter in this case-undergoes a discontinuous change. In effect, in a first order PT, the conversion from one phase to another happens through nucleation and propagation of the true vacuum bubbles. The region separating both phases is called the wall. As the bubble wall sweeps a point in space, the order parameter changes rapidly, leading to a departure from thermal equilibrium.

Preservation of the Baryon Asymmetry

In order to freeze out the produced baryon number, the rate of fermion number non conservation in the broken phase, at temperatures below the bubble nucleation temperature, must be smaller than the rate of expansion of the universe.

The rate, per unit volume, of baryon number violating events depends, at low temperatures $T < T_c$ (more precisely for $M_W << T << M_W/\alpha_W$), on the sphaleron energy:

$$\Gamma = \mu \left(\frac{M_W}{\alpha_W T} \right)^3 M_W^4 \exp \left(\frac{E_{\text{sph}}(T)}{T} \right) , \tag{1}$$

with μ a dimensionless constant and $E_{\rm sph} \sim M_W(T)/\alpha_W$ ($\alpha_W = g^2/4\pi$, with g the $SU(2)_L$ gauge coupling). Comparing this rate with the rate of expansion of the universe $H \sim g_*^{1/2} T^2/m_{Pl}$ [19] pp 60–65 (g_* is the effective number of degrees of freedom and m_{Pl} is the Planck mass), the following bound is found [26]:

$$E_{\rm sph}(T_{\rm nucl})/T_{\rm nucl} > A \;\; ; \;\; A \simeq 35 - 45\,. \tag{2}$$

Here, we assume that the major wash-out is achieved near the nucleation epoch and we therefore consider only the nucleation temperature ($T_{\rm nucl}$).

In principle, this condition can be translated to a bound on the order parameter of the PT, or on the Higgs mass [3]. However, there are a number of approximations and nontrivial steps involved in this procedure. The condition (2) is on the sphaleron energy at the temperature of bubble nucleation and it has to be related to the vacuum expectation value of the Higgs field (*vev*), at critical temperature (T_c). These two temperatures are not exactly the same since, though the quantum tunneling phenomenon starts at T_c, initially the bubbles are not large enough for their volume energy to overcome the surface tension and they shrink. We have to wait for a lower temperature $T_{\rm nucl}$, when bubbles are large enough to grow. Besides, M_H is assumed to be equal to M_W and there are a number of poorly known prefactors involved. In spite of these difficulties, a condition to avoid the sphaleron erasure is found and at present generally accepted:

$$(\phi/T)_{min} \simeq 1.0 - 1.5\,. \tag{3}$$

This bound represents a condition on the order of the PT, requiring a remarkable jump in the Higgs field.

On another hand, the order of the EWPT depends on the mass of all the particles of the theory ($SU(2)_L \times U(1)_Y$ SM) and in particular on the Higgs boson mass M_H, which is at present not known. We only have constraints on it: the current lower bound on the Higgs mass from LEP [27] is: $m_H \gtrsim 114$ GeV.

The effective potential for the Higgs field , at finite temperature, can be written, including the radiative corrections from all the known SM particles:

$$V_{\rm eff} \simeq -\frac{1}{2}(\mu^2 - \alpha T^2)\phi^2 - T\delta\phi^3 + \frac{1}{4}(\lambda - \delta\lambda_T)\phi^4\,, \tag{4}$$

where the coefficients depend on the masses of the heaviest particles, on the temperature and on the *vev* (for details, see e.g. [8]). The cubic term in the effective potential is responsible for the existence of the barrier between the two degenerate vacua at T_c which makes the transition first order.

From the effective potential, the value of (ϕ/T) can be estimated at the critical value T_c, when the two minima of the potential become degenerate: $\phi/T = 2\delta/(\lambda - \delta\lambda_T)$. This is proportional to the inverse of the Higgs mass, and its maximum value is 0.55 for $m_H = 0$. The transition weakens with increasing Higgs mass, a result that is basically in agreement with lattice calculations for the EWPT in the standard model [3]. These values for (ϕ/T) do not overlap with those for the requirement (3) to avoid the sphaleron wash-out.

3 Hypermagnetic Fields and Phase Transitions

For temperatures above the EWPT, the $SU(2)_L \times U(1)_Y$ symmetry is restored, the magnetic fields correspond to the $U(1)_Y$ group instead of to the $U(1)_{em}$ group and they are therefore properly called *hypermagnetic* fields. The only fields able to propagate for long distances are the Abelian vector modes that represent a magnetic field. On the other hand, electric fields [28] as well as non-Abelian fields are screened due to the development of a temperature dependent mass.

The hypercharge field B_Y contains a component of the vector field Z, which becomes massive in the broken phase and is thus screened, such as a magnetic field in a superconductor. The presence of a hypermagnetic field consequently introduces an extra contribution in the pressure term in the symmetric phase, enhancing the difference in free energies between the two phases, making the PT more strongly first order. Recently, it has been shown [8], [7] and [29], using quite different methods (perturbatively, at tree level and at one loop, and non perturbatively, with lattice calculations) that hypermagnetic fields strengthen the PT, although the calculations differ somewhat on the level of strength reached and on the viable range for the Higgs mass and the field value. Reference [8] concludes that for $B_{Yc} \geq 0.33T_c^2$, the bound (3) is preserved. Lattice calculations [29] have shown that, even high magnetic field values do not suffice to obtain a first-order transition for $m_H \geq 80$ GeV.

Unfortunately, in the presence of an external magnetic field, the relation between the *vev* and the sphaleron energy is altered and even if condition (3) is respected, condition (2) may not be fulfilled anymore. In fact, another aspect that needs to be considered is the effect of the magnetic field on the height of the sphaleron barrier, through the coupling with the sphaleron's dipole moment. Reference [10] has found that in this case the barrier is lowered, facilitating the transition between topologically inequivalent vacua. These calculations conclude that there is no value of B_Y that is enough to push the sphaleronic transitions out of thermal equilibrium. (See also [30]).

4 Magnetic Fields in the Universe

Magnetic fields seem to be pervading the entire universe. They have been observed in galaxies, clusters, intracluster medium and high-redshift objects [31]. Estimation of magnetic fields strength -by synchrotron emission and Faraday rotation- require the independent estimation or assumption of the local electron density and the spatial structure of the field. Both quantities are reasonably known for our galaxy, where the average field strength is measured to be $3 - 4\mu G$; various spiral galaxies in our neighborhood present fields that are homogeneous over galactic size, with similar magnetic field intensities [31, 32]. At larger scales, only model dependent upper limits can be established. These limits are also in the few μG range. In the intracluster medium, recent results detect the presence of μG magnetic fields [33, 34]. For intergalactic large scale fields (dissociated from any particular galaxy or cluster), an upper bound of $10^{-9}G$ has been estimated, adopting some reasonable values for the magnetic coherence length [31].

The origin of these fields is nowadays unknown but it is widely believed that two ingredients are needed for their generation: a mechanism for creating the seed fields and a process for amplifying both their amplitude and their coherence scale [9, 35].

Generation of the seed field (magnetogenesis) may be either primordial or associated to the process of structure formation. In the early universe, which is the case of interest here, there are a number of proposed mechanisms that could possibly generate primordial magnetic fields. Among the best suited are first order phase transitions, [36, 37, 38], which provide favorable conditions for magnetogenesis such as charge separation, turbulence and out-of-equilibrium conditions. Local charge separation, creating local currents, can be achieved through the high pressure effect on the different equations of state of baryons and leptons, behind the shock fronts which precede the expanding bubbles. Turbulent flow near the bubble walls is then expected to amplify and freeze the seed field and when two shock fronts collide, turbulence and vorticity -and hence magnetic fields- can be generated to larger scales. Other proposals include bubble wall collisions, which produce phase gradients of the complex order parameter that act as a source for gauge fields [39]. A low expansion velocity of the bubbles wall then allows the magnetic flux generated in the intersection region to penetrate the colliding bubbles.

When interested in larger coherence scales, a plausible scenario is inflation, where super-horizon scale fields are generated through the amplification of quantum fluctuations of the gauge fields. This process needs however a mechanism for breaking conformal invariance of the electromagnetic field [40]. Several possibilities have been proposed, introducing non-minimal coupling of photons to curvature [41], to the dilaton/inflaton field [42] or to fermions [43].

The most promising way to distinguish between primordial and protogalactic fields is searching for their imprint on the cosmic microwave background radiation (CMBR). A homogeneous magnetic field would spoil the universe isotropy, giving rise to a dipole anisotropy in the background radiation; on this basis, COBE results set an upper bound on the present equivalent field strength [44] at the level of 10^{-9}G. On the other hand, primordial magnetic fields affect the wave patterns generating fluctuations in the energy density, producing distortions in the Planckian spectrum [45] and on the Doppler peaks [46]. Here the bounds depend on the coherence scale [45, 47]. The polarization can also be affected by primordial magnetic fields, through depolarization [48] and cross-correlations between temperature and polarization anisotropies [49]. The future CMBR satellite mission PLANCK may reach the required sensitivity for the detection of these last signals.

Although there is at present no conclusive evidence about the origin of magnetic fields, their existence prior to the EWPT epoch cannot certainly be ruled out. We will then work with primordial hypermagnetic fields, homogeneous over bubble scales.

5 CP Violating Fermion Scattering with Hypermagnetic Fields

During the EWPT, the properties of the bubble wall depend on the effective, finite temperature Higgs potential. Under the assumption that the wall is thin and that the phase transition happens when the energy densities of both phases are degenerate, it is possible to find a one-dimensional analytical solution for the Higgs field ϕ called the *kink*. When scattering is not affected by diffusion, the problem of fermion reflection and transmission through the wall can be cast in terms of solving the Dirac equation with a position dependent fermion mass, proportional to the Higgs field [50]. In order to simplify the discussion, let us consider a situation in which we take the limit when the width of the wall approaches zero. In this case, the kink solution becomes a step function $\Theta(z)$, where z is the coordinate along the direction of the phase change [12]. Since the mass of the particles is dictated by its coupling to the Higgs field, in our approximation, the former is given by

$$m(z) = m_0 \Theta(z) \,. \tag{5}$$

In terms of (5), we can see that $z \leq 0$ represents the region outside the bubble, that is the region in the symmetric phase where particles are massless. Conversely, for $z \geq 0$, the system is inside the bubble, that is in the broken phase and particles have acquired a finite mass m_0.

In the presence of an external magnetic field, we need to consider that fermion modes couple differently to the field in the broken and the symmetric phases. Let us first look at the symmetric phase.

For $z \leq 0$, the coupling is chiral. Let

$$\begin{aligned} \Psi_R &= \frac{1}{2}\left(1+\gamma_5\right)\Psi \\ \Psi_L &= \frac{1}{2}\left(1-\gamma_5\right)\Psi \end{aligned} \tag{6}$$

represent, as usual, the right and left-handed chirality modes for the spinor Ψ, respectively. Then, the equations of motion for these modes, as derived from the electroweak interaction Lagrangian, are

$$
\begin{aligned}
(i\partial\!\!\!/ - \frac{y_L}{2}g' A\!\!\!/)\Psi_L - m(z)\Psi_R = 0 \\
(i\partial\!\!\!/ - \frac{y_R}{2}g' A\!\!\!/)\Psi_R - m(z)\Psi_L = 0 \,,
\end{aligned}
\tag{7}
$$

where $y_{R,L}$ are the right and left-handed hypercharges corresponding to the given fermion, respectively, g' the $U(1)_Y$ coupling constant and we take $A^\mu = (0, \mathbf{A})$ representing a, not as yet specified, four-vector potential having non-zero components only for its spatial part, in the rest frame of the wall.

The set of equations (7) can be written as a single equation for the spinor $\Psi = \Psi_R + \Psi_L$ by adding up the former equations

$$
\left\{ i\partial\!\!\!/ - A\!\!\!/ \left[\frac{y_R}{4} g' (1 + \gamma_5) + \frac{y_L}{4} g' (1 - \gamma_5) \right] - m(z) \right\} \Psi = 0 \,. \tag{8}
$$

Hereafter, we explicitly work in the chiral representation of the gamma matrices

$$
\gamma^0 = \begin{pmatrix} 0 & -I \\ -I & 0 \end{pmatrix} \; \boldsymbol{\gamma} = \begin{pmatrix} 0 & \boldsymbol{\sigma} \\ -\boldsymbol{\sigma} & 0 \end{pmatrix} \; \gamma_5 = \begin{pmatrix} I & 0 \\ 0 & -I \end{pmatrix} , \tag{9}
$$

and thus write (8) as

$$
\left\{ i\partial\!\!\!/ - \mathcal{G} A_\mu \gamma^\mu - m(z) \right\} \Psi = 0 \,, \tag{10}
$$

where we have introduced the matrix

$$
\mathcal{G} = \begin{pmatrix} \frac{y_L}{2} g' I & 0 \\ 0 & \frac{y_R}{2} g' I \end{pmatrix} . \tag{11}
$$

We now look at the corresponding equation in the symmetry broken phase. For $z \geq 0$ the coupling of the fermion with the external field is through the electric charge e and thus, the equation of motion is simply the Dirac equation describing an electrically charged fermion in a background magnetic field, namely,

$$
\left\{ i\partial\!\!\!/ - e A_\mu \gamma^\mu - m(z) \right\} \Psi = 0 \,. \tag{12}
$$

For definiteness, let us consider a constant magnetic field $\mathbf{B} = B\hat{z}$ pointing along the $\hat{z}$ direction. In this case, the vector potential $\mathbf{A}$ can only have components perpendicular to $\hat{z}$ and the solution to the above equations factorize, namely

$$
\Phi(t, \boldsymbol{x}) = \zeta(x, y)\Phi(t, z) \,. \tag{13}
$$

We concentrate on the solution describing the motion of fermions perpendicular to the wall, *i.e.*, along the $\hat{z}$ axis and thus effectively treating the problem as the motion of fermions in one dimension. The case where the width of the wall is allowed to become finite has been addressed in [13] and we will briefly present below the results of this more realistic case; whereas the motion of the fermions in three dimensions has been given a solution in [14]. Equations (10) and (12) can be solved analytically. We look for the scattering states appropriate to describe the motion of fermions in the symmetric and broken symmetry phases. For our purposes, these are fermions incident toward and reflected from the wall in the symmetric phase. There are two types of such solutions; those coupled with y_L and those coupled with y_R. For an incident wave coupled with y_L (y_R), the fact that the differential equation mixes up the solutions means that the reflected wave will also include a component coupled with y_R (y_L). In analogy, the solution to (12) is found by looking for the scattering states appropriate for the description of transmitted waves. The solutions are explicitly constructed in [12] to where we refer the reader for details. For the purposes of this work, we proceed to describe how to use these solutions to construct the transmission and reflection probabilities.

In order to quantitatively describe the scattering of fermions, we need to compute the corresponding reflection and transmission coefficients. These are built from the reflected, transmitted and incident currents of each type. Recall that for a given spinor wave function Ψ, the current normal to the wall is given by

$$J = \Psi^\dagger \gamma^0 \gamma^3 \Psi \,. \tag{14}$$

The reflection and transmission coefficients, R and T, are given as the ratios of the reflected and transmitted currents, to the incident one, respectively, projected along a unit vector normal to the wall.

The probabilities for finding a left or a right-handed particle in the symmetric phase after reflection, PR_L, PR_R are given, respectively by

$$PR_L = R_{L\to L} + R_{R\to L} \tag{15}$$

$$PR_R = R_{L\to R} + R_{R\to R} \,, \tag{16}$$

whereas the probabilities for finding a left or a right-handed particle in the symmetry broken phase after transmission, PT_L, PT_R are given, respectively by

$$PT_L = T_{L\to L} + T_{R\to L} \tag{17}$$

$$PT_R = T_{L\to R} + T_{R\to R} \,. \tag{18}$$

Figure 1 shows the probabilities PR_L and PR_R as a function of the magnetic field parametrized as $B = bT^2$ for a temperature $T = 100$ GeV, a fixed $E = 184$ GeV and for a fermion taken as the top quark with a mass $m_0 = 175$ GeV, $y_R = 4/3$, $y_L = 1/3$ and for a value of $g' = 0.344$, as appropriate for the EWPT epoch. Notice that when $b \to 0$, these probabilities approach each other and that the difference grows with increasing field strength. We have considered the top quark since it is assumed to be the heaviest particle in the broken phase, and hence to have the larger Yukawa coupling. The τ-lepton may also be a good candidate.

In the case that we allow for the vacuum expectation value of the Higgs field and $m(z)$ to vary continuously through the wall, we work in the thin wall regime with the kink solution:

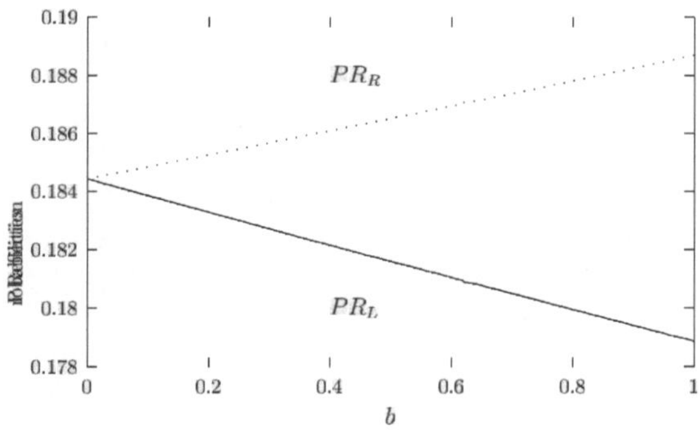

Fig. 1. Probabilities PR_L and PR_R as a function of the magnetic field parametrized as $B = bT^2$ for $T = 100$ GeV, $E = 184$ GeV and a top quark with a mass $m_0 = 175$ GeV, $y_R = 4/3$, $y_L = 1/3$. The value for the U(1)$_Y$ coupling constant is taken as $g' = 0.344$, corresponding to the EWPT epoch.

$$\varphi(x) = 1 + \tanh(x)\,, \tag{19}$$

where the dimensionless position coordinate x is proportional to z. In this case, we have to solve the same equation (8), but with a different $m(z)$ profile. We achieved this with a combination of analytical and numerical methods and we report here only the results for the reflection probabilities.

Figure 2 shows the coefficients $R_{l\to r}$ and $R_{r\to l}$ as a function of the magnetic field parameter $b \equiv \left(\delta T/\sqrt{2\lambda}\right)^{-2} B$, an energy parameter $\epsilon \equiv \left(\delta T/\sqrt{2\lambda}\right)^{-1} E = 7.03$ (slightly larger than the height of the barrier, in order to avoid the exponential damping of the transmitted waves), and the other parameters as in the previous case. Again, notice that, when $b \to 0$, these coefficients approach each other and that the difference grows with increasing field strength. The results are in good agreement with the simplest case. This scheme is more realistic then the former since the height and width of the wall are typically related to each other in such a way that it is not entirely realistic to vary one without affecting the other.

Though not explicitly worked out here, it is easy to convince oneself that when considering the scattering in three dimensions, the quantum mechanical motion of the fermion will include in general the description of its velocity vector with a component perpendicular to the field. In this case, due to the Lorentz force, the particle circles around the field lines maintaining its velocity along the direction of the field. The motion of the particle is thus described as an overall displacement along the field lines superimposed to a circular motion around these lines [14]. These circles are labeled by the principal quantum number. We see that if we have fermions that start off moving by making a nonzero angle with the field lines, all of these trajectories will result at the end in the same overall direction of incidence. Also, since

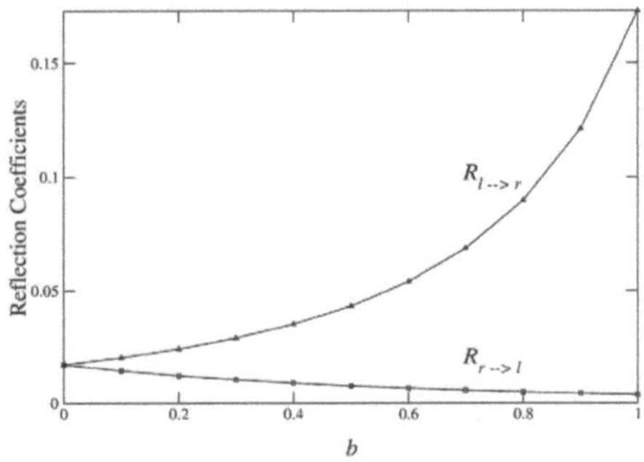

Fig. 2. Coefficients $R_{l\rightarrow r}$ and $R_{r\rightarrow l}$, in the case of a smooth variation of $m(z)$, as a function of the magnetic field parameter b for an energy parameter $\epsilon = 7.03$, $y_R = 4/3$, $y_L = 1/3$. The value for the $U(1)_Y$ coupling constant is taken as $g' = 0.344$, corresponding to the EWPT epoch. The dots represent the computed values.

the fermion coupling with the external field is through its spin, changing the direction of the field exchanges the role of each spin component but since each chirality mode contains both spin orientations, this does not affect the final probabilities and thus the asymmetry is independent of the orientation of the field with respect to the $\hat{z}$ axis.

It is interesting to notice that, with this mechanism, we are not generating a net excess of one type of particle (left- or right- handed) over the other; it is merely a segregation between the two sides of the bubble wall.

We also emphasize that, under the very general assumptions of CPT invariance and unitarity, the total axial asymmetry (which includes contributions both from particles and antiparticles) is quantified in terms of the particle axial asymmetry. Let ρ_i represent the number density for species i. The net densities in left-handed and right-handed axial charges are obtained by taking the differences $\rho_L - \rho_{\bar{L}}$ and $\rho_R - \rho_{\bar{R}}$, respectively. It is straightforward to show [15] that CPT invariance and unitarity imply that the above net densities are given by

$$\begin{aligned} \rho_L - \rho_{\bar{L}} &= (f^s - f^b)(R_{r\rightarrow l} - R_{l\rightarrow r}) \\ \rho_R - \rho_{\bar{R}} &= (f^s - f^b)(R_{l\rightarrow r} - R_{r\rightarrow l})\,, \end{aligned} \qquad (20)$$

where f^s and f^b are the statistical distributions for particles or antiparticles (since the chemical potentials are assumed to be zero or small compared to the temperature, these distributions are the same for particles or antiparticles) in the symmetric and the broken symmetry phases, respectively. From (20), the asymmetry in the axial charge density is finally given by

$$(\rho_L - \rho_{\bar{L}}) - (\rho_R - \rho_{\bar{R}}) = 2(f^s - f^b)(R_{r\rightarrow l} - R_{l\rightarrow r}). \qquad (21)$$

This asymmetry, built on either side of the wall, is dissociated from non-conserving baryon number processes and can subsequently be converted to baryon number in the symmetric phase where sphaleron induced transitions are taking place with a large rate. This mechanism receives the name of *non-local baryogenesis* [4, 15, 51, 52] and, in the absence of the external field, it can only be realized in extensions of the SM where a source of CP violation is introduced ad hoc into a complex, space-dependent phase of the Higgs field during the development of the EWPT [53].

In our case, the relation of this axial asymmetry to CP violation is understood as follows: recall that for instance, in the SM, CP is violated in the quark sector through the mixing between different weak interaction eigenstates to form states with definite mass. However, in the present scenario, no such mixing occurs since we are concerned only with the evolution of a single quark (*e.g.*, the top quark) species. The relation is thus to be found in the dynamics of the scattering process itself and becomes clear once we notice that this can be thought of as describing the mixing of two levels, right- and left-handed quarks coupled to an external hypermagnetic field. When the two chirality modes interact only with the external field, they evolve separately. It is only the scattering with the bubble wall what allows a finite transition probability for one mode to become the other. Since the modes are coupled differently to the external field, these probabilities are different and give rise to the axial asymmetry. CP is violated in the process because, though C is conserved, P is violated and thus is CP.

6 Summary and Outlook

In this work we have given a quantitative outline of the CP violating scattering of fermions off (a simplified picture of) EWPT bubbles in the presence of hypermagnetic fields. This scattering produces an axial asymmetry built on either side of the bubble walls. The origin of this asymmetry is the chiral nature of the fermion coupling to the hypermagnetic field in the symmetric phase. We have shown how to compute reflection and transmission coefficients and also that these differ for left and right-handed incident particles.

Primordial hypermagnetic fields thus provide with a much needed ingredient, namely, additional CP violation, for the possible generation of baryon number during the EWPT. A second ingredient, the strengthening of the order of the phase transition and thus the avoidance of the sphaleron bound seems at the moment a difficult to surmount problem. Nonetheless, it is important to bear in mind that so far, the calculations that provide insight into the effect of the hypermagnetic fields on the order of the EWPT do not account for the non-perturbative effects, cast in the language of resummation, which are otherwise well known to play a very important role for the dynamics of the phase transition in the absence of magnetic fields. Much work is needed in this direction. This is for the future.

Acknowledgments

GP is grateful to Instituto de Astronomía, UNAM, for its kind hospitality during the development of this work. Support for this work has been received in part by DGAPA-UNAM under PAPIIT grant number IN108001 and by CONACyT-México under grant numbers 32279-E and 40025-F.

References

1. A. D. Sakharov, Pis'ma Zh. Eksp. Teor. Fiz. **5**, 32 (1967) [JETP Lett. **5**, 24 (1967)].
2. M.B. Gavela, P. Hernández, J. Orloff, and O. Pène, Mod. Phys. Lett. A **9**, 795 (1994).
3. K. Kajantie, M. Laine, K. Rummukainen, and M. Shaposnikov, Nucl. Phys. B **466**, 189 (1996).
4. A. G. Cohen, D. B. Kaplan and A. E. Nelson, Phys. Lett. B **263**, 86 (1991).
5. M. Carena, M. Quiros and C. E. Wagner, Nucl. Phys. B **524**, 3 (1998).
6. J. M. Cline, *Electroweak phase transition and baryogenesis*, hep-ph/0201286.
7. M. Giovannini and M. E. Shaposhnikov, Phys. Rev. D **57**, 2186 (1998).
8. P. Elmfors, K. Enqvist and K. Kainulainen, Phys. Lett. B **440**, 269 (1998).
9. M. Giovannini, *Primordial Magnetic Fields*, hep-ph/0208152.
10. D. Comelli, D. Grasso, M. Pietroni and A. Riotto, Phys. Lett. B **458**, 304 (1999).
11. M. E. Carrington, Phys. Rev. D **45**, 2933 (1992).
12. A. Ayala, J. Besprosvany, G. Pallares and G. Piccinelli, Phys. Rev. D **64**, 123529 (2001).
13. A. Ayala, G. Piccinelli and G. Pallares, Phys. Rev. D **66**, 103503 (2002).
14. A. Ayala and J. Besprosvany, Nucl. Phys. B **651**, 211 (2003).
15. A. E. Nelson, D. B. Kaplan and A. G. Cohen, Nucl. Phys. B **373**, 453 (1992).
16. M. Trodden, Rev. Mod. Phys. **71**, 1463 (1999).
17. A. Riotto and M. Trodden, Ann. Rev. Nucl. Part. Sci. **49**, 35 (1999).
18. V. A. Rubakov and M. E. Shaposhnivov, Usp.Fiz.Nauk **166**, 493 (1996), hep-ph/9603208.
19. E. W. Kolb and M. S. Turner: *The Early Universe*, (Addison–Wesley Publishing Company 1990).
20. F. W. Stecker: The matter-antimatter asymmetry of the universe. In: *XIVth Rencontres de Blois 2002 on Matter-Antimatter Asymmetry*, ed. by J. Tran Thanh Van, hep-ph/0207323.
21. A. D. Dolgov: Cosmological matter-antimatter asymmetry and antimatter in the universe. In: *XIVth Rencontres de Blois 2002 on Matter-Antimatter Asymmetry*, ed. by J. Tran Thanh Van, hep-ph/0211260.
22. F. R. Klinkhamer and N. S. Manton, Phys. Rev. D **30**, 2212 (1984).
23. G. 't Hooft, Phys. Rev. Lett. **37**, 8 (1976); Phys. Rev. D **14**, 3432 (1976).
24. For a recent review on the subject, see e.g. W. Bernreuther, Lect. Notes Phys. **591**, 237 (2002), hep-ph/0205279.
25. V. A. Kuzmin, V. A. Rubakov and M. E. Shaposhnikov, Phys. Lett. B **155**, 36 (1985).

26. M. E. Shaposhnikov, Nucl. Phys. B **287**, 757 (1987).
27. K. Hagiwara et al., Phys. Rev. D **66**, 010001 (2002), http://pdg.lbl.gov/.
28. M. Le Bellac: *Thermal Field Theory*, (Cambridge University Press, Cambridge 1996) pp 124–127.
29. K. Kajantie, M. Laine, J. Peisa, K. Rummukainen, and M. Shaposnikov, Nucl. Phys. B **544**, 357 (1999).
30. V. Skalozub and V. Demchik, *Can baryogenesis survive in the standard model due to strong hypermagnetic field?*, hep-ph/9909550.
31. The observational situation is discussed in P.P. Kronberg, Rep. Prog. Phys. **57**, 325 (1994); or more recently in J.-L. Han and R. Wielebinski, *Milestones in the Observations of Cosmic Magnetic Fields*, ChJ A & A, **2**, 293 (2002), astro-ph/0209090.
32. R. Beck, A. Brandenburg, D. Moss, A. Shukurov and D. Sokoloff, Annu. Rev. Astron. Astrophys. **34**, 155 (1996).
33. J. A. Eilek and F. N. Owen, Ap. J. **567**, 202 (2002).
34. T. E. Clarke, P. P. Kronberg and H. Böhringer, Ap. J. **547**, L111 (2001).
35. For reviews on the origin, evolution and some cosmological consequences of primordial magnetic fields see: K. Enqvist, Int. J. Mod. Phys. D **7**, 331 (1998); R. Maartens: Cosmological magnetic fields. In: *International Conference on Gravitation and Cosmology*, Pramana **55**, 575 (2000) and references therein; D. Grasso and H.R. Rubinstein, Phys. Rep. **348**, 163 (2001).
36. For quark-hadron PT: J. Quashnock, A. Loeb and D.N. Spergel, Ap. J. **344**, L49 (1989); B. Cheng and A.V. Olinto, Phys. Rev. D **50**, 2421 (1994); G. Sigl, A.V. Olinto and K. Jedamzik, Phys. Rev. D **55**, 4582 (1997).
37. For EWPT: G. Baym, D. Bödeker and L. McLerran, Phys. Rev. D **53**, 662 (1996).
38. For a PT at a critical temperature larger than EW scale: D. Boyanovsky, H. J. de Vega and M. Simionato, *Large scale magnetogenesis from a non-equilibrium phase transition in the radiation dominated era*, Phys. Rev. D **67**, 123505 (2003), hep-ph/0211022.
39. T. Vachaspati, Phys. Lett. B **265**, 258 (1991); T. W. B. Kibble and A. Vilenkin, Phys. Rev. D **52**, 679 (1995); E.J. Copeland, P. M. Saffin and O. Törnqvist, Phys. Rev. D **61**, 105005 (2000).
40. For an overview of the subject, see e.g., A. D. Dolgov, *Generation of magnetic fields in cosmology*, hep-ph/0110293.
41. M. S. Turner and L. M. Widrow, Phys. Rev. D **37**, 2743 (1988).
42. B. Ratra, Ap. J. **391**, L1 (1992); M. Gasperini, M. Giovannini and G. Veneziano, Phys. Rev. Lett. **75**, 3796 (1995); D. Lemoine and M. Lemoine, Phys. Rev. D **52**, 1955 (1995).
43. T. Prokopec, *Cosmological magnetic fields from photon coupling to fermions and bosons in inflation*, astro-ph/0106247.
44. J.D. Barrow, P. Ferreira and J. Silk, Phys. Rev. Lett. **78**, 3610 (1997).
45. K. Jedamzik, V. Katalinić and A. V. Olinto, Phys. Rev. Lett. **85**, 700 (2000).
46. J. Adams, U. H. Danielsson, D. Grasso and H. rubinstein, Phys. Lett. B **388**, 253 (1996).
47. For the effect of stochastic magnetic fields, see A. Mack, T. Kahniashvili and A. Kosowsky, Phys.Rev. D **65**, 123004 (2002), and references therein.
48. D. D. Harari, J. D. Hayward and M. Zaldarriaga, Phys. Rev. D **55**, 1841 (1997); M Giovannini, Phys. Rev. D **56**, 3198 (1997).

49. A. Kosowsky and A. Loeb, Ap. J. **469**, 1 (1996); E. S. Scannapieco and P.G. Ferreira, Phys. Rev. D **56**, 7493 (1997).
50. A. Ayala, J. Jalilian-Marian, L. McLerran and A. P. Vischer, Phys. Rev. D **49**, 5559 (1994).
51. M. Dine, O. Lechtenfield, B. Sakita, W. Fischel and J. Polchinski, Nucl. Phys. B **342**, 381 (1990).
52. M. Joyce, T. Prokopec and N. Turok, Phys. Lett. B **338**, 269 (1994).
53. E. Torrente-Lujan, Phys. Rev. D **60**, 085003 (1999).

Infering Annihilation Channels of Neutralinos in Galactic Halos

Luis G. Cabral–Rosetti[1], Xavier Hernández[2], and Roberto A. Sussman[3]

[1] Instituto de Ciencias Nucleares, UNAM, A.P. 70543, México D.F., 04510, México. luis@nuclecu.unam.mx
[2] Instituto de Astronomía, UNAM, A.P. 70264, México D.F., 04510, México. xavier@astroscu.unam.mx
[3] Instituto de Ciencias Nucleares, UNAM, A.P. 70543, México D.F., 04510, México. sussman@nuclecu.unam.mx

Abstract. Applying the microcanonical definition of entropy to a weakly interacting and self–gravitating neutralino gas, we evaluate the change in the local entropy per particle of this gas between the freeze out era and present day virialized halo structures. An "entropy consistency" criterion emerges by comparing the obtained theoretical entropy per particle of the virialized halos with an empirical entropy per particle given in terms of dynamical halo variables of actual galactic structures. We apply this criterion to the cases when neutralinos are mostly B-inos and mostly Higgsinos, in conjunction with the usual "abundance" criterion requiring that present neutralino relic density complies with $0.2 < \Omega_{\tilde{\chi}_0^1} < 0.4$ for $h \simeq 0.65$. The joint application of both criteria reveals that a much better fitting occurs for the B-ino than for the Higgsino channels, so that the former seems to be a favored channel along the mass range of $150\,\mathrm{GeV} < m_{\tilde{\chi}_0^1} < 250\,\mathrm{GeV}$. These results are consistent with neutralino annihilation patterns that emerge from recent theoretical analysis on cosmic ray positron excess data reported by the HEAT collaboration. The suggested methodology can be applied to test other annihilation channels of the neutralino, as well as other particle candidates of thermal WIMP gas relics.

1 Introduction

There are strong theoretical arguments favoring lightest supersymmetric particles (LSP) as making up the relic gas that forms the halos of actual galactic structures. Assuming that R parity is conserved and that the LSP is stable, it might be an ideal candidate for cold dark matter (CDM), provided it is neutral and has no strong interactions. The most favored scenario [1, 2, 3, 4, 5, 6] considers the LSP to be the lightest neutralino ($\tilde{\chi}_1^0$), a mixture of supersymmetric partners of the photon, Z boson and neutral Higgs boson [2]. Since neutralinos must have decoupled once they were non-relativistic, it is reasonable to assume that they constituted originally a Maxwell-Boltzmann (MB) gas in thermal equilibrium with other components of the primordial cosmic plasma. In the present cosmic era, such a gas is practically collision–less and is either virialized in galactic and galactic cluster halos, in the process of virialization or still in the linear regime for superclusters and structures near the scale of homogeneity[7, 8, 9].

L.G. Cabral–Rosetti, X. Hernández, and R.A. Sussman, Infering Annihilation Channels of Neutralinos in Galactic Halos, Lect. Notes Phys. **646**, 309–320 (2004)
http://www.springerlink.com/

Besides the constraint due to their present abundance as main constituents of cosmic dark matter ($\Omega_{\tilde{\chi}_0^1} \sim 0.3$), it is still uncertain which type of annihilation cross section characterizes these neutralinos. In this paper we present a method that discriminates between different cross sections, based on demanding (together with the correct abundance) that a theoretically estimated entropy per particle matches an empiric estimate of the same entropy, but constructed with dynamic variables of actual halo structures. The application of this "entropy consistency" criterion is straightforward because entropy is a state variable that can be evaluated at equilibrium states, irrespectively of how enormously complicated could be the evolution between each state. In this context, the two fiducial equilibrium states of the neutralino gas are (to a good approximation) the decoupling (or "freeze out") and their present state as a virialized relic gas. Considering simplified forms of annihilation cross sections. the joint application of the abundance and entropy–consistency criteria favors the neutralinos as mainly "B–inos" over neutralinos as mainly "higgsinos". These results are consistent with the theoretical analysis of the HEAT experiment [10, 11, 12] which aims at relating the observed positron excess in cosmic rays with a possible weak interaction between neutralinos and nucleons in galactic halos. The paper is organized as follows. In Sect. 2 we describe the thermodynamics of the neutralino gas as it decouples. Section 3 applies to the post–decoupling neutralino gas the entropy definition of the microcanonical ensemble entropy, leading to a suitable theoretical estimate of the entropy per particle. In Sect. 4 we obtain an empiric estimate of this entropy based on actual halo variables, while in Sect. 5 we examine the consequences of demanding that these two entropies coincide. Section 6 provides a summary of these results.

2 The Neutralino Gas

The equation of state of a non-relativistic MB neutralino gas is [7, 8, 9]

$$\rho = m_{\tilde{\chi}_0^1} n_{\tilde{\chi}_0^1} \left(1 + \frac{3}{2\,x}\right), \qquad p = \frac{m_{\tilde{\chi}_0^1} n_{\tilde{\chi}_0^1}}{x}, \tag{1}$$

$$x \equiv \frac{m_{\tilde{\chi}_0^1}}{T}, \tag{2}$$

where $m_{\tilde{\chi}_0^1}$ and $n_{\tilde{\chi}_0^1}$ are the neutralino mass and number density. Since we will deal exclusively with the lightest neutralino, we will omit henceforth the subscript $_{\tilde{\chi}_0^1}$, understanding that all usage of the term "neutralino" and all symbols of physical and observational variables (*i.e.* Ω_0, m, ρ, n, etc.) will correspond to this specific particle. As long as the neutralino gas is in thermal equilibrium, we have

$$n \approx n^{(\mathrm{eq})} = g \left[\frac{m}{\sqrt{2\,\pi}}\right]^3 x^{-3/2} \exp\left(-x\right), \tag{3}$$

where $g = 1$ is the degeneracy factor of the neutralino species. The number density n satisfies the Boltzmann equation [2, 7]

$$\dot{n} + 3\,H\,n \;=\; -\langle\sigma|\mathrm{v}|\rangle\left[n^2 - \left(n^{(\mathrm{eq})}\right)^2\right], \tag{4}$$

where H is the Hubble expansion factor and $\langle\sigma|\mathrm{v}|\rangle$ is the annihilation cross section. Since the neutralino is non-relativistic as annihilation reactions "freeze out" and it decouples from the radiation dominated cosmic plasma, we can assume for H and $\langle\sigma|\mathrm{v}|\rangle$ the following forms

$$H \;=\; 1.66\,g_*^{1/2}\frac{T^2}{m_p}, \tag{5}$$

$$\langle\sigma|\mathrm{v}|\rangle \;=\; a \;+\; b\langle\mathrm{v}^2\rangle, \tag{6}$$

where $m_p = 1.22\times10^{19}$ GeV is Planck's mass, $g_* = g_*(T)$ is the sum of relativistic degrees of freedom, $\langle\mathrm{v}^2\rangle$ is the thermal averaging of the center of mass velocity (roughly $\mathrm{v}^2 \propto 1/x$ in non-relativistic conditions) and the constants a and b are determined by the parameters characterizing specific annihilation processes of the neutralino (s-wave or p-wave) [2]. The decoupling of the neutralino gas follows from the condition

$$\Gamma \;\equiv\; n\,\langle\sigma|\mathrm{v}|\rangle \;=\; H, \tag{7}$$

leading to the freeze out temperature T_f. Reasonable approximated solutions of (7) follow by solving for x_f the implicit relation [2]

$$x_\mathrm{f} = \ln\left[\frac{0.0764\,m_p\,c_0(2+c_0)\,(a+6\,b/x_\mathrm{f})\,m}{(g_{*\mathrm{f}}\,x_\mathrm{f})^{1/2}}\right], \tag{8}$$

where $g_{*\mathrm{f}} = g_*(T_\mathrm{f})$ and $c_0 \approx 1/2$ yields the best fit to the numerical solution of (4) and (7). From the asymptotic solution of (4) we obtain the present abundance of the relic neutralino gas [2]

$$\Omega_0\,h^2 \;=\; Y_\infty\,\frac{\mathcal{S}_0\,m}{\rho_\mathrm{crit}/h^2} \;\approx\; 2.82\times10^8\,Y_\infty\,\frac{m}{\mathrm{GeV}}, \tag{9}$$

$$\begin{aligned} Y_\infty \;&\equiv\; \frac{n_0}{\mathcal{S}_0} \\ &=\; \left[0.264\,g_{*\mathrm{f}}^{1/2}\,m_p\,m\left\{a/x_\mathrm{f} + 3(b-1/4\,a)/x_\mathrm{f}^2\right\}\right]^{-1}, \end{aligned} \tag{10}$$

where $\mathcal{S}_0 \approx 4000\,\mathrm{cm}^{-3}$ is the present radiation entropy density (CMB plus neutrinos), $\rho_\mathrm{crit} = 1.05\times10^{-5}\,\mathrm{GeV\,cm}^{-3}$.

Since neutralino masses are expected to be in the range of tens to hundreds of GeV's and typically we have $x_\mathrm{f} \sim 20$ so that $T_\mathrm{f} >$ GeV, we can use $g_{*\mathrm{f}} \simeq 106.75$ [3] in equations (8) – (10). Equation (8) shows how x_f has a logarithmic dependence on m, while theoretical considerations [1, 2, 3, 4, 5, 6] related to the minimal supersymetric extensions of the Standard Model (MSSM) yield specific forms for a and b that also depend on m. Inserting into (9)–(10) the specific forms of a and b for each annihilation channel leads to a specific range of m that satisfies the "abundance" criterion based on current observational constraints that require $0.1 < \Omega_0 < 0.3$ and $h \approx 0.65$ [9].

Suitable forms for $\langle\sigma|v|\rangle$ can be obtained for all types of annihilation reactions [2]. If the neutralino is mainly pure B-ino, it will mostly annihilate into lepton pairs through t-channel exchange of right-handed sleptons. In this case the cross section is p-wave dominated and can be approximated by (6) with [3, 13, 14]

$$a \approx 0, \qquad b \approx \frac{8\,\pi\,\alpha_1^2}{m^2\,[1+m_l^2/m^2]^2}, \tag{11}$$

where m_l is the mass of the right-handed slepton ($m_l \sim m$ [3]) and $\alpha_1^2 = g_1^2/4\pi \simeq 0.01$ is the fine structure coupling constant for the $U(1)_Y$ gauge interaction. If the neutralino is Higgsino-like, annihilating into W-boson pairs, then the cross section is s-wave dominated and can be approximated by (6) with [3, 13, 14]

$$b \approx 0, \qquad a \approx \frac{\pi\,\alpha_2^2\,(1-m_W^2/m^2)^{3/2}}{2\,m^2\,(2-m_W^2/m^2)^2}, \tag{12}$$

where $m_W = 80.44$ GeV is the mass of the W-boson and $\alpha_2^2 = g_2^2/4\pi \simeq 0.03$ is the fine structure coupling constant for the $SU(2)_L$ gauge interaction.

In the freeze out era the entropy per particle (in units of the Boltzmann constant k_B) for the neutralino gas is given by [7, 9, 8]

$$s_f = \left[\frac{\rho+p}{n\,T}\right]_f = \frac{5}{2} + x_f, \tag{13}$$

where we have assumed that chemical potential is negligible and have used the equation of state (1). From (8) and (13), it is evident that the dependence of s_f on m will be determined by the specific details of the annihilation processes through the forms of a and b. In particular, we will use (11) and (12) to compute s_f from (8)-(13).

3 The Microcanonical Entropy

After the freeze out era, particle numbers are conserved and the neutralinos constitute a weakly interacting and practically collision–less self–gravitating gas. This gas is only gravitationally coupled to other components of the cosmic fluid. As it expands, it experiences free streaming and eventually undergoes gravitational clustering forming stable bound virialized structures

[9, 8, 15, 16]. The evolution between a spectrum of density perturbations at the freeze out and the final virialized structures is extremely complex, involving a variety of dissipative effects characterized by collisional and collision–less relaxation processes [15, 16, 17]. However, the freeze out and present day virialized structures roughly correspond to "initial" and "final" equilibrium states of this gas. Therefore, instead of dealing with the enormous complexity of the details of the intermediary processes, we will deal only with quantities defined in these states with the help of simplifying but general physical assumptions.

The microcanonical ensemble in the "mean field" approximation yields an entropy definition that is well defined for a self–gravitating gas in an intermediate scale, between the short range and long range regimes of the gravitational potential. This intermediate scale can be associated with a region that is "sufficiently large as to contain a large number of particles but small enough for the gravitational potential to be treated as a constant" [15]. Considering the neutralino gas in present day virialized halo structures as a diluted, non-relativistic (nearly) ideal gas of weakly interacting particles, its microcanonical entropy per particle under these conditions can be given in terms of the volume of phase space [16]

$$s = \ln\left[\frac{(2mE)^{3/2}\,V}{(2\pi\hbar)^3}\right], \tag{14}$$

where V and E are local average values of volume and energy associated with the intermediate scale. For non-relativistic velocities $v/c \ll 1$, we have $V \propto 1/n \propto m/\rho$ and $E \propto m v^2/2 \propto m/x$. In fact, under these assumptions the definition (14), evaluated at the freeze out, is consistent with (3) and (13), and so it is also valid immediately after the freeze out era (once particle numbers are conserved). Since (14) is valid at both the initial and final states, respectively corresponding to the decoupling ($s_{\rm f}$, $x_{\rm f}$, $n_{\rm f}$) and the values ($s^{(\rm h)}$, $x^{(\rm h)}$, $n^{(\rm h)}$) associated with a suitable halo structure, the change in entropy per particle that follows from (14) between these two states is given by

$$s^{(\rm h)} - s_{\rm f} = \ln\left[\frac{n_{\rm f}}{n^{(\rm h)}}\left(\frac{x_{\rm f}}{x^{(\rm h)}}\right)^{3/2}\right], \tag{15}$$

where (13) can be used to eliminate $s_{\rm f}$ in terms of $x_{\rm f}$. Considering present day halo structures as roughly spherical, inhomogeneous and self-gravitating gaseous systems, the intermediate scale of the microcanonical description is an excellent approximation for gas particles in a typical region of $\sim 1\,\mathrm{pc}^3$ within the halo core, near the symmetry center of the halo where the gas density enhancement is maximum but spacial gradients of all macroscopic quantities are negligible [18, 19]. Therefore, we will consider current halo macroscopic variables as evaluated at the center of the halo: $s_c^{(\rm h)}$, $x_c^{(\rm h)}$, $n_c^{(\rm h)}$.

In order to obtain a convenient theoretical estimate of $s_c^{(\rm h)}$ from (15), we need to relate $n_{\rm f}$ with present day cosmological parameters like Ω_0 and h.

Bearing in mind that density perturbations at the freeze out era were very small ($\delta n_{\rm f}/n_{\rm f} < 10^{-4}$, [7, 8, 9]), the density $n_{\rm f}$ is practically homogeneous and so we can estimate it from the conservation of particle numbers: $n_{\rm f} = n_0\,(1+z_{\rm f})^3$, and of photon entropy: $g_{*{\rm f}}\mathcal{S}_{\rm f} = g_{*0}\,\mathcal{S}_0\,(1+z_{\rm f})^3$, valid from the freeze out era to the present for the unperturbed homogeneous background. Eliminating $(1+z_{\rm f})^3$ from these conservation laws yields

$$
\begin{aligned}
n_{\rm f} &= n_0\,\frac{g_{*{\rm f}}}{g_{*0}}\left[\frac{T_{\rm f}}{T_0^{\rm CMB}}\right]^3 \simeq 27.3\,n_0\left[\frac{x_0^{\rm CMB}}{x_{\rm f}}\right]^3, \\
\text{where}\quad x_0^{\rm CMB} &\equiv \frac{m}{T_0^{\rm CMB}} = 4.29\times 10^{12}\,\frac{m}{\rm GeV}
\end{aligned}
\tag{16}
$$

where $g_{*0} = g_*(T_0^{\rm CMB}) \simeq 3.91$ and $T_0^{\rm CMB} = 2.7\,{\rm K}$. Since for present day conditions $n_0/n_c^{\rm (h)} = \rho_0/\rho_c^{\rm (h)}$ and $\rho_0 = \rho_{\rm crit}\,\Omega_0\,h^2$, we collect the results from (16) and write (15) as

$$
\begin{aligned}
s_c^{\rm (h)}|_{\rm th} &= x_{\rm f} + 93.06 + \ln\left[\left(\frac{m}{\rm GeV}\right)^3\frac{h^2\,\Omega_0}{(x_{\rm f}\,x_c^{\rm (h)})^{3/2}}\,\frac{\rho_{\rm crit}}{\rho_c^{\rm (h)}}\right] \\
&= x_{\rm f} + 81.60 + \ln\left[\left(\frac{m}{\rm GeV}\right)^3\frac{h^2\,\Omega_0}{(x_{\rm f}\,x_c^{\rm (h)})^{3/2}}\,\frac{{\rm GeV/cm^3}}{\rho_c^{\rm (h)}}\right],
\end{aligned}
\tag{17}
$$

Therefore, given m and a specific form of $\langle\sigma|{\rm v}|\rangle$ associated with a and b, equation (17) provides a theoretical estimate of the entropy per particle of the neutralino halo gas that depends on the initial state given by $x_{\rm f}$ in (8) and (13), on observable cosmological parameters Ω_0, h and on generic state variables associated to the halo structure.

4 Theoretical and Empiric Entropies

If the neutralino gas in present halo structures strictly satisfies MB statistics, the entropy per particle, $s_c^{\rm (h)}$, in terms of $\rho_c^{\rm (h)} = m\,n_c^{\rm (h)}$ and $x_c^{\rm (h)} = m\,c^2/(k_{\scriptscriptstyle B}\,T_c^{\rm (h)})$, follows from the well known Sackur–Tetrode entropy formula [20]

$$
\begin{aligned}
s_c^{\rm (h)}|_{\rm MB} &= \frac{5}{2} + \ln\left[\frac{m^4\,c^3}{\hbar^3\,(2\pi\,x_c^{\rm (h)})^{3/2}\,\rho_c^{\rm (h)}}\right] \\
&= 94.42 + \ln\left[\left(\frac{m}{\rm GeV}\right)^4\left(\frac{1}{x_c^{\rm (h)}}\right)^{3/2}\frac{{\rm GeV/cm^3}}{\rho_c^{\rm (h)}}\right].
\end{aligned}
\tag{18}
$$

Such a MB gas in equilibrium is equivalent to an isothermal halo if we identify [21]

$$
\frac{c^2}{x^{\rm (h)}} = \frac{k_{\scriptscriptstyle B}\,T^{\rm (h)}}{m} = \sigma^2_{\rm (h)},
\tag{19}
$$

where $\sigma^2_{\rm (h)}$ is the velocity dispersion (a constant for isothermal halos).

However, an exactly isothermal halo is not a realistic model, since its total mass diverges and it allows for infinite particle velocities (theoretically accessible in the velocity range of the MB distribution). More realistic halo models follow from "energy truncated" (ET) distribution functions [16, 21, 22, 23, 24] that assume a maximal "cut off" velocity (an escape velocity). Therefore, we can provide a convenient empirical estimate of the halo entropy, $s_c^{(\mathrm{h})}$, from the microcanonical entropy definition (14) in terms of phase space volume, but restricting this volume to the actual range of velocities (i.e. momenta) accessible to the central particles, that is up to a maximal escape velocity $v_e(0)$. From theoretical studies of dynamical and thermodynamical stability associated with ET distribution functions [22, 23, 26, 24, 27, 28, 25] and from observational data for elliptic and LSB galaxies and clusters [29, 30, 18, 31, 32], it is reasonable to assume

$$v_e^2(0) \;=\; 2\,|\Phi(0)| \;\simeq\; \alpha\,\sigma_{(\mathrm{h})}^2(0), \quad 12 < \alpha < 18, \tag{20}$$

where $\Phi(r)$ is the newtonian gravitational potential. We have then

$$\begin{aligned} s_c^{(\mathrm{h})}\big|_{\mathrm{em}} \;&\simeq\; \ln\left[\frac{m^4\,v_e^3}{(2\pi\hbar)^3\,\rho_c^{(\mathrm{h})}}\right] \\ &=\; 89.17 + \ln\left[\left(\frac{m}{\mathrm{GeV}}\right)^4\left(\frac{\alpha}{x_c^{(\mathrm{h})}}\right)^{3/2}\frac{\mathrm{GeV/cm^3}}{\rho_c^{(\mathrm{h})}}\right], \end{aligned} \tag{21}$$

where we used $x_c^{(\mathrm{h})} = c^2/\sigma_{(\mathrm{h})}^2(0)$ as in (19). As expected, the scalings of (21) are identical to those of (18). Similar entropy expressions for elliptic galaxies have been examined in [33].

Comparison between $s_c^{(\mathrm{h})}$ obtained from (21) and from (17) leads to the constraint

$$\begin{aligned} s_c^{(\mathrm{h})}\big|_{\mathrm{th}} \;&=\; s_c^{(\mathrm{h})}\big|_{\mathrm{em}} \quad \Rightarrow \\ x_{\mathrm{f}} \;&=\; 7.57 \;+\; \ln\left[\frac{(\alpha\,x_{\mathrm{f}})^{3/2}}{h^2\,\Omega_0}\,\frac{m}{\mathrm{GeV}}\right]. \end{aligned} \tag{22}$$

which does not depend on the halo variables $x_c^{(\mathrm{h})}$, $\rho_c^{(\mathrm{h})}$, hence it can be interpreted as the constraint on $s_{\mathrm{f}} = 5/2 + x_{\mathrm{f}}$ that follows from the condition $s_c^{(\mathrm{h})}\big|_{\mathrm{th}} = s_c^{(\mathrm{h})}\big|_{\mathrm{em}}$. Since we can use (9) and (10) to eliminate $h^2\,\Omega_0$, the constraint (22) becomes a relation involving only x_{f}, m, a, b, α. This constraint is independent of (8), which is another (independent) expression for $s_{\mathrm{f}} = 5/2 + x_{\mathrm{f}}$, but an expression that follows *only* from the neutralino annihilation processes. Therefore, the comparison between $s_c^{(\mathrm{h})}\big|_{\mathrm{th}}$ and $s_c^{(\mathrm{h})}\big|_{\mathrm{em}}$, leading to a comparison of two independent expressions for s_{f}, is not trivial but leads to an "entropy consistency" criterion that can be tested on suitable desired values of m, a, b, α. This implies that a given dark matter particle candidate, characterized by m and by specific annihilation channels given by x_{f} through (8), will pass or fail to pass this consistency test independently of the details one assumes regarding the present day dark halo structure. This is so, whether we conduct the consistency test by comparing (8) and (22) or (17) and (21). However, the actual values of $s_c^{(\mathrm{h})}$ for a given halo structure, whether obtained from (21) or from (17), do depend on the precise values of $\rho_c^{(\mathrm{h})}$ and $x_c^{(\mathrm{h})}$. Since the matching of either (8) and (22) or (17) and (21) shows a weak logarithmic dependence on m, the fulfillment of the "entropy consistency" criterion identifies a specific mass range for each dark matter particle. This allows us to discriminate, in favor or against, suggested dark matter particle candidates and/or annihilation channels by verifying if the standard abundance criterion (9) is simultaneously satisfied for this range of masses.

5 Testing the Entropy Consistent Criterion

Since we can write (22) as:

$$\ln(h^2 \Omega_0) = 7.57 - x_{\mathrm{f}} + \ln\left[(\alpha x_{\mathrm{f}})^{3/2} m\right] . \tag{23}$$

this constraint becomes a new estimate of the cosmological parameters $h^2\Omega_0$, given as in terms of a structural parameter of galactic dark matter halos, α, the mass of the neutralino, m, and the temperature of the neutralino gas at freeze out, x_{f}. This last quantity depends explicitly not only on m, but also on its interaction cross section, and hence on the details of its phenomenological physics *viz* (8).

At this point we consider values for the constants a and b that define the interaction cross section of the neutralino, and use (23) to plot Ω_0 as a function of m in GeV's. Using $h = 0.65$ and given the uncertainty range of α, we will obtain not a curve, but a region in the $\Omega_0 - m$ plane. Considering first condition (12), corresponding to Higgsino–like neutralinos, leads to the shaded region in Fig. 1a. On this figure we have also plotted the relation which the abundance criterion (9) yields on this same plane. Firstly, we notice that the mass range that results from our entropy criterion intersects the one resulting from the abundance criterion. However, it is evident that within the observationally determined range of Ω_0 (the horizontal dashed lines 0.2-0.4), there is no intersection between the shaded region and the abundance criterion curve. This implies that both criteria are mutually inconsistent, thus the possibility that Higgsino-like neutralinos make up both the cosmological dark matter and galactic dark matter appears unlikely.

Repeating the same procedure for mainly B–ino neutralinos, (11) yields Fig. 1b. In this case, we can see that the abundance criterion curve falls well within the shaded region defined by the entropy criterion. Although we can not improve on the mass estimate provided by the abundance criterion alone, the consistency of both criteria reveals the B-ino neutralino as a viable option for both the cosmological and the galactic dark matter.

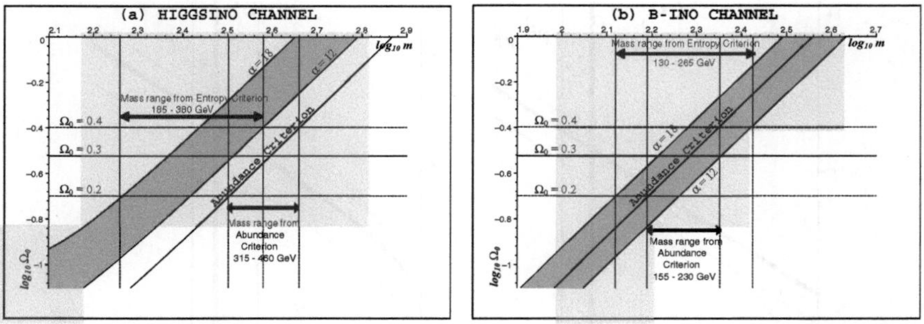

Fig. 1. Figures (a) and (b) respectively correspond to the Higgsino and B-ino channels. The shaded regions display Ω_0 vs m from our entropy criterion (23) with the solid curve giving Ω_0 from the cosmological abundance criterion (9), in all cases for $h = 0.65$. The horizontal dashed lines give current estimates of $\Omega_0 = 0.3 \pm 0.1$. It is evident that only the B-ino channels allow for a simultaneous fitting of both the abundance and the entropy criteria.

It is also interesting to evaluate (21) and (17) for the two cases of neutralino channels: the B-ino and Higgsino, but now considering numerical estimates for $x^{(\mathrm{h})}$ and $\rho^{(\mathrm{h})}$ that correspond to central regions of actual halo structures. Considering terminal velocties in rotation curves we have $v_{\mathrm{term}}^2 \simeq 2\sigma_{(\mathrm{h})}^2(0)$, so that $x_c^{(\mathrm{h})} \simeq 2(c/v_{\mathrm{term}})^2$, while recent data from LSB galaxies and clusters [31, 32, 34, 19, 35] suggest the range of values $0.01\,\mathrm{M}_\odot/\mathrm{pc}^3 < \rho_c^{(\mathrm{h})} < 1\,\mathrm{M}_\odot/\mathrm{pc}^3$. Hence, we will use in the comparison of (17) and (21) the following numerical values: $\rho_c^{(\mathrm{h})} = 0.01\,\mathrm{M}_\odot/\mathrm{pc}^3 = 0.416\,\mathrm{GeV/cm}^3$ and $x_c^{(\mathrm{h})} = 2 \times 10^6$, typical values for a large elliptical or spiral galaxy with $v_{\mathrm{term}} \simeq 300\,\mathrm{km/sec}$ [34, 19, 35].

Figure 2a displays $s_c^{(\mathrm{h})}|_{\mathrm{th}}$ and $s_c^{(\mathrm{h})}|_{\mathrm{em}}$ as functions of $\log_{10} m$, for the halo structure described above, for the case of a neutralino that is mostly Higgsino. The shaded region marks $s_c^{(\mathrm{h})}|_{\mathrm{em}}$ given by (21) for the range of values of α, while the vertical lines correspond to the range of masses selected by the abundance criterion (9) for $\Omega_0 = 0.2, 0.3, 0.4$. The solid curves are $s_c^{(\mathrm{h})}|_{\mathrm{th}}$ given by (17) for the same values of Ω_0, intersecting the shaded region associated with (21) at some range of masses. However, the ranges of coincidence of a fixed (17) curve with the shaded region (21) occurs at masses which correspond to values of Ω_0 that are different from those used in (17), that is, the vertical lines and solid curves with same Ω_0 intersect out of the shaded region. Hence, this annihilation channel does not seem to be favored.

Figure 2b depicts the same variables as Fig. 2a, for the same halo structure, but for the case of a neutralino that is mostly B-ino. In this case, the joint application of the abundance and entropy criteria yield a consistent mass range of $150\,\mathrm{GeV} < m_{\tilde{\chi}_0^1} < 250\,\mathrm{GeV}$), which allows us to favor this annihilation channel as a plausible dark matter candidate, with m lying in

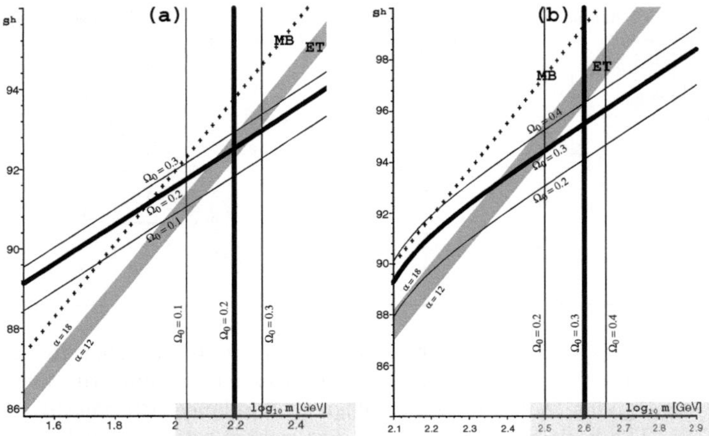

Fig. 2. Figures (a) and (b) respectively correspond to the Higgsino and B-ino channels. The figures display $s_c^{(h)}|_{em}$ from (20)–(21) (gray strip), $s_c^{(h)}|_{th}$ from (17) for $h = 0.65$ and the uncertainty strip $\Omega_0 = 0.3 \pm 0.1$ (thick curves) and $s_c^{(h)}|_{MB}$ from (18) (crosses), all of them as functions of $\log_{10} m$. The vertical strip marks the range of values of m that follow from (9)–(10) for the same values of Ω_0 and h. It is evident that only the B-ino channels allow for a simultaneous fitting of both the abundance and the entropy criteria.

the narrow ranges given by this figure for any chosen value of Ω_0. As noted above, the results of Figs. 1a and 1b are totally insensitive to the values of halo variables, $x_c^{(h)}$ and $\rho_c^{(h)}$, used in evaluating (21) and (17). Different values of these variables (say, for a different halo structure) would only result in a relabeling of the values of $s_c^{(h)}$ along the vertical axis of the figures.

6 Conclusions

We have presented a robust consistency criterion that can be verified for any annihilation channel of a given dark matter candidate proposed as the constituent particle of the present galactic dark matter halos. Since we require that the empirical estimate $s_c^{(h)}|_{em}$ of present dark matter haloes must match the theoretical value $s_c^{(h)}|_{th}$, derived from the microcanonical definition and from freeze out conditions for the candidate particle, the criterion is of a very general applicability, as it is largely insensitive to the details of the structure formation scenario assumed. Further, the details of the present day halo structure enter only through an integral feature of the dark halos, the central escape velocity, thus our results are also insensitive to the fine details concerning the central density and the various models describing the structure of dark matter halos. A crucial feature of this criterion is its direct dependence on the physical details (*i.e.* annihilation channels and mass) of any particle candidate.

Recent theoretical work by E. A. Baltz *et al.* [10] confirmed that neutralino annihilation in the galactic halo can produce enough positrons to make up for the excess of cosmic ray positrons experimentally detected by the HEAT collaboration [11, 12]. Baltz *et al.* concluded that for a boost factor $B_s \sim 30$ the neutralinos must be primarily B-inos with mass around 160 GeV. For a boost factor $30 < B_s < 100$, the gaugino–dominated SUSY models complying with all constraints yield neutralino masses in the range of $150\,\mathrm{GeV} < m_{\tilde{\chi}^1_0} < 400\,\mathrm{GeV}$. On the other hand, Higgsino dominated neutralinos are possible but only for $B_s \sim 1000$ with masses larger than 2 TeV. The results that we have presented in this letter are in agreement with the predictions that follow from [10], as we obtain roughly the same mass range for the B-ino dominated case (see Fig. 1b) and the Higgsino channel is shown to be less favored in the mass range lower than TeV's.

We have examined the specific case of the lightest neutralino for the mostly B-ino and mostly Higgsino channels. The joint application of the "entropy consistency" and the usual abundance criteria clearly shows that the B-ino channel is favored over the Higgsino. This result can be helpful in enhancing the study of the parameter space of annihilation channels of LSP's in MSSM models, as the latter only use equations (8) and (9)–(10) in order to find out which parameters yield relic gas abundances that are compatible with observational constraints [1, 2, 3, 4, 5, 6]. However, equations (8) and (9)–(10) by themselves are insufficient to discriminate between annihilation channels. A more efficient study of the parameter space of MSSM can be achieved by the joint usage of the two criteria, for example, by considering more general cross section terms (see for example [2]) than the simplified approximated forms (11) and (12). This work is currently in progress.

References

1. John Ellis: Summary of DARK 2002. In: *4th International Heidelberg Conference on Dark Matter in Astro and Particle Physics*, Cape Town, South Africa, 4-9 Feb. 2002. [astro-ph/0204059].
2. G. Jungman, M. Kamionkowski and K. Griest: Phys. Rep. **267**, 195 (1996).
3. S. Khalil, C. Muñoz and E. Torrente-Lujan: New Jour. Phys. **4**, 27 (2002) [hep-ph/0202139].
4. Leszek Roszkowski, In: *37th Rencontres de Moriond on Electroweak Interactions and Unified Theories*, Les Arcs, France, 9-16 Mar 2002, [hep-ph/0206178].
5. N. Fornengo, In: *5th International UCLA Symposium on Sources and Detection of Dark Matter and Dark Energy in the Universe (DM 2002)*, Marina del Rey, California, 20-22 Feb 2002, [hep-ph/0206092].
6. J.Ellis et al.: Nuc. Phys. B, **238**, 453 (1984).
7. E.W. Kolb and M.S. Turner: *The Early Universe*, (Addison–Wesley Publishing Co., 1990).

8. T. Padmanabhan: *Structure formation in the universe*, (Cambridge University Press, 1993).
9. J.A. Peacock: *Cosmological Physics*, (Cambridge University Press, 1999).
10. E. A. Baltz, J. Edsjö, K. Freese and P. Gondolo: Phys. Rev. D, **65**, 063511 (2002).
11. HEAT Collaboration, S.W. Barwick et al.: Astrophys. J. Lett., **482**, L191 (1997).
12. HEAT–pbar Collaboration, S. Coutu et al, In: *Proceedings of 27th ICRC, 2001.*
13. T. Moroi, M. Yamaguchi and T. Yanagida: Phys. Lett. B**342**, 105 (1995).
14. Keith A. Olive and Mark Srednicki: Phys. Lett. B**230**, 78 (1989).
15. T. Padmanabhan: Phys. Rep. **188**, 285 (1990).
16. T. Padmanabhan: *Theoretical Astrophysics, Volume I: Astrophysical Processes*, (Cambridge University Press, 2000).
17. C.J. Hogan and J.J. Dalcanton: Phys Rev D, **62**, 063511, (2000).
18. X. Hernandez and G. Gilmore: Mon. Not. Roy. Astron. Soc., **294**, 595 (1998).
19. I.T. Iliev and P.R. Shapiro: Mon. Not. Roy. Astron. Soc., **325**, 468 (2001).
20. R.K. Pathria: *Statistical Mechanics*, (Pergamon Press, 1972).
21. J. Binney and S. Tremaine: *Galactic Dynamics*, (Princeton University Press, 1987).
22. J. Katz, G. Horowitz and A. Dekel: Astrophys. J., **223**, 299 (1978).
23. J. Katz: Mon. Not. Roy. Astron. Soc., **190**, 497 (1980).
24. M Magliocchetti, G. Pugacco and E. Vesperini: Nuovo Cimento B, **112**, 423 (1997).
25. F.A. Rasio, S.L. Shapiro and S.A. Teukolsky: Astrophys. J., **336**, L–63 (1989).
26. H. Cohn: Astrophys. J., **242**, 765 (1980).
27. J. Hjorth and J. Madsen: Mon. Not. Roy. Astron. Soc., **253**, 703 (1991).
28. O.Y. Gnedin and H.S. Zhao: Mon. Not. Roy. Astron. Soc., **333**, 299 (2002).
29. P.J. Young: Astrophys. J., **81**, 807 (1976).
30. W.J.G. de Blok and S.S. McGaugh: Mon. Not. Roy. Astron. Soc., **290**, 533 (1997).
31. C. Firmani et al.: Mon. Not. Roy. Astron. Soc., **315**, L–29 (2000).
32. C. Firmani et al: Mon. Not. Roy. Astron. Soc., **321**, 713 (2001).
33. G.B. Lima Neto, D. Gerbal and I. Márquez: [astro-ph/9905048].
34. P.R. Shapiro and I.T. Iliev: Astrophys. J., **542**, L–1 (2000).
35. I.T. Iliev and P.R. Shapiro: Astrophys. J., **546**, L–5 (2001).

Part III

Braneworlds, Loop Quantum Cosmology

Brane World Cosmology

Kei-ichi Maeda

Dept. of Physics, Waseda University, Shinjuku, Tokyo 169-8555, Japan.
maeda@waseda.jp

Abstract. We first give an overview of several brane models, and show how to deal with a higher-dimensional brane world scenario. We then discuss one formalism in detail, which is the four-dimensional (4D) effective approach. It is applied to the case where gravity is confined in the brane. We then present the effective equations to describe the 4D gravity of a brane world, assuming the Z_2 symmetry.

Applying this formalism to the Randall-Sundrum II model, we find two additional terms in the Einstein equations: One is the quadratic term of the energy-momentum tensor of matter and the other is the 5-dimensional (5D) Weyl curvature term. Although the 4D system of effective equations is not closed and further information from the bulk is required, those terms may provide us a window for the search of extra dimensions. We discuss some effects induced by those terms in cosmology.

Using this formalism, we also analyze other brane models: a model with a bulk dilaton field motivated by the Hořava-Witten model, and one with a bulk Yang-Mills field. The induced gravity brane model proposed by Dvali, Gabadadze and Porrati and its extension are also discussed. In the latter case, we find that the effective cosmological constant on a brane can be extremely reduced in contrast to that of the Randall-Sundrum model even if a bulk cosmological constant and a brane tension are not fine-tuned.

1 Introduction

Gravity is a very interesting interaction. When we discuss great achievements in physics, gravity is often deeply related. Newton unified the gravity on Earth with the force acting between the planets and the Sun as a universal attractive force, which was the dawn of modern science. This universal interaction was extended by Einstein's general relativity. General relativity not only gives a precise description of the gravitational force but also has completely changed the idea of space and time. Gravity is described as a geometrical object, and energy and momentum densities of matter fields deform a spacetime. As a result, a spacetime becomes a dynamic object, which allows us to think about the universe as a whole.

Another great achievement in physics in 20th century is quantum mechanics. This has completely changed our world view at microscopic scales. To understand this microscopic world, particle physics has played a very important role. Among the achievements in particle physics, the unification of

K. Maeda, Brane World Cosmology, Lect. Notes Phys. **646**, 323–356 (2004)
http://www.springerlink.com/

fundamental interactions is one of the most important subjects. All interactions except gravity could be unified by grand unified theories. The matter (fermions) and the forces (bosons) may also be unified by supersymmetry. When we discuss quantization of fields, gravity is again exceptional because it is not renormalizable. We have not so far been able to quantize the gravitational interaction. Therefore, if we could solve those fundamental problems of gravity, we may reach the next revolution in physics. One of the most promising approaches is a superstring theory, or M-theory [1]. Such unified theories are usually formulated in higher dimensions than four. However, the dimension of our spacetime is certainly four. How can one obtain a realistic world from such a higher-dimensional theory? Kaluza and Klein first discussed a 5D spacetime to unify gravity and electromagnetism [2]. In their model, the fifth direction is curled up to very small scales such that our effective 4D spacetime is recovered. From a view point of gravity, a supergravity theory, which may unify all interactions and particles, is just an extension of the Kaluza-Klein theory. The extra dimensions should also be compactified into below 10^{-17} cm because of the constraints placed by high-energy experiments. The size of extra dimensions is usually assumed to be the Planck length 10^{-33} cm.

Then the idea of "superstring" appeared in physics. A string theory has predicted a new object, a *brane*, on which edges of open strings stand [3]. The existence of such natural boundaries changes our world view completely. Particles in the standard model are expected to be confined to a three-dimensional (3D) brane, whereas the gravitons propagate in a higher-dimensional bulk spacetime. This suggests a new perspective in cosmology: a brane world scenario, that is, we live in a brane world, which is a 3D hypersurface in a higher-dimensional spacetime [4]. Such a world view is very interesting because TeV gravity might be realistic and a quantum gravity effect could be observed by a next-generation particle collider [5].

New approaches for extra dimensions proposed by Randall and Sundrum are also very important [6, 7]. In their first paper [6], they proposed a mechanism to solve the hierarchy problem by a small extra dimension, while in their second paper [7], they proposed a single brane model with a positive tension, where 4D Newtonian gravity is recovered at low energies even if the extra dimension is not compact. This mechanism provides us an alternative compactification of extra dimensions.

If the brane world is realistic, we may find some evidence of higher-dimensions in strong gravity phenomena. Here we first present an overview of several brane models in Sect. 2, and summarize the approaches which allow us to deal with a higher-dimensional brane world in Sect. 3. We then discuss a class of the brane models, in which gravity is confined on the brane just as in the Randall-Sundrum second model. In Sect. 4, we derive the effective gravitational equations for the induced 4D metric obtained by projecting the 5D metric onto the brane world-volume [8, 9, 10]. The gravitational action on the brane, which may be induced via quantum effects of matter fields, could be included in the present approach.

We apply our formalism to the Randall-Sundrum second model (RSII) in Sect. 5 [8]. We find two new terms in the "Einstein equations" [11]: One is the quadratic term of the energy-momentum tensor of matter and the other is the 5D Weyl curvature term. Although the 4D system of effective equations are not closed and further information from the bulk is required, here we explore a window to look for extra dimensions in the context of cosmology. The quadratic term of the energy-momentum tensor will change the dynamics of the early universe, while the Weyl curvature term will play a new role in strong gravitational phenomena.

Using this formalism, we also analyze other brane models in Sect. 6 and Sect. 7; a model with a bulk dilaton field [9] motivated by the Hořava-Witten model [12, 13], and one with a bulk Yang-Mills field [14]. The induced gravity brane model proposed by Dvali, Gabadadze and Porrati [15] and the curvature squared inflationary model first proposed by Starobinsky [16, 17, 18] are also discussed [10]. Concluding Remarks follow in Sect. 8.

2 Several Models for a Brane World

String theory predicts a new type of nonlinear structure, which is called a *brane*, a nomenclature created artificially from "membrane." It is a boundary layer on which edges of open strings stand. The idea of a brane with duality plays an important role in a statistical derivation of a black hole entropy. This also suggests a new perspective in cosmology: we are living in a brane world, which is a 3D hypersurface in a higher-dimensional spacetime. In contrast to the familiar Kaluza-Klein theory, our world view appears to be changed completely. In the following, we present a brief overview of several brane world models.

2.1 Brane Models with Large Extra Dimensions

Based on a brane world picture, a new type of Kaluza-Klein cosmology was proposed by Arkani-Hamed, Dimopoulos and Dvali [5]. Ordinary matter fields are confined on the brane which is infinitesimally "thin" mathematically, though it may be thick physically, probably of the order of the Planck length. Compared with this thickness, on the other hand, the extra dimensions where only gravitons propagate could be larger. How large can they be is a question that can be answered only by gravitational experiments. Since the experiments probe the Newtonian gravity to scales above 1 mm, the laws of gravity might be different below this scale.

Suppose that the fundamental theory of gravity is given by the Einstein-Hilbert action in $D(=4+n)$ dimensions and the spacetime is compactified into (4D spacetime)$\times$(n extra dimensions). The action is reduced as:

$$S=\frac{1}{2\pi\kappa_D^2}\int d^Dx\,\mathcal{R}\approx\frac{1}{2\pi\kappa_D^2}V_n\int d^4x\,R=\frac{1}{2\pi\kappa^2}\int d^4x\,R, \tag{1}$$

where V_n is the volume of extra dimensions, while κ_D^2 is a gravitational constant in D-dimensional spacetime. Defining the D-dimensional Planck mass ${}^{(D)}m_{PL}$ by $\kappa_D^2 = \left({}^{(D)}m_{PL}\right)^{-(n+2)}$, we obtain from (1)

$$m_{PL}^2 = \left({}^{(D)}m_{PL}\right)^{(n+2)} \times V_n. \tag{2}$$

In the old Kaluza-Klein idea, the size of extra dimensions should be much smaller than 10^{-17} cm ($\sim$TeV^{-1}) in order to keep them unobservable by high-energy experiments. We have even assumed that the size is nearly as small as the Planck length, the inverse of m_{PL}. If we believe, however, that ordinary matter fields are confined on a brane world, the extra dimensions are not necessarily required to be so small. This might also be connected with a conjecture that the mass scale ${}^{(D)}m_{PL}$ in D-dimensional spacetime at the more fundamental level is as low as $\sim$ TeV, nearly the same as the electroweak mass scale. This removes what is called a "hierarchy problem". Once we accept this idea, we can derive a typical size r_n of the extra dimensions from (2):

$$r_n \sim (V_n)^{1/n} \sim 10^{(30/n)-17}\text{cm}. \tag{3}$$

If $n = 1$, we expect $r_1 \sim 10^{13}$ cm ($\sim$ 1 astronomical unit), which is excluded. If $n = 2$, however, we find $r_2 \sim 1$ mm, precisely the shortest distance only above which the Newtonian law of the inverse-square law has been tested. This is very interesting model because it shows a possibility of the unification scale within reach of our near-future experiments.

2.2 Randall-Sundrum Model

Randall and Sundrum [6, 7] proposed a very simple brane model, in which our brane is identical to a domain wall in 5D anti-de Sitter (AdS) spacetime with a negative cosmological constant ${}^{(5)}\Lambda(<0)$. The 5D spacetime is described by the metric, which is not factorizable as

$$ds^2 = e^{-2|y|/\ell} g_{\mu\nu}(x) dx^\mu dx^\nu + dy^2, \tag{4}$$

where $\ell = \sqrt{-6/{}^{(5)}\Lambda}$ is a typical curvature scale of the AdS spacetime. To find the Minkowski space on the brane ($g_{\mu\nu} = \eta_{\mu\nu}$), the tension λ of the brane must satisfy $|\lambda| = 6/\ell\kappa_5^2$. The "warp" factor $e^{-2|y|/\ell}$, which is rapidly changing in the extra dimension, plays a very important role in contrast to the usual Kaluza-Klein compactification. They discussed two models.

(1) two-brane model (RSI)
In their first model [6], they proposed a mechanism to solve the hierarchy problem by a small extra dimension bounded by two boundary branes, with positive and negative tension, located at $y = 0$ and $y = s$, respectively, as illustrated in Fig. 1.

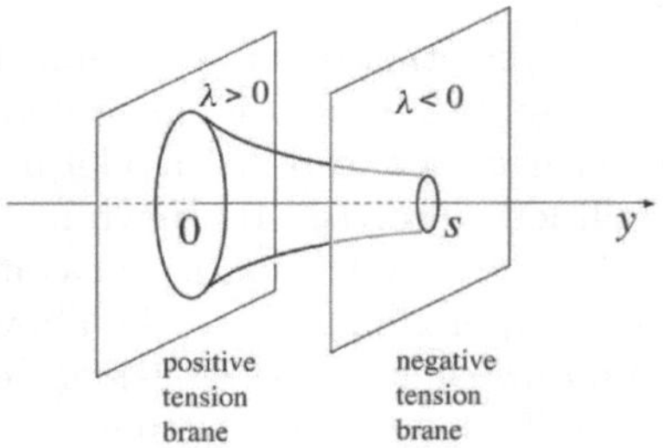

Fig. 1. Randall-Sundrum type I model. We are living in a negative tension brane ($y = s$). The circles describe the warp factor $e^{-2|y|/\ell}$

If we are living in a negative tension brane, we find a solution to the hierarchy problem. We first estimate the 4D Planck mass scale by integrating the 5D action in the 5-th direction as

$$S_g \sim \frac{1}{2\kappa_5^2}\int_0^s dy e^{-2|y|/\ell} \cdot \int d^4x\sqrt{-g}R, \tag{5}$$

where we made an approximate estimate ${}^{(5)}R \sim R$, without solving Einstein's equation in 5 dimensions rigorously. We then find

$$m_{PL}^2 = \frac{1}{\kappa^2} \sim \frac{1}{2\kappa_5^2}\int_0^s dy e^{-2|y|/\ell} = \frac{\ell}{2\kappa_5^2}\left(1 - e^{-2s/\ell}\right). \tag{6}$$

This means that m_{PL} depends weakly on the distance s, as long as $e^{-s/\ell} \ll 1$. To show how the hierarchy problem is resolved, we consider a fundamental Higgs field confined in the visible brane with negative tension. The action is given by

$$S_{\rm vis} = \int d^4x\sqrt{-g_{\rm vis}}\left[g_{\rm vis}^{\mu\nu}D_\mu H^\dagger D_\nu H - \lambda\left(|H|^2 - v_0^2\right)^2\right], \tag{7}$$

where $g_{\mu\nu}^{\rm vis}$ is the 4D components of the 5D metric evaluated at $y = s$, *i.e.* $g_{\mu\nu}^{\rm vis} = e^{-2s/\ell}g_{\mu\nu}$. This, together with redefinition of the Higgs field, $H \to e^{s/\ell}H$, leads to

$$S_{\rm vis}^{\rm eff} = \int d^4x\sqrt{-g}\left[g^{\mu\nu}D_\mu H^\dagger D_\nu H - \lambda\left(|H|^2 - v_{\rm eff}^2\right)^2\right], \tag{8}$$

where $v_{\rm eff} = e^{-s/\ell}v_0$ gives the physical symmetry-breaking energy scale, which could be much smaller than the original energy scale v_0. In fact, if $s/\ell \sim 35$, this produces TeV energy scale from the 4D Planck scale m_{PL}.

(2) one-brane model (RSII)
In their second model [7], on the other hand, we assume to live in the positive-tension brane surrounded by AdS. There is no second brane with negative tension, which is obtained from a two-brane model in the limit of $s \to \infty$. Although hierarchy is still left unsolved, 4D Newtonian gravity is recovered at low energies even if the extra dimension is not compact. This is proved by applying a perturbation approximation to the above solution (4) with a positive-tension Minkowski brane at $y = 0$. Consider perturbation to the 4D components, ${}^{(5)}g_{\mu\nu} = e^{-2|y|/\ell}\eta_{\mu\nu} + h_{\mu\nu}$. By setting $h(x, y) = \hat{\psi}(z)e^{-|y|/2\ell}e^{ipx}$ with $z = \ell(e^{|y|/\ell} - 1)$ and $p^2 = -m^2$, we find the perturbation equations for the graviton as

$$\left[-\frac{1}{2}\partial_z^2 + V(z)\right]\hat{\psi} = m^2\hat{\psi}, \tag{9}$$

where

$$V(z) = \frac{15}{8\ell^2(|z|/\ell + 1)} - \frac{3}{2\ell}\delta(z), \tag{10}$$

is a volcano-shaped potential (see Fig. 2).

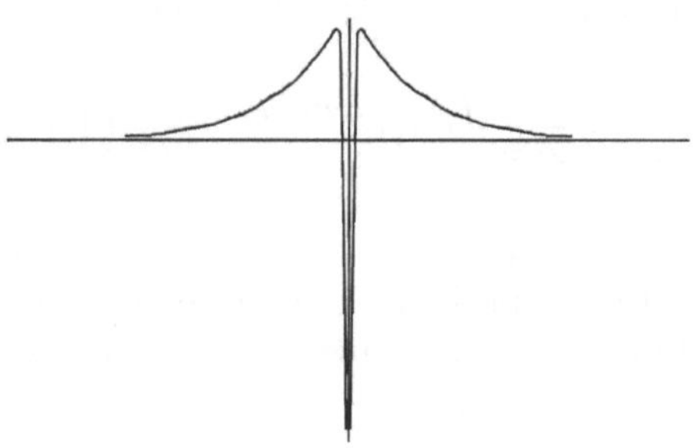

Fig. 2. Volcano-shape potential in the Randall-Sundrum type II model. Since gravity is confined on the brane $\mathcal{B}$, gravity on the brane can be described by the induced metric of $\mathcal{B}$.

This potential confines a massless graviton mode on the brane. As a result, even if the 5-th dimension is not compact, Newtonian gravitational potential is recovered in low energy limit as

$$V(r) \sim G\frac{m_1 m_2}{r}\left(1 + \frac{\ell^2}{r^2}\right), \tag{11}$$

where m_1 and m_2 are masses of two particles on the brane. This "compactification" is completely different from the Kaluza-Klein type compactification.

2.3 Brane Models with Bulk Fields

According to the recent progress in superstring theory, different string theories are connected with each other via dualities, which unify them to the *M theory* in 11 dimensions. Among string theories, the 10-dimensional (10D) $E_8 \times E_8$ heterotic string theory is a strong candidate for our real world because the theory may contain the standard model. Hořava and Witten showed that this heterotic string model is equivalent to an 11-dimensional (11D) realization of M theory compactified on the orbifold $R^{10} \times S^1/Z_2$ [12]. Each gauge field on E_8 is confined to the 10D boundary brane of S^1/Z_2. The 10D spacetime is compactified to $M^4 \times CY^6$, where M^4 and CY^6 are 4D Minkowski spacetime (our Universe) and Calabi-Yau space, respectively. Lukas, Ovrut and Waldram then derived an effective 5D action by dimensional reduction from 11D M-theory [13, 19, 20]. In their reduction, we find a dilaton field, which corresponds to a volume of CY^6 space, as well as a $U(1)$ "gauge" field in the 5D bulk spacetime. Hence, when we reduce our model from most fundamental higher-dimensional spacetime to an effective bulk spacetime, one would usually expect scalar fields associated with many moduli fields as well as "gauge" fields in the gravitational sector, which are allowed to propagate in the 5D "bulk" spacetime. We may find non-Abelian gauge field from some type of dimensional reduction of a unified theory. There is another interesting point in discussing non-Abelian gauge fields in a bulk spacetime. Using a brane structure, new mechanism of spontaneous symmetry breaking of gauge interactions has been proposed [21]. In this picture, the present standard model (SU(3)×SU(2)×U(1)) is obtained on the brane assuming some higher-symmetric gauge interactions such as SU(5) in the bulk.

2.4 Induced Gravity on a Brane

In a brane world, usual matter fields are confined on a brane. We may have to include quantum effects of those matter fields, because we have extra dimensions. This idea is closely related to AdS/CFT correspondence, by which we expect some equivalence between classical dynamics in AdS background and a conformal field theory in its flat Minkowski boundary [22]. We then find modification of gravitational interaction on the brane, which is the so-called *induced gravity*. In particular, when we consider conformal field theory on a brane, we will obtain a trace anomaly term in 4D brane world. Those additional gravitational action on the brane induced via quantum effects of matter fields should be included.

This type of brane model (a flat brane in a 5D Minkowski bulk space) was first discussed by Dvali, Gabadadze and Porrati [15]. This model is interesting because, phenomenologically, 4D Newtonian gravity on a 3-brane is recovered at high energy scale, whereas 5D gravity emerges at low energy scale (see, however, [23]). Applying this model to cosmology, they showed that there exists an exponential expansion of the universe without a cosmological constant. This would be the present acceleration of the universe, if we fine-tuned a 5D Planck mass ($m_5 \sim 100$ MeV).

We also generalize their model to the case with a bulk cosmological constant and a tension of the brane. Assuming the energy scale of the tension is much larger than the 5D Planck mass, we show that the effective cosmological constant on the brane is extremely suppressed in contrast to the RS model even if the cosmological constant and the tension are not fine-tuned [10]. This might explain the present acceleration of the universe. We also study the curvature-squared model motivated by the appearance of trace anomaly [24, 16, 17].

3 Approaches to a Brane World

Recently there has been tremendous interest in a brane world scenario and many works have been done in cosmology as well as in particle physics. Here we classify those into three approaches: (1) 4D effective approach; (2) holographic approach; (3) higher-dimensional approach, which we shall explain those three in order.

3.1 Four-Dimensional Approach

Since a brane world is localized on a hypersurface in a bulk spacetime, this entire spacetime is quite inhomogeneous. Hence, even if we are interested in dynamics of our isotropic and homogeneous 3D universe, this set up is very complicated. Since the system is described by partial differential equations, it may be difficult to solve the basic equations without any additional ansatz [11, 13]. However, because we are living in a brane world, which is simply isotropic and homogeneous, we may not need to know what happens in a complicated bulk spacetime. We are interested only in a brane world. Therefore, if one can construct an effective theory which is applied to our brane world, the analysis of our universe will be very simplified. If gravity is confined just as the RS II model, such construction would be possible because an induced metric describes the brane gravity. Projecting the basic equations in a bulk spacetime onto a brane, we can construct an effective gravitational theory. We will show how to obtain such an effective theory in the following sections.

3.2 Holographic Approach

The holographic principle has been proposed based on the equivalence between classical dynamics of superstring (supergravity) in AdS background spacetime and a conformal field theory on its Minkowski boundary. If this AdS/CFT correspondence is true, when we discuss the dynamics of a brane world, we have to include its quantum effects, because of the existence of extra-dimensions. There are two ways: one where we analyze a high-dimensional classical spacetime, and the other one where we solve 4D effective models including quantum effects. If we adopt the latter approach, including quantum corrections on a brane gravity, we can apply the above 4D approach. Nojiri, Odintsov, and Zerbini [16], and Hawking, Hertog, and Reall [17] studied such models. We will discuss those as an induced gravity brane model in Sect. 7.

3.3 Higher-Dimensional Approach

Although the 4D approach is very effective for some specified situations, most interesting aspects in a brane world may be found in analyzing highly inhomogeneous bulk spacetime. Here we just list up some subjects which should be discussed in higher dimensions.

(1) Stability and perturbations

Since a bulk spacetime with branes is inhomogeneous, it is not trivial whether such a spacetime is stable or not. In fact, two branes in the RS I model are marginally stable because spacetimes with two branes at any distance are allowed. In order to explain the hierarchy problem, we have to fix the distance around $s \sim 35\ell$. Such a spacetime should be realized as a stable spacetime [25]. It is also very important to study perturbation equations of a brane world in the bulk spacetime, when we analyze density fluctuations in brane cosmology [26].

(2) Higher-dimensional black hole and brane black hole

The black hole solution is important from two points of view: One is in the context of a brane cosmology. Our universe may be regarded just as a domain wall in a bulk black hole spacetime [27, 14]. The other view point is related to the TeV gravity. If a fundamental unification scale is about TeV energy scale, a black hole could be formed in a particle-collider. Then we may have a chance to find quantum gravity effects in current high-energy experiments [28]. There is another interesting topic, which is very fundamental [29]. We know that various types of black hole spacetimes become possible in higher dimensions. In particular, a topology of a event horizon is not always S^n, but it could be a torus or other type of geometry. Many questions on higher-dimensional black holes are still open.

(3) Brane collision and new scenario of the early universe

One of typical phenomena in brane dynamics is a collision of branes. With this idea, one interesting scenario for the early universe model has been proposed, the so called *ekpyrotic universe* [30, 31]. Starting with two parallel branes, we find a brane collision, which initiates the *Big Bang*. It could provide an alternative to an inflationary scenario. Although there may be so far some serious troubles in density perturbations [32], we should pursue such possibilities furthermore, including the detail analysis of collision mechanism and its dissipation process [33].

(4) Creation of a brane world

We also have to discuss how the universe appears in a brane world scenario. The conventional quantum cosmology may be applied to a spacetime with a brane. Some instanton solution might describe what kind of a brane world is created quantum mechanically [34]. We may have another scenario for creation of our universe, which is based on an instability of a brane system. The system with D-brane and anti D-brane will decay by collision and then more stable lower-dimensional brane world (which could be our universe) is formed [35]. This may provide an alternative for creation of our brane world, although we still have to discuss about how such a system with two parallel branes appears in the beginning.

4 The Effective Gravitational Equations on a Brane World

Now we discuss how to describe our brane world by an effective 4D theory. We consider a 5D bulk spacetime with one 3-brane, on which we assume that gravity is confined. We derive the effective 4D gravitational equations, which describe our 4D brane world.

Suppose that the 4D brane world $(M, g_{\mu\nu})$ is located at a hypersurface $(\mathcal{B}(X^A) = 0)$ in the 5D bulk spacetime $(\mathcal{M}, {}^{(5)}g_{AB})$, which coordinates are described by X^A $(A = 0, 1, 2, 3, 5)$. We assume the Einstein-Hilbert action in the 5D bulk spacetime. The action discussed here is

$$S = S_{\text{bulk}} + S_{\text{brane}} \tag{12}$$

where

$$S_{\text{bulk}} = \int d^5X \sqrt{-{}^{(5)}g} \left[\frac{1}{2\kappa_5^2} {}^{(5)}R + {}^{(5)}L_{\text{m}} \right] \tag{13}$$

and

$$S_{\text{brane}} = \int_{\mathcal{B}=0} d^4x \sqrt{-g} \left[\frac{1}{\kappa_5^2} K^{\pm} + L_{\text{brane}}(g_{\alpha\beta}, \psi) \right], \tag{14}$$

κ_5^2 is the 5D gravitational constant, ${}^{(5)}R$ and ${}^{(5)}L_{\text{m}}$ are the 5D scalar curvature and the matter Lagrangian in the bulk, respectively. x^μ $(\mu = 0, 1, 2, 3)$ are the induced 4D brane world coordinates on the brane, $K^{\pm}$ is the trace of extrinsic curvature on either side of the brane [36, 37]. $L_{\text{brane}}(g_{\alpha\beta}, \psi)$ is the effective 4D Lagrangian, which is given by a generic functional of the brane metric $g_{\alpha\beta}$ as well as matter field ψ because we may include an additional gravitational interaction induced on a brane via quantum effects.

The 5D Einstein equations in the bulk are

$$ {}^{(5)}G_{AB} = \kappa_5^2 \left({}^{(5)}T_{AB} + \tau_{AB}\, \delta(\mathcal{B}) \right) , \tag{15} $$

where

$$ {}^{(5)}T_{AB} \equiv -2 \frac{\delta^{(5)} L_{\rm m}}{\delta^{(5)} g^{AB}} + {}^{(5)}g_{AB}{}^{(5)}L_{\rm m} \tag{16} $$

is the energy-momentum tensor of bulk matter fields, while $\tau_{\mu\nu}$ is the "effective" energy-momentum tensor localized on the brane which is defined by

$$ \tau_{\mu\nu} \equiv -2 \frac{\delta L_{\rm brane}}{\delta g^{\mu\nu}} + g_{\mu\nu} L_{\rm brane} \,. \tag{17} $$

The $\delta(\mathcal{B})$ denotes the localization of brane contributions. The basic equations in the brane world is obtained by projection of the variables onto the 3-brane, because we assume that gravity is confined on the brane. The induced 4D metric is $g_{AB} = {}^{(5)}g_{AB} - n_A n_B$, where n_A is the spacelike unit-vector field normal to the brane hypersurface M.

We first project the 5D Riemann tensor onto the brane spacetime as

$$ {}^{(5)}R_{MNRS}\; g^M_{\ A}\, g^N_{\ B}\, g^R_{\ C}\, g^S_{\ D} = {}^{(4)}R_{ABCD} - K_{AC}K_{BD} + K_{AD}K_{BC} \tag{18} $$

$$ {}^{(5)}R_{MNRS}\; g^M_{\ A}\, g^N_{\ B}\, g^R_{\ C}\, n^S = 2 D_{[A} K_{B]C} \tag{19} $$

$$ {}^{(5)}R_{MNRS}\; g^M_{\ A}\, g^R_{\ C}\, n^N\, n^S = -\pounds_n K_{AC} + K_{AB}\, K^B_{\ C} \,, \tag{20} $$

where the extrinsic curvature of M is denoted by $K_{MN} = g^A_{\ M} g^B_{\ N} \nabla_A n_B$, D_M is the covariant differentiation with respect to g_{MN}, and $\pounds_n$ denotes the Lie derivative in the n-direction. The first equation is called the Gauss equation. Combining (18) and (20), and contracting it and (19), we find

$$ {}^{(5)}R_{MN}\; g^M_{\ A}\; g^N_{\ B} = -\pounds_n K_{AB} - K K_{AB} + 2 K_{AC}\, K^C_{\ B} + {}^{(4)}R_{AB} \tag{21} $$

$$ {}^{(5)}R_{RS}\; n^S g^R_{\ M} = D_N K^N_{\ M} - D_M K \,. \tag{22} $$

The last equation is called the Codazzi equation.

To find the effective "Einstein equations" of the induced metric, we use (18) and (22). Contracting the Gauss equation (18), we find the effective 4D equations as

$$ G_{\mu\nu} = \frac{2\kappa_5^2}{3} \left[{}^{(5)}T_{RS}\; g^R_{\ \mu} g^S_{\ \nu} + g_{\mu\nu} \left({}^{(5)}T_{RS}\; n^R n^S - \frac{1}{4} {}^{(5)}T \right) \right] $$

$$ + K K_{\mu\nu} - K_\mu^{\ \lambda} K_{\nu\lambda} - \frac{1}{2} g_{\mu\nu} \left(K^2 - K^{\alpha\beta} K_{\alpha\beta} \right) - E_{\mu\nu}, \tag{23} $$

where

$$ E_{\mu\nu} \equiv {}^{(5)}C_{ARBS}\; n^A n^B g^R_{\ \mu}\; g^S_{\ \nu} \tag{24} $$

is some projected components of the 5D Weyl curvature. Here we have used the 5D Einstein equations (15).

In (23), we have so far three unknown variables; ${}^{(5)}T_{AB}$, $E_{\mu\nu}$, and $K_{\mu\nu}$. Although the first two variables are described by bulk information, we can determine the extrinsic curvature $K_{\mu\nu}$ from brane information as follows.

We assume that the brane is infinitely thin. The singular behavior in 5D bulk spacetime appears in the 5D Einstein equations (15) as a delta function. Inserting (15) into (21), we find that the extrinsic curvature must be discontinuous on the brane. Then, integrating (21) in the n-direction, we obtain the so-called Israel's junction condition [38],

$$[g_{\mu\nu}]_{\pm} = 0 \quad \text{and} \quad [K_{\mu\nu}]_{\pm} = -\kappa_5^2\left(\tau_{\mu\nu} - \frac{1}{3}g_{\mu\nu}\tau\right), \tag{25}$$

where we define the difference between values evaluated on the + or − side of the brane by $[X]_{\pm} \equiv X^+ - X^-$. Because of the Z_2-symmetry, the extrinsic curvature of the brane is uniquely determined in terms of $\tau_{\mu\nu}$ as

$$K^+_{\mu\nu} = -K^-_{\mu\nu} = -\frac{1}{2}\kappa_5^2\left(\tau_{\mu\nu} - \frac{1}{3}g_{\mu\nu}\tau\right). \tag{26}$$

In what follows, we omit the indices ± below for brevity.

Substituting (26) into (23), we obtain the gravitational equations on the 3-brane in the form,

$$G_{\mu\nu} = \frac{2\kappa_5^2}{3}\left[{}^{(5)}T_{RS}\; g^R_{\ \mu} g^S_{\ \nu} + g_{\mu\nu}\left({}^{(5)}T_{RS}\; n^R n^S - \frac{1}{4}{}^{(5)}T\right)\right] + \kappa_5^4 \pi_{\mu\nu} - E_{\mu\nu}, \tag{27}$$

where

$$\pi_{\mu\nu} = -\frac{1}{4}\tau_{\mu\alpha}\tau_\nu^{\ \alpha} + \frac{1}{12}\tau\tau_{\mu\nu} + \frac{1}{8}g_{\mu\nu}\tau_{\alpha\beta}\tau^{\alpha\beta} - \frac{1}{24}g_{\mu\nu}\tau^2. \tag{28}$$

Together with (15) and (26), the Codazzi equation (22) gives

$$D_\nu \tau_\mu^{\ \nu} = -2\,{}^{(5)}T_{RS}\, n^R g^S_{\ \mu}. \tag{29}$$

Equations (27), (28) and (29) give the effective gravity theory on the brane. If the brane Lagrangian $L_{\rm brane}$ contains only a matter field ψ, $\tau_{\mu\nu}$ is just the energy-momentum tensor of matter field, and then the gravity is described by the 4D Einstein tensor in (27). Then we recover the Einstein gravitational theory in the 4D brane world. We will discuss this case in detail in the next section. However, if $L_{\rm brane}$ includes some additional contribution to gravity such as an induced gravity on the brane, the effective energy-momentum tensor $\tau_{\mu\alpha}$ gives modification of gravitational interaction in the effective theory. We will analyze such cases in Sect. 6 and Sect. 7 .

5 Randall-Sundrum Type II Brane World Model

5.1 The "Einstein Equations" in the RS II Model

First we study the RS II brane world model in our approach [8]. Since this compactification is completely different from conventional Kaluza-Klein picture, we may find much difference from the Einstein gravity. Following Randall and Sundrum [7], we assume a negative cosmological constant ${}^{(5)}\Lambda(<0)$ in a bulk spacetime and a positive tension $\lambda(>0)$ of the brane. We also assume matter field ($T_{\mu\nu}$) on the brane to discuss cosmology. From (27) with

$$\tau_{\mu\nu} = -\lambda g_{\mu\nu} + T_{\mu\nu}\,. \tag{30}$$

we find the effective 4D gravitational equations as

$$G_{\mu\nu} = -\Lambda g_{\mu\nu} + 8\pi G_N T_{\mu\nu} + \kappa_5^4 \Pi_{\mu\nu} - E_{\mu\nu} \quad , \tag{31}$$

where

$$\Lambda = \frac{1}{2}\left({}^{(5)}\Lambda + \frac{1}{6}\kappa_5^4\lambda^2\right) , \tag{32}$$

$$G_N = \frac{\kappa_5^4\lambda}{48\pi} , \tag{33}$$

$$\Pi_{\mu\nu} = -\frac{1}{4}T_{\mu\alpha}T_{\nu}{}^{\alpha} + \frac{1}{12}TT_{\mu\nu} + \frac{1}{8}g_{\mu\nu}T_{\alpha\beta}T^{\alpha\beta} - \frac{1}{24}g_{\mu\nu}T^2\,. \tag{34}$$

It resembles the conventional Einstein equations. The Codazzi equation (29) now implies the conservation law for the matter,

$$D_\nu T_\mu{}^\nu = 0\,. \tag{35}$$

The existence of non-zero Newton's gravitational constant G_N strongly relies on the presence of the tension λ. If $\lambda < 0$, although we have the wrong sign of G_N, it does not mean anti-gravity. We have to be careful to conclude our result in this case because gravity is not confined on the brane. Gravity may not only be described by the induced metric on the brane, but a gravitational effect from the bulk, which is described by $E_{\mu\nu}$, should be included. (see the analysis for two-brane model by Garriga and Tanaka [39].)

The $\Pi_{\mu\nu}$ term, which is quadratic in $T_{\mu\nu}$, could play a very important role, especially when the energy density is very high as in the early universe [11]. In addition, (31) contains a new term, $E_{\mu\nu}$, which is a part of the 5D Weyl tensor and carries information of the gravitational field outside the brane. $E_{\mu\nu}$ satisfies the trace free condition , $E^\mu{}_\mu = 0$, which is from the property of the Weyl tensor. Since we have the contracted Bianchi identities ($D^\nu G_{\mu\nu} = 0$), with the energy-momentum conservation (35), we find one additional constraint equation on $E_{\mu\nu}$ as

$$D^\nu E_{\mu\nu} = \kappa_5^4 D^\nu \Pi_{\mu\nu}(T_{\alpha\beta})\,. \tag{36}$$

This equation, however, does not fix $E_{\mu\nu}$ completely. Hence the effective gravitational equations on the brane (31) are not closed. We have to solve the gravitational field in the bulk as well in generic cases. However, in the case of cosmology, it is not the case. We discuss it in the next subsection.

5.2 Cosmology as a Window to Extra Dimensions

Using the effective equations on the brane, we shall discuss some applications. Since the Randall-Sundrum model is completely different from conventional Kaluza-Klein theory, we expect new aspects in strong gravitational phenomena. In fact, we found two new terms, the quadratic term in the energy-momentum tensor and the 5D Weyl curvature. In this subsection, we discuss on cosmology.

Suppose we have a perfect fluid in the brane, where the energy-momentum tensor is given by $T^{\mu\nu} = \rho u^\mu u^\nu + P h^{\mu\nu}$, where $h^{\mu\nu} = g^{\mu\nu} + u^\mu u^\nu$ with u^μ being a timelike fluid four-velocity. Using the energy-momentum conservation, we find $D_\nu \Pi^{\mu\nu} = \frac{1}{6}(\rho + P) h^{\mu\nu} D_\nu \rho$. This means that $D_\nu \Pi^{\mu\nu} = 0$ if the matter density is spatially uniform, which is usually assumed in cosmology.

Assuming Friedmann-Robertson-Walker (FRW) spacetime on the brane, (36) for $E_{\mu\nu}$ turns out to be

$$\dot{E}_{00} + 4\frac{\dot{a}}{a} E_{00} = 0, \tag{37}$$

leading to $E_{00} = \mathcal{E}_0 / a^4$ where $\mathcal{E}_0$ is an integration constant. Appart from the fact that we need extra information about the bulk spacetime to determine $\mathcal{E}_0$, our system becomes closed. The Friedmann equation is now

$$H^2 + \frac{k}{a^2} = \Lambda + \frac{8\pi G_N}{3}\rho + \frac{\kappa_5^4}{36}\rho^2 - \frac{\mathcal{E}_0}{a^4}, \tag{38}$$

where $H = \dot{a}/a$ is the Hubble parameter, a is the scale factor of the universe, and k is a curvature constant taking either value of 0, or ± 1. The third and last terms on the right hand side have again appeared in the brane cosmology. The ρ^2-term comes from the quadratic term $\Pi_{\mu\nu}$, while the last one corresponds to the so-called "dark radiation" [11]. Since these two terms do not appear in the conventional big bang cosmology, such terms could be a window to the extra dimensions.

Here we list up some possible cosmological consequence due to those new terms in order.

(1) The effects of ρ^2 term

Since the energy density decreases as the universe expands, the quadratic term must be very important in the early stage of the universe. In fact it becomes dominant when

$$\rho > \rho_{\rm cr} \equiv 12\frac{m_5^6}{m_{PL}^2} = 2\lambda\,, \tag{39}$$

where $m_5 = \kappa_5^{-2/3}$.

(1-i) If matter field is described by the perfect fluid with the equation of state $P = (\gamma - 1)\rho$, the energy momentum conservation of matter fluid implies $\rho = \rho_0 a^{-3\gamma}$, resulting in the Friedmann equation written by the equation for the scalar factor a as

$$\frac{1}{2}\dot{a}^2 + U(a) = 0\,,$$

$$\text{with} \quad U(a) = -\frac{\Lambda}{2}a^2 - \frac{\rho_0}{6m_{PL}^2}a^{-3\gamma+2} - \frac{\rho_0^2}{72m_5^6}a^{-6\gamma+2} + \frac{\mathcal{E}_0}{6}a^{-2} + \frac{k}{2}. \tag{40}$$

If $\Lambda = 0$ and $\mathcal{E}_0 = 0$, the expansion law of the flat universe is given by $a \propto t^{1/3\gamma}$ in the ρ^2-term dominant stage, which eventually comes to the conventional expansion phase ($a \propto t^{2/3\gamma}$) after the linear term dominates.

(1-ii) If a scalar field ϕ is confined on the brane, its dynamics will be drastically changed. The power-law expansion of the universe is found for the potential $V(\phi) = \mu^6\phi^{-2}$. The expansion law of the universe is $a \propto t^p$ with $p = \frac{1}{6}[1 + \frac{1}{8}(\mu/m_5)^6]$. Then, if $\mu > 40^{1/6}m_5 (\approx 1.85 m_5)$, we find a power-law inflationary solution. For the potential $V(\phi) = \mu^{\alpha+4}\phi^{-\alpha}$ ($\alpha > 0$), we also find some interesting behaviors. If $\alpha < 2$, we have an inflationary solution as $a \propto \exp[H_0 t^{(2-\alpha)/2}]$, where H_0 is a constant determined by μ, m_5 and α. While, for $\alpha > 2$, the perfect fluid (if it exists) will eventually dominate for any initial conditions. The density parameter of the scalar field decreases as $\Omega_\phi \propto a^{-2(\alpha-2)/(\alpha+2)}$ in the radiation-dominated stage. This may provide us a natural initial condition for a quintessence scenario [40].

(1-iii) Inflation could be modified if the quadratic term is dominant during inflation. The quadratic term makes inflation stronger [41]. The preheating mechanism is also changed [42]. It is important to study the density perturbations, which is beyond the present approach.

(2) The effect of "dark radiation" $E_{\mu\nu}$

(2-i) In the case of $\mathcal{E}_0 < 0$, the term of $E_{\mu\nu}$ behaves just as radiation, which is constrained by a successful nucleosynthesis [11]. If inflation occurs, however, this radiation is suppressed by the exponential expansion. Then, this window will be closed unless $E_{\mu\nu}$ is produced again after inflation.

(2-ii) If $\mathcal{E}_0 > 0$, we have "negative" radiation, which changes the dynamics of the universe completely. For the universe filled by the perfect fluid, it is easy to analyze it by (40). If $0 \le \gamma < 2/3$, the universe has no initial singularity. Even if the universe is contracting initially, it will bounce and eventually expand. For $\gamma = 0$, we find the analytic solution as

$$a^2 = \frac{1}{2H_0}\left[k + \sqrt{k^2 + 2H_0\mathcal{E}_0}\cosh(2H_0 t)\right] \quad \to \quad e^{2H_0 t}, \tag{41}$$

where $H_0^2 = \rho_0/(3m_{PL}^2) + \rho_0^2/(36m_5^6)$. Inflation will occur via non-singular universe.

If $2/3 \le \gamma < 4/3$, which includes the case of dust fluid, the large value of $\mathcal{E}_0$ guarantees the existence of a bounce solution. For the case with $4/3 \le \gamma \le 2$, which includes the case of radiation fluid or stiff matter, we find a bounce solution only for the $k = -1$ universe and large $\mathcal{E}_0$. In the case of positive $\mathcal{E}_0$, although all models are not always non-singular, it seems to have a tendency of singularity avoidance.

Why we have "radiation" in the beginning of the universe in addition to the conventional radiation fluid ? In order to understand this, we re-interpret the present cosmological model as follow: Suppose we have 5D Schwarzschild-AdS spacetime

$$ds_5^2 = -\left(k + \frac{r^2}{\ell^2} - \frac{M}{r^2}\right)dt^2 + \left(k + \frac{r^2}{\ell^2} - \frac{M}{r^2}\right)^{-1} dr^2 + r^2 d\Sigma_k^2, \tag{42}$$

which is a solution of the bulk spacetime in the present model. Put a spherically symmetric singular thin shell with a radius a, and throw the outside spacetime away. Then, preparing two sets of such a spacetime, we identify each shell, which guarantees Z_2 symmetry. It turns out that the motion of the shell is exactly the same as the previous Friedmann equation (38) [27]. We then find $E_{00} = -M/a^4$. This means that $E_{\mu\nu}$ is the 5D tidal force by the "mass" M of a black hole in the bulk spacetime. Since the tidal force is proportional to a^{-4} in 5 dimensions, we realize that the radiation-like behavior is just coming from its dimensionality. If we impose the condition that a regular horizon exists for the Schwarzschild-AdS spacetime, M is positive for $k = 0$ or 1, which implies $\mathcal{E}_0 <$. However, for the open universe ($k = -1$), the condition is $M > -\ell^2/4$ and then $\mathcal{E}_0$ could be positive. This domain wall interpretation of the brane universe is extended to the model with the bulk Yang-Mills fields (see Sect. 6.2).

6 Brane Model with a Bulk Field

6.1 Brane Model with a Dilaton Field [9]

It is easy to apply our effective 4D approach to the model with a bulk dilaton field Φ, which is motivated by the Hořava-Witten model [12]. The matter Lagrangians in the bulk and on the brane are now

$${}^{(5)}L_{\rm m} = -\left[\frac{1}{2}(\nabla\Phi)^2 + V(\Phi)\right] \quad \text{and} \quad L_{\rm brane} = L_{\rm matter} - \lambda(\Phi), \tag{43}$$

where $V(\Phi)$ and $\lambda(\Phi)$ are a potential of the dilaton and a tension of the brane, respectively. Equation (27) can be rewritten using the bulk energy-momentum tensor, ${}^{(5)}T_{AB}(\Phi)$, and $\tau_{\mu\nu} = -\lambda(\Phi)g_{\mu\nu} + T_{\mu\nu}$, as

$$G_{\mu\nu} = \frac{2{\kappa_5}^2}{3}\left[T^{(\phi)}_{\mu\nu} + \Delta T^{(\phi)}_{\mu\nu}\right] + 8\pi G_N(\phi)T_{\mu\nu} + \kappa_5^4 \Pi_{\mu\nu} - E_{\mu\nu}\,, \tag{44}$$

where $\phi(x)$ is the value of the dilaton field Φ on the brane,

$$T^{(\phi)}_{\mu\nu} = D_\mu\phi D_\nu\phi - g_{\mu\nu}\left[\frac{1}{2}(D\phi)^2 + U_{eff}\right], \tag{45}$$

$$\Delta T^{(\phi)}_{\mu\nu} = \frac{1}{4}g_{\mu\nu}\left[U_{eff} - \frac{1}{2}(D\phi)^2\right]. \tag{46}$$

$$8\pi G_N = \frac{\kappa_5^4}{6}\lambda(\phi)\,, \tag{47}$$

with

$$U_{eff} = \left[V + \frac{1}{6}\kappa_5^2\lambda^2 - \frac{1}{8}\left(\frac{d\lambda}{d\phi}\right)^2\right] \tag{48}$$

and $\Pi_{\mu\nu}$ is given by (34)

Choosing a Gaussian normal coordinate χ such that the hypersurface $\chi = 0$ coincides with our brane ($\mathcal{B} = 0$), we expand the dilaton field Φ near the brane as

$$\Phi = \phi(x) + \Phi_1(x)|\chi| + \frac{1}{2}\Phi_2(x)\chi^2 + \mathcal{O}(\chi^3), \tag{49}$$

and, inserting this into

$$\Box\Phi + K\pounds_n\Phi + {\pounds_n}^2\Phi - \frac{dV}{d\Phi} - \frac{d\lambda}{d\Phi}\delta(\mathcal{B}) = 0, \tag{50}$$

we find the jump condition for a dilaton field as

$$\Phi_1 = \frac{1}{2}\frac{d\lambda}{d\phi}\,, \tag{51}$$

and the equation for ϕ as

$$\Box\phi - \frac{dU_{eff}}{d\phi} = -\Delta\Phi_2\,, \tag{52}$$

where

$$\Delta\Phi_2 \equiv \Phi_2 - \frac{1}{4}\frac{d\lambda}{d\phi}\frac{d^2\lambda}{d\phi^2}\,. \tag{53}$$

Introducing a vector field $J_\mu = \Delta\Phi_2 \cdot D_\mu\phi$, we find that (52) yields

$$D^\nu T^{(\phi)}_{\mu\nu} = -J_\mu\,. \tag{54}$$

Thus we regard J_μ as the energy-momentum lost from the scalar field on the brane to the bulk.

In order to find 4D Minkowski spacetime on the brane, U_{eff} should vanish at some point ($\phi = \phi_0$), i.e.

$$V + \frac{\kappa_5^2}{6}\lambda^2 = \frac{1}{8}\left(\frac{d\lambda}{d\phi}\right)^2 \tag{55}$$

An example is provided by the 5D effective action obtained by Lukas, Ovrut and Waldram [13] from a dimensional reduction of an 11D Hořava-Witten model, in which

$$V = \frac{\alpha_0^2}{6\kappa_5^2}e^{-2\sqrt{2}\kappa_5\Phi} \quad \text{and} \quad \lambda = \frac{\sqrt{2}\alpha_0}{{\kappa_5}^2}e^{-\sqrt{2}\kappa_5\Phi} \tag{56}$$

where α_0 is a constant. In this case, U_{eff} vanishes for any value of ϕ. This fact ($U_{eff} \equiv 0$) is also true for a supersymmetric model [43], in which V and λ are given by a superpotential $W(\Phi)$ as

$$2\kappa_5^2 V = \frac{1}{4\kappa_5^2}\left(\frac{dW}{d\Phi}\right)^2 - \frac{1}{3}W^2$$
$$\kappa_5^2\lambda(\phi) = W(\phi)\,. \tag{57}$$

It is also possible to find FRW cosmological solutions on the brane with time-dependent scale factor, $a(t)$, and scalar field, $\phi(t)$, where t is cosmic proper time, if we know, or make some assumption about the energy transfer from brane to bulk, $J_0 = \dot{\phi}\,\Delta\Phi_2$.

The scalar field equation of motion (52) is

$$\ddot{\phi} + 3H\dot{\phi} = \frac{dU_{eff}}{d\phi} + \Delta\Phi_2\,, \tag{58}$$

while the Friedmann equation is

$$H^2 + \frac{k}{a^2} = \frac{2{\kappa_5}^2}{9}\left(\rho_\phi + \Delta\rho_\phi\right) + \frac{1}{3}E^0{}_0\,. \tag{59}$$

The energy density of the scalar field ρ_ϕ and the effect from the scalar field in the bulk $\Delta\rho_\phi$ are given by

$$\rho_\phi = -T^0_{\ 0} = \frac{1}{2}\dot{\phi}^2 + U_{eff} \quad \text{and} \quad \Delta\rho_\phi = -\Delta T^0_{\ 0} = -\frac{1}{4}\rho_\phi \tag{60}$$

The equation for $E^0{}_0$ is given by (44) with the contracted Bianchi identity as

$$\dot{E}^0{}_0 + 4HE^0{}_0 = \frac{2{\kappa_5}^2}{3}\left[\frac{1}{4}\dot{\rho}_\phi - J_0\right], \tag{61}$$

where we have used $E^\mu{}_\mu = 0$ and $\Delta T_{\mu\nu} = (\rho_\phi/4)g_{\mu\nu}$.

Here we give two simple examples:

(1) No energy transfer from a bulk ($J_0 = 0$)

When there is no energy transfer from brane to bulk, i.e. $J_0 = 0$, we find a closed set of equations (58–61) for the dilaton-vacuum universe, for a given U_{eff}. Equations (58) and (61) can then be integrated, if the bulk and brane potentials obey the generalized Randall-Sundrum condition given in (55), so that $U_{eff} = 0$. Equation (58) can be simply integrated to give $\dot{\phi} = C_\phi / a^3$, where C_ϕ is an integration constant, and (61) can be integrated to give

$$E^0{}_0 = \frac{{\kappa_5}^2 C_\phi^2}{4a^6} - \frac{\mathcal{E}_0}{a^4}, \tag{62}$$

where $\mathcal{E}_0$ is another integration constant. Inserting those solutions into (59), we find

$$H^2 + \frac{k}{a^2} = \frac{{\kappa_5}^2 C_\phi^2}{6a^6} - \frac{\mathcal{E}_0}{3a^4}. \tag{63}$$

This is the same as the standard Friedmann equation with stiff matter and radiation, which is easily integrated [9].

(2) energy transfer from a bulk ($J_0 \propto \dot{\phi}^3$)

Next we discuss the case with an energy transfer between the scalar field on the brane, $\phi(t)$, and the radion field b, described by $J_0 \propto (\dot{b}/b)\rho_\phi$. At the same time the expansion of the bulk metric is itself determined by the local density. In order to obtain separable solutions for the bulk metric, Lukas et al [13] require some ansatz in the bulk, corresponding to $J_0 \propto \dot{\phi}^3$ on the brane. Then we assume that the energy transfer is given by

$$J_0 = -\sqrt{2}\Gamma \kappa_5 \dot{\phi}^3, \tag{64}$$

where Γ is a constant, then (58), with $dU_{eff}/d\phi = 0$, can be integrated to give $\dot{\phi} = C_\phi a^{-3} e^{-\sqrt{2}\Gamma\kappa_5\phi}$. The remaining equations (61) and (59) can be integrated if we make the power-law ansatz

$$a \propto |t|^p \quad \text{and} \quad e^{\sqrt{2}\kappa_5\phi} \propto |t|^q. \tag{65}$$

Equation (61) yields

$$E^0_0 = -\frac{(4p-1)q^2}{8(2p-1)}|t|^{-2} - \mathcal{E}_0|t|^{-4p}. \tag{66}$$

We find

$$p = p_{(\pm)} = \frac{3(2\Gamma^2+1) \pm 2\sqrt{3}\Gamma\sqrt{3\Gamma^2+1}}{3(8\Gamma^2+3)} \tag{67}$$

$$q = q_{(\pm)} = \frac{2[\Gamma \mp \sqrt{3(3\Gamma^2+1)}]}{8\Gamma^2+3}. \tag{68}$$

For $\Gamma = 1$ we recover the solutions of Lukas et al [13] with

$$p_{(\pm)} = \frac{3}{11}\left(1 \pm \frac{4\sqrt{3}}{9}\right) \quad \text{and} \quad q_{(\pm)} = \frac{2}{11}\left(1 \mp 2\sqrt{3}\right). \tag{69}$$

We recover the stiff-matter dominated solution with $p_{(\pm)} = 1/3, q_{(\pm)} = \mp 2/\sqrt{3}$ for $\Gamma = 0$. Over the entire range $-\infty < \Gamma < \infty$, $p_{(+)}$ changes monotonically from 0 to 1/2, while $q_{(+)}$ always takes negative values from zero to zero, having its minimum value $-\sqrt{6}/2$ at $\Gamma = -\sqrt{6}/12$.

6.2 Brane Model with Bulk Yang-Mills Field [14]

Next we study the 5D Einstein-Yang-Mills system with a cosmological constant and discuss a brane universe model. In the brane world cosmology [11], a higher-dimensional black hole solution plays an important role. Our universe is just a domain wall expanding in the black hole background spacetime [27]. The black hole mass gives a contribution as dark radiation through its tidal force. Hence, a higher-dimensional black hole or a globally regular solution with a cosmological constant is now a very important subject. However, from a view point of brane cosmology, although a black hole singularity is covered by a horizon, the universe can pass through the horizon and hit on the singularity, which is the end of our world. If a string/M theory is fundamental, such a singularity should not exist. Then, if we can construct some non-singular object in the bulk spacetime, it might be a manifestation of singularity avoidance immanent in a fundamental theory. In four dimensions, Bartnik and McKinnon found a particle-like solution as a globally regular spacetime in a spherically symmetric Einstein-Yang-Mills system with SU(2) gauge group [44]. Soon after, a colored black hole with a nontrivial non-Abelian structure was also found [45]. These solutions were also extended to those in the system with a cosmological constant [46, 47, 21]. From stability analysis, it turns out that the solutions with zero or positive cosmological constant are unstable [48], while those with negative cosmological constant are stable [49]. Since a negative cosmological constant is naturally expected in a brane world scenario just as the RS model [6, 7], the above fact becomes very important. Then, we study a nontrivial particle-like solution or black hole solution in five dimensions with a cosmological constant [14].

Here, assuming that non-Abelian gauge field appears in 5D bulk spacetime, we first discuss a spherically symmetric Einstein-Yang-Mills system in 5-dimensions. The action we discuss here is given by

$$S = \frac{1}{2}\int d^5x\sqrt{-g_5}\left[\frac{1}{\kappa_5^2}(R - 2^{(5)}\Lambda) - \frac{1}{8\pi g^2}\mathrm{Tr}\mathbf{F}^2\right], \tag{70}$$

where g is a gauge coupling constant. $\mathbf{F} = F_{\mu\nu}dx^\mu \wedge dx^\nu$ is a field strength of the gauge field, which is described by the vector potential $\mathbf{A} = A_\mu dx^\mu$ as

$$F_{\mu\nu} = \partial_\mu A_\nu - \partial_\nu A_\mu - [A_\mu, A_\nu]. \tag{71}$$

Defining a fundamental mass scale of the gauge field by $m_g = g^{-2}$, we normalize the present system by a typical length scale given by $(8\pi m_g/m_5^3)^{1/2} = (8\pi\kappa_5^2/g^2)^{1/2}$.

A spherically symmetric and static spacetime is given by

$$ds^2 = \left[-f(r)e^{-2\delta(r)}dt^2 + \frac{dr^2}{f(r)} + r^2 d\Omega_3^2\right], \tag{72}$$

where

$$f(r) = 1 - \frac{\mu(r)}{r^2} + \epsilon\frac{r^2}{\ell^2}. \tag{73}$$

We set ${}^{(5)}\Lambda = -6\epsilon/\ell^2$ with $\epsilon = 0$ or ± 1, corresponding to the signature of ${}^{(5)}\Lambda$. We call μ a 'mass' function.

We find the following solutions:

(1) "electrically" charged solutions

If we take only an "electric" part of the field, we find the Reissner-Nordstrom type solution such as

$$\mu = \mathcal{M} - \frac{2Q^2}{3r^2}, \quad \delta = 0, \quad A = -\frac{Q}{r^2}. \tag{74}$$

(2) "magnetically" charged solutions

If the "magnetic" part of the gauge field appears, the gauge potentials is given by one scalar function $w(r)$ as

$$A_t^a = 0, \quad A_r^a = 0, \tag{75}$$

$$A_\psi^a = (0, 0, w), \tag{76}$$

$$A_\theta^a = (w\sin\psi, -\cos\psi, 0), \tag{77}$$

$$A_\varphi^a = (\cos\psi\sin\theta, w\sin\psi\sin\theta, -\cos\theta), \tag{78}$$

resulting in the Einstein-Yang-Mills equation of the present system as

$$\mu' = 2r\left[fw'^2 + \frac{(1-w^2)^2}{r^2}\right], \tag{79}$$

$$\delta' = -\frac{2}{r}w'^2, \tag{80}$$

$$\frac{1}{r}(rfe^{-\delta}w')' + \frac{2}{r^2}e^{-\delta}w(1-w^2) = 0. \tag{81}$$

(2-i) analytic solutions

The above differential equations (79-81) have two analytic solutions. One analytic solution is

$$w = \pm 1, \quad \mu = \mathcal{M}, \quad \delta = 0, \tag{82}$$

which corresponds to the Schwarzschild or the Schwarzschild-AdS (or dS) spacetime, which properties are well known.

Another analytic solution is given by

$$w = 0, \quad \mu = \mathcal{M} + 2\ln r, \quad \delta = 0. \tag{83}$$

In the 4D spacetime, this type of solution describes the Reissner-Nordstrom type geometry with a magnetic charge. In the 5D spacetime, $2\ln r$ term appear in the mass function μ. Although μ diverges, the metric itself approaches that of well-known symmetric spacetime for each ϵ, i.e. the Minkowski, de Sitter and anti de Sitter one. There is no singularity except for the origin which is covered by an event horizon. We regard this spacetime as a localized object (a magnetically charged black hole).

(2-ii) non-trivial particle solution and black holes

Like in the four dimensional case [44, 45, 49], we can find non-trivial structure of self-gravitating Yang-Mills field. We obtain those solutions numerically. There are two types of solutions; a particle solution and a black hole. Here, we show only a particle-like solution in the case of $\epsilon = 0$ or 1 ($^{(5)}\Lambda \leq 0$).

Imposing regularity at the origin $r = 0$, we find non-singular particle-like solutions. For the case of $\epsilon = 0$, the spacetime approaches to Minkowski as $r \to \infty$ for $b_{\min}(\approx -0.635607) < b < 0$, where an expansion parameter b is defined by $w \approx 1 + br^2$ near $r \approx 0$. We show the numerical result in Fig. 3.

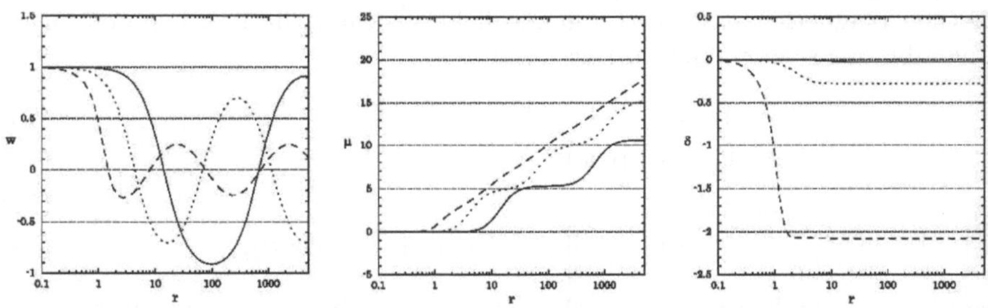

Fig. 3. The potential function $w(r)$, the mass function $\mu(r)$ and the metric function $\delta(r)$ for a particle-like solution with $\epsilon = 0$. The solid, dotted and dash lines depict those for $b = -0.01$, -0.1 and -0.5, respectively.

The potential function w is oscillating between ± 1 and the mass function μ is increasing without bound just as a step function. There is no finite mass particle-like solution. The mass function increases as $\ln r$ asymptotically like the analytic solution (83). The periodic steps in μ in terms of $\ln r$ are caused by infinite set of t'Hooft-Polyakov instantons in the radial direction [14].

For the case of ϵ=1, we also find a regular solution for $b_{\min}$<b<0. $b_{\min}$ depends on ℓ, and decreases as ℓ decreases. For example, $b_{\min}\approx -0.644036$ for ℓ=10, $b_{\min}\approx -1.105002$ for ℓ=1. We show the numerical result in Fig. 4.

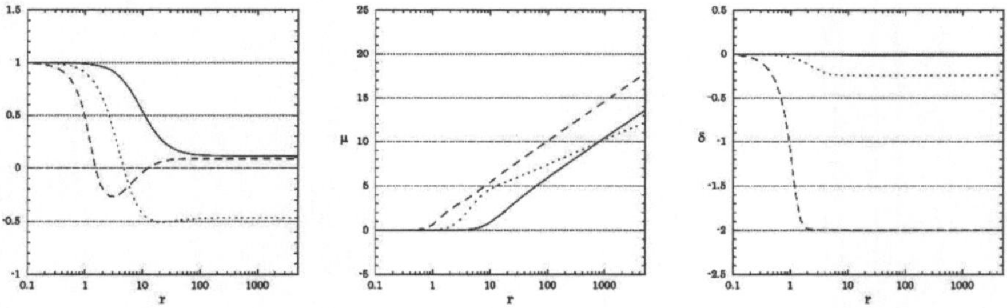

Fig. 4. The potential function $w(r)$, the mass function $\mu(r)$ and the metric function $\delta(r)$ for a particle-like solution with $\epsilon = 1$. The solid, dotted and dash lines depict those for $b = -0.01, -0.1$ and -0.5, respectively.

In this case, the potential w does not oscillate, and converge to some value w_∞, then the number of node is finite. The mass function increases monotonically as

$$\mu \to 2(1 - w_\infty^2)^2 \ln r \tag{84}$$

as $r \to \infty$.

Universe as a domain wall in a non-singular spacetime

Using this non-singular solution, we can construct the oscillating universe without any singularity, neither in a bulk nor on a brane. The Friedmann equation is given by

$$H^2 + \frac{f(a)}{a^2} = \left[\frac{\kappa_5^2}{3}(\lambda + \rho)\right]^2 . \tag{85}$$

We can rewrite (85) as

$$\begin{aligned} &\frac{1}{2}\dot{a}^2 + U(a) = 0 \\ &\text{with} \quad U(a) = \frac{1}{2}\left[1 - \frac{\mu(a)}{a^2} - \frac{\Lambda}{3}a^2\right] . \end{aligned} \tag{86}$$

In Fig. 5, we depict the potential $U(a)$ with $\Lambda = 0$, and show how the universe oscillates.

7 Models with Induced Gravity on a Brane

Finally, we study models with an induced gravity on the brane due to quantum corrections. Since the matter fields are confined on the brane, if we take into account of quantum effects of matter fields, the gravitational action on the brane will be modified. Here we shall discuss two models: First one is

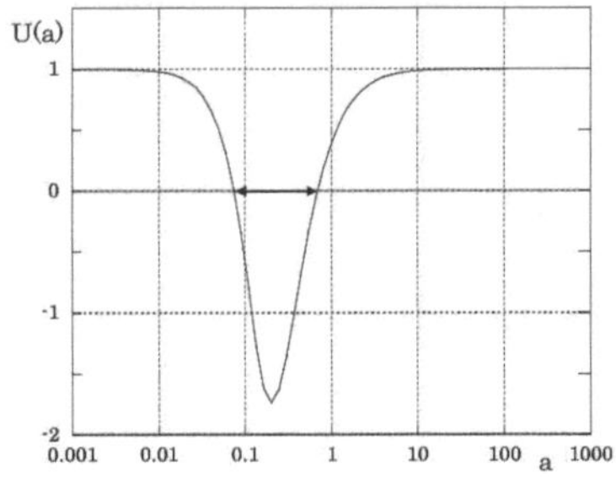

Fig. 5. The bouncing universe in non-singular bulk spacetime. The bulk spacetime is described by a particle-like solution with a negative cosmological constant. The brane universe oscillates between two scale factors as shown by the solid line with arrows.

the model of a brane world proposed by Dvali, Gabadadze, and Porrati [15]. Secondly, if we have conformally invariant fields on the brane, trace anomaly appears naturally in 4D brane world via quantum effects. This type of curvature squared term was first discussed by Starobinsky in his inflationary scenario [18].

We consider the brane Lagrangian

$$L_{\rm brane} = -\lambda + L_{\rm m} + \frac{\mu^2}{2} R + \beta R^2. \tag{87}$$

It is just a simple version of the R^2 inflationary model. If we set $\beta = 0$, the action gives the Dvali-Gabadadze-Porrati's model, or its generalization. When we take into account quantum effects such as a trace anomaly, the other curvature terms should also be included in the basic equations as quantum corrections of the energy-momentum tensor. Although we can apply the present approach to such a model [24], here we study only the model with the Lagrangian (87).

In order to find the basic equations on the brane, we just calculate the "energy-momentum" tensor of the brane $\tau_{\mu\nu}$ by the definition (17) from the Lagrangian (87). We find

$$\begin{aligned} \tau^\mu{}_\nu &= -\lambda \delta^\mu{}_\nu + T^\mu{}_\nu - \mu^2 G^\mu{}_\nu \\ & -4\beta \left[R G^\mu{}_\nu - \nabla^\mu \nabla_\nu R + \Box R \delta^\mu{}_\nu + \frac{1}{4} R^2 \delta^\mu{}_\nu \right] \end{aligned} \tag{88}$$

7.1 Dvali-Gabadadze-Porrati's Induced Gravity Model [10]

Setting $\beta = 0$, we find the effective equations as

$$\begin{aligned}\left(1+\frac{\lambda}{6}\kappa_5^4\mu^2\right)G_{\mu\nu}+\kappa_5^4\mu^2\mathcal{K}_{\mu\nu\rho\sigma}(T_{\alpha\beta})G^{\rho\sigma}+\Lambda g_{\mu\nu}\\ =\frac{\lambda}{6}\kappa_5^4T_{\mu\nu}+\kappa_5^4\left[\Pi_{\mu\nu}+\mu^4\pi^{(G)}_{\mu\nu}\right]-E_{\mu\nu},\end{aligned}\tag{89}$$

where

$$\begin{aligned}\mathcal{K}_{\mu\nu\rho\sigma}&=\frac{1}{4}\left(g_{\mu\nu}T_{\rho\sigma}-g_{\mu\rho}T_{\nu\sigma}-g_{\nu\sigma}T_{\mu\rho}\right)\\ &\quad+\frac{1}{12}\Big[T_{\mu\nu}g_{\rho\sigma}+T\left(g_{\mu\rho}g_{\nu\sigma}-g_{\mu\nu}g_{\rho\sigma}\right)\Big],\end{aligned}\tag{90}$$

$$\Lambda=\frac{1}{2}\left[{}^{(5)}\Lambda+\frac{1}{6}\kappa_5^4\lambda^2\right],\tag{91}$$

$$\pi^{(G)}_{\mu\nu}=-\frac{1}{4}G_{\mu\alpha}G_{\nu}^{\ \alpha}+\frac{1}{12}GG_{\mu\nu}+\frac{1}{8}g_{\mu\nu}G_{\alpha\beta}G^{\alpha\beta}-\frac{1}{24}g_{\mu\nu}G^2.\tag{92}$$

The Codazzi equation is now $D^\nu\tau_{\mu\nu}=0$, which implies the energy momentum conservation, i.e. $D^\nu T_{\mu\nu}=0$ because of the contracted Bianchi identity.

Now we discuss the FRW universe with a perfect fluid $P=(\gamma-1)\rho$. Since the spacetime is isotropic and homogeneous, we can show $D^\nu\pi_{\mu\nu}=0$ following [8], which implies $D^\nu E_{\mu\nu}=0$.

The basic equations (27) are written as

$$3X=\frac{1}{2}{}^{(5)}\Lambda+E^0_{\ 0}+\frac{\kappa_5^4}{12}\left(\lambda+\rho-3\mu^2X\right)^2,\tag{93}$$

$$\begin{aligned}&\left[1+\frac{\kappa_5^4}{6}\mu^2\left(\lambda+\rho-3\mu^2X\right)\right]Y\\ &\quad=-\frac{2}{3}E^0_{\ 0}-\frac{\kappa_5^4}{12}(\rho+P)\left(\lambda+\rho-3\mu^2X\right),\end{aligned}\tag{94}$$

where

$$X\equiv H^2+\frac{k}{a^2},\qquad Y\equiv\dot{H}-\frac{k}{a^2}.\tag{95}$$

The equation for $E^0_{\ 0}$ is reduced to $\dot{E}^0_{\ 0}+4HE^0_{\ 0}=0$. This equation is the same as the dark radiation in the case of the RS model [8], and is easily solved as $E^0_{\ 0}=\mathcal{E}_0/a^4$, where $\mathcal{E}_0$ is just an integration constant.

We now have to solve one equation (93), which is a quadratic equation with respect to X, we rewrite it in the form of the conventional Friedmann equation as

$$H^2+\frac{k}{a^2}=\frac{1}{3}\Lambda^{(\epsilon)}_{\rm eff}+\frac{8\pi G^{(\epsilon)}_{\rm eff}}{3}\left(\rho+\rho^{(\epsilon)}_{\rm DR}\right),\tag{96}$$

where

$$\Lambda_{\rm eff}^{(\epsilon)} = \frac{\rho_0}{\mu^2}\left(1+\epsilon\mathcal{A}_0\right), \qquad 8\pi G_{\rm eff}^{(\epsilon)} = \frac{1}{\mu^2}\left[1+\epsilon\mathcal{F}(\rho,a)\right],$$
$$8\pi G_{\rm eff}^{(\epsilon)}\rho_{\rm DR}^{(\epsilon)} = -\epsilon\mathcal{F}(\rho,a)\frac{\mathcal{E}_0}{a^4}, \tag{97}$$

with

$$\mathcal{F}(\rho,a) = \frac{2\eta}{\mathcal{A}_0+\mathcal{A}(\rho,a)}. \tag{98}$$

ρ_0, η, $\mathcal{A}$ and $\mathcal{A}_0$ are defined by

$$\rho_0 = m_\lambda^4 + 6\frac{m_5^6}{\mu^2}, \qquad \eta = \frac{6m_5^6}{\rho_0\mu^2} \quad (0<\eta\le 1),$$
$$\mathcal{A} \equiv \left[\mathcal{A}_0^2 + \frac{2\eta}{\rho_0}\left(\rho - \mu^2\frac{\mathcal{E}_0}{a^4}\right)\right]^{\frac{1}{2}}, \quad \mathcal{A}_0 = \sqrt{1-2\eta\frac{\mu^2\Lambda}{\rho_0}}. \tag{99}$$

ϵ denotes either $+1$ or -1. The choice of the sign of ϵ also has a geometrical meaning as shown by Deffayet [50].

Using the above expression, we discuss the evolution of the universe. $\Lambda_{\rm eff}^{(\epsilon)}$ acts as a cosmological constant in each branch. The effective gravitational "constant" $G_{\rm eff}^{(\epsilon)}$ changes in the history of the universe. As ρ decreases from ∞ to zero (and a increases from 0 to ∞),

$$8\pi G_{\rm eff}^{(\epsilon)} \;:\; \frac{1}{\mu^2} \to \frac{1}{\mu^2}\left(1+\frac{\epsilon\eta}{\mathcal{A}_0}\right). \tag{100}$$

In particular, in the negative branch ($\epsilon=-1$), if $\eta>\eta_{\rm cr}$, where

$$\eta_{\rm cr} \equiv -\mu^2\frac{\Lambda}{\rho_0} + \sqrt{1+\left(\mu^2\frac{\Lambda}{\rho_0}\right)^2}, \tag{101}$$

$G_{\rm eff}^{(-)}$ vanishes at some density and becomes negative below that density. In this case, $\eta\le 1$ implies $\eta_{\rm cr}<1$, which requires $\Lambda>0$. Although the effective gravitational constant becomes negative, it does not naively mean the observed "Newtonian gravitational constant" is negative. The expansion of the universe first slows down after this critical point, and then approaches some constant given by $\Lambda_{\rm eff}^{(-)}(>0)$. This cosmological model could be interesting because the expansion gets slow in some period of the universe and then it might help a structure formation process.

Equation (94) is also rewritten in the conventional form as

$$\dot{H} - \frac{k}{a^2} = -4\pi G_{\rm eff}^{(\epsilon)}\left(\gamma_{\rm eff}^{(\epsilon)}\rho + \gamma_{\rm DR}\rho_{\rm DR}^{(\epsilon)}\right), \tag{102}$$

where

$$\gamma_{\rm eff}^{(\epsilon)} = \gamma \left[1 - \frac{\epsilon\eta(\rho - \mu^2\mathcal{E}_0/a^4)\mathcal{F}(\rho,a)}{\rho_0\mathcal{A}(\rho,a)[\mathcal{A}(\rho,a)+\mathcal{A}_0+2\epsilon\eta]} \right], \gamma_{\rm DR} = \frac{2}{3}\left[1+\frac{\mathcal{A}_0}{\mathcal{A}(\rho,a)}\right]. \quad (103)$$

$\gamma_{\rm eff}^{(\epsilon)}$ and $\gamma_{\rm DR}$ denote the effective adiabatic indexes of matter fluid and of "dark radiation". We depict the behavior of the effective adiabatic indexes $\gamma_{\rm eff}^{(\epsilon)}$ when ρ (or $1/a$) changes from ∞ to 0 in Fig. 6. This behavior is interesting because the "effective" negative pressure ($\gamma_{\rm eff}^{(+)} < 1$) can be obtained during the evolution of the universe from standard matter fluid such as dust ($\gamma = 1$). Although $\gamma_{\rm eff}^{(-)}$ diverges at some density, (102) is not singular because $G_{\rm eff}^{(-)}$ vanishes at the same density.

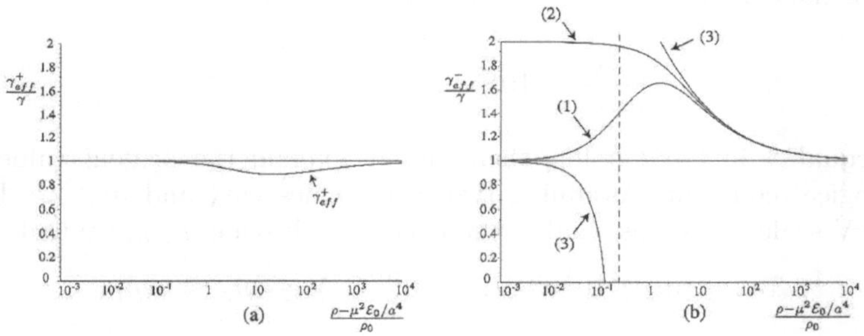

Fig. 6. The behavior of $\gamma_{\rm eff}^{(\pm)}$. On the left ((a)), we show one typical example for positive branch ($\epsilon = +1$) , while, on the right ((b)), we depict the figures for negative branch (($\epsilon = -1$)) in the following three typical cases; (1) $\eta < \eta_{cr}$, (2) $\eta = \eta_{cr}$ and (3) $\eta > \eta_{cr}$. For an expanding universe, the universe evolves from the right hand side to the left in the figures.

Here we discuss the late time behavior of the universe ($\rho \ll \rho_0$ and $a \to \infty$), and focus on the value of a cosmological constant. In this limit, if we take positive branch ($\epsilon = 1$), since $\Lambda_{\rm eff}^{(+)} = \rho_0\left(1+\mathcal{A}_0\right)/\mu^2$ (> 0), we find an inflationary expansion in the late stage of the universe if $\Lambda < \rho_0^2/(12m_5^6)$. The case of $\Lambda = 0$ corresponds to the original Dvali et al's model. The present gravitational constant in the Friedmann equation, which is given by $8\pi G_N^{(+)}$, becomes larger than that in the early stage ($1/\mu^2$).

If we take negative branch ($\epsilon = -1$), in the case of $\Lambda = 0$, we have zero cosmological constant ($\Lambda_{\rm eff}^{(-)} = 0$) on the brane, which is the conventional Friedmann equation with dark radiation. The gravitational constant becomes smaller than that in the early stage.

If $0 < \Lambda \le \rho_0^2/(12m_5^6)$, however, we expect a positive cosmological constant on the brane, which could be very small. Suppose that $\lambda \gg m_5^6/\mu^2$ ($\eta \ll 1$). We then approximate the cosmological constant in the Friedmann equation as

$$\Lambda_{\text{eff}}^{(-)} \approx \eta\Lambda \approx \frac{6m_5^6}{\lambda\mu^2}\Lambda \ll \Lambda. \tag{104}$$

This means that the 4D cosmological constant is suppressed in the Friedmann equation from its proper value (Λ). Hence, we might have a possibility to explain the tiny value of the present cosmological constant, of which observational limit is $\Lambda_{\text{eff}}^{(-)}/m_{PL}^2 \lesssim 10^{-120}$. In the RS model, Λ is fine-tuned to zero, but in more realistic brane models such as the Hořava-Witten model, the 4D cosmological constant may automatically vanish if a supersymmetry is preserved (see Sect. 6.1). In the present universe, however, supersymmetry must be broken, and then we expect that non-zero value of Λ is created by the SUSY breaking, which may occur at 1 TeV. This gives $\Lambda/m_{PL}^2 \sim (1\text{TeV}/m_{PL})^4 \sim 10^{-60}$.

Then the constraint by observation ($\Omega_\Lambda \lesssim 1$) is now

$$\frac{m_\lambda}{m_{PL}} \gtrsim 10^{15} \times \left(\frac{m_5}{m_{PL}}\right)^{3/2}. \tag{105}$$

If the equality in (105) holds, then we can explain the present value of a cosmological constant. Assuming two mass scales (m_λ and m_5) are larger than TeV scale as well as smaller than the Planck scale m_{PL}, we find

$$1\ \text{TeV} \lesssim m_5 \lesssim 10^8\ \text{GeV}, \qquad 10^{10}\ \text{GeV} \lesssim m_\lambda \lesssim m_{PL}. \tag{106}$$

7.2 the R^2 Inflationary Model

Next we discuss cosmology in the case with R^2-term. For the FRW universe with a perfect fluid, from (88), we have

$$\tau^0{}_0 = -(\lambda+\rho) - (\mu^2 + 4\beta R)G^0{}_0 - \beta R^2 + 12\beta H\dot{R} \tag{107}$$
$$\tau^i{}_j = (P-\lambda)\delta^i{}_j - (\mu^2 + 4\beta R)G^i{}_j - \beta R^2\delta^i{}_j + 4\beta(\ddot{R} + 2H\dot{R})\delta^i{}_j\,. \tag{108}$$

We also find

$$\pi^0{}_0 = -\frac{1}{12}\left(\tau^0{}_0\right)^2 \quad \pi^i{}_j = \frac{1}{12}\tau^0{}_0\left(\tau^0{}_0 - 2\tau^1{}_1\right)\delta^i{}_j. \tag{109}$$

Inserting the above equations into (27), we obtain the "Friedmann" equation on the brane:

$$\beta\left[\dot{Y} + 3HY - \frac{Y}{2H}\left(Y + \frac{2k}{a^2}\right)\right]$$
$$= \frac{1}{12H}\left[-\frac{\mu^2}{2}\left(H^2 + \frac{k}{a^2}\right) + \frac{\lambda+\rho}{6} \pm \frac{1}{\kappa_5^2}\sqrt{H^2 + \frac{k}{a^2} + \frac{1}{\ell^2} - \frac{E_0^0}{3}}\right], \tag{110}$$

where $Y = \dot{H} - k/a^2$. Note that another dynamical equation becomes trivial if we assume the energy-momentum conservation of matter field, just as in the conventional Einstein system.

We set $k = 0$, $E^0_{\ 0} = 0$ and $\Lambda = 0$ for brevity. We also consider only the vacuum case, i.e. $\rho = 0$. Then we find the following dynamical equation:

$$\ddot{H} = -3H\dot{H} + \frac{\dot{H}^2}{2H} - \frac{d}{dH}U^{(\pm)}_{eff}, \tag{111}$$

where the "effective potential" $U^{(\pm)}_{eff}$, which describes the dynamics of the universe, is defined by

$$U^{(\pm)}_{eff}(H) = \int dH \left(\frac{\mu^2}{24\beta} H - \frac{\lambda}{72\beta H} \left[1 \pm \sqrt{1 + \frac{36H^2}{\kappa_5^4 \lambda^2}} \right] \right). \tag{112}$$

It is integrated as

$$U^{(+)}_{eff} = \frac{\mu^2 H^2}{48\beta} - \frac{\lambda}{72\beta} \left[2 \ln H + \sqrt{1 + \frac{36H^2}{\kappa_5^4 \lambda^2}} - \ln \left| 1 + \sqrt{1 + \frac{36H^2}{\kappa_5^4 \lambda^2}} \right| \right]. \tag{113}$$

$$U^{(-)}_{eff} = \frac{\mu^2 H^2}{48\beta} + \frac{\lambda}{72\beta} \left[\sqrt{1 + \frac{36H^2}{\kappa_5^4 \lambda^2}} - \ln \left| 1 + \sqrt{1 + \frac{36H^2}{\kappa_5^4 \lambda^2}} \right| \right]. \tag{114}$$

The dynamical equation is written in the following form:

$$\frac{dE^{(\pm)}}{dt} = \frac{\dot{H}^2}{2H} \left(\dot{H} - 6H^2 \right) \tag{115}$$

where the "energy" $E^{(\pm)}$ is defined by

$$E^{(\pm)} = \frac{1}{2}\dot{H}^2 + U^{(\pm)}_{eff} \tag{116}$$

Then, if we discuss the expanding universe ($H > 0$), the "energy" $E^{(\pm)}$ is increasing for $\dot{H} > 6H^2$, while decreasing for $\dot{H} < 6H^2$.

If $\mu \neq 0$ (Fig. 7), we have the asymptotic behavior of the "potential" as $U^{(\pm)}_{eff} \approx \mu^2 H^2/(48\beta)$ for large value of H. Assuming a slow-rolling condition and setting $H \approx H_0 + Y_0 t$, which describes the R^2-inflation found by Starobinsky [18], we obtain $Y_0 = -\mu^2/(72\beta)$. To satisfy the slow rolling condition, we require $H_0^2 \gg \mu^2/(72\beta)$. For the "potential" $U^{(+)}_{eff}$, we find one minimum, which gives another inflationary solution. The Hubble constant H_1 is given by

$$H_1^2 = \frac{2}{3\mu^2} \left[\lambda + \frac{6}{\kappa_5^4 \mu^2} \right] \tag{117}$$

This is a second inflation, which might correspond to the present acceleration of the universe.

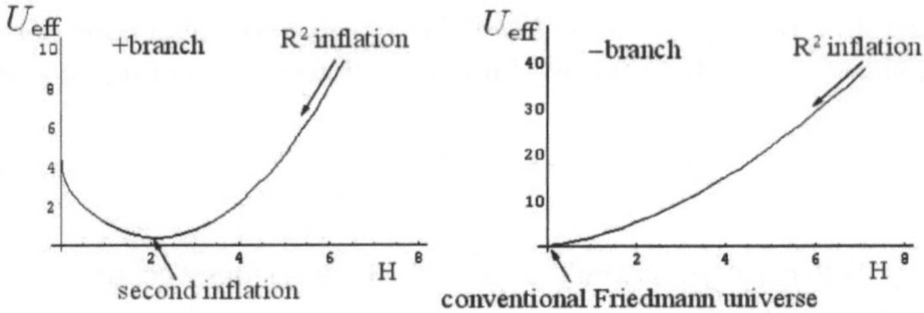

Fig. 7. The potential $U_{eff}^{(\pm)}$ for $\mu \neq 0$.

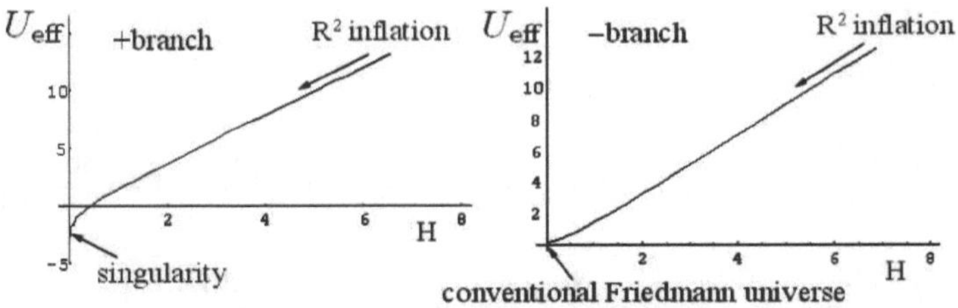

Fig. 8. The potential $U_{eff}^{(\pm)}$ for $\mu = 0$.

If $\mu = 0$ (Fig. 8), which corresponds to the model only with the trace anomaly, the asymptotic behaviors of the potential is changed to be

$$U_{\text{eff}}^{(\pm)} \approx \mp \frac{H}{12\kappa_5^2 \beta} \quad \text{as} \quad H \to \infty. \tag{118}$$

Then, in order for the universe to reach the present small-H stage, $\beta < 0$ is required for $U_{\text{eff}}^{(+)}$, while $\beta > 0$ for $U_{\text{eff}}^{(-)}$. The slow-rolling solution is given by $H = H_0 + Y_0 t$ where $Y_0 = \pm 1/(36\beta H_0)$. The slow-rolling condition implies that $H_0^3 \gg 1/(\kappa_5^2 |\beta|)$.

In the 4D spacetime, if we have only the R^2 term, the universe expands exponentially forever, and inflation will never end. However, in the present brane scenario, the inflation will finish because of the existence of the Einstein-Hilbert action in 5D bulk spacetime. After the inflationary stage, we find either the conventional Hubble expansion or a curvature singularity, instead of the second inflationary stage.

8 Concluding Remarks

We have first given an overview of several models for a brane world, and then listed up how to deal with such a higher-dimensional spacetime. Three approaches are introduced: (1) 4D effective approach, (2) holographic approach, and (3) higher-dimensional approach. We have discussed the first approach in detail. We derive the effective 4D gravitational equations in 5 dimensions, (27), assuming that gravity is confined on a brane. This approach yields the most general form of the 4D gravitational field equations for a brane world observer whatever the form of the bulk metric, in contrast to the usual Kaluza-Klein type dimensional reduction which relies on taking a particular form for the bulk metric in order to integrate over the extra dimensions. The price to be paid for such generality, is that the brane world observer may be subject to influences from the bulk, in particular gravitational waves, which are not constrained by local brane quantities, i.e., the set of 4D equations does not form a closed system in general. Nonetheless, when a brane is localized, the energy-momentum tensor on the brane is sufficient to determine the extrinsic curvature of the brane, and together with the local induced metric, this strongly constrains the brane world gravity. The effect of the bulk gravity can be described by the projected 5D Weyl tensor as well as the energy-momentum tensor of bulk matter fields. It turns out that this 5D object may provide us a window to see the extra dimensions. We then studied the Randall-Sundrum model in detail and applied our formalism to cosmology. The effective theory is reduced to the conventional Einstein equations in the low-energy limit. There appear two new terms: One is a quadratic term of the energy-momentum tensor of matter field and the other is the 5D Weyl curvature term. We discussed about some effects of those new terms on cosmological models.

We also studied other brane models, i.e. the model with a bulk dilaton field motivated by the Hořava-Witten model, the models with bulk Yang-Mills field, the induced gravity brane model by by Dvali, Gabadadze and Porrati, and the R^2 inflationary model based on trace anomaly quantum correction. Note that the action of brane gravity may be modified by quantum effect (or via the AdS/CFT correspondence).

Although our 4D approach is very efficient for some specified situations, the analysis of an entire higher-dimensional spacetime is definitely required in order to clarify the brane world scenario. We hope that those attempts will reveal new physics in the early universe.

References

1. J. Polchinski, *String Theory I & II* (Cambridge Univ. Press, Cambridge, 1998).
2. Th. Kaluza, Sitzungsber. Preuss. Akad. Wiss. Berlin, Phys. Math. Klasse, 966 (1921); O. Klein, Z. Phys. **37**, 895 (1926).
3. J. Dai, R. G. Leigh and J. Polchinski, Mod. Phys. Lett. A **4**, 2073 (1989); J. Polchinski, Phys. Rev. Lett. **75**, 4724 (1995).
4. For earlier work on this topic, see
   K. Akama, Lect. Notes Phys. **176**, 267 (1982) [hep-th/0001113]; V. A. Rubakov and M. E. Shaposhnikov, Phys. Lett. B**152**,136 (1983); M. Visser, Phys. Lett. B **159**, 22 (1985); M. Gogberashvili, Mod. Phys. Lett. A **14**, 2025(1999);hep-ph/9908347.
5. N. Arkani-Hamed, S. Dimopoulos and G. Dvali, Phys. Lett. B **429**, 263 (1998); I. Antoniadis, N. Arkani-Hamed, S. Dimopoulos and G. Dvali, Phys. Lett. B **436**, 257 (1998); N. Arkani-Hamed, S. Dimopoulos and G. Dvali, Phys. Rev. D **59**, 086004 (1999); N. Arkani-Hamed, S. Dimopoulos, N. Kaloper, J. March-Russell, Nucl. Phys. B **567**, 189 (2000).
6. L. Randall and R. Sundrum, Phys. Rev. Lett. **83**, 3370 (1999).
7. L. Randall and R. Sundrum, Phys. Rev. Lett. **83**, 4690 (1999).
8. T. Shiromizu, K. Maeda and M. Sasaki, Phys. Rev. D **62**, 024012 (2000); M. Sasaki, T. Shiromizu and K. Maeda, Phys. Rev. D **62**, 024008 (2000).
9. K. Maeda and D. Wands, Phys. Rev. D **62**, 124009 (2000).
10. K. Maeda, S. Mizuno and T. Torii, Phys. Rev. D **68**, 024033 (2003) [gr-qc/0303039]; K. Maeda, Prog. Theor. Phys. Suppl. **148**, 59 (2003).
11. P. Binétruy, C. Deffayet and D. Langlois, Nucl. Phys. B **565**, 269 (2000); N. Kaloper, Phys. Rev. D **60**, 123506 (1999) ; C. Csaki, M. Graesser, C. Kolda and J. Terning Phys. Lett. B **462**, 34 (1999) ; T. Nihei, Phys. Lett. B **465**, 81 (1999) ; P. Kanti, I. I. Kogan, K. A. Olive and M. Prospelov, Phys. Lett. B **468**, 31 (1999); J. M. Cline, C. Grojean and G. Servant, Phys. Rev. Lett. **83**, 4245 (1999); P. Binétruy, C. Deffayet, U. Ellwanger and D. Langlois, Phys. Lett. B **477**, 285 (2000) ;
    S. Mukohyama, T. Shiromizu and K. Maeda, Phys. Rev. D **62**, 024028 (2000).
12. P. Hořava and E. Witten, Nucl. Phys. B **460**, 506 (1996); ibid **475**, 94 (1996)
13. A. Lukas, B. A. Ovrut, K.S. Stelle and D. Waldram, Phys. Rev. D **59**, 086001 (1999); H.S. Reall, Phys. Rev. D **59**, 103506 (1999); A. Lukas, B. A. Ovrut and D. Waldram, Phys. Rev. D**60**, 086001 (1999).
14. N. Okuyama and K. Maeda, Phys.Rev.D **67** (2003) 104012.
15. G. Dvali, G. Gabadadze, and M. Porrati, Phys. Lett. B **485**, 208 (2000); G. Dvali and G. Gabadadze, Phys. Rev.D **63**, 065007 (2001); G. Dvali, G. Gabadadze, and M. Shifman, Phys. Rev. D **67**, 044020 (2003) [hep-th/0202174].
16. S. Nojiri, S. D. Odintsov, and S. Zerbini, Phys. Rev. D **62**, 064006 (2000); S. Nojiri, and S. D. Odintsov, Phys. Lett. B **484**, 119 (2000).
17. S.W. Hawking, T. Hertog, and H.S. Reall, Phys. Rev.D **62**, 043501 (2001);ibid. **63**, 083504 (2001); S.W. Hawking and T. Hertog, Phys. Rev. D **66**, 123509 (2002).
18. A.A. Starobinsky, Phys. Lett. B **91**, 99 (1980).
19. C. S. Chan, P. L. Paul and H. Verlinde, Nucl. Phys. B **581**, 156 (2000) [hep-th/0003236].
20. M. J. Duff, J. T. Liu and K. S. Stelle, J. Math. Phys. **42**, 3027 (2001) [hep-th/0007120].

21. M. Kubo, C. S. Lim and H. Yamashita, Mod. Phys. Lett. A **17**, 2249 (2002) [hep-ph/0111327].
22. J. M. Maldcena, Adv. Theor. Math. Phys. **2**, 231 (1998)
23. T. Tanaka, Phys. Rev. D **69**, 024001 (2004) [gr-qc/0305031].
24. S. Mizuno, K. Maeda and T. Torii, in preparation.
25. W.D. Goldberger and M.B. Wise, Phys. Rev. Lett. **83**,4922 (1999).
26. S. Mukohyama, Phys. Rev. D **62**, 084015 (2000); H. Kodama, A. Ishibashi and O. Seto, Phys. Rev. D **62**, 064022 (2000); R. Maartens, Phys. Rev. D **62**, 084023 (2001); C. Gordon and R. Maartens, Phys. Rev. D **63**, 044022 (2001); D. Langlois, Phys. Rev. D **62**, 126012 (2000); C. van de Bruck, M. Dorca, R. H. Brandenberger and A. Lukas, Phys. Rev. D **62**, 123515 (2000); K. Koyama and J. Soda, Phys. Rev. D **62**, 123502 (2000); D. Langlois, R. Maartens, M. Sasaki and D. Wands, Phys. Rev. D **63**, 084009 (2001).
27. P. Kraus, JHEP **9912**, 011 (1999). D. Ida, JHEP **0009**, 014 (2000)
28. S. Dimopoulos, G. Landsberg, Phys. Rev. Lett. **87**, 161602 (2001); A. Chamblin, G.C. Nayak, Phys. Rev. D **66**, 091901 (2002); S. B. Giddings, S. Thomas, Phys. Rev. D**65**, 056010 (2002).
29. A. Chamblin, S. W. Hawking, and H. S. Reall, Phys. Rev. D **61**, 065007 (2000); R. Emparan, G. T. Horowitz, and R. C. Myers, JHEP, **01**, 007 (2000); T. Wiseman, Class. Quant. Grav., **20**, 1177 (2003); H. Kudoh, T. Tanaka, and T. Nakamura, Phys. Rev. D**68**, 024035 (2003) [hep-th/0301089]; R. Emparan and H.S. Reall, Phys. Rev. Lett. **88**, 101101 (2002).
30. J. Khoury, B.A. Ovrut, P.J. Steinhardt, N. Turok, Phys. Rev. D**64**, 123522 (2001); J. Khoury, B.A. Ovrut, N. Seiberg, P.J. Steinhardt, N. Turok, Phys.Rev. D**65** 086007 (2002); P.J. Steinhardt, N. Turok, Phys. Rev. D **65** 126003, (2002).
31. see also R. Kallosh, L. Kofman, A. Linde, Phys. Rev. D **64**, 123523 (2001).
32. D.H. Lyth, Phys. Lett. B **524**, 1 (2002).
33. D. Langlois, K. Maeda and D. Wands, Phys. Rev. Lett. **88**, 181301 (2002).
34. J. Garriga and M. Sasaki, Phys. Rev. D **62**, 043523 (2000).
35. A. Sen, JHEP 0204 (2002) 048: A. Sen, Mod. Phys. Lett. A **17**, 1797 (2002); G. Dvali and A. Vilenkin, Phys. Rev. D **67**, 046002 (2003).
36. G. W. Gibbons and S. W. Hawking, Phys. Rev. **D15**, 2752 (1977).
37. H. A. Chamblin and H. S. Reall, Nucl. Phys. B **562**, 133 (1999).
38. W. Israel, Nuovo Cim. B **44**, 1 (1966).
39. J. Garriga and T. Tanaka, Phys. Rev. Lett. **84**, 2778 (2000).
40. K. Maeda, Phys. Rev. D **64**, 123525 (2001); S. Mizuno and K. Maeda, Phys. Rev. D**64**, 123521 (2001); S. Mizuno, K. Maeda and K. Yamamoto, Phys. Rev. D **67**, 023516 (2003).
41. R. Maartens, D. Wands, B. Bassett and I. Heard, Phys. Rev. D **62**, 041301 (2000).
42. S. Tsujikawa, K. Maeda and S. Mizuno, Phys. Rev. D **63**, 123511(2001).
43. M. Cvetic, H. Lu and C. N. Pope, Class. Quant. Grav. **17**, 4867 (2000); C. Csaki, J. Erlich, C. Grojean and T. Hollowood, Nucl. Phys. B **584** (2000), V. Barger, T. Han, T. Li, J. D. Lykken and D. Marfatia, Phys.Lett. B **488**, 97 (2000); S. S. Gubser, Adv. Theor. Math. Phys. **4**, 679 (2002).
44. R. Bartnik and J. McKinnon, Phys. Rev. Lett. **61**, 141 (1988).
45. P. Bizon, Phys. Rev. Lett. **64**, 2844 (1990).
46. M. S. Volkov, N. Straumann, G. Lavrelashvili, M. Heusler and O. Brodbeck, Phys. Rev. D **54**, 7243 (1996).

47. T. Torii, K. Maeda and T. Tachizawa, Phys. Rev. D **52**, R4272 (1995).
48. O. Brodbeck, M. Heusler, G. Lavrelashvili, N. Straumann and M. S. Volkov, Phys. Rev. D **54**, 7338 (1996).
49. J. Bjoraker and Y. Hosotani, Phys. Rev. Lett. **84**, 1853 (2000); E. Winstanley, Class. Quantum Grav. **16**, 1963 (1999).
50. C. Deffayet, Phys. Lett. B **502**, 199 (2001).

Inflation and Braneworlds

James E. Lidsey

Astronomy Unit, School of Mathematical Sciences, Queen Mary, University of London, Mile End Road, LONDON, E1 4NS, UK. J.E.Lidsey@qmul.ac.uk

Abstract. An introductory review of the Randall-Sundrum type II braneworld scenario is presented, with emphasis on the relationship between the density and gravitational wave perturbations that are generated during inflation. The implications of relaxing the reflection symmetry in the fifth dimension are considered. The effects of including a Gauss-Bonnet combination of higher-order curvature invariants in the bulk action are briefly discussed.

1 Introduction

A unified description of the origin and very early evolution of the universe that is consistent both with our understanding of unified field theory and astrophysical observations is one of the primary goals of particle cosmology. A synthesis of these two disciplines provides a unique window to high energy physics that would otherwise be inaccessible to any form of terrestrial experiment.

From the observational side, recent years have witnessed rapid advances in the quality and availability of high precision data from numerous cosmic microwave background (CMB) and high redshift surveys. This is resulting in ever more stringent constraints on models of the early universe and the trend is certain to continue in light of the anticipated data that will become available in the near future. Specifically, recent measurements from the Wilkinson Microwave Anisotropy Probe (WMAP) [1] are entirely consistent with a universe that has a total density that is very close to the critical density, implying that the curvature of the universe is very close to spatial flatness [2]. On the other hand, there is by now considerable evidence from a variety of sources – including the CMB power spectrum, galaxy clustering statistics, peculiar velocities, the baryon mass fraction in galaxy clusters and Lyman–α forest data – that the density of clumped baryon and non–baryonic matter can be no more than 30% of the critical density. Moreover, spectral and photometric data from high redshift surveys of type Ia supernovae [3] indicate that the expansion of the universe may be accelerating at the present epoch, thereby requiring the existence of some form of exotic 'dark energy' or 'quintessence' field that contributes the remaining 70% of the total energy density.

The popular explanation of this diverse set of observations is the *inflationary scenario* [4], whereby the universe underwent an epoch of very rapid,

J.E. Lidsey, Inflation and Braneworlds, Lect. Notes Phys. **646**, 357–379 (2004)
http://www.springerlink.com/

accelerated expansion sometime before the electroweak phase transition. Inflation is presently the cornerstone of modern, early universe cosmology. (For a review, see, e.g., [5]). Not only is inflation able to resolve the horizon and flatness problems of the hot, big bang model, it also provides the mechanism for generating the primordial density perturbations necessary for galaxy formation [6]. In the simplest class of inflationary models, the accelerated expansion is driven by the potential energy arising through the self–interactions of a single quantum scalar field, referred to as the *inflaton* field and denoted ϕ. If the potential is sufficiently flat and smooth, the field is able to slowly roll towards the minimum of its potential. In this case, the kinetic energy of the field is subdominant and its pressure becomes sufficiently negative for the strong energy condition of General Relativity to become violated. Inflation ends as the field reaches its ground state and the hot big bang model is recovered through a reheating process.

It is now widely believed that the observed large–scale structure in the universe evolved through the process of gravitational instability from density perturbations that were generated *quantum mechanically* during the inflationary expansion. In single field inflationary models, the perturbations are predicted to be adiabatic, nearly scale–invariant and Gaussian distributed. Moreover, inflation results in an effectively flat universe. The current CMB data, most notably from WMAP [2, 7, 8, 9], supports these predictions whilst simultaneously providing strong constraints on such models [10, 11]. In particular, an anti–correlation between the temperature and polarization E–mode maps of the CMB on degree scales has been detected by WMAP [8], thereby providing strong evidence for correlations on length scales beyond the Hubble radius [12].

Despite the success of inflationary cosmology in passing these key observational tests, there is presently no canonical theory for explaining the origin of the inflaton field. Consequently, it is imperative to establish that inflation can arise generically within the context of unified field theory. Superstring theory has emerged as the leading candidate for such a theory of the fundamental interactions, including gravity. Developments over recent years towards a non–perturbative formulation of the theory have indicated that the five, anomaly–free, supersymmetric perturbative string theories – known respectively as types I, IIA, IIB, SO(32) heterotic and $E_8 \times E_8$ heterotic – represent different limits of a more fundamental theory referred to as M–theory. (For reviews, see, e.g., [13, 14]). M–theory was originally defined as the strong coupling limit of the type IIA superstring [15]. However, since its infra–red (low–energy) limit is eleven–dimensional supergravity, it must be more than another theory of superstrings [15, 16].

Given this change of perspective, it is crucial to study the cosmological consequences of string/M–theory [17]. Supersymmetry implies that a consistent quantum string theory can only be formulated if spacetime is higher–dimensional. Given that these extra dimensions are not observed, some mechanism is required to ensure that they remain undetected. One possibility is that the dimensions are compactified through the Kaluza–Klein mechanism. In this case, tests of quantum electrodynamics limit the size of the extra dimensions to be less than 10^{-17} cm.

On the other hand, a key theoretical development has been the realization that the standard model fields (quarks, electrons, photons, etc.) may be confined to a four–dimensional domain wall or 'membrane' that is embedded in a higher–dimensional space (referred to as the bulk). This picture has developed following the discovery that the quantum dynamics of D–branes can be described by open strings whose ends are fixed on the brane [18]. In string theory, branes are static, solitonic configurations extending over a number of spatial, tangential dimensions. Thus, a 0–brane may be viewed as a pointlike particle or a black hole, a 1–brane represents a string, a 2–brane a membrane, and so forth. In this picture, our observable, four–dimensional universe is interpreted as a 3–brane. The spatial dimensions tangential to our 3–brane describe our familiar three–dimensional space of length, width and height. The only long–range interaction that propagates in the bulk dimensions is gravitational. In this case, corrections to Newton gravity necessarily arise, but the weak nature of gravity implies that any modifications can not presently be observed below scales of 1 mm. In principle, therefore, the extra dimensions may extend over scales that are many orders of magnitude larger than previously thought possible and, depending on the model, may even be infinite in extent. This paradigm shift in our understanding of the observable universe is referred to as the *braneworld scenario*.

The radical proposal, therefore, is that *our universe is a brane embedded in a higher–dimensional space*. The implications for cosmology, and for our understanding of the inflationary scenario in particular, are significant and there is presently a high level of active research in this field. Broadly speaking, the key objectives from the astrophysical and cosmological perspectives are:

- To determine the nature of cosmological solutions that are possible in braneworld scenarios, to investigate their asymptotic early– and late–time behaviours, and to uncover important differences and similarities between braneworld scenarios and conventional cosmologies based on Einstein gravity.
- To establish the conditions whereby inflation may occur, both in the arena of the early universe and at the present epoch (quintessence scenarios) and to determine whether inflation is more or less generic in this new paradigm.
- To investigate the production of scalar (density), vector (electromagnetic) and tensor (gravitational wave) perturbations during braneworld inflation.
- To develop cosmological tests of inflationary braneworld scenarios and determine whether the perturbations generated are compatible with limits imposed by the CMB power spectrum and large–scale structure observations.

Ultimately, such a programme will yield unique information on the dimensionality of the universe.

2 Types of Braneworlds

It is impossible in a talk of this nature to fully review the vast body of work in this field. The emphasis from a cosmological point of view has focused on models consisting of a single brane or of two or more parallel branes; for early papers see, e.g., [19]. These configurations are shown schematically in Fig. 1. From an historical point of view, a significant development was the interpretation by Hořava and Witten of the strongly coupled limit of the $E_8 \times E_8$ heterotic string as M–theory compactified on the eleven–dimensional orbifold $R^{10} \times S^1/Z^2$ [20]. The weakly coupled limit of this string theory then corresponds to the limit where the radius of the circle (as parametrized by the value of the dilaton field) tends to zero. The orbifold S^1/Z_2 may be viewed as the segment of the real line bounded by two fixed points on the circle, such that the orientation of the circle is reversed by the Z_2 transformation, $y \to -y$. Gravitational anomalies are cancelled by placing the two sets of E_8 gauge supermultiplets on each of the ten–dimensional orbifold fixed planes. An effective five–dimensional theory may then be derived by compactifying on an appropriate Calabi–Yau surface [21].

Cosmological solutions admitted in this theory were found and analyzed [22]. In particular, models where the branes approach and move away from each other were found and interpreted in terms of the pre–big bang scenario, an earlier string–inspired inflationary scenario driven by the kinetic energy of a scalar field [23]. Recently, the idea of interpreting the big bang in terms of brane collisions has been advocated through the 'ekpyrotic' scenario [24]. In these models, the brane dynamics is reduced to an effective four–dimensional theory, where a scalar field parametrizes the brane separation.

A further key development was the proposal that the hierarchy problem of particle physics (namely the problem of understanding why the weak scale is so much smaller than the Planck scale) could be alleviated if the volume of the extra dimensions were to be made sufficiently large [25]. In general, the four–dimensional Planck scale, m_4, is related to the $(4+n)$–dimensional Planck scale, M, through the relationship $m_4^2 = M^{n+2}V_n$, where V_n is the volume of the compact space. In the model of [25], the extra dimensions were assumed to be topologically equivalent to a n–torus and a single brane configuration was considered. Pursuing such ideas further, Randall and Sundrum considered two parallel branes, with equal and opposite tension, embedded in five–dimensional Anti–de Sitter (AdS) space, with a Z_2 reflection symmetry imposed in the fifth dimension [26]. In this model, the weak scale is generated from a larger (Planck) scale through an exponential hierarchy arising directly from the five–dimensional AdS geometry.

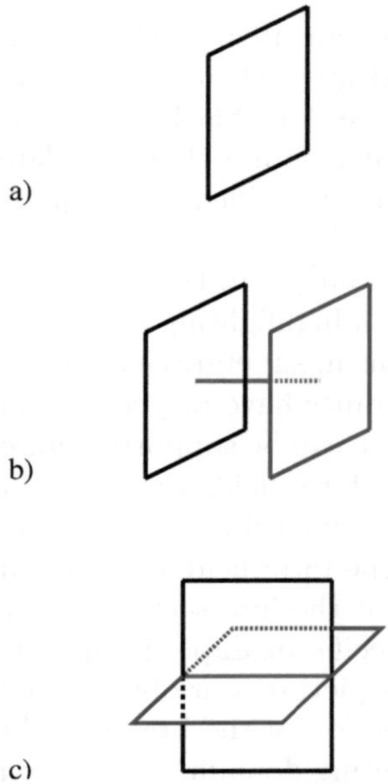

Fig. 1. Illustrating possible braneworld configurations: (a) a single brane embedded in a higher–dimensional space; (b) two parallel branes with equal and opposite tension; (c) intersecting branes.

Perhaps of more interest from a cosmological viewpoint is the second Randall–Sundrum 'type II' (RSII) scenario, consisting of a single brane embedded in five–dimensional AdS space [27]. Formally, this model may be interpreted in terms of the first Randall–Sundrum model, where the negative tension brane is taken to infinity. The bulk space may also contain a black hole. In this case it corresponds to the five–dimensional Schwarzschild–AdS solution. The RSII model is interesting because it is simple enough for analytical results to be derived, yet is sufficiently rich for new physics to be uncovered. Indeed, one of the key areas of interest in theoretical physics at present is focused towards the 'holographic principle' [28]. In short, the holographic principle implies that the number of degrees of freedom associated with gravitational dynamics is determined by the boundary of the spacetime rather than by its volume. This follows naturally from the idea that the state of maximal entropy for a given volume, V, is determined by the size of the largest black hole that can be contained within V [28]. The AdS/Conformal Field Theory (CFT) correspondence provides a realization of this principle within the context of string theory by establishing a duality between semi–classical d–dimensional gravity in AdS space and a quantum CFT located on its boundary [29]. It has recently become apparent that the AdS/CFT correspondence is closely related to braneworld cosmology and the RSII scenario in particular [30]. More specifically, one may view the RSII braneworld as being dual to a CFT (with an ultra–violet cut off) that is coupled to gravity on the brane. Remarkably, when one identifies the entropy, mass and Hawking temperature of the AdS black hole with the entropy, energy and temperature of the CFT, it is found that the Cardy entropy formula of the CFT coincides precisely with the Friedmann equation when the brane passes the black hole event horizon [31].

We focus on the RSII scenario in the remainder of this talk. Before proceeding, however, it is worth highlighting an alternative class of 'intersecting' braneworlds. Configurations in supergravity theories that describe the intersection of two or more p–branes have played a prominent role in advances in string theory [32]. A p–brane can be supported, for example, when the components of the antisymmetric form fields that arise in the string and M–theory effective actions have non–trivial flux over the compactifying manifold. Depending on the degree of the form field and the nature of the internal space, such a model may represent the intersection of two higher–dimensional branes (See Fig. 1c). One specific example is the intersection of two 5–branes over a 3–brane [32]. In this picture, our observable four–dimensional universe corresponds to the intersection of the branes. Solutions representing curved (time–dependent), intersecting domain walls were recently found [33].

3 The Randall–Sundrum Type II Braneworld

In the Randall–Sundrum type II scenario, a co–dimension one brane is embedded in five–dimensional AdS space with a Z_2 reflection symmetry imposed on the bulk. The action is given by

$$S = \int_{\mathcal{M}} d^5x\sqrt{G}\left[2M^3\hat{R} - \Lambda\right] + \int_{\partial\mathcal{M}} d^4x\sqrt{h}\,(\lambda + L_{\rm matter})\,, \tag{1}$$

where $\hat{R}$ is the Ricci curvature scalar of the bulk spacetime, $\mathcal{M}$, with metric G_{AB}, Λ is the five–dimensional (negative) cosmological constant, $G \equiv \det G_{AB}$, h is the determinant of the metric induced on the boundary of $\mathcal{M}$, λ is the tension of the brane, $L_{\rm matter}$ is the Lagrangian density of the matter on the brane and M is the five–dimensional Planck mass.

The bulk solution has a metric of the form [27]

$$ds^2 = e^{-2k|y|}\left(-dt^2 + dx_3^2\right) + dy^2 \tag{2}$$

for a given constant, k. This geometry is non–factorizable due to the presence of the exponential warp factor, in contrast to the standard Kaluza–Klein compactification schemes based on the periodic boundary conditions. Consequently, the fifth dimension (y) may extend to infinity. Imposing four–dimensional Poincaré invariance on the brane world–volume such that the metric corresponds to flat Minkowski spacetime requires the bulk cosmological constant to be fine–tuned with the brane tension [27]:

$$\Lambda = -24M^3k^2, \qquad \lambda = 24M^3k. \tag{3}$$

The remarkable feature of this model is that the graviton equation of motion admits a zero energy ground state solution that is localized around the domain wall. This ground state is naturally interpreted (by a four–dimensional observer) as the four–dimensional, massless spin–2 graviton. A continuum of massive states also arises in the spectrum and these lead to corrections to the form of the Newton potential. However, these corrections fall as the cube of the distance, r [27]:

$$V(r) \approx \frac{G_N m_1 m_2}{r}\left(1 + \frac{1}{k^2r^2}\right) \tag{4}$$

and, if the warping of the bulk geometry is sufficiently strong (i.e. the constant k is sufficiently large), these massive states are suppressed near the brane and are therefore harmless. This indicates that *the curvature of the five–dimensional world effectively determines the four–dimensional physics.* The cosmology of the RSII scenario arises due to the motion of the brane through the bulk space. An observer confined to the surface of the brane interprets such motion in terms of cosmic expansion or contraction [34, 35, 36]. The 'Friedmann' equation describing the cosmic dynamics may be derived within the context of the thin wall formalism of (five–dimensional) General Relativity. Since we are interested primarily in late–time inflationary dynamics, we focus on the simplest case where the world–volume of the brane corresponds to the spatially flat, Friedmann–Robertson–Walker (FRW) metric and consider a pure AdS bulk. The effect of a bulk black hole on the four–dimensional brane dynamics is formally equivalent to that of a relativistic perfect fluid contribution to the energy–momentum tensor and so is rapidly redshifted away by the accelerated motion of the brane.

It is convenient to work with the five–dimensional metric expressed in static coordinates:

$$ds^2 = G_{AB}dx^A dx^B = -\frac{r^2}{L^2}dt^2 + \frac{L^2}{r^2}dr^2 + r^2 dE_3^2, \tag{5}$$

where the constant L is related to the bulk cosmological constant. The induced metric on the wall then has the desired form:

$$\begin{aligned} h_{AB} &= G_{AB} + n_A n_B \\ ds_4^2 &= -d\tau^2 + a^2(\tau)dE_3^2, \end{aligned} \tag{6}$$

where n^A is the unit normal vector to the brane. Cosmic time as measured on the brane is parametrized by τ, defined such that

$$d\tau^2 = \frac{r^2}{L^2}dt^2 - \frac{L^2}{r^2}dr^2 \tag{7}$$

and the radial coordinate of the brane in the bulk space determines the scale factor, $a(\tau)$:

$$r = r\left[a(\tau)\right] . \tag{8}$$

The effective Friedmann equation is then derived directly from the Israel junction conditions [37]:

$$K_{AB} = -4\pi G_5 \left(T_{AB} - \frac{1}{3} T h_{AB} \right) , \tag{9}$$

where G_5 is the five–dimensional Newton constant. These conditions relate the energy–momentum tensor, T_{AB}, of the matter confined on the brane directly to the brane's extrinsic curvature, $K_{AB} \equiv h^C{}_{(A} h_{B)}{}^D \nabla_C n_D$. Note that we have taken into account the Z_2 symmetry in this expression and that $T \equiv T^A_A$ is the trace of the energy–momentum.

Conservation of energy–momentum on the brane then follows as a direct consequence of the Codazzi equation:

$$\nabla_B K^B_A - \nabla_A K = \hat{R}_{BC} G^B_A n^C . \tag{10}$$

It is straightforward to verify that for the case of a pure AdS bulk geometry, the right–hand side of (10) is identically zero. Thus, substitution of the Israel junction conditions (9) into (10) implies conservation of energy–momentum on the brane:

$${}^{(4)}\nabla_\mu T^{\mu\nu} = 0. \tag{11}$$

We will further assume that the energy–momentum tensor of the matter on the brane is given by the perfect fluid form

$$T^A_B|_{\rm brane} = \delta(y) {\rm diag}(-\rho, p, p, p, 0), \tag{12}$$

where ρ and p represent the energy density and pressure, respectively. Hence, we recover the standard expression of conventional cosmology:

$$\dot{\rho} + 3\frac{\dot{a}}{a}(\rho + p) = 0, \tag{13}$$

where a dot denotes $d/d\tau$.

The spatial components of the Israel junction conditions (9), as given by

$$K_{ij} = -\sqrt{\frac{1}{L^2} + \frac{\dot{a}^2}{a^2}} \delta_{ij}, \tag{14}$$

are now sufficient to derive the effective Friedmann equation. (The time–time components of (9) provide no new information for the model we are considering). We therefore deduce that [38, 39, 40, 34, 36, 41]

$$\frac{\dot{a}^2}{a^2} = \left(\frac{4\pi G_5 \rho}{3} \right)^2 - \frac{1}{L^2}. \tag{15}$$

Although the quadratic dependence of the Friedmann equation (15) may appear to be inconsistent with the Hubble expansion, and particularly with constraints form primordial nucleosynthesis, we must recall that the vacuum brane has a tension, λ. This implies that the total matter content on the brane can effectively be separated into two components, the dynamical matter, ρ_B, and the tension. Substituting into (15) then implies that

$$H^2 = \frac{8\pi G_4}{3}\rho_B\left(1+\frac{\rho_B}{2\lambda}\right) + \left(\frac{4\pi G_5 \lambda}{3}\right)^2 - \frac{1}{L^2}, \tag{16}$$

where $G_4 \equiv 4\pi\lambda G_5^2/3$. The constant terms are then cancelled by imposing the fine–tuning condition (3), resulting in a Friedmann equation of the form [38, 39, 40, 34, 36, 41]

$$H^2 = \frac{8\pi}{3m_4^2}\rho\left[1+\frac{\rho}{2\lambda}\right], \tag{17}$$

where the subscript 'B' is dropped for notational simplicity and we define the four–dimensional Planck mass, $m_4 \equiv G_4^{-1/2}$. The standard form of the Friedmann equation is recovered at low energy scales, $\rho \ll \lambda$, whereas the dependence on the energy density is modified to a quadratic form at high energies, $\rho \gg \lambda$.

We now consider the implications of this term for inflationary cosmology.

4 Braneworld Inflation

4.1 Scalar Field Dynamics

Equations (13) and (17) are sufficient to fully determine the cosmic dynamics on the brane once an equation of state has been specified for the matter sources. In what follows, we assume that the brane matter consists of a single scalar field that is confined to the brane and is self–interacting through a potential, $V(\phi)$. The conservation equation (13) then implies that

$$\ddot{\phi} + 3H\dot{\phi} + V' = 0, \tag{18}$$

where a prime denotes differentiation with respect to the scalar field. We further assume the slow–roll approximation, $\dot{\phi}^2 \ll V$ and $|\ddot{\phi}| \ll H|\dot{\phi}|$. Equation (18) then simplifies to $3H\dot{\phi} \approx -V'$.

The slow–roll parameters, $\epsilon \equiv -\dot{H}/H^2$ and $\eta \equiv V''/(3H^2)$, may then be written in the form [42]

$$\epsilon \simeq \frac{m_4^2}{4\pi}\left(\frac{V'}{V}\right)^2\left[\frac{1+V/\lambda}{(2+V/\lambda)^2}\right] \tag{19}$$

$$\eta \simeq \frac{m_4^2}{8\pi}\left(\frac{V''}{V}\right)\left[\frac{2\lambda}{2\lambda+V}\right] \tag{20}$$

and inflation occurs for $\epsilon < 1$. Self–consistency of the slow–roll approximation requires that $\max\{\epsilon, |\eta|\} \ll 1$. The number of e–foldings of inflationary expansion that occur when the scalar field rolls from some value, ϕ, to the value, ϕ_e, corresponding to the end of inflation is given by

$$N \equiv \ln a = \int_t^{t_e} dt\, H \approx -\frac{8\pi}{m_4^2} \int_\phi^{\phi_e} \frac{V}{V'} \left(1 + \frac{V}{2\lambda}\right) d\phi. \tag{21}$$

The effect of the brane corrections is to enhance the value of the Hubble parameter relative to what it would be for a given pure Einstein gravity model of the same energy density [42]. This introduces additional friction on the scalar field and further resists its motion down the potential, thereby enabling a steeper class of potentials to support inflation. This is the basis behind the steep inflationary scenario [43]. The quadratic correction relaxes the condition for slow–roll inflation in the RSII scenario relative to the corresponding condition for the standard model. Generically, steep inflation proceeds in the region of parameter space where $\rho \gg \lambda$ and naturally comes to an end when $\rho \approx \lambda$, since the conventional cosmological dynamics is recovered in this regime.

4.2 Density Perturbations

We now consider the generation of scalar and tensor perturbations in RSII inflation. Since many of the issues of perturbation theory in conventional inflationary models have already been covered in lectures at this school, we omit detailed discussions here and focus instead on the differences that arise between the two scenarios. We employ the normalization conventions of [44].

We begin by recalling that the scalar perturbations generated during inflation that is driven by a single, self–interacting scalar field are *adiabatic*. The curvature perturbation on uniform density hypersurfaces is then given by $\zeta = H\delta\phi/\dot{\phi}$ and is determined by the scalar field fluctuation, $\delta\phi$, on spatially flat hypersurfaces [6]. Conservation of energy–momentum implies that ζ is conserved on large scales, a result that is independent of the specific form of the gravitational physics [45]. This implies that the amplitude of a mode when it re–enters the Hubble radius after inflation is related to the curvature perturbation by $A_S^2 = 4\langle\zeta^2\rangle/25$, where the right–hand side is evaluated when the mode with comoving wavenumber, k, goes beyond the Hubble radius during inflation, i.e., when

$$k(\phi) = a_e H(\phi) \exp[-N(\phi)], \tag{22}$$

where a subscript 'e' denotes values at the end of inflation and $N = \int dt H(t)$ corresponds to the number of e–foldings of inflationary expansion that elapse between the time when the scale crosses the Hubble radius and the end of inflation [cf. (21)]. Finally, the Gibbons–Hawking temperature of de Sitter space determines the magnitude of the field fluctuation, $\langle\delta\phi^2\rangle = H^2/(4\pi^2)$, and we therefore deduce that the scalar perturbation amplitude has the form

$$A_S^2 = \frac{1}{25\pi^2} \frac{H^4}{\dot{\phi}^2}\bigg|_{k=aH} \tag{23}$$

This is given in terms of the potential by [42]

$$A_S^2 = \frac{512\pi}{75m_4^6} \frac{V^3}{V'^2} \left(1 + \frac{V}{2\lambda}\right)^3 \tag{24}$$

after substitution of the Friedmann equation (17). We see that the amplitude is enhanced over that of the standard scenario by the bracketed term.

4.3 Gravitational Waves

Although to first–order the gravitational waves decouple from the matter, the calculation of the tensor perturbation spectrum is more involved in braneworld cosmology because the perturbations extend into the bulk. In this subsection we review the method of Langlois, Maartens and Wands [46]. To proceed analytically, it is necessary to assume pure de Sitter expansion on the brane and this is a good approximation if the inflation field is slowly rolling. It proves convenient to express the perturbed, five–dimensional metric in the form

$$ds_5^2 = \mathcal{A}^2[-dt^2 + a^2(\delta_{ij} + E_{ij})dx^i dx^j] + dy^2, \tag{25}$$

where E_{ij} represents the perturbations. The warp factor is given by

$$\mathcal{A} = (H/\alpha)\sinh[\alpha(y_h - |y|)], \tag{26}$$

where the Cauchy horizons, $g_{00}(\pm y_h) = 0$, are located at $y = \pm y_h$, and the constant $\alpha = \kappa_4/\kappa_5 = (-\Lambda/6)^{1/2}$ is determined by the bulk cosmological constant, Λ. Here and throughout, $\kappa_4^2 \equiv 8\pi m_4^{-2}$ and $\kappa_5^2 \equiv 8\pi M^{-3}$.

The standard approach is to expand the metric perturbations as a Fourier series. In this case, and assuming that any anisotropic stresses are negligible, the linearly perturbed junction conditions (9) reduce to

$$\frac{dE}{dy}\bigg|_{y=0} = 0, \tag{27}$$

where $E(t, y; \boldsymbol{k})$ denotes the amplitude of the modes. Assuming a pure de Sitter expansion of the brane world–volume allows us to separate the corresponding gravitational wave equation of motion and then expand the amplitude into eigenmodes such that $E(t, y; \boldsymbol{k}) = \int dm \varphi_m(t; \boldsymbol{k}) \mathcal{E}_m(y)$, where $\varphi_m(t; \boldsymbol{k})$ and $\mathcal{E}_m(y)$ depend on the world–volume and bulk coordinates, respectively, and m represents the separation constant. It can then be shown that the solution for the zero mode ($m = 0$) is determined in full generality up to a quadrature [46]:

$$\mathcal{E}_0 = C_1 + C_2 \int^y dy' \frac{1}{\mathcal{A}^4(y')}, \tag{28}$$

where $C_{1,2}$ are constants. In general, if a given mode diverges at the Cauchy horizon, it can not form part of the spectrum of orthonormal modes that constitute the basis of the Hilbert space for the quantum field. (Heuristically, this is because such a mode would produce an infinite contribution to the action and so it would cost too much energy to excite it). However, we must specify $C_2 = 0$ to satisfy the boundary condition (27) and this removes the divergent part of the zero–mode. Thus, the physically relevant solution for the zero–mode is $\mathcal{E}_0 = C_1$. The non–zero modes are not excited – modes where $m < 3H/2$ remain divergent at the Cauchy horizon even when (27) is satisfied and modes satisfying $m > 3H/2$ remain in the vacuum state during inflation [46].

The zero–mode, φ_0, remains constant on super–Hubble radius scales, as in the four–dimensional scenario. The amplitude of the quantum fluctuation in this mode is then calculated by deriving an effective, five–dimensional action for the tensor perturbations and integrating over the fifth dimension. This results in a four–dimensional action that corresponds formally to a massless scalar field propagating in a FRW universe. The standard four–dimensional analysis may then be employed to determine the amplitude if the action is normalized appropriately when integrating over the fifth dimension. This requires that

$$2\int_0^{y_h} dy C_1^2 \mathcal{A}^2 = 1 \tag{29}$$

and implies that $C_1 = \sqrt{\alpha} F(x)$, where

$$\frac{1}{F^2} = \sqrt{1+x^2} - x^2 \sinh^{-1}\left(\frac{1}{x}\right) \tag{30}$$

and $x \equiv H/\alpha$. Finally, the tensorial amplitude follows once each polarization state is interpreted as a quantum field propagating in a time–dependent potential [46]:

$$A_T^2 = \left.\frac{\kappa_4^2}{50\pi^2} H^2 F^2\right|_{k=aH} . \tag{31}$$

The effects of the brane modifications are parametrized in terms of the 'correction' function F. In the low–energy limit ($\rho \ll \lambda, x \ll 1$), $F \approx 1$, whereas $F^2 \approx [27H^2 m_4^2/(16\pi\lambda)]^{1/2}$ in the high–energy limit.

4.4 The Consistency Equation

Since the scalar field slowly rolls down its potential, the amplitudes of the perturbations are not precisely scale–invariant. These variations are parametrized in terms of the spectral indices, or *tilts*, of the spectra and are defined by

$$n_S \equiv 1 + d\ln A_S^2/d\ln k, \qquad n_T \equiv d\ln A_T^2/d\ln k \tag{32}$$

for the scalar and tensor perturbations, respectively. The scalar spectral index may be expressed in terms of the slow–roll parameters:

$$n_S - 1 = -6\epsilon + 2\eta. \tag{33}$$

In the high energy limit ($\rho \gg \lambda, x \gg 1$), the tilts are given in terms of the potential and its first two derivatives by

$$n_S - 1 \approx -\frac{m_4^2 \lambda}{2\pi V}\left[3\frac{V'^2}{V^2} - \frac{V''}{V}\right] \tag{34}$$

$$n_T \approx -\frac{3m_4^2}{4\pi}\frac{\lambda V'^2}{V^3}. \tag{35}$$

It is well known that since the scalar and tensor perturbations share a common origin through the inflaton potential, $V(\phi)$, it is possible to relate them in a way that is independent of the functional form of the potential. (For a review, see, e.g., [44]). This relationship is known as the *consistency equation* and, to lowest–order in the slow–roll approximation, determines the relative amplitudes of the tensor and scalar perturbations directly in terms of the tilt of the gravitational wave spectrum:

$$\frac{A_T^2}{A_S^2} = -\frac{1}{2}n_T. \tag{36}$$

Since it is independent of the potential, (36) represents a powerful test of single–field inflationary models and, in principle, failure to satisfy such a constraint could be employed to rule out such a class of models. At present, the contribution of tensor perturbations to the large–angle CMB power spectrum is constrained to be no more than 30 % and, in practice, it will be very difficult, if not impossible, to measure the tilt of the tensor spectrum to a sufficient level of accuracy. Nevertheless, a cosmological background of gravitational waves could be detected through their contribution to the B–mode (curl) of the CMB polarization [47] and interest is growing in this possibility in light of the recent detections of polarization in the CMB [48, 8]. In any case, even in the event that such a detection is not made, the consistency relation remains important because it removes a free parameter (usually chosen to be n_T) when determining the best–fit models to the data.

Given the importance of the consistency equation, it is clearly of interest to determine the form of the corresponding relations in braneworld cosmologies [49, 50]. We have seen that in the case of the RSII scenario, the amplitudes of the perturbations are modified by the brane effects and these modifications become progressively more important at higher energy scales. Consequently, it is to be anticipated that the form of the consistency equation should reflect these differences. If so, this would provide a potentially observable test of RSII inflation.

In the standard scenario, the consistency equation (36) is derived by first differentiating the tensorial spectrum with respect to comoving wavenumber,

k, and then relating a given scale to the corresponding value of the inflaton field through (22). Any dependence on the first derivative of the inflaton potential may then be eliminated by substituting for the scalar perturbation amplitude and any remaining dependence on the inflaton potential itself may be removed by substituting for the tensor perturbation amplitude.

In principle, an identical approach could be followed to derive the consistency equation in RSII inflation. However, given the complicated form of the amplitudes, this is algebraically very difficult (but not impossible) to accomplish. In view of this, we adopt a more elegant approach and proceed by defining a pair of new variables [50]

$$b \equiv \frac{1}{2} \sinh^{-1} x \tag{37}$$

$$\beta \equiv \kappa_4 \frac{d\phi}{dN}, \tag{38}$$

where x is defined after (30). This implies that the Friedmann equation (17) and scalar field equation (18) reduce to a first–order, non–linear system of differential equations:

$$\dot{b} = -\left(\frac{3\kappa_4^2}{8\lambda}\right)^{1/2} \dot{\phi}^2 \tag{39}$$

$$\beta = -\left(\frac{8\lambda}{3}\right)^{1/2} \frac{b'}{H}. \tag{40}$$

Moreover, the correction function (30) arising in the gravitational wave amplitude (31) depends only on the single variable, b, and may be expressed in terms of a single quadrature:

$$\frac{1}{F^2} = -4 \sinh^2 2b \int \frac{db}{\sinh^3 2b}, \tag{41}$$

whereas the scalar perturbations (23) depend on β:

$$A_S^2 = \frac{\kappa_4^2}{25\pi^2} \frac{H^2}{\beta^2}. \tag{42}$$

We are now in a position to derive the form of the consistency equation in this scenario. By employing the definitions (22), (37) and (38) and substituting (31), (40) and (42) into (41), we find that

$$\frac{1}{A_T^2} = 2 \int \frac{d \ln k}{A_S^2}. \tag{43}$$

Thus, differentiation with respect to comoving wavenumber recovers the consistency equation [49, 50]

$$\frac{A_T^2}{A_S^2} = -\frac{1}{2} n_T. \tag{44}$$

Remarkably, the form of the consistency equation is *identical* to that of standard, single–field inflation. This is particularly surprising given that the gravitational physics is manifestly different in the two scenarios. Formally, this degeneracy between the consistency equations arises because the combination of observable parameters in (44) is independent of the brane tension, but it is not immediately transparent from (24), (30) and (31) why this should be so.

5 Asymmetric Braneworld Inflation

We now proceed in this section to consider an extension of the RSII scenario where the Z_2 reflection symmetry in the bulk dimension is no longer imposed. This implies that the brane may be embedded in five–dimensional AdS space where the value of the cosmological constant differs on either side of the brane (Fig. 2). Similar analyses to those summarized in Sects. 3 and 4 may be followed to derive the Friedmann equation and the inflationary perturbation spectra. Here, we omit many of the details and simply highlight the main results.

The spatial components of the junction conditions reduce to [34, 51, 52]

$$\left(\alpha_+ + H^2\right)^{1/2} + \left(\alpha_- + H^2\right)^{1/2} = \frac{\kappa_5^2 \rho}{3}, \tag{45}$$

where

$$\alpha_\pm \equiv -\kappa_5^2 \Lambda_\pm / 6 \tag{46}$$

and $\Lambda_\pm$ are the bulk cosmological constants either side of the brane. Dimensional reduction relates the four– and five–dimensional Newton constants:

$$\frac{\kappa_5^2}{\kappa_4^2} = \frac{1}{2}\left(\frac{1}{\sqrt{\alpha_+}} + \frac{1}{\sqrt{\alpha_-}}\right). \tag{47}$$

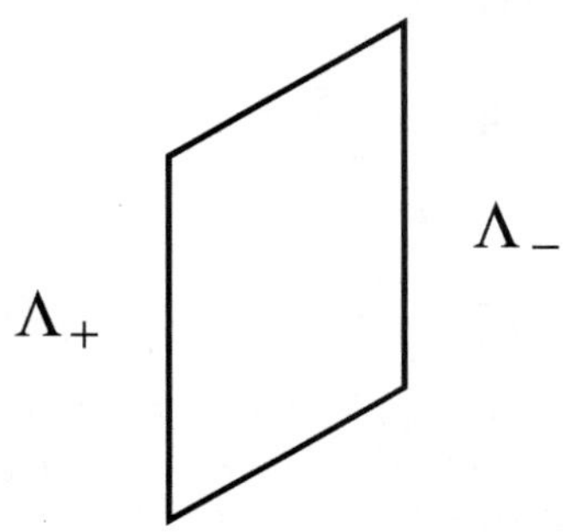

Fig. 2. In the asymmetric RSII braneworld scenario, there is no reflection symmetry imposed on the bulk dimension and the cosmological constant may take different values on either side of the brane.

Solving (45) yields the Friedmann equation [34, 51, 52]:

$$H^2 = \frac{\kappa_5^4 \rho^2}{36} - \frac{1}{2}(\alpha_- + \alpha_+) + \frac{9}{4\kappa_5^4 \rho^2}(\alpha_- - \alpha_+)^2 \tag{48}$$

and, as in the symmetric scenario, a brane tension may be introduced in order to recover the required linear dependence on the energy density at low energy scales. Relaxing the Z_2 symmetry results in the appearance of the third term on the right–hand side of (48). Note how this term is proportional to ρ^{-2} and becomes negligible at high energies. (However, this does not imply that such a term diverges at very low energies, since the density as shown here represents the total energy density on the brane. This consists of both the matter contributions as well as the brane tension, λ). It is worth remarking that the Friedmann equation (48) exhibits an infra–red/ultra–violet duality, in the sense that it is invariant under the transformation of the energy density, $\kappa_5^2 \rho \leftrightarrow 9|\alpha_+ - \alpha_-|/(\kappa_5^2 \rho)$.

Since the bulk space on either side of the brane is AdS, the Gauss–Codazzi equation (10) once more implies that energy–momentum is conserved on the brane and (13) therefore remains valid. This is important because it implies that the argument of Sect. 4.2 may be employed once more to determine the amplitude of the scalar perturbations. Consequently, the amplitude is given by (23), although its specific dependence on the inflaton potential is altered from that of the symmetric scenario due to the additional term arising in the Friedmann equation (48). Indeed, substituting (48) and the scalar field equation (18) into (23) implies that [50]

$$A_S^2 = \frac{9}{25\pi^2} \frac{1}{V'^2} \left[\frac{\kappa_5^4 (V+\lambda)^2}{36} - \frac{1}{2}(\alpha_- + \alpha_+) \right. \\ \left. + \frac{9}{4\kappa_5^4 (V+\lambda)^2}(\alpha_- - \alpha_+)^2 \right]^3 \Bigg|_{k=aH} \tag{49}$$

The tensor spectrum for this model has been calculated in [50] by extending the method of [46]. The result is

$$A_T^2 = \frac{\kappa_5^2}{50\pi^2} H^2 J^2 \Bigg|_{k=aH} \tag{50}$$

where

$$\frac{1}{J^2} = \frac{1}{2\sqrt{\alpha_-} F^2(x_-)} + \frac{1}{2\sqrt{\alpha_+} F^2(x_+)}, \tag{51}$$

the functional form of $F = F(x_\pm)$ is given by (30) and $x_\pm \equiv H/\sqrt{\alpha_\pm}$. At low energy scales ($x_\pm \ll 1$), we find that $J \to \kappa_4/\kappa_5$, implying that the standard expression is recovered in this limit, as expected.

The consistency equation can be derived in this model [50]. Given the results of Sect. 4, the simplest approach is to assume *a priori* that the consistency equation has the same form as that of (36) and to then verify that this is indeed the case. Let us therefore substitute the scalar and tensor perturbations, (23) and (50), directly into (36). We find that

$$\frac{d\ln(HJ)}{dH}\frac{dH}{dN} = -\frac{\kappa_5^2}{18}\frac{J^2V'^2}{H^4}, \tag{52}$$

where we have employed the slow–roll approximation in the form $d\ln k \approx d\ln a = dN$. Noting that the scalar field equation, $3H\dot{\phi} = -V'$, may be expressed in the somewhat unconventional form

$$\frac{dH}{dN} = -\frac{dH}{dV}\frac{V'^2}{3H^2} \tag{53}$$

then allows us to simplify (52) through substitution of (53) [50]:

$$H^4\frac{dH}{dV}\frac{d(HJ)^{-2}}{dH} = -\frac{\kappa_5^2}{3}. \tag{54}$$

Equation (54) represents a *necessary* condition for the consistency equation (36) to hold. It should be emphasized that this condition applies to any (single field) braneworld scenario, where energy–momentum is conserved on the brane. In general, a given braneworld model may be characterized by the functional form of its Friedmann equation, i.e., by the dependence of the Hubble parameter on the inflaton potential, $H = H(V)$. Once this relation has been established, (54) may then be interpreted as a constraint that must be satisfied by the correction function, $J = J(H)$, if the consistency equation is to remain degenerate. To illustrate this, consider the standard scenario, where $H \propto V^{1/2}$. We see immediately that the left–hand side of (54) is constant when $J = 1$.

It is possible, after some lengthy algebra, to confirm that (54) is indeed satisfied for the asymmetric RSII scenario when the correction to the tensor spectrum takes the form given by (51) and we therefore conclude that even in this generalized, asymmetric RSII model, the consistency equation remains degenerate [50]:

$$\frac{A_T^2}{A_S^2} = -\frac{1}{2}n_T. \tag{55}$$

Before concluding this section, we briefly consider the braneworld model of [53], where the brane is embedded in a five–dimensional bulk space with a stabilized radius. This model differs from the RSII scenario in that the Friedmann equation, and therefore the scalar perturbation spectrum, remain unmodified. However, the presence of the fifth dimension becomes apparent through a correction to the tensor spectrum. Thus, one would certainly expect the consistency equation to be modified in this model. However, surprisingly, this is not the case – the correction to the gravitational wave amplitude is given by $J^2 = (1 - \alpha H^2)^{-2}$, where α is a constant with a numerical value determined by the radius of the extra dimension [53]. One may readily deduce that a correction of this form satisfies (54) and, moreover, represents the general solution to (54) when the Friedmann equation has the standard form. Thus, we now know of four classes of inflationary cosmologies, modelled on different gravitational physics, where the consistency equation remains robust.

Such a degeneracy implies that the task of identifying the correct inflationary model through observations will be more difficult. On the other hand, although the consistency equation may be interpreted as a prediction of single field inflation, it should be emphasized that it is a prediction relating the *primordial* perturbations. In particular, in the above analyses we have neglected the influence of the bulk space on the subsequent evolution of the perturbations. This is equivalent to assuming that the projection of the five–dimensional Weyl tensor vanishes to linear order. More generally, however, the backreaction of the bulk will perturb the bulk space away from conformal invariance and generate a non–trivial Weyl tensor in five dimensions. This results in a non–local energy–momentum source in the gravitational field equations when projected down to four dimensions [54]. As a result, the background dynamics is altered. The subsequent evolution of the perturbations is difficult to determine in general, because the system of equations is not closed, although it is expected that it will be model–dependent to some extent.

A further assumption that we have made is that the field is rolling sufficiently slowly down its potential. This assumption can be relaxed in the standard scenario by working to the 'next–to–leading' order in the slow–roll approximation. In this regime, it has been shown that the consistency equation (36) receives modifications [55]:

$$n_T = -2\frac{A_T^2}{A_S^2}\left[1 - \frac{A_T^2}{A_S^2} + (1 - n_S)\right] . \tag{56}$$

The question that naturally arises, therefore, is whether the corresponding consistency equations in the RSII scenarios receive similar corrections or whether the degeneracy can be lifted by moving away from the slow–roll approximation. The answer to this and related questions must be left for future work.

6 Gauss–Bonnet Braneworld Cosmology

As well as developing the framework for testing braneworld inflation through a confrontation with observations, another important task is to enhance the connection of the scenario with string/M–theory. One approach towards this goal is to include combinations of higher–order curvature invariants in the bulk action [56, 57, 58, 59, 60, 61, 62, 63]. Within the context of the AdS/CFT correspondence, such terms arise as next–to–leading order corrections in the $1/N$ expansion of the CFT [64]. The Gauss–Bonnet combination,

$\hat{R}^2 - 4\hat{R}_{ab}\hat{R}^{ab} + \hat{R}_{abcd}\hat{R}^{abcd}$, is of particular relevance, given that it is the unique combination in five dimensions that results in second–order field equations in the metric and it also appears as a leading–order quantum correction in the heterotic string theory action [65, 66, 67].

The extension of the RSII scenario to include this term is presently attracting attention. The five–dimensional field equations admit Schwarzschild–AdS space as a solution [66, 68] and the Friedmann equation may be derived through a variety of methods. These include generalizing Birkhoff's theorem [59], varying the boundary terms in the action [60], or by employing the formalism of differential forms [61]. When the bulk space is Z_2 symmetric, the Friedmann equation takes the form [59, 60, 61]

$$H^2 = \frac{c_+ + c_- - 2}{8\alpha} , \tag{57}$$

where

$$c_{\pm} = \left\{ \left[\left(1 + \frac{4}{3}\alpha\Lambda \right)^{3/2} + \frac{\alpha}{2}\kappa_5^4\rho^2 \right]^{1/2} \pm \sqrt{\frac{\alpha}{2}}\kappa_5^2\rho \right\}^{2/3} , \tag{58}$$

the bulk cosmological constant is Λ and $\alpha > 0$ represents the Gauss–Bonnet coupling constant. As in the models discussed above, conservation of energy–momentum on the brane follows directly from the Gauss–Codazzi equations.

Despite the rather complicated form of (57), it is possible to make progress analytically by introducing a new variable, r [63]:

$$\rho \equiv \left(\frac{2b}{\alpha\kappa_5^4} \right)^{1/2} \sinh r \tag{59}$$

and defining the constant

$$b \equiv \left(1 + \frac{4}{3}\alpha\Lambda \right)^{3/2} . \tag{60}$$

Substituting (59) and (60) into (58) then implies that

$$c_{\pm} = b^{1/3} \exp(\pm 2r/3) \tag{61}$$

and, as a result, the Friedmann equation (57) simplifies considerably [63]:

$$H^2 = \frac{1}{4\alpha} \left[b^{1/3} \cosh\left(\frac{2r}{3} \right) - 1 \right] . \tag{62}$$

It may be verified that the Friedmann equation (62) exhibits a quadratic dependence on the total energy density in the low energy limit corresponding to the RSII model. At sufficiently high energies, however, the dependence scales as $H^2 \propto \rho^{2/3}$. The condition for inflation to proceed in this regime is simply that the pressure of the matter be negative, $p < 0$. More precisely, slow–roll parameters may be introduced and, in the slow–roll limit where the inflaton potential dominates the brane tension, λ, they are given by [63]

$$\epsilon = \left(\frac{2\lambda}{\kappa_4^2}\frac{V'^2}{V^3}\right)\left[\frac{2b^{2/3}}{27}\frac{\sinh(2r/3)\tanh r\sinh^2 r}{\left[b^{1/3}\cosh(2r/3)-1\right]^2}\right], \tag{63}$$

$$\eta = \left(\frac{2\lambda}{\kappa_4^2}\frac{V''}{V^2}\right)\left[\frac{2b^{1/3}}{9}\frac{\sinh^2 r}{b^{1/3}\cosh(2r/3)-1}\right]. \tag{64}$$

The terms in the square brackets parametrize the effects of the Gauss–Bonnet contribution. These are monotonically decreasing functions of r and tend to unity from above as $\{r, \alpha\} \to 0$. In this limit, the slow–roll parameters reduce to those of the RSII model. It follows, therefore, that the Gauss–Bonnet contribution tightens the constraints for inflation to proceed relative to the RSII scenario.

A further consequence of introducing a Gauss–Bonnet term is that the spectrum of perturbations is altered. For example, in the high–energy limit where $H^2 \propto \rho^{2/3}$, we find that $A_S^2 \propto H^4/\dot{\phi}^2 \propto H^6/V'^2 \propto (V/V')^2$. Thus, for the case of an exponential potential, the spectrum is pushed very close to a scale–invariant form [63]. This is interesting given that potentials of this nature generically arises in a number of particle physics inspired settings.

Finally, it would be of interest to determine whether the degeneracy of the inflationary consistency equation is lifted by introducing a Gauss–Bonnet term into the bulk action. To date, the gravitational wave spectrum in this model has yet to be determined. The calculation of the spectrum is more involved than that of the RSII scenario, because the linearly perturbed junction conditions must be employed to impose the necessary boundary conditions on the perturbations.

7 Concluding Remark

To summarize, inflationary cosmology based on Randall–Sundrum braneworlds remains a rich environment for future work. The scenario has already revealed unexpected surprises and surely has more to offer. It is well motivated from a string theoretic perspective in view of its close relationship with the AdS/CFT correspondence. It is sufficiently simple to provide a framework for performing analytical calculations and thereby making observational predictions. Thus, it provides a unique window into higher–dimensional physics.

Acknowledgements

JEL is supported by the Royal Society. It is a pleasure to thank G. Huey and N. Nunes with whom various results presented in Sects. 4 and 5 and Sect. 6 were derived.

References

1. C. L. Bennett et al.: Astrophys. J. Suppl. **148**, 1 (2003) [astro-ph/0302207]; G. Hinshaw et al.:Astrophys. J. Suppl. **148**, 135 (2003), [astro-ph/0302217].
2. D. N. Spergel et al.: Astrophys. J. Suppl. **148**, 175 (2003), [astro-ph/0302209].
3. B. P. Schmidt et al., Astrophys. J. **507**, 46 (1998); A. G. Riess et al. [Supernova Search Team Collaboration], Astron. J. **116**, 1009 (1998); S. Perlmutter et al. [Supernova Cosmology Project Collaboration], Astrophys. J. **517**, 565 (1999).
4. A. A. Starobinsky, Phys. Lett. B **91**, 99 (1980); A. H. Guth, Phys. Rev. D**23**, 347 (1981); A. Albrecht and P. J. Steinhardt, Phys. Rev. Lett. **48**, 1220 (1982); S. W. Hawking and I. G. Moss, Phys. Lett. B **110**, 35 (1982); A. D. Linde, Phys. Lett. B **108**, 389 (1982); A. D. Linde, Phys. Lett. B**129**, 177 (1983).
5. A. R. Liddle and D. H. Lyth, *Cosmological Inflation and Large-Scale Structure* (Cambridge University Press, Cambridge, 2000).
6. V. Mukhanov and G. Chibisov, Pis'ma Zh. Eksp. Teor. Fiz. **33**, 549 (1981) [JETP Lett. **33**, 532 (1981), astro-ph/0303077]; A. H. Guth and S. Y. Pi, Phys. Rev. Lett. **49**, 1110 (1982); S. W. Hawking, Phys. Lett. B **115**, 295 (1982); A. D. Linde, Phys. Lett. B **116**, 335 (1982); A. A. Starobinsky, Phys. Lett.B **117**, 175 (1982); A. A. Starobinsky, Sov. Astron. Lett. **9**, 302 (1983); J. M. Bardeen, P. J. Steinhardt, and M. S. Turner, Phys. Rev. D **28**, 679 (1983); D. H. Lyth, Phys. Rev. D **31**, 1792 (1985).
7. H. V. Peiris et al., Astrophys. J. Suppl. **148**, 213 (2003) [astro-ph/0302225].
8. A. Kogut et al., Astrophys. J. Suppl. **148**, 161 (2003) [astro-ph/0302213].
9. E. Komatsu *et al.*, Astrophys. J. Suppl. **148**, 119 (2003) [astro-ph/0302223].
10. S. L. Bridle, A. M. Lewis, J. Weller, and G. Efstathiou, Mon. Not. Roy. Astron. Soc. **342**, L72 (2003) [astro-ph/0302306].
11. S. Dodelson and L. Hui, Phys. Rev. Lett. **91**, 131301 (2003) [astro-ph/0305113]; W. H. Kinney, E. W. Kolb, A. Melchiorri, and A. Riotto, astro-ph/0305130; A. R. Liddle and S. M. Leach, Phys. Rev. D **68**, 103503 (2003) [astro-ph/0305263].
12. W. Hu and M. White, Astrophys. J. **479**, 568 (1997); D. N. Spergel and M. Zaldarriaga, Phys. Rev. Lett. **79**, 2180 (1997).
13. J. Polchinski, *String Theory* (Cambridge University Press, Cambridge, 1998).
14. C. V. Johnson, *D–Branes* (Cambridge University Press, Cambridge, 2003).
15. E. Witten, Nucl. Phys.B **443**, 85 (1995).
16. P. Townsend, Phys. Lett. B **350**, 184 (1995).
17. J. E. Lidsey, D. Wands, and E. J. Copeland, Phys. Rep. **337**, 343 (2000); F. Quevedo, Class. Quant. Grav. **19**, 5721 (2002) [hep-th/0210292]; M. Gasperini and G. Veneziano, Phys. Rep. **373**, 1 (2003).
18. J. Polchinski, Phys. Rev. Lett. **75**, 4724 (1995).
19. K. Akama, Lect. Notes Phys. **176**, 267 (1982), hep-th/0001113; V. A. Rubakov and M. E. Shaposhnikov, Phys. Lett. B **125**, 136 (1983); M. Visser, Phys. Lett. B **159**, 22 (1985).
20. P. Hořava and E. Witten, Nucl. Phys. B **460**, 506 (1996); P. Hořava and E. Witten, Nucl. Phys. B **475**, 94 (1996).
21. A. Lukas, B. A. Ovrut, K. S. Stelle, and D. Waldram, Phys. Rev. D **59**, 086001 (1999).
22. K. Benakli, Int. J. Mod. Phys. D **8**, 153 (1999); K. Benakli, Phys. Lett. B **447**, 51 (1999); A Lukas, B. A. Ovrut, and D. Waldram, Phys. Rev. D **60**, 086001 (1999); H. S. Reall, Phys. Rev. D **59**, 103506 (1999); J. E. Lidsey, Class. Quant. Grav. **17**, L39 (2000).

23. M. Gasperini and G. Veneziano, Astropart. Phys. **1**, 317 (1991).
24. J. Khoury, B. A. Ovrut, P. J. Steinhardt, and N. Turok, Phys. Rev. D **64**, 123522 (2001).
25. N. Arkani–Hamed, S. Dimopoulos, and G. Dvali, Phys. Lett. B **429**, 263 (1998); I. Antoniadis, N. Arkani–Hamed, S. Dimopoulos, and G. Dvali, Phys. Lett. B **436**, 257 (1998).
26. L. Randall and R. Sundrum, Phys. Rev. Lett. **83**, 3370 (1999).
27. L. Randall and R. Sundrum, Phys. Rev. Lett. **83**, 4690 (1999).
28. G. 't Hooft, gr-qc/9310026; L. Susskind, J. Math. Phys. **36**, 6337 (1995).
29. J. M. Maldacena, Adv. Theor. Math. Phys. **2**, 231 (1998); E. Witten, Adv. Theor. Math. Phys. **2**, 505 (1998); S. Gubser, I. Klebanov, and A. Polyakov, Phys. Lett. B **428**, 105 (1998); O. Aharony, S. Gubser, J. Maldacena, H. Ooguri, and Y. Oz, Phys. Rep. **323**, 183 (2000).
30. S. W. Hawking, T. Hertog, and H. S. Reall, Phys. Rev. D **62** 043501 (2000); S. Nojiri, S. D. Odintsov, and S. Zerbini, Phys. Rev. D **62** 064006 (2000); S. Nojiri and S. Odintsov, Phys. Lett. B **484**, 119 (2000); L. Anchordoqui, C. Nunez, and K. Olsen, J. High Energy Phys. **10**, 050 (2000); S. Nojiri and S. Odintsov, Phys. Lett. B **494**, 135 (2000); S. Gubser, Phys. Rev. D **63**, 084017 (2001); T. Shiromizu and D. Ida, Phys. Rev. D **64**, 044015 (2001).
31. E. Verlinde, hep-th/0008140; I. Savonije and E. Verlinde, Phys. Lett. B **507**, 305 (2001).
32. R. Argurio: Brane Physics in M-theory, PhD thesis, Universite Libre de Bruxelles, [hep-th/9807171]; J. P. Gauntlett, hep-th/9705011.
33. J. E. Lidsey, Phys. Rev. D **64**, 063507 (2001).
34. P. Kraus, J. High Energy Phys. **12**, 011 (1999); D. Ida, J. High Energy Phys. **09**, 014 (2000).
35. H. A. Chamblin and H. S. Reall, Nucl. Phys. B **562**, 133 (1999); H. A. Chamblin, M. J. Perry, and H. S. Reall, J. High Energy Phys. **09**, 014 (1999).
36. C. Barcelo and M. Visser, Phys. Lett. B **482**, 183 (2000).
37. W. Israel, Nuovo Cim. B **44**, 1 (1966).
38. P. Binétruy, C. Deffayet, and D. Langlois, Nucl. Phys. B **565**, 269 (2000); P. Binétruy, C. Deffayet, U. Ellwanger, and D. Langlois, Phys. Lett. B **477**, 285 (2000).
39. E. E. Flanagan, S. -H. Tye, and I. Wasserman, Phys. Rev. D **62**, 044039 (2000).
40. J. M. Cline, C. Grojean, and G. Servant, Phys. Rev. Lett. **83**, 4245 (1999); C. Cśaki, M. Graesser, C. Kolda, and J. Terning, Phys. Lett. B **462**, 34 (1999).
41. T. Shiromizu, K. Maeda, and M. Sasaki, Phys. Rev. D **62**, 024012 (2000).
42. R. Maartens, D. Wands, B. Bassett, and I. Heard, Phys. Rev. D **62**, 041301 (2000).
43. E. J. Copeland, A. R. Liddle, and J. E. Lidsey, Phys. Rev. D **64**, 023509 (2001).
44. J. E. Lidsey, A. R. Liddle, E. W. Kolb, E. J. Copeland, T. Barreiro, and M. Abney, Rev. Mod. Phys. **69**, 373 (1997).
45. D. Wands, K. A. Malik, D. H. Lyth, and A. R. Liddle, Phys. Rev. D **62**, 043527 (2000).
46. D. Langlois, R. Maartens, and D. Wands, Phys. Lett. B **489**, 259 (2000).
47. M. Kamionkowski, A. Kosowsky, and A. Stebbins, Phys. Rev. Lett. **78**, 2058 (1997); U. Seljak and M. Zaldarriaga, Phys. Rev. Lett. **78**, 2054 (1997).
48. J. Kovac et al., Nat. **420**, 772 (2002).
49. G. Huey and J. E. Lidsey, Phys. Lett. B **514**, 217 (2001).

50. G. Huey and J. E. Lidsey, Phys. Rev. D **66**, 043514 (2002).
51. H. Stoica, S. -H. Tye, and I. Wasserman, Phys. Lett. B **482**, 205 (2000).
52. H. Collins and B. Holdom, Phys. Rev. **D62**, 105009 (2000); N. Deruelle and T. Dolezel, Phys. Rev. D **62**, 103502 (2000); P. Bowcock, C. Charmousis, and R. Gregory, Class. Quant. Grav. **17**, 4745 (2000); W. B. Perkins, Phys. Lett. B **504**, 28 (2001).
53. G. F. Giudice, E. W. Kolb, J. Lesgourgues, and A. Riotto, Phys. Rev. D **66**, 083512 (2002).
54. C. Gordon and R. Maartens, Phys. Rev. D **63**, 044022 (2001).
55. E. J. Copeland, E. W. Kolb, A. R. Liddle, and J. E. Lidsey, Phys. Rev. D **49**, 1840 (1994).
56. S. Nojiri and S. D. Odintsov, J. High Energy Phys. **07**, 049 (2000); M. Giovannini, Phys. Rev. D **63**, 064011 (2001); S. Mukohyama, Phys. Rev. D **63**, 104025 (2001); G. Kofinas, J. High Energy Phys. **08**, 034 (2001); S. Nojiri, S. D. Odintsov, and S. Ogushi, Int. J. Mod. Phys. A **16**, 5085 (2001); S. Nojiri, S. D. Odintsov, and S. Ogushi, Phys. Rev. D **65**, 023521 (2002).
57. J. E. Kim, B. Kyae, and H. M. Lee, Phys. Rev. D **62**, 045013 (2000); N. Deruelle and T. Dolezel, Phys. Rev. D **62**, 103502 (2000); I. Low and A. Zee, Nucl. Phys. B **585**, 395 (2000); O. Corradini and Z. Kakushadze, Phys. Lett. B **494**, 302 (2000); J. E. Kim, B. Kyae, and H. M. Lee, Nucl. Phys. B **582**, 296 (2000); Erratum–*ibid* **591**, 587 (2000); J. E. Kim and H. M. Lee, Nucl. Phys. B **602**, 346 (2001); B. Abdesselam and N. Mohammedi, Phys. Rev. D **65**, 084018 (2002); C. Germani and C. F. Sopuerta, Phys. Rev. Lett. **88**, 231101 (2002); J. E. Lidsey, S. Nojiri, and S. D. Odintsov, J. High Energy Phys. **06**, 026 (2002).
58. N. E. Mavromatos and J. Rizos, Phys. Rev. D **62**, 124004 (2000); I. P. Neupane, J. High Energy Phys. **09**, 040 (2000); I. P. Neupane, Phys. Lett. B **512**, 137 (2001); K. A. Meissner and M. Olechowski, Phys. Rev. Lett. **86**, 3708 (2001); Y. M. Cho, I. Neupane, and P. S. Wesson, Nucl. Phys. B **621**, 388 (2002).
59. C. Charmousis and J. Dufaux, Class. Quant. Grav. **19**, 4671 (2002).
60. S. C. Davis, Phys. Rev. D **67**, 024030 (2003).
61. E. Gravanis and S. Willison, Phys. Lett. B **562**, 118 (2003) [hep-th/0209076].
62. P. Binetruy, C. Charmousis, S. C. Davis, and J. Dufaux, Phys. Lett. B **544**, 183 (2002).
63. J. E. Lidsey and N. Nunes, Phys. Rev. D **67**, 103510 (2003), [astro-ph/0303168].
64. A. Fayyazuddin and M. Spalinski, Nucl. Phys. B **535**, 219 (1998); O. Aharony, A. Fayyazuddin, and J. Maldacena, J. High Energy Phys. **07**, 013 (1998).
65. B. Zwiebach, Phys. Lett. B **156**, 315 (1985); A. Sen, Phys. Rev. Lett. **55**, 1846 (1985); R. R. Metsaev and A. A. Tseytlin, Nucl. Phys. B **293**, 385 (1987).
66. D. G. Boulware and S. Deser, Phys. Rev. Lett. **55**, 2656 (1985).
67. N. Deruelle and J. Madore, Mod. Phys. Lett. A **1**, 237 (1986); N. Deruelle and L. Farina–Busto, Phys. Rev. D **41**, 3696 (1990).
68. R. G. Cai, Phys. Rev. D **65**, 084014 (2002).

Creation of Brane Universes

Rubén Cordero[1] and Efraín Rojas[2]

[1] Departamento de Física, Escuela Superior de Física y Matemáticas del I.P.N. Unidad Adolfo López Mateos, Edificio 9, 07738 México, D.F., Mexico. cordero@fis.cinvestav.mx

[2] Facultad de Física e Inteligencia Artificial, Universidad Veracruzana Sebastián Camacho 5; Xalapa, Veracruz; 91000, Mexico. efrojas@uv.mx

Abstract. The presence of a 4-form field in a fixed background spacetime can give rise to the creation of brane universes, that mimic the Swinger pair production. By means of a canonical quantum approach, we study the birth of a brane universe involving its intrinsic curvature. The nucleation probability for the brane is calculated taking into account both an instanton method and a WKB approximation. We discuss some cosmological implications resulting from the model.

1 Introduction

One of the most fundamental questions of human history is: "Where did it all come from?" Standard cosmology does not have a convincing answer, for this reason a new description is necessary. Cosmologists during a long time have believed that quantum cosmology can shed light on this question [1, 2, 3, 4] but some issues are still controversial, e.g. the lack of an intrinsic time variable in the theory [5], the problem of cosmological boundary conditions [6], to mention some. Among the several ideas that try to give a possible answer to the fundamental question, the so-called Brane World Scenarios (BWS) [7, 8] became a promising way to understand the birth and then the evolution of our universe. Based in the proposal that our universe can be understood as a 4-dimensional spacetime object embedded in an N-dimensional spacetime, $N > 4$, the main physical idea behind BWS is that matter fields are confined to a 3-dimensional space (brane) while gravitational fields can extend into a higher-dimensional space (bulk), where graviton can travel into the extra dimensions. BWS, besides resolving the hierarchy problem, have been applied to a great diversity of situations such as dark matter/energy, quintessence, cosmology, inflation and particle physics. Furthermore, BWS have been the motivation for other related applications of embedding theory such as generation of internal symmetries, quantum gravity and alternative Kaluza-Klein theories [9, 10, 11, 12, 13]. In the context of cosmology, there are predictions that could be tested by astronomical observations, this constitutes one of the several reasons why BWS is so attractive [14].

This new approach should be able to reproduce all known features of a gravitational theory. Even so, there are some questions that puzzle cosmologists: the origin of these branes (universes) still is a matter of research. In

R. Cordero and E. Rojas, Creation of Brane Universes, Lect. Notes Phys. **646**, 381–399 (2004)
http://www.springerlink.com/

these brane world programs, gravity on the brane can be recovered by compactifying the extra dimensions [7] or by introducing an AdS background spacetime [8]. However, Dvali, Gabadadze and Porrati [15] (DGP) showed that, even in an asymptotically Minkowski bulk, 4-dimensional gravity can be recovered if one includes a brane curvature term in the action. Furthermore, DGP considered the Z_2 reflection symmetry with respect to the brane finding that gravity is 4-dimensional on scales smaller than a certain scale and that it is 5-dimensional on larger distances [16, 17]. It is noteworthy to mention that the reflection symmetry is not the only possibility in these models. In this regard, several works have been devoted to antisymmetric cases [18, 19, 20, 21, 22, 23, 24, 25], for instance, when the brane is coupled to a 4-form field [23]. In a pioneering work, Brown and Teitelboim worked out the process of membrane creation by an antisymmetric field motivated by Schwinger process of pair creation induced by the presence of an electric field [26]. Garriga [27] has also studied the creation of membranes for this field in dS background. Other authors have been interested in brane world creation in AdS spacetime or in other particular situations [28, 29, 30, 31, 32] but, to our knowledge, nobody has been devoted to the nucleation of Brane World Universes (BWU) induced by a 4-form field and a brane curvature term included in the action. Mostly, BWS are studied in AdS/dS as well as in empty (Minkowski) backgrounds.

In this contribution we are going to discuss the nucleation of BWU with a curvature term induced by a 4-form field in a dS background spacetime. We get the Friedman like equation when 5-dimensional gravity is fixed and perform geometric Hamiltonian analysis in order to obtain, by means of canonical quantization, the corresponding Wheeler-DeWitt equation. The setup for the induced brane production is as follows. There is an external homogeneous field that produces a brane. Then, the natural question is: What is the probability of such a process? In the present paper we calculate the creation probability for a brane universe embedded in a de Sitter space, produced by a 4-form potential gauge field in the same way that the standard electromagnetic potential bears to a charged particle. Within quantum analysis we shall use the WKB approximation getting the same results by the instanton method. We could try to answer the question: Is our universe one of the more probable universes produced in this model? or Is it a very special one? Parameters of this model must be constrained by cosmological requirements like nucleosynthesis [23].

The work is organized as follows. In Sect. 2 we present the equations of motion of a brane with matter and a curvature term that lives in AdS/dS or Minkowski bulk when there is no Z_2 symmetry. We show that under a certain limit the former equations describe a brane interacting with a 4-form field in a fixed background. A geometric Hamiltonian approach is done in Sect. 3, where the fundamental canonical structure is obtained and the canonical constraints are listed. The next step is to specialize the general canonical analysis to the case of a spherical 3-brane floating in dS_5 background spacetime which is the topic of Sect. 4. The last section provides the preamble to obtain the WdW equation in the canonical quantization context, which is done in Sect. 5. The creation probability is calculated in Sect. 6 by two methods, the first one is an instanton approach and the other one is a WKB approach for barrier tunneling of the WdW equation. Finally in Sect. 7 we present our conclusions as well as some perspectives on this work.

2 The Model

As we mention before, BWS is based on the proposal that our universe can be thought as a 4-dimensional spacetime object embedded in a fixed background spacetime of dimension N, $(N > 4)$. The main physical idea behind BWS is that matter fields are confined to a 3-dimensional space (brane) while gravitational fields can extend into the bulk. The extra dimensions maybe larger or even infinite and also there exists the possibility of detecting them, which is another reason why BWS is so interesting.

Strong motivation for BWS comes from string/M-theory where the general idea of a brane world appears naturally [33]. For example, some kind of branes are able to carry matter fields like D-branes.

In these theories, low energy physics is effectively 4-dimensional due to localization of matter on the brane. However, gravity is multidimensional and the corresponding action has the bulk curvature term

$$S_G^{(N)} = \frac{M_{(N)}^{N-2}}{2} \int d^N x \sqrt{-g}^{(N)} \mathcal{R}. \tag{1}$$

where M_N is the bulk Planck mass. One way to recover 4-dimensional gravity is, as proposed in [7], to neglect the brane tension and to consider compact extra dimensions. In this approach the 4-dimensional gravity is mediated by the graviton zero mode and the metric is independent of extra dimensions. The integration on the extra coordinates is straightforward and the effective 4-dimensional action is

$$S_G^{(4)} = \frac{M_{(N)}^{N-2}}{2} V_{ed} \int d^4 \xi \sqrt{-\gamma}\, \mathcal{R}. \tag{2}$$

where $V_{ed} \sim R^d$ is the volume of extra dimensions. We have now a relation between 4-dimensional Planck mass and bulk Planck mass (up to a numerical factor)

$$M_{(4)} = M_{(N)} (M_{(N)} R)^{d/2}. \tag{3}$$

In fact, this can be used to address the hierarchy problem: the big difference between the Planck scale and the electroweak scale. Now we can accept that the fundamental gravity scale is of the same order as the electroweak scale, $M_{(N)} \sim 1$Tev. From this point of view, the hierarchy between $M_{(4)}$ and

M_{EW} is a consequence of the large size of extra dimensions, i.e. gravity is very weak because it spreads into the extra dimensions. For $d = 2$ and $M_{(N)} \sim 30$Tev ($M_{(N)} \sim 1$Tev is excluded by astrophysics and cosmology), the size of $R \sim 1 - 10\mu m$ has motivated the search for deviations from Newton's law in the scale of micro-meter [34].

Also, the 4D gravity on the brane can be recovered by introducing a large negative cosmological constant in the bulk

$$S_m^{(N)} = \int d^N x \sqrt{-G} \Lambda, \tag{4}$$

that causes the bulk space to warp, confining low-energy gravitons to the brane [8].

Very recently, DGP [15] have pointed out that 4D gravity can be recovered even in an asymptotically Minkowski bulk, provided that one includes the brane curvature action (2). Assuming a 5-dimensional bulk and a Z_2 symmetry of reflections with respect to the brane, they found that gravity on the brane is effectively 4D on scales $r << r_0$, with

$$r_0 = \frac{M_{(4)}^2}{2M_{(5)}^3}, \tag{5}$$

and becomes 5D on larger scales. Analysis of cosmological solutions with a Robertson-Walker metric on the brane indicates the same crossover scale (5) in this class of models [16].

The DGP model can be extended in several directions. One possible extension is to lift the requirement of Z_2 symmetry. This symmetry is certainly a necessary condition and actually cannot be enforced in 5D models where the brane is coupled to a 4-form field, so that the 5D cosmological constant is different on the two sides of the brane. Brane world cosmology without Z_2 symmetry has been discussed by a number of authors [18, 19, 20, 21, 22, 23, 24, 25], but in most of this work, the brane curvature term (2) has not been included in the action. Some effects of including the brane curvature term have been discussed in [25].

The effective action that we are interested in is the corresponding to a 3-brane with an intrinsic curvature term considered from its worldsheet and no Z_2 symmetry in the presence of a fixed background spacetime $\mathcal{M}$. We assume the 3-brane immersed in $\mathcal{M}$ by means of the embedding $x^\mu = X^\mu(\xi^a)$ where x^μ are coordinates on the bulk and ξ^a are coordinates on the worldsheet. For concreteness, we consider the following action ($N = 5$),

$$S = \int \sqrt{-g} \left(\frac{1}{2k} {}^{(5)}\mathcal{R} + \mathcal{L}_m \right) + \int \sqrt{-\gamma} \left(\frac{1}{2k'} \mathcal{R} - L_m \right), \tag{6}$$

where $\mathcal{L}_m$ and $L_m = \rho_v$ stand for matter Lagrangians for the bulk and the brane, respectively, and we have absorbed the differentials d^5x, $d^4\xi$, throughout the paper in the integral sign, for simplicity. In our case, we will consider those as cosmological constants. The constants $k = M_{(5)}^{-3}$ and $k' = M_{(4)}^{-2}$, where $M_{(4)}$ and $M_{(5)}$ are the brane Planck and bulk masses, respectively. The corresponding equations of motion for the brane are [19],

$$[K]\gamma_{ab} - [K_{ab}] = k\widetilde{T}_{ab}, \tag{7}$$

$$\widetilde{T}^{ab} < K_{ab} > = [\mathcal{T}_{nn}], \tag{8}$$

$$\nabla_a(T^a{}_b) = -[\mathcal{T}_{bn}]. \tag{9}$$

where $K_{ab} = -g_{\mu\nu}n^\mu D_a X^\nu{}_b$ is the extrinsic curvature of the brane, n^μ stands for the normal vector to the worldsheet, $X^\mu{}_a = \partial_a X^\mu$ denote the tangent vectors to the worldsheet and $D_a = X^\mu{}_a D_\mu = e^\mu{}_a D_\mu$ where D_μ is the covariant derivative compatible with $g_{\mu\nu}$. γ_{ab} denotes the worldsheet metric. $\mathcal{T}_{an} = (\mathcal{T}_{bulk})_{\mu\nu}e^\mu{}_a n^\nu$ and $\mathcal{T}_{nn} = (\mathcal{T}_{bulk})_{\mu\nu}n^\mu n^\nu$ are projections of the bulk energy-momentum tensor. $\widetilde{T}_{ab} = T_{ab} - \frac{1}{k'}\left(\mathcal{R}_{ab} - \frac{1}{2}\gamma_{ab}\mathcal{R}\right)$ and T_{ab} is the brane energy-momentum tensor. The square and angular brackets represent the difference and the average of the corresponding embraced quantity, on the two sides of the brane, respectively, i.e., $[K_{ab}] = K^+_{ab} - K^-_{ab}$ and $< K_{ab} >= \frac{1}{2}(K^+_{ab} + K^-_{ab})$, where the superscripts '+' and '-' denote the exterior and interior for a spherical brane, respectively.

Taking into account that the bulk energy momentum tensor has the form

$$\mathcal{T}^{\pm}_{\mu\nu} = -k^{-1}\Lambda^{\pm}g_{\mu\nu}, \tag{10}$$

and by means of the generalized Birkhoff theorem, the 5-dimensional FRW metric can be written as

$$dS_5^2 = -A_\pm d\tau^2 + A_\pm^{-1}da^2 + a^2 d\Omega_3^2\,, \tag{11}$$

where

$$A_\pm = \kappa - \frac{\Lambda^\pm}{6}a^2 - \frac{2\mathcal{M}^\pm}{M_{(5)}^3 a^2}\,, \tag{12}$$

and $d\Omega_3^2$ denotes the metric of a 3-sphere, a is the cosmic scale factor and $\mathcal{M}^\pm$ is the mass. Furthermore, in the cosmic time gauge the 4-dimensional metric on the brane reduces to

$$dS_4^2 = -dt^2 + a^2 d\Omega_3^2. \tag{13}$$

Using the junction conditions, and due to isotropy and homogeneity in (11), the matter can be parametrized completely via a perfect fluid brane energy-momentum tensor

$$T^a{}_b = \mathrm{diag}(-\rho, P, P, P), \tag{14}$$

so the relevant equations of motion for the model read

$$\left(\dot{a}^2 + A_-\right)^{1/2} - \left(\dot{a}^2 + A_+\right)^{1/2} = \frac{ka}{3}\left(\rho - \frac{3(\dot{a}^2+1)}{k'a^2}\right), \tag{15}$$

$$\dot{\rho} + 3\frac{\dot{a}}{a}(\rho + P) = 0. \tag{16}$$

The equation on the second line, represents the energy-momentum conservation on the brane. The former system was discussed in [35] where several interesting cases were treated. Suppose now $\mathcal{M}^- = 0$, $\rho = \text{const}$, and consider at the same time the limits of fixed bulk gravity, $M_{(5)} \to \infty$ and, $\Lambda^+ \to \Lambda^-$ but restricted to the following relation

$$\mathrm{Lim}_{(M_{(5)},\Lambda^+)\to(\infty,\Lambda^-)}(\Lambda^+ - \Lambda^-)M_{(5)}^3 = \alpha, \tag{17}$$

so, expanding the second term of the LHS of Eq. (15), this equation transforms to

$$\left(\frac{\rho}{3} - M_{(4)}^2\frac{\dot{a}+1}{a^2}\right)\left(\frac{\dot{a}+1}{a^2} - \frac{\Lambda}{6}\right)^{1/2} = \frac{\alpha}{12} + \frac{\mathcal{M}}{a^4}. \tag{18}$$

In order to get the Friedman like equation we define Υ through the relation

$$\frac{\dot{a}+1}{a^2} \equiv \frac{\rho}{3M_{(4)}^2}\Upsilon \equiv H^2\Upsilon. \tag{19}$$

Note that Υ is only a function of a and it is a solution of the following relation

$$M_{(4)}^4(1-\Upsilon)^2\left(\Upsilon - \frac{\Lambda}{6H^2}\right) = H^{-6}\left(\frac{\alpha}{12} + \frac{\mathcal{M}}{a^4}\right)^2. \tag{20}$$

As we will see below, this approach is equivalent to a brane interacting with a 4-form field and propagating in a fixed background spacetime.

3 Hamiltonian Approach

The Hamiltonian framework has been a fundamental tool in the study of the dynamics of field theories and a preliminary step towards canonical quantization of physical theories. Canonical quantization is the oldest and most conservative approach to quantization and we will use it to study quantum cosmology for our model. To carry out the previous goal, we must begin by casting the theory in a canonical form, then we shall proceed to its quantization.

To begin with, we are going to mimic the well known ADM procedure for canonical gravity to get a hamiltonian description of the brane. We shall assume that the worldsheet m admits a foliation, i.e., we will begin with a time like 4-manifold m of the topology $\Sigma \times R$, equipped with a metric γ_{ab}, such that m is an outcome of the evolution of a space like 3-manifold Σ_t, representing "instants of time", each of which is diffeomorphic to Σ. Then we shall identify the several geometric quantities inherent to the hypersurface Σ_t. The ADM decomposition of the action, computation of the momenta as well as the recognition of the constraints, are the next stages. Before going on, we would like to glimpse onto the ADM decomposition of some important geometric quantities defined on the branes in our geometrical approach.

3.1 Embedding Theory

Consider a brane, Σ, of dimension d whose worldsheet, m, is an oriented timelike manifold living in a N-dimensional arbitrary fixed background spacetime M with metric $g_{\mu\nu}$. For hamiltonian purposes, we shall foliate the worldsheet m in spacelike leaves Σ_t.

Employing geometry of surfaces, as well as novelty variational techniques developed in [36, 37], we can write the Gauss-Weingarten equations associated with the embedding of Σ_t in M ($x^\mu = X^\mu(u^A)$), i.e., the gradients of the Σ_t basis $\{\epsilon^\mu{}_A, \eta^\mu, n^\mu{}_i\}$. These spacetime vectors can be decomposed with respect to the adapted basis to Σ_t, as

$$\mathcal{D}_A \epsilon^\mu{}_B = -\Gamma^\mu_{\alpha\beta}\,\epsilon^\alpha{}_A \epsilon^\beta{}_B + k_{AB}\,\eta^\mu - K^i_{AB}\,n^\mu{}_i \tag{21}$$

$$\mathcal{D}_A \eta^\mu = k_{AB}\,\epsilon^{\mu\,B} - K_A{}^i\,n^\mu{}_i \tag{22}$$

$$\tilde{\mathcal{D}}_A n^{\mu\,i} = K^i_{AB}\,\epsilon^{\mu\,B} - K_A{}^i\,\eta^\mu \tag{23}$$

where $\Gamma^\alpha_{\beta\gamma}$ are the Christoffel coefficients of the background manifold and $K_A{}^i$ is a piece of the generalized extrinsic twist potential and both k_{AB} and K^i_{AB} are the extrinsic curvatures of Σ_t associated with the normals η^μ and $n^\mu{}_i$, respectively. $\mathcal{D}_A$ denotes the covariant derivative adapted to Σ_t and $\tilde{\mathcal{D}}_A$ is the covariant derivative that preserves invariance under rotations of the normals $n^\mu{}_i$, i.e., $\tilde{\mathcal{D}}^i_A = \mathcal{D}^i_A - \omega^{ij}_A\,n_j$. In a similar way, we can write the Gauss-Weingarten equations associated with the embedding of Σ_t in the worldsheet m, ($x^a = X^a(u^A)$), i.e., the gradients of the Σ_t basis $\{\epsilon^a{}_A, \eta^a\}$. These worldsheet vectors can be decomposed with respect to the adapted basis to Σ_t, as

$$\nabla_A \epsilon^a{}_B = \gamma^C_{AB}\,\epsilon^a{}_C + k_{AB}\,\eta^a \tag{24}$$

$$\nabla_A \eta^a = k_{AB}\,\epsilon^{a\,B}\,, \tag{25}$$

where ∇_A is the gradient along the tangent basis, i.e., $\nabla_A = \epsilon^a{}_A \nabla_a$, where ∇_a is the covariant derivative compatible with γ_{ab} $\quad(\mu,\nu = 0,1,2,\ldots,N-1;$ $a,b = 0,1,\ldots,d$ and $A,B = 1,2,\ldots,d)$.

The time vector field, written in terms of the adapted basis of a leaf Σ_t, is given by

$$t^\mu = \dot{X}^\mu = N\eta^\mu + N^A\,\epsilon^\mu{}_A\,, \tag{26}$$

which represents the flow of time throughout spacetime. Furthermore, from (26) note that the following relations hold:

$$N = -g_{\mu\nu}\eta^\mu \dot{X}^\nu \qquad \text{and} \qquad N^A = g_{\mu\nu}h^{AB}\epsilon^\mu{}_B \dot{X}^\nu .$$

3.2 Model ADM Decomposed

Following the results obtained in [39, 40, 41], we are going to display the standard procedure. We start by considering the action

$$S = \frac{k_1}{2}\int_m \sqrt{-\gamma}\,(\mathcal{R} + \Lambda_b) + \frac{k_2}{4!}\int_m \sqrt{-\gamma} A_{\mu\nu\rho\sigma}\epsilon^{\mu\nu\rho\sigma} \,, \tag{27}$$

where $\mathcal{R}$ is the Ricci scalar curvature of the worldsheet m, $k_1 = M^2_{(4)}$ and $\Lambda_b = -2\rho_v/M^2_{(4)}$ being the cosmological constant on the brane; $A_{\mu\nu\rho\sigma}$ is a gauge 4-form Ramond-Ramond field onto the bulk, $\mu, \nu = 0, 1, \ldots, N-1$; $\epsilon^{\mu\nu\rho\sigma}$ is an antisymmetric bulk tensor which can be expressed in terms of the worldsheet Levi-Civita tensor as $\epsilon^{\mu\nu\rho\sigma} = \epsilon^{abcd} e^\mu{}_a e^\nu{}_b e^\rho{}_c e^\sigma{}_d$, where $a, b = 0, 1, 2, 3$. k_2 is the coupling constant between the brane and the antisymmetric tensor.

Taking into account the Gauss-Codazzi relations for the embedding of Σ_t in m, Eqs. (24) and (25), we obtain an equation involving the extrinsic and intrinsic curvatures, up to a divergence term,

$$\mathcal{R} = R + (k_{AB}k^{AB} - k^2) \,, \tag{28}$$

where R denotes the intrinsic curvature[3] of Σ_t which does not have any dependence of the velocity and k_{AB} its extrinsic curvature associated with the unit timelike normal η^μ, given by

$$\begin{aligned} k_{AB} &= -g_{\mu\nu}\eta^\mu(\mathcal{D}_A\epsilon^\nu{}_B + \Gamma^\nu_{\alpha\beta}\epsilon^\alpha{}_A\epsilon^\beta{}_B) \\ &:= -g_{\mu\nu}\eta^\mu\tilde{\mathcal{D}}_A\epsilon^\nu{}_B \,. \end{aligned} \tag{29}$$

Besides (29), in Σ_t we have another curvature tensor associated with the i-th unit normal $n^{\mu\, i}$

$$K^i_{AB} = -g_{\mu\nu}n^{\mu\, i}\tilde{\mathcal{D}}_A\epsilon^\nu{}_B \,, \tag{30}$$

where $g_{\mu\nu}$ denotes the background spacetime metric and $i = 1, 2, \ldots, N-d$; $A, B = 1, 2, 3$. Note that the configuration space consists of the embedding functions X^μ for the brane, instead of 3-metrics as is customary in the ADM approach to general relativity.

In order to simplify the computations below, we use the next relations where the velocities appear explicitly

$$\begin{aligned} \kappa_{AB} &= N\,k_{AB} \\ &= -g_{\mu\nu}\dot{X}^\mu\tilde{\mathcal{D}}_A\epsilon^\nu{}_B \,. \end{aligned} \tag{31}$$

For canonical purposes it will be useful also the time derivative

$$\frac{\partial N}{\partial \dot{X}^\mu} = -\eta_\mu = -\, g_{\mu\nu}\eta^\nu \,. \tag{32}$$

As before, we will need the derivatives of the extrinsic curvature

[3] We will adhere to Wald's convention concerning the definitions of Riemannian curvature, namely, $2\nabla_{[a}\nabla_{b]}t^c = -\mathcal{R}_{abd}{}^c t^d$ [42]

$$\frac{\partial \kappa_{AB}}{\partial \dot{X}^\mu} = -g_{\mu\nu}\tilde{\mathcal{D}}_A \epsilon^\nu{}_B \qquad (33)$$
$$= -k_{AB}\,\eta_\mu + K^i_{AB}\,n_{\mu\,i},$$

where in the second line on the RHS we have used the Gauss-Weingarten equations (21).

The ADM decomposed action (27) now looks like

$$S = \int_{\Sigma_t}\int_R \frac{k_1}{2} N\sqrt{h}\,[\bar{R} + k_{AB}k^{AB} - k^2] + \int_{\Sigma_t}\int_R \frac{k_2}{3!}\,A_{\mu\nu\rho\sigma}\dot{X}^\mu \epsilon^\nu{}_A \epsilon^\rho{}_B \epsilon^\sigma{}_C\,\varepsilon^{ABC} \qquad (34)$$

where we have defined $\bar{R} := R + \Lambda_b$, h is the determinant of the hypersurface metric h_{AB} and ε^{ABC} is the Σ_t Levi-Civita antisymmetric symbol.

3.3 Primordial Tensor

We define for convenience the following symmetric tensor that is independent of the velocities

$$\begin{aligned}\Theta^\mu{}_\nu &:= (h^{AB}h^{CD} - h^{AC}h^{BD})\,\tilde{\mathcal{D}}_A\epsilon^\mu{}_B \tilde{\mathcal{D}}_C\epsilon_{\nu\,D}\\ &= (k^2 - k_{AB}k^{AB})\,\eta^\mu\eta_\nu - (kL^i - K^i_{AB}k^{AB})\,n^\mu{}_i\eta_\nu\\ &\quad - (kL^i - K^i_{AB}k^{AB})\,\eta^\mu n_{\nu\,i} + (L^iL^j - K^i_{AB}K^{AB\,j})\,n^\mu{}_i n_{\nu\,j},\end{aligned} \qquad (35)$$

where L^i denotes the trace of the curvature K^i_{AB}, i.e., $L^i = h^{AB}K^i_{AB}$. This tensor will keep track of the dynamics of the theory as we will see below. The tensor (35) was previously defined in [38] where a Hamiltonian analysis for geodetic brane gravity was performed. We will have in mind some ideas of the classical approach developed there.

Some important properties of the tensor (35) are the following

$$\begin{aligned}\Theta^\mu{}_\alpha\epsilon^\alpha{}_A &= 0\,,\\ \Theta^\mu{}_\alpha\dot{X}^\alpha &= -N(k^2 - k_{AB}k^{AB})\,\eta^\mu + N(kL^i - K^i_{AB}k^{AB})\,n^\mu{}_i,\\ g_{\mu\nu}\dot{X}^\mu\Theta^\nu{}_\alpha\dot{X}^\alpha &= N^2(k^2 - k_{AB}k^{AB}).\end{aligned}$$

We shall adopt the notation $\dot{X}\cdot\Theta\cdot\dot{X} := g_{\mu\nu}\dot{X}^\mu\Theta^\nu{}_\alpha\dot{X}^\alpha$ throughout the paper. Taking advantage of the previous results we are able to rewrite the Lagrangian density,

$$\mathcal{L} = \frac{k_1}{2}N\sqrt{h}\left[\bar{R} - \frac{1}{N^2}\,\dot{X}\cdot\Theta\cdot\dot{X}\right] + \frac{k_2}{3!}A_{\mu\nu\rho\sigma}\dot{X}^\mu\epsilon^\nu{}_A\epsilon^\rho{}_B\epsilon^\sigma{}_C\,\varepsilon^{ABC}. \qquad (36)$$

Using the tensor (35), the momenta associated to the embedding functions are

$$
\begin{aligned}
P_\mu &= \frac{\partial \mathcal{L}}{\partial \dot{X}^\mu} \\
&= -\frac{k_1}{2}\sqrt{h}\left\{\left[\bar{R} + \frac{1}{N^2}\,\dot{X}\cdot\Theta\cdot\dot{X}\right]\eta_\mu + \frac{2}{N}\Theta_{\mu\nu}\dot{X}^\nu\right\} \\
&\quad + \frac{k_2}{3!}\,A_{\mu\alpha\beta\gamma}\,\bar{\varepsilon}^{\alpha\beta\gamma},
\end{aligned} \tag{37}
$$

where we have defined the Σ_t tangent tensor $\bar{\varepsilon}^{\mu\nu\rho} = \varepsilon^{ABC}\epsilon^\mu{}_A\epsilon^\nu{}_B\epsilon^\rho{}_C$ with normalization $\bar{\varepsilon}^{\mu\nu\rho}\bar{\varepsilon}_{\mu\nu\rho} = 3!$.

3.4 Canonical Constraints

A natural question to ask in an invariant reparametrization theory is: What are the primary constraints? This is part of the fundamental structure of constrained field theories. According to the standard Dirac-Bergmann algorithm, we will get the constraints from the momenta (37). It is convenient for the computation to define the matrix $\Psi^\mu{}_\nu := \Theta^\mu{}_\nu - \lambda g^\mu{}_\nu$ where $\lambda(x)$ is not a dynamical field that is gauge dependent [38] and that should to be find. If we assume that the form of momenta has the following pattern,

$$
P_\mu = -\sqrt{h}k_1\,(\Theta - \lambda\,g)_{\mu\nu}\,\eta^\nu + \frac{k_2}{3!}\,A_{\mu\alpha\beta\gamma}\,\bar{\varepsilon}^{\alpha\beta\gamma}\,, \tag{38}
$$

we are free to compare both expressions (37) and (38) to get a condition to be satisfied

$$
\bar{R} + \eta\cdot\Theta\cdot\eta + 2\lambda = 0\,. \tag{39}
$$

This expression will transform in a primary constraint after we express it in terms of phase space variables.

Profitably is the introduction of the field $\lambda(x)$ since we can solve Eq.(38) for the timelike unit normal vector

$$
\eta^\mu = \frac{-1}{\sqrt{h}k_1}\,\left(\Psi^{-1}\right)^\mu{}_\alpha g^{\alpha\beta}\mathcal{P}_\beta\,, \tag{40}
$$

where we have defined $\mathcal{P}_\mu = P_\mu - \frac{k_2}{3!}\,A_{\mu\alpha\beta\gamma}\,\bar{\varepsilon}^{\alpha\beta\gamma}$, but we have to increase the number of constraints as we will see below. Inserting this form of the unit time-like vector in the relation (39) we obtain the main scalar primary constraint. In a similar way, inserting η^μ in its square relation, $g(\eta,\eta) = -1$, we determine another scalar constraint.

The complete set of primary constraints are

$$
\begin{aligned}
C_0 &= \mathcal{P}\cdot(\Psi^{-1})\cdot\mathcal{P} + h\lambda_0 k_1^2 = 0\,, &(41)\\
\mathcal{C}_0 &= \mathcal{P}\cdot(\Psi^{-2})\cdot\mathcal{P} + hk_1^2 = 0\,, &(42)\\
C_A &= \mathcal{P}_\mu\epsilon^\mu{}_A = 0\,, &(43)\\
\mathcal{C}_\lambda &= P_\lambda = 0\,, &(44)
\end{aligned}
$$

where we have defined $\lambda_0 = \lambda + \bar{R}$. The third constraint is the usual constraint of parametrized theories while the last one came from the fact that λ is not a dynamical field, i.e., its time derivative does not appear in the Lagrangian. It is worth to mention that the constraint $\mathcal{C}_0$ is a byproduct of C_0 taking advantage of the identity $\partial(\Psi^{-1})^\mu{}_\nu/\partial\lambda = (\Psi^{-2})^\mu{}_\nu$.

4 Brane Universe Floating in a de Sitter Space

The main idea in this section is to adapt the previous dynamical description to the case of a spherical brane immersed in a specific background spacetime in order to apply the quantum approach to our BW model.

Consider a 3-dimensional spherical brane evolving in a de Sitter 5-dimensional background spacetime, $dS_5^2 = -A_\pm\, d\tau^2 + A_\pm^{-1}\, da^2 + a^2 d\Omega_3^2$, where $A_\pm$ is given by (12). The worldsheet generated by the motion of the brane can be described by the following embedding

$$x^\mu = X^\mu(\tau,\chi,\theta,\phi) = \begin{pmatrix} t(\tau) \\ a(\tau) \\ \chi \\ \theta \\ \phi \end{pmatrix} . \tag{45}$$

The line element induced on the worldsheet is given by

$$ds^2 = (-A_\pm \dot{t}^2 + A_\pm^{-1}\dot{a}^2)\, d\tau^2 + a^2\, d\chi^2 + a^2 \sin^2\chi\, d\theta^2 + a^2 \sin^2\chi \sin^2\theta\, d\phi^2 , \tag{46}$$

where the dot stands for derivative with respect to cosmic time τ. For convenience in notation we define $\Delta = -A_\pm \dot{t}^2 + A_\pm^{-1}\dot{a}^2$. The frequently appealed cosmic gauge will be set up by $\Delta = -1$.

In order to evaluate the extrinsic curvature tensors involved in our approach, (29) and (30), we need the orthonormal Σ_t basis

$$\eta^\mu = \frac{1}{\sqrt{-\Delta}}\left(\dot{t}, \dot{a}, 0, 0, 0\right) , \qquad n^\mu = \frac{1}{\sqrt{-\Delta}}\left(A_\pm^{-1}\,\dot{a}, A_\pm\,\dot{t}, 0, 0, 0\right) .$$

The only non null components for the extrinsic curvatures are

$$\begin{aligned}
k_{\chi\chi} &= \frac{a\dot{a}}{(-\Delta)^{1/2}} & K_{\chi\chi} &= \frac{a\dot{t}}{(-\Delta)^{1/2}} A_\pm \\
k_{\theta\theta} &= \frac{a\dot{a}}{(-\Delta)^{1/2}} \sin^2\chi & K_{\theta\theta} &= \frac{a\dot{t}}{(-\Delta)^{1/2}} A_\pm \sin^2\chi \\
k_{\phi\phi} &= \frac{a\dot{a}}{(-\Delta)^{1/2}} \sin^2\chi \sin^2\theta & K_{\phi\phi} &= \frac{a\dot{t}}{(-\Delta)^{1/2}} A_\pm \sin^2\chi \sin^2\theta .
\end{aligned}$$

It is a straightforward task to compute the tensor (35) for the present case, which give us

$$(\Theta)^{\mu}{}_{\nu} = \begin{pmatrix} 0 & 0 & 0 \\ 0 & \frac{6}{a^2} A_{\pm} & 0 \\ 0 & 0 & 0_{3\times 3} \end{pmatrix}_{5\times 5} . \tag{47}$$

The next step is to compute the matrix Ψ so, in order to know Ψ it is necessary to evaluate λ. It is easily calculated from the relation (39) to give

$$\lambda = -\frac{1}{2a^2}\left(6 + \Lambda_b a^2 + \frac{6\dot{a}^2}{(-\Delta)}\right). \tag{48}$$

This seems to contradict the functional dependence previously assumed for this field, but we are free to implement a trick to convert the velocity dependence to the right form by means of the generalized evolution equation, $(\dot{a}^2+1)/a^2 = \Upsilon H^2$, avoiding any misunderstanding.

We turn now to compute a first integral for our specific model. This is done using (37) by setting up P_0 proportional to the brane energy, $P_0 := 3E\Phi = 3E(\sin^2\chi \sin\theta)$. Furthermore, since we have a homogeneous isotropic space in (46), we can invoke the typical value $A_{0\chi\theta\phi} = \frac{F}{4}a^4\Phi$ for the gauge field, which is supported by some kind of cosmological solutions [23, 43], where F is a constant and the corresponding gauge independent field tensor $F_{\mu\nu\rho\delta\gamma} = 5\nabla_{[\mu}A_{\nu\rho\delta\gamma]}$ is expressed in terms of it $F_{\mu\nu\rho\delta\gamma} = F\epsilon_{\mu\nu\rho\delta\gamma}$. Explicitly, we have

$$P_0 = \frac{3k_1 a\dot{t}\Phi A_{\pm}}{\sqrt{-\Delta}}\left(1 + \frac{\Lambda_b}{6}a^2 + \frac{\dot{a}^2}{(-\Delta)}\right) + \frac{k_2 F}{4}a^4\Phi . \tag{49}$$

Now, taking into account the generalized evolution equation with Λ_b being the cosmological constant on the brane, we find the desired result

$$E = M_{(4)}^2 a^4 H^3 \left(\Upsilon - \frac{\Lambda}{6H^2}\right)^{1/2}(\Upsilon - 1) + \frac{k_2 F}{12}a^4 , \tag{50}$$

where Λ is the cosmological constant living in the bulk appearing in Eq. (12) and we have used the cosmic gauge in the last step. Note that (50) is in agreement with Eq. (20), then confirming the equivalence with the limiting process developed in Sect. 2.

5 Quantum Brane Cosmology

From the point of view of quantum cosmology, the whole universe is treated quantum mechanically and it is described by a wave function. DeWitt [44] introduced this quantum approach to cosmology long time ago, but recently it has attracted a lot of interest. In the original 4-dimensional quantum cosmology, a small closed universe can be spontaneously nucleate out from nothing [1, 2, 3, 45], where nothing refers to the absence of matter, space and time.

Unlike ordinary quantum mechanics, where boundary conditions for the wave function are specified by the physical setup external to the system, in

4-dimensional quantum cosmology there is nothing external. In fact, there is a debate about the right boundary conditions for the wave function of the universe [46].

However, there is an important difference due to the presence of an embedding space. We can think that our universe was a small nearly spherical brane nucleating in de Sitter space time induced by a totally antisymmetric field. We believe that tunneling boundary condition is the right one, in agreement with the idea that a tunneling process was the mechanism involved in the nucleation of the universe.

We turn now in this section to develop the quantum description for our specific problem. The canonical quantization procedure is well known and it just remains to apply it to our case.

We shall set $P_\mu \to -i\frac{\delta}{\delta X^\mu}$ in such a way that scalar constraints (41) and (42) transform into quantum equations

$$\left(-i\frac{\delta}{\delta X^\mu} - p_{A\,\mu}\right)(\Psi^{-1})^{\mu\nu}\left(-i\frac{\delta}{\delta X^\nu} - p_{A\,\nu}\right)\psi = -h\lambda_0 k_1^2\,\psi\,, \tag{51}$$

$$\left(-i\frac{\delta}{\delta X^\mu} - p_{A\,\mu}\right)(\Psi^{-2})^{\mu\nu}\left(-i\frac{\delta}{\delta X^\nu} - p_{A\,\nu}\right)\psi = -hk_1^2\,\psi\,, \tag{52}$$

where we have defined $p_{A\,\mu} := k_2\,A_{\mu\alpha\beta\gamma}\bar{\varepsilon}^{\alpha\beta\gamma}/3!$.

For the universe we are considering, (45), it is necesary to know explicitly the matrix Ψ. Taking into account Eq. (47) as well as Eq. (48), it is a straightforward task to compute the full expression for Ψ,

$$(\Psi)^{\mu\nu} = \begin{pmatrix} -\frac{1}{2a^2A_\pm}\left[6 + \Lambda_b a^2 + \frac{6\dot{a}^2}{(-\Delta)}\right] & 0 & 0 & 0 \\ 0 & \frac{A_\pm}{2a^2}\left[6 + \Lambda_b a^2 + \frac{6\dot{a}^2}{(-\Delta)} + 12A_\pm\right] & 0 & 0 \\ 0 & 0 & \frac{1}{2a^4}\left[6 + \Lambda_b a^2 + \frac{6\dot{a}^2}{(-\Delta)}\right] & 0 \\ 0 & 0 & 0 & M_{2\times 2} \end{pmatrix} \tag{53}$$

The previous matrix, in the cosmic gauge, reduces to a more manageable form

$$(\Psi)^{\mu\nu} = \begin{pmatrix} 3H^2A_\pm^{-1}(1-\Upsilon) & 0 & 0 & 0 \\ 0 & 3A_\pm a^{-2}[-H^2a^2(1-\Upsilon) + 2A_\pm] & 0 & 0 \\ 0 & 0 & -3a^{-2}H^2(1-\Upsilon) & 0 \\ 0 & 0 & 0 & N_{2\times 2} \end{pmatrix}, \tag{54}$$

where $M_{2\times 2}$ and $N_{2\times 2}$ denote 2×2 diagonal matrices.

Taking the embedding (45) and having in mind the matrix (54) in the cosmic gauge, we are able to get the inverse matrix

$$(\Psi^{-1})^\mu{}_\nu \equiv \begin{pmatrix} A & 0 & 0 \\ 0 & B & 0 \\ 0 & 0 & N_{3\times 3}^{-1} \end{pmatrix} = \begin{pmatrix} \frac{-1}{3H^2(1-\Upsilon)} & 0 & 0 \\ 0 & \frac{a^2}{3[-H^2a^2(1-\Upsilon)+2A_\pm]} & 0 \\ 0 & 0 & N_{3\times 3}^{-1} \end{pmatrix}, \tag{55}$$

in such a way that (51) and (52) transform into the pair of relations

$$-A_{\pm}^{-1}A\widetilde{P}_0^2\,\psi + A_{\pm}B\widetilde{P}_1^2\,\psi = -h\lambda_0 k_1^2\,\psi, \tag{56}$$

$$-A_{\pm}^{-1}A^2\widetilde{P}_0^2\,\psi + A_{\pm}B^2\widetilde{P}_1^2\,\psi = -hk_1^2\,\psi, \tag{57}$$

where we introduced the notation $\widetilde{P}_\mu = -i\frac{\delta}{\delta X^\mu} - p_{A\ \mu}$. Taking into account the value $\lambda_0 = 3\left[-H^2(1+\Upsilon) + \frac{2}{a^2}\right]$ expressed in the cosmic gauge, the couple of quantum relations can be rewritten as,

$$\widetilde{P}_0^2\,\psi = k_1^2(3\Phi)^2a^8H^6(1-\Upsilon)^2\left(\Upsilon - \frac{\Lambda}{6H^2}\right)\psi, \tag{58}$$

$$\widetilde{P}_1^2\,\psi = -k_1^2(3\Phi)^2a^2\frac{(1-H^2\Upsilon a^2)[H^2a^2(1-\Upsilon) - 2 + \frac{\Lambda a^2}{3}]^2}{(1-\frac{\Lambda a^2}{6})^2}\psi\,. \tag{59}$$

At this time, we are more interested in identifying the potential governing the dynamics of our model instead of solving exactly the WdW equation so we propose the wave function to be of separable form, $\psi(t,a) = \psi_1(t)\Psi(a)$. The WdW equation acquires the form

$$-\frac{\partial^2\Psi}{\partial a^2} = \frac{a^2M_{(4)}^4\left[2 - \frac{\Lambda a^2}{3} + (\Upsilon-1)\,H^2a^2\right]^2(-1+\Upsilon H^2a^2)}{\left(1-\frac{\Lambda a^2}{6}\right)^2}\Psi, \tag{60}$$

accompanied by the energy equation

$$\left(E - \frac{k_2F}{12}a^4\right)^2 = H^6a^8M_{(4)}^4(1-\Upsilon)^2\left(\Upsilon - \frac{\Lambda}{6H^2}\right), \tag{61}$$

where we redefined the momenta $\widetilde{P}_0 \to (3\Phi)P_\mu$ and assumed $\psi_1 = e^{-iEt}$.

6 Nucleation Rate

At this stage, we are ready to compute the creation probability for the universe to be created. Some simplifications are necessary since the general problem itself is hard to solve.

From WdW equation (60), the potential is easily read off

$$V(a) = \frac{a^2M_{(4)}^4[2 - \frac{\Lambda a^2}{3} + (\Upsilon-1)H^2a^2]^2(1-\Upsilon H^2a^2)}{(1-\frac{\Lambda a^2}{6})^2}\,. \tag{62}$$

Note that this is a very hard expression to work out if one is interested in the general integration, specially if, as in the cosmological context, creation probability is the desired calculation. Taking into account the tunneling boundary conditions, the probability nucleation [47] is written in terms of the potential extracted from the WdW equation, namely,

$$\mathcal{P} \sim e^{-2\int_{a_l}^{a_r}\sqrt{V}da}\,. \tag{63}$$

In order to get some interesting results from the quantum approach, we shall consider some special cases.

6.1 Case A

If $E = 0$ from Eq. (61) then Υ is just a constant given by

$$\frac{(k_2F/12M_{(2)}^2)^2}{H^6} = (1-\Upsilon)^2(\Upsilon - \frac{\Lambda}{6H^2}) \,. \tag{64}$$

and the probability rate in this case is

$$\mathcal{P} \sim e^{\frac{4((\Upsilon-1)-\Lambda/3H^2)}{\Upsilon\Lambda}+2(\Upsilon-1)H^2(\frac{6}{\Lambda})^2[1-\frac{1}{X}\tan^{-1}X]}, \tag{65}$$

where $X^2 = (\frac{\Lambda}{6H^2})^2\left(\Upsilon - \frac{\Lambda}{6H^2}\right)^{-1}$. If $k_2F, \Lambda << H^2$ to first order the probability rate is

$$\mathcal{P} \sim e^{-\frac{4}{3H^2}+\frac{16k_2F}{15H^5}} \,. \tag{66}$$

This means that it is more probable to create a universe when $k_2F > 0$ than if $k_2F < 0$. We will comment it below.

Now, we would like to calculate the probability nucleation using the instanton method.

The corresponding Euclidean action in de Sitter bulk can be found by complexifying the temporal coordinate and keeping the field strength $F_{\mu\nu\rho\delta\gamma}$ fixed

$$S_{(E)} = \int_m \sqrt{-\gamma}\left(-\frac{M_{(2)}^2}{2}\mathcal{R} + \rho_v\right) + \frac{k_2}{4!}\int_m \sqrt{-\gamma}A_{\mu\nu\rho\sigma}\epsilon^{\mu\nu\rho\sigma} \,. \tag{67}$$

In Euclidean space we have now closed worldsheets that split the deSitter background spacetime of radius $H_{dS}^{-1} = (\Lambda/6)^{-1/2}$ in two regions. This is the basic geometry of the instanton calculation.

Following [27], by using Stoke's theorem we can transform (67) to an instanton action that involves a volume of the spacetime enclosed by the brane

$$S_{(E)} = \int_m \sqrt{-\gamma}\left(-\frac{M_{(2)}^2}{2}\mathcal{R} + \rho_v\right) - k_2F\int_v \sqrt{-g} \,. \tag{68}$$

For spherical worlsheets the former action is expressed through the radius R_0 of the brane

$$S_{(E)} = \left(\rho_v - \frac{12M_{(4)}^2}{R^2}\right)S_4(R_0) - k_2FV_4(R_0), \tag{69}$$

where

$$S_{(4)} = \frac{8\pi^2}{3}R_0^4, \tag{70}$$

is the surface of a worldsheet of radius R_0, and

$$V_4 = \pi^2 H_{dS}^{-5}\phi_0 - \frac{\pi^2 H_{dS}^{-4}}{R_0}(1-R_0H_{dS})^{1/2}\left(1+\frac{2}{3}R_0\right), \tag{71}$$

is the volume enclosed by the brane of radius R_0 and $\sin(\phi_0) = R_0 H_{dS}$. Extremizing (69) we find that the radius of the Euclidean brane is a solution of

$$M_{(4)}^2 H^3 \left(\Upsilon - \frac{\Lambda}{6H^2}\right)^{1/2} (1-\Upsilon) = \frac{k_2 F}{12}, \tag{72}$$

where $\Upsilon \equiv H_{dS}^2 (R_0 H)^{-2}$. The resulting Euclidean action is

$$S_{(E)} = -6\pi^2 M_{(4)}^2 \left\{ \frac{4\left[(\Upsilon-1) - \frac{\Lambda}{3H^2}\right]}{\Upsilon\Lambda} + 2(\Upsilon-1)\left(\frac{6H}{\Lambda}\right)^2 \left[1 - \frac{1}{X}\tan^{-1} X\right] \right\}, \tag{73}$$

and the nucleation probability $\mathcal{P} \sim e^{-S_{(E)}}$ is in agreement with (65) modulo a normalizing factor. We now turn back to the meaning of equation (66). The behavior of strength field $F_{\mu\nu\rho\delta\gamma}$ is the key in this process. When $k_2 > 0$ the value of the field decreases with respect to the original one in the region inside and it corresponds to the screening membrane discussed in [27]. In contrast, when $k_2 < 0$ the field increases its value, corresponding it to the antiscreening membrane and, as it is expected, it is less probable to produce such a Universe. This situation resembles phenomena of vacuum decay, where ordinary transition from false to true vacuum corresponds to $k_2 > 0$, and the decay of true vacuum, by means of false vacuum bubbles, corresponds to $k_2 < 0$ and $k_2 F$ represents the difference in energy density between the false and true vacuum.

6.2 Case B

We proceed to calculate an approximate expression for the nucleation rate at first order, when both E and F are small. The potential is

$$V(a) = 4a^2(1 - H^2a^2 - EH - k_2FHa^4) \tag{74}$$

and the nucleation probability is

$$\mathcal{P} \sim e^{-\frac{4}{3H^2} + EH^{-1} + \frac{16k_2F}{15H^5}} \tag{75}$$

in complete agreement with (66) when E vanishes.

The potential for case A, is plotted in Fig. 1 and the corresponding one for the case B in Fig. 2. Using this kind of plots for the potential and the analytic expressions for the nucleation rate, we can deduce that creation probability is enhanced when the nucleation process takes place in the de Sitter background spacetime with small radius H_{dS}^{-1}.

7 Conclusions

We have calculated the nucleation probability of brane world universes induced by a totally antisymmetric tensor living in a dS fixed background

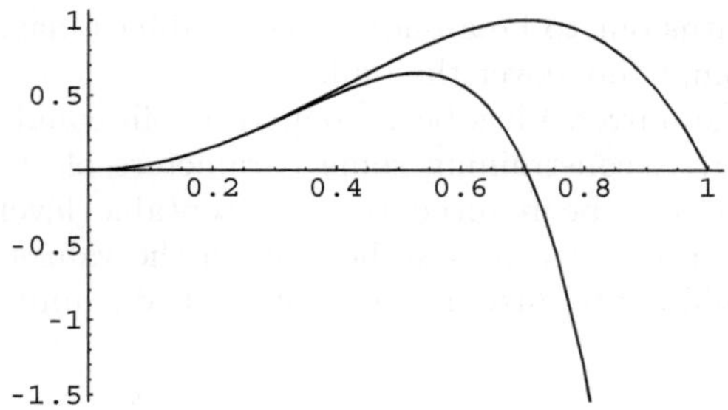

Fig. 1. Potential for case A. In this case $E = 0$ and $k = k_2 F$ taking the values: $k = 0$ (Einstein case) for the upper curve and $k \neq 0$ for the lower curve.

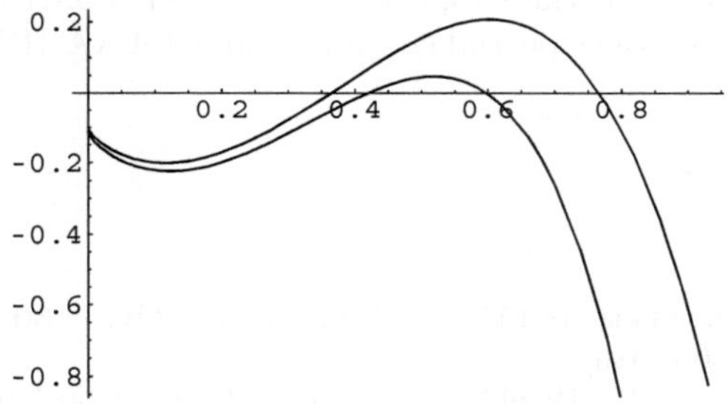

Fig. 2. Potential for case B. In this case $E \neq 0$ and the background is a de Sitter space. $k = 0$ for the upper curve and $k \neq 0$ for the lower curve.

spacetime. This was done by means of canonical quantum approach where the Wheeler-DeWitt equation was found. Besides, we determined, for one specific case, the nucleation rate by computing the corresponding instanton. When the energy of the brane is zero, $\mathcal{E} = 0$, in the bulk space and the coupling constant of the brane with the antisymmetric field, k_2, is positive, the creation probability is enhanced with respect to the situation when there is no interaction of the brane with the 4-form. For $k_2 < 0$ the nucleation rate decreases as it is expected. This situation resembles phenomena of vacuum decay, where ordinary transition from false to true vacuum corresponds to $k_2 > 0$, and the decay of true vacuum by means of false vacuum bubbles corresponds to $k_2 < 0$. Furthermore, $k_2 F$ represents the difference in energy density between the false and true vacuum. For large expansion rate of the de Sitter bulk we observed an increasing nucleation rate. At this point we ask ourselves about possible brane collisions, and what is the most important factor. The branes will be driven apart by the exponential expansion of the bulk reducing brane collision but at the same time, there is an increase in nucleation rate. We expect now that the problem of old inflationary model of the universe turns out to be an advantage: bubbles may not be produced fast enough to completely cover the bulk.

Once the brane universe has been created it still could be hit by stealth branes [35], however, constraining some parameters of the model the rate of brane collisions can be reduced to an acceptable level. We think that cosmological constraints can impose bounds on the values of k_2F and with this value one could try to answer the question: Is our universe very special?

Acknowledgements

The authors wish to thank Keichi Maeda for useful comments. This work was supported in part from CONACyT under grant CO1-41639. We also thank SNI-México for partial support. ER thanks PROMEP-SEP for partial support. RC acknowledges partial support from COFAA-IPN and CGPI-IPN Grant 20030642.

References

1. A. Vilenkin, Phys.Lett. B **117**, 25 (1982), Phys. Rev. D **30**, 509 (1984); Phys. Rev. D **33**, 3560 (1986).
2. J. B. Hartle and S. W. Hawking, Phys. Rev. D **28**, 2960 (1983).
3. A. D. Linde, Lett. Nuovo Cimento **39**, 401 (1984).
4. V. A. Rubakov, *Phys. Lett. B* **148**, 280 (1984).
5. K. V. Kuchař, *The Problem of Time in Canonical Quantization*, in *Conceptual Problems of Quantum Gravity*, Eds. A. Ashtekar and J. Stachel, Birkhäuser, Boston, (1991).
6. J. B. Hartle, *Quantum Cosmology: Problems for the* 21st *Century*,in Procc. of the 11th Nishinomiya-Yukawa Symp, World Scientific, Singapure (1998); gr-qc/9701022.
7. N. Arkani-Hamed, S. Dimopoulos and G. Dvali, Phys. Lett. B **429**, 263 (1998); I. Antoniadis, N. Arkani-Hamed, S. Dimopoulos and G. Dvali, Phys. Lett. B **436**, 257 (1998).
8. L. Randall and R. Sundrum, Phys. Rev. Lett **83**, 3370 (1999); 4690 (1999).
9. T. Regge and C. Teitelboim, in *Proc. Marcel Grossman*, p. 77 (Trieste, 1975).
10. M. D. Maia and W. Mecklemburg, J. Math. Phys. **25** 3047 (1984); M. D. Maia, Phys. Rev. D **31** 262 (1985).
11. M. Pavsic, Phys. Lett. A **116** 1 (1986).
12. M. Pavsic and V. Tapia, *Resource Letter on Geometrical Results for Embeddings and Branes*, gr-qc/0010045;
13. V. A. Rubakov, *Large and Infinite Extra Dimensions*, hep-ph/0104152;
14. *Supernova Cosmology Project*, http://www.supernova.lbl.gov
15. G. Dvali, G. Gabadadze and M. Porrati, Phys. Lett. B **484**, 112 (2000).
16. C. Deffayet, Phys. Lett. B **502**, 199 (2001).
17. G. Dvali, G. Gabadadze, Phys. Rev. D **63**, 065007 (2001).
18. P. Binetruy, C. Defayyet, and D. Langlois Nucl. Phys. B **565**, 269 (2000).

19. D. Ida, JHEP **0009**, 014 (2000).
20. N. Deruelle and T. Dolezel, Phys. Rev. D **62**, 103502 (2001).
21. P. Bowcock, C. Charmousis and R. Gregory, Class. Quant. Grav. **17**, 4745 (2000).
22. A. C. Davis, S. C. Davis, W. B. Perkins and I. R. Vernon, Phys.Lett B **504**, 254 (2001).
23. B. Carter and J.P. Uzan, Nucl. Phys B **606**, 45 (2001); R. Battye, B. Carter, A. Mennim and J.P. Uzan, Phys. Rev. D **64**, 124007 (2001); B. Carter, J. P. Uzan, R. A. Battye and A. Mennim, Class. Q. Grav. **18** 4871 (2001).
24. L. Anchordoqui, C. Nuñez and K. Olsen , JHEP **0010**, 050 (2000).
25. H. Collins and B. Holdom, Phys. Rev. D **62**, 105009 (2000); **62**, 124008 (2000).
26. J. D. Brown and C. Teitelboim, Nucl. Phys. B **297**, 787 (1988).
27. J. Garriga Phys. Rev. D **49**, 6327 (1994).
28. J. Garriga and M. Sasaki, Phys. Rev. D **62**, 043523 (2000).
29. A. Gorsky and K. Selivanov, Phys. Lett. B **485**, 271 (2000).
30. S. Nojiri and S.D. Odintsov, JHEP **12**, 033 (2001); S. Nojiri and S. D. Odintsov and A.S Ogushi, Int. J. of Mod. Phys. A **17** 4809 (2002).
31. M. Bouhmadi-López, P. F. González-Díaz and A. Zhuk, Class. Quant. Grav. **19**, 4863 (2002).
32. K. Koyama and J. Soda, Phys. Lett. B **483**, 432 (2000).
33. P. Horava and E. Witten, Nucl. Phys. B **460**, 506 (1996).
34. J. C. Long, A.B. Churnside and J.C. Price, Gravity Experiment Below 1 Millimeter in Procc. IX Marcel Grossmann Conf. Rome (2000), hep-ph/0009062.
35. R. Cordero and A. Vilenkin, Phys. Rev. D **65** 083519 (2002) .
36. R. Capovilla and J. Guven, Phys. Rev. D **51** 6736 (1995); Phys. Rev. D **57** 5158 (1998).
37. J. Guven, Phys. Rev. D **48** 5562 (1993); Phys. Rev. D **48** 4604 (1993).
38. D. Karasik and A. Davidson, Phys. Rev. D **67** 064012 (2003); Mod. Phys. Lett. A **13** 2187 (1998).
39. R. Cordero, and E. Rojas, Phys. Lett. **470** 45 (1999).
40. R. Capovilla, J. Guven and E. Rojas Nucl. Phys. Proc. Suppl. **88** 337 (2000).
41. R. Cordero, and E. Rojas Int. Journal of Mod. Phys. A **17** 73 (2002).
42. R. M. Wald, *General Relativity*, (University of Chicago, Chicago, 1986).
43. A. Kehagias and E. Kiritsis, JHEP **9911** 022 (1999).
44. B. S. DeWitt, *Phys. Rev.* **160**, 1113 (1967).
45. A. Vilenkin, Phys. Rev. D **30**, 509 (1984); Phys. Rev. D **33**, 3560 (1986).
46. A. Vilenkin, Phys. Rev. D **58**, 067301 (1998);
47. A. Vilenkin, Phys. Rev. D **50**, 2581 (1994).

The Scalar Field Dark Matter Model: A Braneworld Connection

Tonatiuh Matos[1], Luis Arturo Ureña-López[2], Miguel Alcubierre[3], Ricardo Becerril[4], Francisco S. Guzmán[5], and Darío Núñez[3]

[1] Centro de Investigación y de Estudios Avanzados del I.P.N, A.P. 14-740, 07000 D.F., México. tmatos@fis.cinvestav.mx
[2] Instituto de Física de la Universidad de Guanajuato, A.P. E-143, 37150, León, Guanajuato, México. lurena@fisica.ugto.mx
[3] Instituto de Ciencias Nucleares, Universidad Nacional Autónoma de México, A.P. 70-543, 04510 México, D.F., México. malcubi@nuclecu.unam.mx, nunez@nuclecu.unam.mx
[4] Instituto de Física y Matemáticas, Universidad Michoacana, Edif. C-3, Ciudad Universitaria 58040, Morelia, Michoacán, México. becerril@ifm1.ifm.umich.mx
[5] Max Planck Institut für Gravitationsphysik, Albert Einstein Institut, Am Müehlenberg 1, 14476 Golm, Germany. guzman@aei-potsdam.mpg.de

Abstract. This work is a review for the progress of the scalar field dark matter (SFDM) hypothesis. Here we outline a possible brane world model justifying the hypothesis of the scalar field origin for the dark matter (DM). This model contains two branes, on one of the branes lives the matter of the standard model of particles but on the other one, only spin-0 fundamental interactions are present. In this model these spin-0 fields are the DM and maybe the dark energy (DE). Thus, DM and DE interact only through the gravitational force with the matter brane. Starting with this hypothesis, we write a Lagrangian where the dark matter is of scalar origin with a cosh scalar field potential. The scalar field with a cosh potential behaves exactly in the same way as dust at cosmological scales. In this sense the scalar field mimics very well cold dark matter (CDM). The free parameters of the Lagrangian can be fixed using cosmological observations. After fixing all the free parameters of the model, we found that a scalar field fluctuation collapses forming objects with a preferred mass of $\sim 10^{12} M_{\odot}$. Nevertheless, at galactic scales there exist some strong differences with the CDM paradigm. The scalar field contains an effective pression which avoids the collapse of a scalar field fluctuation, implying that scalar field objects like galaxies contain a flat density profile in the center. This implies that the SFDM paradigm could resolve the cusp density profile problem of galaxies.

1 Introduction

The clarification of the origin of the dark matter in the universe is doubtless one of the most important challenges for theoretical physics and cosmology at the present time. Several models have been proposed to get a possible answer, but at some point they seem to have serious flaws. At the present time, the best model we have is the so called Lambda Cold Dark Matter

T. Matos, L.A. Ureña-López, M. Alcubierre, R. Becerril, F.S. Guzmán, and D. Núñez, The Scalar Field Dark Matter Model: A Braneworld Connection, Lect. Notes Phys. **646**, 401–420 (2004)
http://www.springerlink.com/

model (ΛCDM) [1, 2, 3], which can be considered as the standard cosmological model. The most recent observations of the WMAP satellite [2] confirm that stars and dust (made of baryons) represent something like 4% of the whole energy density of the universe; this amount of baryonic matter is in concordance with the limits imposed by nucleosynthesis. The new measurements of the neutrino mass indicate that neutrinos contribute with less than 0.2% of the whole matter. As the WMAP observations also confirm that our universe is flat, then we should admit that most of the matter content of the cosmos is of unknown nature. In other words, the matter component has the following contributions, $\Omega_{\rm M} = \Omega_b + \Omega_\nu + \cdots + \Omega_{\rm CDM} \sim 0.04 + \Omega_{\rm CDM} \sim 0.27$, where $\Omega_{\rm CDM}$ represents the *cold dark matter* (CDM) part of the contributions, which implies $\Omega_{\rm CDM} \sim 0.23$. Even at this point, we are still missing matter as required for a flat universe. The next breakpoint came with the observations of supernovae type Ia [4], which indicate an actual accelerated expansion of the cosmos. The simplest explanation is to introduce the existence of some vacuum energy (also generically called *dark energy*), in the form of a cosmological constant Λ. Then, the ΛCDM model considers a flat universe ($\Omega_\Lambda + \Omega_{\rm M} \approx 1$) containing 96% of unknown matter.

The dark matter candidates in the ΛCDM model are the WIMP's, (Weakly Interacting Massive Particles), most of them being supersymmetric particles. WIMP's behave as dust particles, i.e., as a (cold) pressureless fluid, and then they are very appropriate for the cosmological scales. However, CDM seems not to be completely appropriate at galactic scales. High resolution numerical simulations show that dust fluids follow the Navarro-Frenk-White (NFW) density profile [5] $\varrho_{\rm NFW} \sim 1/[r(r+b)^2]$ for galaxies, which is singular at $r \to 0$. This fact disagrees with recent observations on LSB galaxies [6, 7], showing that the density profile of the dark matter in the center of galaxies seems to be flat $\varrho_{\rm obs}(r \to 0) \sim$ const. rather than cusp. Also, CDM seems to predict an excess of subgalactic structure [8, 9], an excess that has not been detected at all.

There are some arguments stating that such disagreements are not fundamental. For example, it could be that actual telescopes have not enough resolution to see the dark matter density cusp in the center of galaxies due to baryonic dust or something else. Also, the excess of subgalactic structure could be made only of CDM, and then we cannot see them because they do not have baryonic matter at all [10]. Maybe, it is a matter of time that we will find the right CDM physics and the disagreements will disappear. However, it can also be argued that these disagreements between theory and observations could be because we do not have the right candidate for CDM. This is the reason we propose to look for other candidates which could give us all the successful predictions of ΛCDM at cosmological scales, and the right behavior at galactic scales.

As a motivation for the present work, let us start with the following reasoning. In the actual status of our understanding of the universe, there is an apparent asymmetry in the kind of interactions that take part in Nature. The known fundamental interactions are either spin-1, or spin-2. Electromagnetic, weak and strong interactions are spin-1 interactions, while gravitational interactions are spin-2. Of course, this could be just a coincidence. Nevertheless, we know that the simplest particles are the spin-0 ones. The asymmetry lies in the fact that there is no spin-0 fundamental interactions. Why did Nature forget to use spin-0 fundamental interactions? On the other hand, we know from the success of the ΛCDM model that two fields currently take the main role in the Cosmos, the *dark matter* and the *dark energy*. Recently, we have indeed proposed that dark matter is a scalar field, that is, a spin-0 fundamental interaction. This is the so called *Scalar Field Dark Matter* (SFDM) hypothesis [11, 12, 13, 14]. If true, this hypothesis could solve the problem of the apparent asymmetry in our picture of nature.

Let us go further. All unification theories which try to unify gravitation with the rest of the interactions always contain scalar fields. For instance, in brane theory there is a scalar field, the radion, that must be stabilized, and then this scalar field must be endowed somehow with a scalar field potential with a minimum. This point discards that the scalar field could be identified with the dark energy, because a dark energy scalar field (the so called Quintessence), must be running away when the dark energy dominates [15, 16, 17], its potential must be exponential–like, without a minimum. On the other hand, we know that the scalar field with a scalar field potential containing a minimum sometimes behaves as a dust fluid [6]. From the ΛCDM model, we expect the dark matter to behave as a dust fluid at cosmic scales [1]. Then, the radion or some scalar field of this theory could be identified with the dark matter.

Second, it is known that a exponential–like scalar field potential fits very well the constraints from big bang nucleosynthesis, due to its (so called) tracker solutions [15, 16, 18], and then fine tunning can be avoided. The most simple example of a potential with both exponential behavior and a minimum is a cosh potential.

And third, the anisotropies of the CMB suggest that the universe should have had an early inflationary epoch in order to solve the horizon and the flatness problem, among others. Recently, braneworld scenarios [19, 20], where a 3-brane hypersurface represents our observable universe, have attracted a lot of interest [21, 22, 23, 24, 25, 26, 27, 28]. Originally proposed to solve the hierarchy problem, this kind of ideas have been applied to cosmology where they can be confronted to astronomical observations. In *brane cosmology*, the Friedmann equation (see (12) below) contains an extra quadratic density term, which permits inflation to occur even in the cases not permitted within standard cosmology [29]. As the universe expands, the quadratic term of the density in the Friedmann equation becomes negligible and then inflation comes to an end while at the same time the dynamics of stan-

[6] The existence of a minimum is a necessary condition but not a sufficient one, see [17]

dard cosmology is recovered. Furthermore, it is possible for the inflaton field to survive and become part of the dark matter at late times [29, 30, 31, 32].

This paper is organized as follows. In Sect. 2 we propose a consistent model of inflationary dark matter within the so called braneworld model. In Sect. 3 we study the standard cosmology for the surviving radion field, and in Sect. 4 we discuss the behavior of this scalar field at galactic level and the issue of structure formation under the SFDM hypothesis. Finally in Sect. 5 we give some conclusions.

2 Scalar Field Matter from Brane Cosmology

A series of theories and models like the superstring theory, have shown certain success as fundamental theories. The main feature they show is the existence of many scalar fields called dilatons, inflatons, radions, etc., depending on the type of fields they are coupled to. Nevertheless, fundamental scalar fields have not been observed in nature, and then this fundamental theories should postulate a mechanism to decouple those scalar fields at some point in the history of the universe.

Using the previous analysis, we start with a model which contains two branes [33], as in the Randal-Sundrum (RS) model. Like in the RS model, we suppose that in one of the branes lives the matter of the standard model of particles, in this brane the fundamental interactions are all of them of spin-1. The difference is that we suppose that in the other brane lives a set of fundamental spin-0 interactions which interact only through the gravitational interaction on the bulk with the first brane. Of course, we suppose that in this second brane, with only spin-0 fundamental interactions, lives the dark matter and maybe the dark energy, if it is modelled by Quintessence. This could explain why we are not able to see the dark matter, because it is living in the other brane, but we are able to feel it, because it acts through the gravitational interactions with our brane (see [33] for a extended explanation). This is also the reason why we cannot see spin-0 fundamental interactions in the universe. They determine the structure and maybe the behavior of the universe, but they live in the other brane, we cannot detect them, we can only feel them through their gravitation. The universe is then, a higher-dimensional manifold, which contains two three-dimensional submanifolds, i.e., two branes. In one of them there are only spin-0 fundamental interactions while in the other one there are only spin-1 fundamental interactions. In the higher-dimensional manifold the fundamental interaction is the gravitational one, i.e., the spin-2 interaction. Our brane, with only spin-1 fundamental interactions, lives very close to the other brane, feeling its presence, but not being able to see it at all. Then, in our brane, the dark matter can be modelled effectively using a scalar feel Lagrangian in the main action. In the next section we see that this model contains some very nice features.

2.1 Brane World Scalar Field Dynamics

The effective action describing the scalar field dynamics in the brane world paradigm can be obtained from the following Lagrangian

$$S = \int dx^4 dy \sqrt{-G}\,(R + \Lambda) + \int dx^4 \sqrt{-g}\left(\lambda_b - \frac{1}{2}(\partial\Phi)^2 - V(\Phi)\right) , \quad (1)$$

where

$$V(\Phi) = V_0\left[\cosh(\sqrt{\kappa_0}\lambda\Phi) - 1\right] , \quad (2)$$

is a convenient potential for the scalar field and λ_b is the brane tension. The 5-dimensional metric can be written as

$$dS_5^2 = -A_\pm d\tau^2 + A_\pm^{-1} da^2 + a^2 d\Omega_3^2 , \quad (3)$$

$$A_\pm = \kappa_l - \frac{\Lambda^\pm}{6} a^2 - \frac{2\mathcal{M}^\pm}{M_{(5)}^3 a^2} , \quad (4)$$

being κ_l an integrations constant. In the cosmic time gauge the 4-dimensional inherited metric of the brane is

$$dS_4^2 = -dt^2 + a^2 d\Omega_3^2 . \quad (5)$$

The relevant equations of motion for the model are the following

$$\left(\dot{a}^2 + A_-\right)^{1/2} - \left(\dot{a}^2 + A_+\right)^{1/2} = \frac{\kappa_5 a \rho}{3} , \quad (6)$$

$$\dot{\rho} + 3\frac{\dot{a}}{a}(\rho + P) = 0 . \quad (7)$$

where κ_5 is the five dimensional gravitational constant. The last equation represents the energy-momentum conservation on the brane. Here P is the total pressure and ρ is the total energy density. If we assume that $\Phi = \Phi(t)$, the scalar field energy and pressure are expressed respectively as

$$\varrho_\Phi = \frac{1}{2}\dot{\Phi}^2 + V(\Phi) , \quad (8)$$

$$p_\Phi = \frac{1}{2}\dot{\Phi}^2 - V(\Phi) . \quad (9)$$

These equations can be cast as

$$\dot{a}^2 + \frac{1}{2}(A_+ + A_-) = \frac{3}{4}\kappa_5^2 a^2 \varrho^2 + \left(\frac{3}{2\kappa_5 a \varrho}\right)^2 (A_+ - A_-)^2 , \quad (10)$$

where $\rho = \varrho_\Phi + \lambda_b$. For Z_2 symmetry the former equation can be written in the following way

$$H^2 = \frac{\dot{a}^2}{a^2} = \frac{-\kappa_l^2}{a^2} + \frac{3}{2}\kappa_5^2\lambda\varrho_\Phi(1+\varrho_\Phi/2\lambda_\mathrm{b}) + \frac{\Lambda}{6} + \frac{3}{4}\kappa_5^2\lambda_\mathrm{b}^2 + \frac{2\mathcal{M}}{M_{(5)}^3 a^4} . \quad (11)$$

For flat geometry ($\kappa_l = 0$) and vanishing $\mathcal{M}$ we get

$$H^2 = \frac{3}{2}\kappa_5^2\lambda_\mathrm{b}\varrho_\Phi(1+\varrho_\Phi/2\lambda_\mathrm{b}) + \frac{\Lambda}{6} . \quad (12)$$

Einstein cosmology in four-dimensions is recovered when the brane tension is significantly higher than energy density $\lambda_\mathrm{b} \gg \varrho_\Phi$. The quadratic correction is important at high energies and means that the expansion rate is increased with respect to the standard Einstein cosmology. Analysis of the dynamics of this system has been done in [36, 29, 30, 31, 32, 38].

If we were interested in a zero effective cosmological constant Λ we would require fine-tuning, but the changes to the former analysis would be negligible.

2.2 Inflation in the Braneworld Scenario

The appealing feature of the braneworld model described above is that it inspired new ideas about inflationary cosmology, which would have taken place in the high-energy regime of (12) corresponding to the early universe. Under quite general conditions, there is an extra-friction induced on the KG equation due to the quadratic density term in (12). This allows for the existence of inflationary solutions for *steep* scalar field potentials, which are otherwise not capable of supporting inflation in standard cosmology[29, 30, 31].

We assume that the scalar (inflaton) field Φ was initially displaced from the global minimum of its cosh potential, $\kappa_0\lambda^2\Phi^2 \gg 1$, which is a reasonable assumption if we think of the high energy scales of the very early universe. In this limit, the potential (2) can be approximated by an exponential function. Braneworld inflation driven by such a potential has been studied in [29, 31, 32].

Recalling the main results, the COBE normalization of the cosmic microwave background (CMB) power spectrum relates the value of the brane tension to the scalar field self–coupling such that $\lambda_\mathrm{b}^{1/4}\lambda^{3/2} \approx 10^{15}$ GeV. For the favored value of the latter, as implied by (21) below, we deduce that

$$\lambda_\mathrm{b} \simeq \left(6 \times 10^{-7} M_4\right)^4 = 2.88 \times 10^{51}\,\mathrm{GeV}^4 . \quad (13)$$

For these given values of $\{\lambda, \lambda_\mathrm{b}\}$, the magnitude of the potential energy at the end of inflation is $V_\mathrm{end} \simeq (3.2 \times 10^{-6} M_4)^4 = 2.33 \times 10^{54}\,\mathrm{GeV}^4$ and this implies that $\Phi_\mathrm{end} \approx 2M_4$, thus justifying the exponential approximation during the inflationary era.

An important inflationary parameter is the spectral index n_s of the scalar fluctuation spectrum, which is given by to be [31]

$$n_{\rm s} = 1 - \frac{4}{N+1} = 0.94 \,, \tag{14}$$

where $N \approx 70$ is the number of e-foldings that elapse between the epoch that a given observable mode crosses the Hubble radius during inflation and the end of the inflationary epoch. Remarkably, the tilt of the scalar perturbation spectrum in this scenario is *uniquely* determined by the number of e–foldings and is *independent* of the parameters in the potential (2). Furthermore, the amplitude of the primordial gravitational wave spectrum, A_T, relative to that of the density perturbations, A_S, can be estimated as [29]

$$r = 4\pi \frac{A_T^2}{A_S^2} \simeq 0.4 \tag{15}$$

implying after COBE normalization that $A_T^2 \approx 1.7 \times 10^{-10}$. This ratio is also independent of the model's parameters and is within the projected sensitivity of the Planck satellite. It provides a potentially powerful test of the model.

Inflation ends when the quadratic corrections to the Friedmann equation (12) become negligible. Due to the steep nature of its potential, the inflaton then behaves as a massless field, where its energy density redshifts as $\varrho_\Phi \propto a^{-6}$. For the reheating of the universe after inflation, we can think of different mechanisms. One first option would be reheating via gravitational particle production due to the time–variation of the gravitational field [37].

However, there are some caveats of the model that should be mentioned. First, the tilted spectral index (14) is at variance with current observations of the CMB anisotropies [2], which favor a flat spectral index $n_{\rm s} \simeq 1$. Second, there is an excess of gravitational waves produced between the end of inflation and the expected recovering of standard cosmology that the latter is not recovered at all. After inflation we have a gravitational waves dominated universe, which seems to be a typical feature of these models[31].

Fortunately, these caveats can be cured if we consider the existence of another scalar field within the so called *curvaton* hypothesis [38]. Such another scalar field is not rare since most super theories consider the existence of more than one scalar field. The curvaton field takes the main role for the predictions, and then both a flat $n_{\rm s}$ and a low density of gravitational waves can be achieved. Moreover, the universe is reheated through the decay of the curvaton field, which is a more efficient reheating mechanism than gravitational particle production.

Notice then that, even within the curvaton hypothesis, the decay of the inflaton field after inflation is not a necessary condition for reheating to occur. This implies that the inflaton field can survive until late times to play other roles in the evolution of the universe, like being part of the missing (dark) matter revealed by observations. In particular, we describe here how the inflaton field associated to the potential (2) could become the CDM at late times[30, 31, 32].

3 Scalar Field Dark Matter in the Cosmological Context

As we mentioned before, the CDM model has become a paradigm for the missing matter responsible for the formation of the large scale structure in the universe, from galaxies to clusters of galaxies. In this respect, we can determine the values of the free parameters of a scalar field dark matter model by comparing its evolution to CDM. Such comparison should be done also at the level of linear perturbations. The latter are important, since we should recover the successful picture of structure formation of CDM, in which primordial perturbations grow by means of gravitational instability.

As mentioned in a previous section, after inflation, we expect to recover the standard Big Bang scenario in which nucleosynthesis can proceed successfully, as suggested by observations and the Standard Model of Particles. This occurs during the epoch of radiation domination, in which the scalar field dark matter evolves with an effective exponential potential. The coexistence of the two fluids, radiation and scalar field, is restricted by a successful nucleosynthesis scenario. The ratio of the scalar field (ϱ_Φ) and radiation (ϱ_γ) energy densities at that time should be [18]

$$\frac{\varrho_\Phi}{\varrho_\gamma} = \frac{4}{\lambda^2 - 4} < 0.2 \, . \tag{16}$$

Observe that the ratio is a constant determined uniquely by the self-coupling λ, that indicates that the scalar field evolves, along its attractor solution, as radiation.

After that, the scalar field reaches the minimum of its potential, and starts to oscillate around it so rapidly that the universe only feels the average energy density and pressure of the field. In fact it can be shown that

$$< \varrho_\Phi > \simeq \varrho_{\mathrm{CDM}} \, , \tag{17}$$

and then the scalar field evolves now as pressureless matter, $< p_\Phi > \simeq 0$ [17, 14, 39, 12, 40, 41]. In order to have a scalar field dark matter model equivalent to CDM at cosmic scales, it is required that the free parameters of the cosh potential obey the following relation,

$$\kappa_0 V_0 \simeq \frac{1.7}{3} \left(\lambda^2 - 4\right)^3 \left(\frac{\Omega_{0\mathrm{CDM}}}{\Omega_{0\gamma}}\right)^3 \Omega_{0\mathrm{CDM}} H_0^2 \, , \tag{18}$$

where $\Omega_{0\mathrm{i}}$ is the current density parameter of the ith-matter component.

Equation (18) shows the allowed values of $\{\lambda \, , V_0\}$ to have an appropriate model of scalar dark matter, one that can mimic the standard CDM model. Up to now, the parameters of the cosh potential cannot be determined uniquely, but more information is necessary. As we shall see below, such information can be taken from the theory of linear perturbations.

3.1 Damping of the Scalar Power Spectrum

As more and new cosmological observations are appearing, it is always a must to check whether the correspondence between SFDM and CDM is preserved at the level of linear perturbations, where CDM is very successful.

When a detailed analysis of scalar perturbations is made [12, 40], it is found that the processed mass power spectrum for the SFDM, in the linear regime, is much the same as the CDM spectrum. Actually, the two spectra are related through the formula[39]

$$P_\Phi(k) = \left[\frac{\cos(x^3)}{1+x^8}\right]^2 P_{\rm CDM}(k) \,, \tag{19}$$

where $x = k/k_c$, with k the wave number of the different cosmological scales involved in the analysis. k_c corresponds to a cutoff scale below which linear perturbations are highly suppressed, and whose value is given by

$$k_c \simeq 1.3\,\lambda\,\sqrt{\lambda^2-4}\,\frac{\Omega_{0\rm CDM}}{\sqrt{\Omega_{0\gamma}}}\,H_0 \,. \tag{20}$$

If we take a cut-off of the mass power spectrum at $k_c = 4.5\,h\,{\rm Mpc}^{-1}$[8], we find that

$$\begin{aligned} \lambda &\simeq 20.3, \\ V_0 &\simeq \left(3.0\times10^{-27}\,m_{\rm Pl} \simeq 36.5\,{\rm eV}\right)^4, \\ m_\Phi &\simeq 9.1\times10^{-52}\,m_{\rm Pl} \simeq 1.1\times10^{-23}\,{\rm eV}. \end{aligned} \tag{21}$$

All parameters of potential (2) are now completely determined.

We would like to mention that the results shown in (21) are not unchangeable, since the cutoff scale k_c could take a larger value; in fact, the larger the cutoff wavenumber value the more similar the SFDM is to the CDM model. However, it is worth to mention that the free parameters of the particular model (2) can be determined from cosmological observations only. As we shall see below, the values in (21) seem to be the appropriate ones for the model to fulfill the requisites of non-linear structure formation.

4 Scalar Field Dark Matter and Structure Formation

At galactic scale, the CDM paradigm predicts a density profile which corresponds to the Navarro-Frenk-White (NFW) profile [5] given by

$$\varrho_{NFW} = \frac{\varrho_0}{(r/r_0)(r/r_0+1)^2} \,, \tag{22}$$

where r is the radial coordinate and r_0 is a parameter. However, this profile seems to have some differences with the inferred profiles from observations on

LSB galaxies[6]. The evidence points to the fact that, in the central regions, galaxies prefer to follow an almost constant density profile. The point is that the NFW model has the support of N-body simulations and has proved to be consistent with observations at cosmic scales, whereas the new observations have no cosmic counterpart that could support them, and it is here where the SFDM shows some advantages with respect to NFW as shown below [9].

The formation of galaxies through gravitational collapse of dark matter is not an easy problem to understand. A good model for galaxy formation has to take into account all the observed features of real galaxies. For example, it seems that many disk galaxies contain a black hole in their center, but others do not [42]. Typical galaxies are spiral, elliptical or dwarf galaxies (irregular galaxies may be galaxies still evolving). In most spiral and elliptical galaxies the luminous matter extends to $\sim 10-30\,\mathrm{kpc}$, and the total content of matter (including dark matter) is of the order of $10^{10}-10^{12}\,\mathrm{M}_\odot$, with about 10 times more dark matter than luminous component. The central density profile of the dark matter in galaxies should not be cusp[43]. Even though the luminous matter is only a small fraction of the total amount of matter in galaxies, it plays an important role in galaxy formation and stability, etc.

Let us now draw the general picture for the gravitational collapse of a scalar field. It is known that the final stage of a collapsed scalar field could be a massive object made of scalar field particles in quantum coherent states, like boson stars (for a complex scalar field) or oscillatons (for a real scalar field) [44, 45]. It is thus important to investigate whether the scalar field would collapse to form structures of the size of galaxies and provide the correct properties of any galactic dark matter candidate, like growing rotation curves and appropriate dark matter distribution functions. Also, we need to know which are the conditions that must be imposed on the scalar field particles.

For instance, axions are massive scalar particles with no self–interaction. In order for axions to be an essential component of the dark matter content of the Universe, their mass should be $m \sim 10^{-5}$ eV. With this axion mass, the scalar field collapses forming compact objects with masses of order of $M_{\mathrm{crit}} \sim 0.6\, m_{\mathrm{Pl}}^2/m \sim 10^{-6}\,\mathrm{M}_\odot$ [44, 45], which corresponds to objects with the mass of a planet. Since the amount of dark matter in galaxies is ten times bigger than the luminous matter, we would need tenths of millions of such objects around the solar system, which is clearly not the case. Also, these candidates behave just like standard CDM, and then they cannot explain the observed scarcity of dwarf galaxies and the smoothness of the galactic-core matter densities. That is, they should behave as objects made of baryonic matter do, except that they do not emit light; in such case this type of stars should be distributed as luminous matter is. In one word, such scalar field stars have been proposed as dark matter simply because they could account for the total amount of matter in the whole universe, without taking care of the local effect like rotation curves.

Let us return to the SFDM hypothesis for a real scalar field Φ (for a complex counterpart picture, see [46]). If a galaxy is an oscillaton, i.e., an oscillating soliton object, it must correspond to coherent scalar oscillations around the minimum of the scalar potential (2). The scalar field Φ and the metric coefficients (considering the spherically symmetric case) would be time-dependent, though it has been shown that such a configuration can be stable, non-singular and asymptotically flat [44, 47, 48, 49, 50]. For the collapse of a real scalar field, the critical value for the mass of an oscillaton (the maximum mass for which a stable configuration exists) will depend on the mass of the boson. Roughly speaking, if we take $m_\Phi = 10^{-23}\,\mathrm{eV}$ (see (21)), and use the formula for the critical mass of the oscillaton corresponding to a scalar field with a Φ^2 potential (i.e. just a mass term), we expect the critical mass to be [44, 45, 50]

$$M_{\mathrm{crit}} \sim 0.6\,\frac{m_{\mathrm{Pl}}^2}{m_\Phi} \sim 10^{12}\,\mathrm{M}_\odot\,. \tag{23}$$

which means that the critical mass of the cosmological model depicted in Sect. 3 is of the order of magnitude of the dark matter content of a standard galactic halo.

The scenario of galactic formation we assume is the following. A sea of scalar field particles fills the Universe and forms localized primordial fluctuations that could collapse to form stable objects, which we will interpret as the dark matter halos of galaxies. The equations governing the evolution of such scalar galaxy halos correspond to the Einstein-Klein-Gordon (EKG) system, which we show below applied to the simplest case of spherical symmetry.

The spherically symmetric line element is written in the form

$$ds^2 = -\alpha^2(r,t)dt^2 + a^2(r,t)dr^2 + r^2 d\Omega^2\,, \tag{24}$$

with $\alpha(r,t)$ the lapse function and $a^2(r,t)$ the radial metric function, and where we have chosen the polar-areal slicing condition (i.e. we force the line element to have the above form at all times, so that the area of a sphere with $r = R$ is always equal to $4\pi R^2$). This choice of gauge will force the lapse function $\alpha(r,t)$ to satisfy an ordinary differential equation in r (see below). The energy momentum tensor of the scalar field is

$$T_{\mu\nu} = \Phi_{,\mu}\Phi_{,\nu} - \frac{1}{2}g_{\mu\nu}\left[\Phi^{,\sigma}\Phi_{,\sigma} + 2V(\Phi)\right]\,. \tag{25}$$

We now introduce the first order variables $\Psi = \Phi_{,r}$ and $\Pi = a\Phi_{,t}/\alpha$, with which the Hamiltonian constraint becomes

$$\frac{a_{,r}}{a} = \frac{1-a^2}{2r} + \frac{\kappa_0 r}{4}\left[\Psi^2 + \Pi^2 + 2a^2 V\right]\,, \tag{26}$$

and the polar-areal slicing condition takes the form:

$$\frac{\alpha_{,r}}{\alpha} = \frac{a_{,r}}{a} + \frac{a^2-1}{r} - \kappa_0 r a^2 V\,. \tag{27}$$

All other components of Einstein's equations either vanish, or are a consequence of the last two equations. The Klein-Gordon (KG) equation now reads

$$\Phi_{,t} = \frac{\alpha}{a}\Pi\,, \tag{28}$$

$$\Pi_{,t} = 3\frac{\partial}{\partial(r^3)}\left(\frac{r^2\alpha\Psi}{a}\right) - a\alpha\,\frac{\mathrm{d}V}{\mathrm{d}\Phi} \tag{29}$$

$$\Psi_{,t} = \left(\frac{\alpha\Pi}{a}\right)_{,r}\,. \tag{30}$$

Equations (26-30) form the complete set of differential equations to be solved numerically.

The study of the evolution of oscillatons is a very interesting field by itself and its description would deserve more space in this text [44, 47, 48, 49, 50]. However, we will focus our attention on the results related to the SFDM hypothesis and the non-linear collapse of density perturbations.

Fortunately, the collapse of scalar fluctuations under realistic conditions does not require the solution of the relativistic EKG system. We can make use of some approximations to show the promising features of the SFDM at galactic level.

4.1 The Flat Space-Time Approximation

In order to show some properties of scalar halos, we start by taking the simplest approximation, that is, the flat space-time approximation. A scalar field object (an oscillaton) in flat space-time contains a flat central density profile, as it seems to be the case in galaxies [47, 51]. In order to see this, we study a massive oscillaton with potential $V = \frac{1}{2}m_\Phi^2\Phi^2$ in the Minkowski background. Even though it is not a solution to the Einstein equations as we are neglecting the gravitational force provoked by the scalar field, the solution is analytic and helps us to understand some features that appear in the non-flat oscillatons.

Using spherical coordinates, the KG equation in the Minkowski spacetime reads

$$\Phi'' + \frac{2}{r}\Phi' - m_\Phi^2\Phi = \ddot{\Phi} \tag{31}$$

where over-dot denotes $\partial/\partial t$ and prime $\partial/\partial r$. The exact general solution for the scalar field Φ is

$$\Phi(t,r) = \frac{\mathrm{e}^{\pm ikr}}{r}\mathrm{e}^{\pm i\omega t} \tag{32}$$

from which we obtain the dispersion relation $k^2 = \omega^2 - m_\Phi^2$. For $\omega > m_\Phi$ the solution is non-singular and vanishes at infinity. We will restrict ourselves to this case. It is more convenient to use trigonometric functions and to write the particular solution in the form

$$\Phi(t,x) = \Phi_0 \frac{\sin(x)}{x} \cos(\omega t) \tag{33}$$

where $x = kr$. It oscillates in harmonic manner in time. The scalar field spreads over all space, i.e., it is not confined to a finite region, as we are neglecting the gravitational interactions.

The analytic expression for the scalar field energy density derived from (33) is

$$\varrho_\Phi = \frac{1}{2}{\Phi_0}^2 \left\{ \left[\left(\frac{x\,\cos(x) - \sin(x)}{x^2} \right)^2 - \frac{\sin^2(x)}{x^4} \right] k^4 \cos^2(\omega\, t) + \frac{\omega^2 k^2 \sin^2(x)}{x^2} \right\} \tag{34}$$

which oscillates with a frequency $2\omega t$. Observe that close to the central regions of the object, the density of the oscillaton behaves like

$$\varrho_\Phi \sim \frac{1}{2}{\Phi_0}^2 k^2 \left[\omega^2 - k^2 \cos^2(\omega\, t) \right] + O(x^2) \tag{35}$$

which implies that the density is almost constant in the central regions, i.e. when $x \to 0$ the central density oscillates around a fixed value.

On the other hand, the asymptotic behavior when $x \to \infty$, is such that $\varrho_\Phi \sim 1/x^2$, i.e., far away from the center, the flat oscillaton density profile behaves like that of an isothermal halo sphere (IHS): the mass function oscillates around $M \sim x$ at large distances from the center. Nevertheless, if the gravitational interaction is taken into account, the object must be confined[47] and the integrated mass of the scalar field gives a finite value. In Fig. 1 we show the comparison between the NFW, the IHS and a flat oscillaton for the same galaxy. Observe that the flat oscillaton and NFW profiles remain very

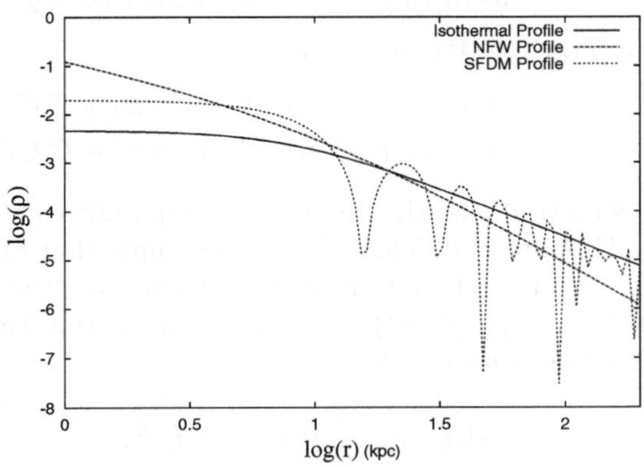

Fig. 1. Comparison between the energy density profile for the Scalar Field Dark Matter model with the NFW and the Isothermal models. The parameters for the isothermal model are $\rho_{\rm ISO} = 0.3/(r^2 + 8^2)$ and for the NFW profile are $\rho_{\rm NFW} = 10/(r(r+8)^2)$. The parameter used for the SFDM model are $k = 0.2$, $\sqrt{\kappa_0}\Phi_0 = 5 \times 10^{-2}$ in 33.

similar up to 100 kpc, then the flat oscillaton profile starts to follow the $1/r^2$ behavior, as the IHS one. The parameters used in the figures, correspond to a middle size galaxy.

As simple as the flat-spacetime approximation can be, it still shows that the density profile at the center of a scalar halo should be flat, as suggested by observations in galaxies. The self–gravity of the latter, if taken into account, will just change the profile at large distances from the center [47].

4.2 The Formation of Scalar Field Galaxy Halos

When the scalar field fluctuations reaches the non-linear regime, the scalar field collapses in a different way as the standard CDM. In a normal dust collapse, as for example in CDM, there is in principle nothing to avoid that the dust matter collapses all the time. There is only a radial gravitational force that provokes the collapse, and to stop it, one needs to invoke some virialization phenomenon.

In the scalar field paradigm the collapse is different. The radial and angular pressures are two natural components of the scalar field which prevent the collapse, avoiding the cusp density profiles in the centers of the collapsed objects (for details, see [50]). This is the main difference between the normal dust collapse and the SFDM one. The pressures play an important role in the SFDM equilibrium. In order to see this, as galaxies have an almost flat background, the Newtonian approximation of the EKG system (26-30) should be sufficient to understand them. This time, we will fully incorporate the influence of gravity in the process [52].

The weak field limit of the EKG equations arises when $\alpha^2 - 1, a^2 - 1, \sqrt{8\pi G}\,\Phi < 10^{-3}$ [45, 47]. We start by writing the scalar field and the metric coefficients in terms of the Newtonian fields ψ, U, U_2, A, A_2 as

$$\sqrt{8\pi G}\,\Phi = \mathrm{e}^{-\mathrm{i}\tau}\psi(\tau, x) + \mathrm{C.C.}\,, \tag{36}$$

$$\alpha^2 = 1 + 2U(\tau, x) + \mathrm{e}^{-2\mathrm{i}\tau}U_2(\tau, x) + \mathrm{C.C.}\,, \tag{37}$$

$$a^2 = 1 + 2A(\tau, x) + \mathrm{e}^{-2\mathrm{i}\tau}A_2(\tau, x) + \mathrm{C.C.}\,, \tag{38}$$

where we have also introduced the dimensionless quantities $\tau = mt$, $x = mr$. Notice that only U, A are real fields. Next, we assume that all the new fields are of order $\mathcal{O}(\varepsilon^2) \ll 1$, and that their derivatives obey the standard post-Newtonian rules $\partial_\tau \sim \varepsilon\partial_x \sim \mathcal{O}(\varepsilon^4)$. Therefore, considering the leading order terms only, the EKG equations now read

$$i\partial_\tau\psi = -\frac{1}{2x}\partial_x^2(x\psi) + U\psi\,, \tag{39}$$

$$\partial_x^2(xU) = x\psi\psi^*\,, \tag{40}$$

$$\partial_x U_2 = -x\psi^2\,. \tag{41}$$

In addition, $A(\tau, x) = x\partial_x U$ and $A_2 \sim \mathcal{O}(\varepsilon^4)$, that is, the metric coefficient g_{rr} can be taken plainly as time-independent.

It should be stressed here that the whole dynamics of the EKG system is contained in (39) and (40), which are the so-called Schrödinger–Newton (SN) equations[45, 46, 47, 53, 54, 55]; which also stand for the post-Newtonian expansion with complex scalar fields [45]. For example, under appropriate boundary conditions, stationary solutions of (39), (40) and (41) are in turn the so-called Newtonian oscillatons[47]. Indeed, (41) only arises for real scalar fields and represents the particular oscillatory behavior of the metric for oscillatons [44, 47, 49, 50].

For the initial data, we take the profile

$$\psi_i(0, x < x_0) = \psi_0 \frac{\sin(x/\sigma)}{(x/\sigma)}, \tag{42}$$

$$\psi_i(0, x > x_0) = \psi_1 \frac{\exp\left[-\sqrt{2|U_0|\sigma^2 - 1}\,(x/\sigma)\right]}{x}.$$

where $(x_0/\sigma) \simeq (n+1)\pi < \sqrt{2|U_0|}x_0$, where $n = 0, 1, 2, ...$ labels the number of nodes of the initial profile, and $x_0 = mR_0$ with R_0 is the physical size of the initial matter fluctuation. ψ_1 is evaluated from the continuity condition of the radial function at $x = x_0$. This profile is the simplest we can imagine that depicts the expected characteristics of a realistic scalar fluctuation: a confined wave function with nodes. More details can be found in[52].

We mention here that SN system is invariant under the scaling transformation

$$\{\tau, x, U, U_2, \psi\} \to \left\{\beta^{-2}\hat{\tau}, \beta^{-1}\hat{x}, \beta^2\hat{U}, \beta^2\hat{U}_2, \beta^2\hat{\psi}\right\} \tag{43}$$

where β is an arbitrary parameter. By means of (43) and an appropriate β, the collapse of our fluctuation can be reduced to the study of a conveniently sized configuration concerning *hat*-quantities only. Once the hat-configuration has been evolved, we apply the inverse transformation to recover the physical (no-hat) quantities.

We focus now on the numerical solution to the SN equations. Once the initial profile $\psi_i(0, x)$ is given, the Poisson equation (40) is integrated with a second order accurate upwind method inwards under the boundary condition $U(\tau, x_f) = GM(\tau, x_f)/x_f$ (being x_f the last point of our spherical grid), which is valid at each time slice according to the Newton theorems regarding spherical objects. The next scalar configuration is determined by solving the Schrödinger equation (39) using a second order finite differences fully implicit Cranck-Nicholson method [56]. The procedure is then repeated forward in time.

The accuracy depends on the time step $\Delta\tau$ and the grid resolution Δx, which should be chosen appropriately to assure that $|\Delta\psi|/|\psi| \ll 1$ in a time step. Thus, we should comply with both $\Delta\tau/(\Delta x)^2 \leq 1$ and $[\Delta\tau/(\Delta x)^2]|1 + U(\Delta x)^2| < 1$. The former is the condition applied to a free wave function, and the latter takes into account the presence of a potential in (39).

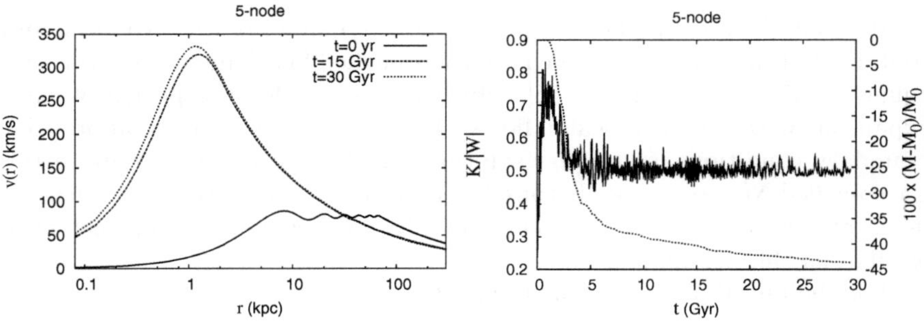

Fig. 2. Evolution of a 5-node initial scalar fluctuation of the form (42), see the text for details. Left: Profiles of the rotation curves $v(r) = \sqrt{GM(r)/r}$ at different times. Right: Evolution of the ratio Kinetic Energy/|Gravitational Potential Energy| (solid line), which shows the virialization of the two systems, and the total integrated mass M (dotted line). It is worth to notice the so called gravitational cooling after free fall of the system.

In Fig. 2 we show the run of an initial 5–node scalar fluctuation for a single scalar halo, whose initial physical parameters are $M_0 = 10^{11}\ M_\odot$ and $R_0 = 70$ kpc. The corresponding scale parameter is $\beta^2 = 6.38\times10^{-6}$, and then $\hat{\sigma} = 14.6$. The grid spacing is $\Delta\hat{x} = 0.25$ with the boundary at $\hat{x}_f = 1250$. The time step is $\Delta\hat{\tau} = 3 \times 10^{-2}$, and the run was followed up to a physical time of $T_0 = 30$ Gyr.

We see in Fig. 2 that the initial rotation curve profile is almost flat, but such flatness is lost at late times during the collapse. The scalar configuration rapidly virializes and forms a giant Newtonian oscillaton, almost as large as a realistic dwarf galaxy. The study of the SN system can be extended to the 3-dimensional case where the inclusion of a matter fluid is made through the r.h.s. of the Poisson equation (40). This issue will be reported elsewhere.

4.3 Supermassive Black Holes

It is worth to mention here that we can study the interaction between a scalar halo and a supermassive black hole at its center [57], as recent observations indicate is the case in real galaxies [42].

In this respect, for the SFDM hypothesis to survive, it should be shown that the central black hole does not represent a serious threat for the existence of a scalar halo. This is done, in first approximation, by solving the KG equation for a scalar field living in a fixed Schwarzschild background. This approximation is valid since the Schwarzschild radius of a typical supermassive black hole is smaller compared to the Compton length of a scalar particle of dark matter[7], $r_S \ll m_\Phi^{-1}$. The manifestation of the self-gravity of the scalar halo occurs at larger scales than m_Φ^{-1}.

[7] $r_S \simeq 10^{-13}\,(M_{bh}/M_\odot)$ pc.

From approximate analytical solutions of the KG equation, we can determine the fraction of incoming particles absorbed by the black hole (Γ), and the corresponding accretion rate of scalar matter. The results are, respectively,

$$\Gamma = 4\pi \left(m_\Phi r_\mathrm{S}\right)^3 \ , \tag{44}$$

$$\frac{\mathrm{d}M_\Phi}{\mathrm{d}t} = 4\pi \left(\Phi_0 m_\Phi r_\mathrm{S}\right)^2 \ , \tag{45}$$

where Φ_0 is the value of the scalar field (in units of $\kappa_0^{-1/2}$) near the horizon.

For typical values of the parameters involved, we get that $\Gamma \sim 10^{-20}$ and an accretion rate of around $10^{-14}\, M_\odot\, \mathrm{y}^{-1}$. This simple calculation shows that a scalar galaxy halo and a supermassive black hole can coexist.

5 Conclusions

In principle, it is possible to consider a braneworld model with an adequate Lagrangian in order to construct a consistent inflaton–dark matter scenario. The price we have to pay is that the inflaton remains after inflation and behaves as dark matter at the present epoch.

After inflation, the scalar dark matter we studied reproduces the success of the standard model of CDM, which seems to be correct at cosmological scales. We would like to stress that the SFDM gives the same results as the CDM model for both the homogeneous and the linearly perturbed cosmological evolutions.

On the non-linear realm of perturbations, we can track the evolution of scalar matter fluctuations through the use of numerical General Relativity. For the case of real scalar fields, scalar fluctuations collapse to form objects, called oscillatons, with a preferred mass of $\sim 10^{12}\,\mathrm{M}_\odot$. These objects are virialized and stable during periods of time longer than the actual age of the universe. Moreover, their density profiles are completely regular, i.e., without cusp centers, in a similar way as for real galaxies. This is encouraging enough [49, 58], to continue considering the scalar field as a good candidate to be the dark matter in the universe.

As a caution remark, we would like to mention that even if it is possible to identify the inflaton with the scalar field dark matter within braneworld models, this might not be necessarily the case. Despite the feasibility of the braneworld models, the results presented for epochs after inflation are still valid even if the dark matter scalar field were not fundamental. We should wait for more accurate observations to restrict the free parameters of the SFDM model and its relation to an inflationary epoch.

In conclusion, we have proposed a simple and unified model of inflationary dark matter, where the same scalar field provides the origin for the primordial spectrum of density perturbation (which are produced quantum–

mechanically) and later plays an important role evolving them to form the cosmological structures we observe today.

Acknowledgements

We would like to thank Aurelio Espíritu for technical suport. This work was partly supported by CONACyT México, under grants 32138-E and 34407-E and by a bilateral project DFG-CONACyT 444 MEX-13/17/0-1. The runs were made using the computer cluster in the Laboratorio de Supercómputo Astrofísico at Cinvestav-IPN, México.

References

1. N. A. Bahcall, J. P. Ostriker, S. Perlmutter and P. J. Steinhardt, Science **284**, 1481-1488. V. Sahni, A. Starobinsky, Int. J. Mod. Phys. D **9**, 373-444 (2000).
2. D. N. Spergel *et al*, `astro-ph/0302209`. J. L. Sievers *et al*, `astro-ph/0205387`.
3. A. R. Liddle and D. H. Lyth, *Cosmological inflation and large-scale structure* (Cambridge University Press, 2000). T. Padmanabhan, *Structure formation in the universe* (Cambridge University Press, 1993).
4. S. Perlmutter *et al. ApJ* **517** (1999)565. A. G. Riess *et al.*, *Astron.J.* **116** (1998)1009-1038.
5. J. Navarro, C. S. Frenk and S. S. M. White. ApJ **490**, 493 (1997).
6. W. J. G. de Blok, S. S. MacGaugh, A. Bosma and V. C. Rubin, ApJ **552**, L23 (2001). S. S. MacGaugh, V. C. Rubin and W. J. G. de Blok, ApJ **122**, 2381 (2201). W. J. G. de Blok, S. S. MacGaugh and V. C. Rubin, ApJ **122**, 2396 (2001).
7. J. J. Binney and N. W. Evans, *MNRAS* **327** L27 (2001). Blais-Ouellette, C. Carignan, and P. Amram, E-print astro-ph/0203146. C. M. Trott and R. L. Wesbster, E-print astro-ph/0203196. P. Salucci, F. Walter and A. Borriello, E-print astro-ph/0206304.
8. M. Kamionkowski and A. R. Liddle, Phys. Rev. Lett. **84**, 4525 (2000).
9. D. N. Spergel and P. J. Steinhardt, Phys. Rev. Lett. **84**, 3760 (2000); B. D. Wandelt, R. Dave, G. R. Farrar, P. C. McGuire, D. N. Spergel, and P. J. Steinhardt, astro-ph/0006344; C. Firmani, E. D'Onghia, V. Avila-Reese, G. Chincarini, and X. Hernández, MNRAS **315**, L29 (2000); C. Firmani, E. D'Onghia, G. Chincarini, X. Hernández, and V. Avila-Reese, MNRAS **321**, 713 (2001); M. Kaplinghat, L. Knox, and M. S. Turner, Phys. Rev. Lett. **85**, 3335 (2000).
10. N. Trentham, O. Moeller and E. Ramirez-Ruiz, MNRAS **322** 658 (2001).
11. F. S. Guzmán and T. Matos, Class. Quant. Grav. **17**, L9-L16 (2000). E-print gr-qc/9810028.
12. T. Matos and L. A. Ureña-López, Class. Quantum Grav. **17**, L75 (2000).
13. H. Dehnen and B. Rose, Astrophys. Sp Sci. **207** (1993) 133-144. H. Dehnen, B. Rose and K. Amer, Astrophys. Sp Sci. **234** (1995) 69-83.
14. P. J. E. Peebles, Astrophys. J. **534**, L127 (2000); J. Goodman, New Astron. **5**, 103 (2000). M. C. Bento, O. Bertolami, R. Rosenfeld, and L. Teodoro, Phys. Rev. D **62**, 041302 (2000); O. Bertolami, M. C. Bento, and R. Rosenfeld, astro-ph/0111415. A. Riotto and I. Tkachev, Phys. Lett. B **484**, 177 (2000).

15. C. Wetterich, Nucl. Phys. B **302**, 668 (1988); B. Ratra and P. J. E. Peebles, Phys. Rev. D **37**, 3406 (1988); E. J. Copeland, A. R. Liddle, and D. Wands, Phys. Rev. D **57**, 4686 (1998). R. R. Caldwell, R. Dave, and P. J. Steinhardt, Phys. Rev. Lett. **80**, 1582 (1998); L. Wang, R. R. Caldwell, J. P. Ostriker, and P. J. Steinhardt, Astrophys. J. **503**, 17 (2000).
16. L. A. Ureña-López and T. Matos, Phys. Rev. D **62**, 081302 (2000).
17. V. Sahni and L. Wang, Phys. Rev. D **62**, 103517 (2000).
18. P. G. Ferreira and M. Joyce, Phys. Rev. D **58**, 023503 (1998).
19. N. Arkani-Hamed, S. Dimopoulos and G. Dvali, *Phys. Lett.* **B 429**, 263 (1998); I. Antoniadis, N. Arkani-Hamed, S. Dimopoulos and G. Dvali, Phys. Lett. **B436**, 257 (1998).
20. L. Randall and R. Sundrum, *Phys. Rev. Lett* **83**, 3370 (1999); 4690 (1999).
21. G. Dvali, G. Gabadadze and M. Porrati, *Phys. Lett.* **B 484**, 112 (2000).
22. C. Deffayet, *Phys. Lett.* **B 502**, 199 (2001).
23. G. Dvali, G. Gabadadze, *Phys. Rev. D* **63**, 065007 (2001).
24. P. Binetruy, C. Defayyet, and D. Langlois *Nucl. Phys.* **B 565**, 269 (2000).
25. D. Ida, *JHEP* **0009**, 014 (2000).
26. N. Deruelle and T. Dolezel, *Phys. Rev. D* **62**, 103502 (2001).
27. P. Bowcock, C. Charmousis and R. Gregory, *Class. Quant. Grav.* **17**, 4745 (2000).
28. A. C. Davis, S. C. Davis, W. B. Perkins and I. R. Vernon, *Phys.Lett* **B 504**, 254 (2001).
29. E. J. Copeland, A. R. Liddle, and J. E. Lidsey, Phys. Rev. D **64**, 023509 (2001).
30. G. Huey and J. Lidsey, Phys. Lett. **B514**, 217 (2001).
31. V. Sahni, M. Sami, and T. Souradeep, Phys.Rev. D **65**, 023518 (2002).
32. J.E. Lidsey, T. Matos and L.A. Ureña-López, *Phys. Rev. D* **66** 023514 (2002).
33. H. García-Compeán, R. Cordero and T. Matos. To be published.
34. C. P. Burgess, A. de la Macorra, I. Maksymyk and F. Quevedo, *Phys. Lett.* **B 410** 181 (1997).
35. A. R. Frey and A. Mazumdar, *Phys. Rev. D* **67**, 046006 (2003). E-print hep-th/0210254.
36. R. Maartens, D. Wands, B. A. Basset, and I. P. C. Heard, Phys. Rev. D **62**, 041301 (2000).
37. L. H. Ford, Phys. Rev. D **35**, 2955 (1987); L. P. Grishchuk and Y. V. Sidorov, Phys. Rev. D **42**, 341 (1990); B. Spokoiny, Phys. Lett. B **315**, 40 (1993).
38. D. H. Lyth, and D. Wands, Phys. Lett. B**524**, 5 (2002). D. H. Lyth, C. Ungarelli, and D. Wands, E-print astro-ph/0208055. N. Bartolo and A. R. Liddle, Phys. Rev. D **65**, 121301(R) (2002). A. R. Liddle and L. A. Ureña-López, E-print astro-ph/0302054.
39. W. Hu, R. Barkana, and A. Gruzinov, Phys. Rev. Lett. **85**, 1158 (2000).
40. T. Matos and L. A. Ureña-López, Phys. Rev. D **63**, 063506 (2001).
41. T. Matos and L. A. Ureña-López, Phys. Lett. B **538**, 246 (2002).
42. L. Ferrarese, E-print astro-ph/0207050, astro-ph/0207050.
43. B. Moore, Nature **370**, 629 (1994). A. Burkert, ApJ **477**, L25 (1995). J. A. Tyson, G. P. Kochanski and I. P. Dell'Antonio, ApJ **498**, L107 (1998).
44. E. Seidel and W. Suen, Phys. Rev. Lett. **66**, 1659 (1991). E. Seidel and W. Suen, Phys. Rev. Lett. **72**, 2516 (1994).
45. E. Seidel and W-M. Suen, Phys. Rev. D **42**, 384 (1990); J. Balakrishna, E. Seidel, and W-M. Suen, Phys. Rev. D **58**, 104004 (1998).

46. A. Arbey, J. Lesgourgues, and P. Salati, Phys. Rev. D **64**, 123528 (2001); *ibid*, **65**, 083514 (2002); astro-ph/0301533.
47. L. A. Ureña-López, Class. Quantum Grav. **19**, 2617 (2002); L. A. Ureña-López, T. Matos, and R. Becerril, Class. Quantum Grav. **19**, 6259 (2002).
48. S. H. Hawley and M. W. Choptuik, Phys. Rev. **D 67**, 024010 (2003).
49. M. Alcubierre, F. S. Guzmán, T. Matos, D. Núñez, L. A. Ureña-López and P. Wiederhold. Class. Quantum Grav. **19**, 5017 (2002).
50. M. Alcubierre, R. Becerril, F. S. Guzmán, T. Matos, D. Núñez, and L. A. Ureña-López, E-print gr-qc/0301105.
51. T. Matos and D. Núñez, E-print astro-ph/0303455.
52. F. Siddhartha Guzmán and L.A. Ureña-López, Phys. Rev. D, in press. E-print astro-ph/0303440.
53. S. J. Sin, Phys. Rev. D **50**, 3650 (1994); S. U. Ji and S. J. Sin, Phys. Rev. D **50**, 3655 (1994).
54. J. W. Lee and I. G. Koh, Phys. Rev. D **53**, 2236 (1996).
55. R. Friedberg, T. D. Lee, and Y. Pang, Phys. Rev. D **35**, 3640 (1987). I. M. Moroz, R. Penrose, and P. Tod, Class. Quantum Grav. **15**, 2733 (1998); P. Tod and I. M. Moroz, Nonlinearity **12**, 201 (1999).
56. W. H. Press, S. A. Teukolsky, W. T. Watterling and B. P. Flannery, *Numerical Recipes in Fortran.* Cambridge University Press, 1992.
57. L. A. Ureña-López and A. R. Liddle, Phys. Rev. D **66**, 083005 (2002).
58. T. Matos and F. S. Guzmán, Class. Quantum Grav. **18**, 5055 (2001).

Cosmological Applications of Loop Quantum Gravity

Martin Bojowald[1] and Hugo A. Morales-Técotl[2,3]

[1] Center for Gravitational Physics and Geometry, The Pennsylvania State University, University Park, PA 16802, USA. bojowald@gravity.psu.edu
[2] Departamento de Física, Universidad Autónoma Metropolitana Iztapalapa, A.P. 55-534 México D.F. 09340, México hugo@xanum.uam.mx
[3] Associate member of AS-ICTP Trieste, Italy.

Abstract. After a brief introduction to classical and quantum gravity we discuss applications of loop quantum gravity in the cosmological realm. This includes the basic formalism and recent results of loop quantum cosmology, and a computation of modified dispersion relations for quantum gravity phenomenology. The presentation is held at a level which does not require much background knowledge in general relativity or mathematical techniques such as functional analysis, so as to make the article accessible to graduate students and researchers from other fields.

1 Introduction

According to general relativity, not only the gravitational field but also the structure of space and time, the stage for all the other fields, is governed by the dynamical laws of physics. The space we see is not a fixed background, but it evolves on large time scales, even to such extreme situations as singularities where all of space collapses into a single point. At such a point, however, energy densities and tidal forces diverge; all classical theories break down, even general relativity itself. This implies that general relativity cannot be complete since it predicts its own breakdown. Already for a long time, it has been widely expected that a quantum theory of general relativity would cure this problem, providing a theory which can tell us about the fate of a classical singularity.

Most of the time, quantum gravity has been regarded as being far away from observational tests. In such a situation, different approaches would have to be judged purely on grounds of internal consistency and their ability to solve conceptual problems. Those requirements are already very restrictive for the quantization of a complicated theory as general relativity, to the extent that in all the decades of intense research not a single completely convincing quantum theory of gravity has emerged yet, even though there is a number of promising candidates with different strengths. Still, the ultimate test of a physical theory must come from a confrontation with observations of the real world. For quantum gravity, this means observations of effects which happen at the smallest scales, the size of the Planck length $\ell_{\rm P} \approx 10^{-32}$cm.

M. Bojowald and H.A. Morales-Técotl, Cosmological Applications of Loop Quantum Gravity, Lect. Notes Phys. **646**, 421–462 (2004)
http://www.springerlink.com/

In particular in the light of recent improvements in precision cosmology, the cosmological arena seems to be most promising for experimental tests. This is fortunate since also many conceptual issues arise in the cosmological setting where the universe is studied as a whole. Examples are the singularity problem mentioned above and the so-called problem of time which we will address later. Therefore, one can use the same methods and approximations to deal with conceptual problems and to derive observational consequences.

One of the main approaches to quantum gravity is based on a canonical quantization of general relativity, which started with the formal Wheeler–DeWitt quantization and more recently evolved into quantum geometry. Its main strength is its background independence, i.e. the metric tensor which describes the geometry of space is quantized as a whole and not split into a background and a dynamical part. Since most familiar techniques of quantum field theory rely on the presence of a background, for this ambitious approach new techniques had to be invented which are often mathematically involved. By now, most of the necessary methods have been developed and we are ready to explore them in simple but physically interesting situations.

Being the quantization of a complicated, non-linear field theory, quantum gravity cannot be expected to be easily understood in full generality. As always in physics, one has to employ approximation techniques which isolate a small number of objects one is interested in without taking into account all possible interactions. Prominent examples are symmetric models (which are usually called minisuperspaces in the context of general relativity) and perturbations of some degrees of freedom around a simple solution. This opens the possibility to study the universe as a whole (which is homogeneous and isotropic at large scales) as well as the propagation of a single particle in otherwise empty space (where complicated interactions can be ignored).

These two scenarios constitute the two main parts of this article. In the context of the first one (Sect. 5) we discuss the basic equations which govern the quantum evolution of an isotropic universe and special properties which reflect general issues of quantum gravity. We then analyze these equations and see how the effects of quantum geometry solve and elucidate important conceptual problems. Quantum gravity effects in these regimes also lead to modifications of classical equations of motion which can be used in a phenomenological analysis. A different kind of phenomenology, related to the propagation of particles in empty space, is discussed in Sect. 6. In this context cosmological scales are involved for many proposals of observations, and so they fit into the present scheme.

Both settings are now at a stage where characteristic effects have been identified and separated from the complicated, often intimidating technical foundation. This is a natural starting point for phenomenological analyzes and opens a convenient port of entry for beginners to the field.

The article is intended to describe the basic formalism to an extent which makes it possible to understand the applications without requiring too much background knowledge (the presentation cannot be entirely background independent, though). The general framework of quantum geometry is reviewed briefly in Sect. 4 after recalling facts about general relativity (Sect. 2) and the Wheeler–DeWitt quantization (Sect. 3). For the details we provide a guide to the literature including technical reviews and original papers.

2 General Relativity

General relativity is a field theory for the metric $g_{\mu\nu}$ on a space-time M which determines the line element[4] $\mathrm{d}s^2 = g_{\mu\nu}(x)\mathrm{d}x^\mu \mathrm{d}x^\nu$. The line element, in turn specifies the geometry of space-time; we can, e.g., measure the length of a curve $C\colon \mathbb{R} \to M, t \mapsto x^\mu(t)$ using

$$\ell(C) = \int \mathrm{d}s = \int \sqrt{g_{\mu\nu}(x(t))\dot{x}^\mu(t)\dot{x}^\nu(t)}\,\mathrm{d}t\,.$$

2.1 Field Equations

While a space-time can be equipped with many different metrics, resulting in different geometries, only a subclass is selected by Einstein's field equations of general relativity which are complicated non-linear partial differential equations with the energy density of matter as a source.

They can be understood as giving the dynamical evolution of a space-like geometry in a physical universe. Due to the four-dimensional covariance, however, in general there is no distinguished space-like slice which could be used to describe the evolution. All possible slices are allowed, and they describe the same four-dimensional picture thanks to symmetries of the field equations.

Selecting a slicing into space-like manifolds, the field equations take on different forms and do not show the four-dimensional covariance explicitly. However, such a formulation has the advantage that it allows a canonical formulation where the metric q_{ab} only of space-like slices plays the role of coordinates of a phase space, whose momenta are related to the time derivative of the metric, or the extrinsic curvature $K_{ab} = -\frac{1}{2}\dot{q}_{ab}$ of a slice [1]. This is in particular helpful for a quantization since canonical quantization techniques become available. The momentum conjugate to the metric q_{ab} is related to the extrinsic curvature by

$$\pi^{ab} = -\tfrac{1}{2}\sqrt{\det q}\,(K^{ab} - q^{ab}K^c_c)$$

where indices are raised by using the inverse q^{ab} of the metric. The dynamical field equation, the analog of Einstein's field equations, takes the form of a

[4] In expressions with repeated indices, a summation over the allowed range is understood unless specified otherwise. We use greek letters $\mu, \nu, \ldots$ for space-time indices ranging from zero to three and latin letters from the beginning of the alphabet, $a, b, \ldots$ for space indices ranging from one to three.

constraint,[5] the Hamiltonian constraint

$$8\pi G\sqrt{\det q}\,(q_{ac}q_{bd}+q_{ad}q_{bc}-q_{ab}q_{cd})\pi^{ab}\pi^{cd}-\frac{1}{16\pi G}\sqrt{\det q}\;\,{}^3\!R(q)$$
$$+\sqrt{\det q}\,\rho_{\rm matter}(q)=0 \qquad (1)$$

where G is the gravitational constant, ${}^3R(q)$ the so-called Ricci scalar of the spatial geometry (which is a function of the metric q), and $\rho_{\rm matter}(q)$ is the energy density of matter depending on the particular matter content (it depends on the metric, but not on its momenta in the absence of curvature couplings).

The complicated constraint can be simplified slightly by transforming to new variables [2], which has the additional advantage of bringing general relativity into the form of a gauge theory, allowing even more powerful mathematical techniques. In this reformulation, the canonical degrees of freedom are a densitized triad E^a_i which can be thought of as giving three vectors labelled by the index $1\leq i\leq 3$. Requiring that these vectors are orthonormal defines a metric given by

$$q_{ab}=\sqrt{|\det E^c_j|}(E^{-1})^i_a(E^{-1})^i_b\,.$$

Its canonical conjugate is the Ashtekar connection

$$A^i_a=\Gamma^i_a-\gamma K^i_a \qquad (2)$$

where Γ^i_a is the spin connection (given uniquely by the triad such that $\partial_aE^b_i+\epsilon_{ijk}\Gamma^j_aE^b_k=0$) and K^i_a is the extrinsic curvature. The positive Barbero–Immirzi parameter γ also appears in the symplectic structure together with the gravitational constant G

$$\{A^i_a(x),E^b_j(y)\}=8\pi\gamma G\delta^b_a\delta^i_j\delta(x,y) \qquad (3)$$

and labels equivalent classical formulations. Thus, it can be chosen arbitrarily, but the freedom will be important later for the quantum theory. The basic variables can be thought of as a "vector potential" A^i_a and the "electric field" E^a_i of a gauge theory, whose gauge group is the rotation group SO(3) which rotates the three triad vectors: $E^a_i\mapsto\Lambda^j_iE^a_j$ for $\Lambda\in$ SO(3) (such a rotation does not change the metric).

Now, the Hamiltonian constraint takes the form [3]:

$$|\det E^c_l|^{-1/2}\epsilon_{ijk}F^i_{ab}E^a_jE^b_k-2(1+\gamma^2)|\det E^c_l|^{-1/2}K^i_{[a}K^j_{b]}E^a_iE^j_b$$
$$+8\pi G\sqrt{|\det E^c_l|}\,\rho_{\rm matter}(E)=0 \qquad (4)$$

where F^i_{ab} the curvature of the Ashtekar connection, and the matter energy density $\rho_{\rm matter}(E)$ now depends on the triad via the metric.

[5] Note that this requires a relation between the basic fields in every point of space; there are infinitely many degrees of freedom and infinitely many constraints.

2.2 Approximations

Given the complicated nature of the field equations, one has to resort to approximation schemes in order to study realistic situations. In the case of gravity, the most widely used approximations are:

- Assume symmetries. This simplifies the field equations by eliminating several degrees of freedom and simplifying the relations between the remaining ones. In a cosmological situation, for instance, one can assume space to be homogeneous such that the field equations reduce to ordinary differential equations in time.
- Perturbations around a simple known solution. One can, e.g., study a small amount of matter, e.g. a gravitational wave or a single particle, and its propagation in Minkowski space. To leading order, the back reaction of the geometry, which changes due to the presence of the particle's energy density, on the particle's propagation can be ignored.
- Asymptotic regimes with boundary conditions. In many situations it is possible to isolate interesting degrees of freedom by looking at boundaries of space-time with special boundary conditions capturing the physical situation. It can then be possible to ignore interactions with the bulk degrees of freedom which simplifies the analysis. This strategy is most widely used in the context of black hole physics, in its most advanced form with isolated horizon conditions; see, e.g. [4].

The first two approximation schemes and their applications in quantum geometry will be discussed on Sects. 5 and 6, respectively. Since the last one so far does not have many cosmological applications, it will not be used here. It does have applications in quantum geometry, however, in the calculation of black hole entropy [5]. In this section we only illustrate the first one in the context of isotropic cosmology.

2.3 Cosmology

In the simplest case of a cosmological model we can assume space to be isotropic (looking the same in all its points and in all directions) which implies that one can choose coordinates in which the line element takes the form

$$\mathrm{d}s^2 = -\mathrm{d}t^2 + a(t)^2((1-kr^2)\mathrm{d}r^2 + r^2(\mathrm{d}\vartheta^2 + \sin\vartheta\mathrm{d}\varphi^2)) \tag{5}$$

with the scale factor $a(t)$ (the evolving "radius" of the universe). The constant k can take the values $k = 0$ for a spatially flat model (planar), $k = 1$ for a model with positive spatial curvature (spherical), and $k = -1$ for a model with negative spatial curvature (hyperbolic). Einstein's field equations restrict the possible behavior of $a(t)$ in the form of the Friedmann equation [6]

$$\left(\frac{\dot{a}}{a}\right)^2 = \frac{16\pi}{3} G\,\rho(a) - \frac{k}{a^2}\,. \tag{6}$$

Since also the matter density $\rho(a)$ enters, we can find $a(t)$ only if we specify the matter content. Common choices are "dust" with $\rho(a) \propto a^{-3}$ or "radiation" with $\rho(a) \propto a^{-4}$ (due to an additional red-shift factor), which describe the matter degrees of freedom collectively. After choosing the matter content, we just need to solve an ordinary differential equation. For radiation in a spatially flat universe, e.g., all solutions are given by $a(t) \propto \sqrt{t-t_0}$ where t_0 is an integration constant.

In a more complicated but also more fundamental way one can describe the matter by using additional matter fields[6] which enter via their Hamiltonian (or total energy). This results in a system of coupled ordinary differential equations, one for the scale factor and others for the matter fields. A common example in cosmology is a scalar ϕ which has Hamiltonian

$$H_\phi(a) = \tfrac{1}{2}a^{-3}p_\phi^2 + a^3 W(\phi) \tag{7}$$

with its potential W and the scalar momentum $p_\phi = a^3\dot{\phi}$. Note that it is important to keep track of the a-dependence in cosmology since a is evolving; in the usual formulas for Hamiltonians on Minkowski space a does not appear.

The Friedmann equation is now given by (6) with energy density $\rho(a) = H_\phi(a)/a^3$. Now, the right hand side depends explicitly on ϕ and p_ϕ which both depend on time. Their evolution is given by the Hamiltonian equations of motion

$$\dot{\phi} = \{\phi, H_\phi\} = p_\phi/a^3 \tag{8}$$

$$\dot{p}_\phi = \{p_\phi, H_\phi\} = -a^3 W'(\phi)\,. \tag{9}$$

By using the first equation one can transform the second one into a second order equation of motion for ϕ:

$$\ddot{\phi} = -3\,\dot{a}a^{-1}\,\dot{\phi} - W'(\phi) \tag{10}$$

which in addition to the usual force term from the potential has a friction term proportional to the first derivative of ϕ. The friction is strongest for a rapid expansion.

When we come close to $a = 0$, the kinetic term usually dominates and even diverges when $a = 0$. This is problematic and leads to the singularity problem discussed in the following subsection. However, the divergence occurs only when $p_\phi \neq 0$ for small a, so one could try to arrange the evolution of the scalar such that the divergence is avoided. In addition to suppressing the diverging kinetic term, we have the additional welcome fact that $p_\phi \approx 0$ implies $\phi \approx \phi_0 = \text{const}$. The right hand side of the Friedmann equation then becomes constant, $(\dot{a}/a)^2 \approx \Lambda = (16\pi G/3)W(\phi_0)$ for $k = 0$. Its solutions are given by $a(t) \propto \exp(\sqrt{\Lambda}t)$ which describes an *accelerated* expansion, or

[6] In a homogeneous model, matter "fields" are also described by a finite number of parameters only, e.g. a single one for a scalar ϕ.

inflation. Though motivated in a different way here, inflation is deemed to be an important ingredient in cosmological model building, in particular for structure formation.

Unfortunately, however, it is very difficult to arrange the evolution of the scalar in the way described here; for – in addition to introducing a new field, the inflaton ϕ – it requires very special scalar potentials and also initial values of the scalar. A common choice is a quadratic potential $W(\phi) = \frac{1}{2}m\phi^2$ (e.g., for chaotic inflation) which requires a very small m (a very flat potential) for inflation to take place long enough, and also a huge initial value ϕ_0 pushing it up to Planck values. There is a plethora of models with intricate potentials, all requiring very special choices.

Inflation in general is the term for accelerated expansion [7], i.e. $\ddot{a} > 0$. It is not necessarily of the exponential form as above, but can be parameterized by different ranges of the so-called equation of state parameter w which needs to be less than $w < -\frac{1}{3}$ for inflation. It can be introduced by a phenomenological a-dependence of the energy density,

$$\rho(a) \propto a^{-3(w+1)} . \tag{11}$$

Note, however, that this is in general possible only with a-dependent w except for special cases. Solutions for a (with $k = 0$) are then of the form

$$a(t) \propto \begin{cases} (t-t_0)^{2/(3w+3)} & \text{for } -1 < w < -\frac{1}{3} \text{ (power-law inflation)} \\ \exp(\sqrt{\Lambda}t) & \text{for } w = -1 \text{ (standard inflation)} \\ (t_0-t)^{2/(3w+3)} & \text{for } w < -1 \text{ (super-inflation)} \end{cases} \tag{12}$$

where t_0 is an initial value (replaced by $\Lambda = (16\pi/3)G\rho$ with the constant energy density ρ for standard inflation). Note in particular that super-inflation (also called pole-law inflation) can be valid only during a limited period of time since otherwise a would diverge for $t = t_0$. While these possibilities add more choices for model building, they share with standard inflation that they are difficult to arrange with scalar potentials.

2.4 Singularities

Trying to suppress the kinetic term has led us to introduce inflation as an ingredient in cosmological models. Can it lead to a regular evolution, provided we manage to arrange it in some way? The answer is no, for the following intuitive reason: We can get p_ϕ to be very small by making special choices, but it will not be exactly zero and eventually the diverging a^{-3} will win if we only go close enough to $a = 0$. In the end, we always have to face the singularity problem illustrated by the simple solution $a(t) \propto \sqrt{t-t_0}$ for radiation: $a(t_0) = 0$ such that all of space collapses to a single point (any length of a space-like curve at t_0 measured with the line element (5) is zero) and the energy density diverges. The most dooming consequence is that the evolution breaks down: We cannot set up an initial value problem at t_0 and evolve to values of t smaller than t_0. The theory does not tell us what happens beyond t_0. This consequence is a general property of general relativity which cannot be avoided. We used the symmetric situation only for purposes of illustration, but the singularity problem remains true for any solution [8]. There will always be points which can be reached in a finite amount of time, but we will not be able to know anything as to what happens beyond such a point. General relativity cannot be complete since it predicts situations where it breaks down.

This is the classical situation. Can it be better in a quantum theory of gravity? In fact, this has been the hope for decades, justified by the following motivation: The classical hydrogen atom is unstable, but we know well that quantum mechanics leads to a ground state of finite energy $E_0 = -\frac{1}{2}m_e e^4/\hbar^2$ which cures the instability problem. One can easily see that this is the only non-relativistic energy scale which can be built from the fundamental parameters purely for dimensional reasons. In particular, a non-zero $\hbar$ is necessary, for in the classical limit $\hbar \to 0$ the ground state energy diverges leading to the classical instability. As an additional consequence we know that the existence of a non-zero $\hbar$ leads to discrete energies.

In gravity the situation is similar. We have its fundamental parameter G from which we can build a natural length scale[7] $\ell_P = \sqrt{8\pi G\hbar}$, the Planck length. It is very tiny and becomes important only at small scales, e.g. close to classical singularities where the whole space is small. Where the Planck length becomes important we expect deviations from the classical behavior which will hopefully cure the singularity problem. In the classical limit, $\hbar \to 0$, the Planck length becomes zero and we would get back the singularity. Completing our suggestions from the hydrogen atom, we also expect discrete lengths in a quantum theory of gravity, the explicit form of which can only be concluded from a precise implementation.

3 Wheeler–DeWitt Quantum Gravity

As discussed, one can bring general relativity into a canonical formulation where the metric q_{ab} and its momenta π^{ab} play the role of phase space coordinates (infinitely many because they depend on the points of space), together with possible matter degrees of freedom and their momenta. This allows us to perform a canonical quantization (see, e.g., [9]) by representing quantum states as functionals $\Psi(q_{ab}, \phi)$ of the metric and matter fields, corresponding to a metric representation. The metric itself then acts as a multiplication operator, and its conjugate π^{ab} by a functional derivative $\hat{\pi}^{ab} = -i\hbar\partial/\partial q_{ab}$. These are the basic operators from which more complicated ones can be constructed.

[7] Sometimes the Planck length is defined as $\sqrt{G\hbar}$.

3.1 The Wheeler–DeWitt Equation

In a canonical formulation of general relativity, the dynamics is determined by a constraint equation, (1) in the variables used here. Replacing q_{ab} and π^{ab} by the respective operators yields a complicated constraint operator $\hat{H}_{\mathrm{ADM}}$ acting on a wave function Ψ. Since the classical expression must vanish, only states Ψ are allowed which are annihilated by the constraint operator, i.e. they have to fulfill the Wheeler–DeWitt equation $\hat{H}_{\mathrm{ADM}}\Psi = 0$. Since the constraint is quadratic in the momenta, this is a second order functional differential equation. However, it is only formal since it contains products of functional derivatives which have to be regularized in a way which does not spoil the properties of the theory, in particular its background independence. Such a regularization is complicated because the classical constraint is not even a polynomial in the basic fields, and so far it has not been done successfully in the ADM formulation.

There is another apparent difficulty with the constraint equation: It is supposed to give us the dynamics, but there is no time dependence at all, and no time derivative part as in a Schrödinger equation. This is a general property of theories as general relativity which are invariant under four-dimensional coordinate transformations. We do not have an absolute notion of time, and thus it cannot appear in the basic evolution equation. Classically, we can introduce a time parameter (coordinate time t), but it just serves to parameterize classical trajectories. It can be changed freely by a coordinate transformation. In the quantum theory, which is formulated in a coordinate independent way, coordinate time cannot appear explicitly. Instead, one has to understand the evolution in a relational way: there is no evolution with respect to an absolute time, but only evolution of all the degrees of freedom with respect to each other. After all, this is how we perceive time. We build a clock, which is a collection of matter degrees of freedom with very special interactions with each other, and observe the evolution of other objects, degrees of freedom with weak interactions with the clock, with respect to its progression. Similarly, we can imagine to select a particular combination of matter or metric degrees of freedom as our clock variable and re-express the constraint equation as an evolution equation with respect to it [10, 11]. For instance, in a cosmological context we can choose the volume of space as *internal time* and measure the evolution of matter degrees of freedom with respect to the expansion or contraction of the universe. In general, however, a global choice of a time degree of freedom which would allow us to bring the full Wheeler–DeWitt equation into the form of an evolution equation, is not known; this is the problem of time in general relativity.

Due to the complicated regularization and interpretational issues, applications of the full Wheeler–DeWitt equation have been done only at a formal level for semiclassical calculations.

3.2 Minisuperspaces

In order to study the theory explicitly, we again have to resort to approximations. A common simplification of the Wheeler–DeWitt formalism is the reduction to minisuperspace models where the space is homogeneous or even isotropic. Therefore, the metric of space is specified by a finite number of parameters only – only the scale factor a in the isotropic case. While this is similar in spirit to looking for symmetric classical solutions as we did in Sect. 2, there is also an important difference: If we want the symmetry to be preserved in time we need to restrict the time derivative of the metric, i.e. its canonical conjugate, in the same symmetric form. This is possible classically, but in quantum mechanics it violates Heisenberg's uncertainty relations for the excluded degrees of freedom. Minisuperspace models do not just give us particular, if very special exact solutions as in the classical theory; their results must be regarded as approximations which are valid only under the assumption that the interaction with the excluded parameters is negligible.

An isotropic minisuperspace model has the two gravitational parameters a and its conjugate $p_a = 3a\dot{a}/8\pi G$ together with possible matter degrees of freedom which we simply denote as ϕ and p_ϕ. Using a Schrödinger quantization of the momenta acting on a wave function $\psi(a,\phi)$, the Friedmann equation (6) is quantized to the Wheeler–DeWitt equation

$$\frac{3}{2}\left(-\frac{1}{9}\ell_{\mathrm{P}}^4 a^{-1}\frac{\partial}{\partial a}a^{-1}\frac{\partial}{\partial a}+k\right)a\psi(a,\phi)=8\pi G\hat{H}_\phi(a)\psi(a,\phi) \qquad (13)$$

with matter Hamiltonian $\hat{H}_\phi(a)$. This equation is not unique due to ordering ambiguities on the left hand side. Here, we use the one which is related to the quantization derived later. Without fixing the ordering ambiguity, consequences derived from the equation are ambiguous [12].

The Wheeler–DeWitt equation quantizes the dynamical classical equation and thus should describe the quantum dynamics. As described before, in an isotropic model we can select the scale factor a as an internal time; evolution of the matter fields will then be measured not in absolute terms but in relation to the expansion or contraction of the universe. Interpreting a as a time variable immediately brings (13) to the form of a time evolution equation, albeit with an unconventional time derivative term.

An unresolvable problem of the Wheeler–DeWitt quantization, however, is that it is still singular. Energy densities, all depending on the multiplication operator a^{-1} are still unbounded, and the Wheeler–DeWitt equation does not tell us what happens at the other side of the classical singularity at $a = 0$. Instead, the point of view has been that the universe is "created" at $a = 0$ such that initial conditions have to be imposed there. DeWitt [10] tried to combine both problems by requiring $\psi(0) = 0$ which can be interpreted as requiring a vanishing probability density to find the universe at the singularity. However, this very probability interpretation, which is just taken over from quantum mechanics, is not known to make sense in a quantum cosmological context. Furthermore, at the very least one would also need appropriate fall-off conditions for the wave function since otherwise we can still get arbitrarily close to the singularity. Appropriate conditions are not known, and it is not at all clear if they could always be implemented. The worst problem is, however, that DeWitt's initial condition is not well-posed in more general models where its only solution would vanish identically.

DeWitt's condition has been replaced by several proposals which are motivated from different intuitions [13, 14]. However, they do not eliminate the singularity; with them the wave function would not even vanish at $a = 0$ in the isotropic case. They accept the singularity as a point of creation.

Thus, the hope motivated from the hydrogen atom has not materialized. The isotropic universe model is still singular after quantizing. Do we have to accept that singularities of gravitational systems cannot be removed, not even by quantization? If the answer would be affirmative, it would spell severe problems for any desire to describe the real world by a physical theory. It would mean that there can be no complete description at all; any attempt would stop at the singularity problem.

Fortunately, the answer turns out not to be affirmative. We will see in the following sections that singularities are removed by an appropriate quantization. Why, then, is this not the case for the Wheeler–DeWitt quantization? One has to keep in mind that there is no mathematically well-defined Wheeler–DeWitt quantization of full general relativity, and systematic investigations of even the formal equations are lacking. What one usually does instead is merely a quantum mechanical application in a simple model with only a few degrees of freedom. There is no full theory which one could use to see if all quantization steps would also be possible there. General relativity is a complicated theory and its quantization can be done, if at all, only in very special ways which have to respect complicated consistency conditions, e.g. in the form of commutation relations between basic operators. In a simple model, all these problems can be brushed over and consistency conditions are easily overlooked. One hint that this in fact happened in the Wheeler–DeWitt quantization is the lacking discreteness of space. We expected that a non-zero Planck length in quantum gravity would lead to the discreteness of space. While we did see the Planck length in (13), there was no associated discreteness: the scale factor operator, which is simply the multiplication operator a, still has continuous spectrum.

After the discussion it should now be clear how one has to proceed in a more reliable way. We have to use as much of the full theory of quantum gravity as we know and be very careful to use only techniques in our symmetric models which can also be implemented in the full theory. In this way, we would respect all consistency conditions and obtain a faithful model of the full theory. Ideally, we would even start from the full theory and define symmetric models there at the level of states and operators.

By now, we have good candidates for a full theory of quantum gravity, and in the case of quantum geometry [15, 16, 17] also a procedure to define symmetric models from it [18]. We will describe the main results in the following two sections.

4 Quantum Geometry

While the Wheeler–DeWitt quantization is based on the ADM formulation whose basic variables are the metric and extrinsic curvature of a spatial slice, quantum geometry is a newer approach based on Ashtekar's variables. Quantization, in particular of a non-linear field theory, is a delicate step whose success can depend significantly on the formulation of a given classical theory. Classical theories can usually be formulated in many different, equivalent ways, all being related by canonical transformations. Not all of these transformations, however, can be implemented unitarily at the quantum level which would be necessary for the quantum theories to be equivalent, too. For instance, when quantizing one has to choose a set of basic variables closed under taking Poisson brackets which are promoted unambiguously to operators in such a way that their Poisson brackets are mapped to commutation relations. There is no unique prescription to quantize other functions on phase space which are not just linear functions of the basic ones, giving rise to quantization ambiguities. In quantum mechanics one can give quite general conditions for a representation of at least the basic variables to be unique (this representation is the well-known Schrödinger quantization). However, such a theorem is not available for a field theory with infinitely many degrees of freedom such that even the basic variables cannot be quantized uniquely without further conditions.

One can often use symmetry requirements together with other natural conditions in order to select a unique representation of the basic variables, e.g. Poincaré invariance for a field theory on Minkowski space as a background [19]. For general relativity, which is background independent, it has recently been proven in the context of quantum geometry that diffeomorphism invariance, i.e. invariance under arbitrary deformations of space, can replace Poincaré invariance in strongly restricting the class of possible representations [20]. It is clear that those precise theorems can only be achieved within a theory which is mathematically well-defined. The Wheeler–DeWitt quantization, on the other hand, does not exist beyond a purely formal level and it is unknown if it can give a well-defined quantum representation of the ADM variables at all. In any case, it is based on basic variables different from the ones quantum geometry is based on so that any representation it defines would likely be inequivalent to the one of quantum geometry.

From the beginning, quantum geometry was striving for a mathematically rigorous formulation. This has been possible because it uses Ashtekar's variables which bring general relativity into the form of a gauge theory. While not all standard techniques for quantizing a gauge theory can be applied (most of them are not background independent), new powerful techniques for a *background independent* quantization have been developed [21, 22, 23]. This was possible only because the space of connections, which is the configuration space of quantum geometry, has a structure much better understood than the configuration space of the Wheeler–DeWitt quantization, namely the space of metrics.

We do not describe those techniques here and instead refer the interested reader to the literature where by now several technical reviews are available [17, 24]. In this section, instead, we present an intuitive construction which illustrates all the main results.

4.1 Basic Operators and States

As usually in gauge theories (for instance in lattice formulations), one can form holonomies as functions of connections for all curves $e\colon [0,1] \to \Sigma$ in a manifold Σ,

$$h_e(A) = \mathcal{P}\exp\left(\int_e A_a^i(e(t))\,\dot{e}^a(t)\tau_i \mathrm{d}t\right) \in \mathrm{SU}(2) \tag{14}$$

where $\dot{e}^a$ is the tangent vector to the curve e and $\tau_i = -\frac{\mathrm{i}}{2}\sigma_i$ are generators of the gauge group SU(2) in terms of the Pauli matrices. The symbol $\mathcal{P}$ denotes path ordering which means that the non-commuting su(2) elements in the exponential are ordered along the curve. Similarly, given a surface $S\colon [0,1]^2 \to \Sigma$ we can form a flux as a function of the triads,

$$E(S) = \int_S E_i^a(y) n_a(y) \tau^i \mathrm{d}^2 y \tag{15}$$

where n_a is the co-normal[8] to the surface S. Holonomies and fluxes are the basic variables which are used for quantum geometry, and they represent the phase space of general relativity faithfully in the sense that any two configurations of general relativity can be distinguished by evaluating holonomies and fluxes in them.

One can now prove that the set of holonomies and fluxes is closed under taking Poisson brackets and that there is a representation of this Poisson algebra as an operator algebra on a function space. Moreover, using the action of the diffeomorphism group on Σ, which deforms the edges and surfaces involved in the above definitions, this representation is the unique covariant one. Note that unlike the functional derivatives appearing in the Wheeler-DeWitt quantization, these are well-defined operators on an infinite dimensional Hilbert space. Note in particular that holonomies are well-defined as operators, but *not* the connection itself. A Wheeler–DeWitt quantization, on the other

[8] The co-normal is defined as $n_a = \frac{1}{2}\epsilon_{abc}\epsilon^{de}(\partial x^b/\partial y^d)(\partial x^c/\partial y^e)$ without using a background metric, where x^a are coordinates of Σ and y^d coordinates of the surface S.

hand, regards the extrinsic curvature, related to the connection, as one of the basic fields and would try to promote it to an operator. This is not possible in quantum geometry (and it is not known if it is possible at a precise level at all) which demonstrates the inequivalence of both approaches. The fact that only the holonomies can be quantized can also be seen as one of the consistency conditions of a full theory of quantum gravity mentioned earlier. In a minisuperspace model one can easily quantize the isotropic extrinsic curvature, which is proportional to p_a/a. However, since it is not possible in the full theory, the model departs from it already at a very basic level. A reliable model of a quantum theory of gravity should implement the feature that only holonomies can be quantized; we will come back to this issue later.

We did not yet specify the space of functions on which the basic operators act in the representation of quantum geometry. Understandably, a full definition involves many techniques of functional analysis, but it can also be described in intuitive terms. As mentioned already, it is convenient to define the theory in a connection representation since the space of connections is well-understood. We can then start with the function **1** which takes the value one in every connection and regard it as our ground state.[9] The holonomies depend only on the connection and thus act as multiplication operators in a connection formulation [25]. Acting with a single holonomy $h_e(A)$ on the state **1** results in a state which depends on the connection in a non-trivial way, but only on its values along the curve e. More precisely, since holonomies take values in the group SU(2), we should choose an SU(2)-representation, for instance the fundamental one, and regard the matrix elements of the holonomy in this representation as multiplication operators. This can be done with holonomies along all possible curves, and also acting with the same curve several times. Those operators can be regarded as basic creation operators of the quantum theory. Acting with holonomies along different curves results in a dependence on the connection along all the curves, while acting with holonomies along the same curve leads to a dependence along the curve in a more complicated way given by multiplying all the fundamental representations to higher ones. One can imagine that the state space obtained in this way with all possible edges (possibly intersecting and overlapping) in arbitrary numbers is quite complicated, but not all states obtained in this way are independent: one has to respect the decomposition rules of representations. This can all be done resulting in a basis of states, the so-called spin network states [26]. Furthermore, they are orthonormal with respect to

[9] Note that we do not call it "vacuum state" since the usual term "vacuum" denotes a state in which matter is unexcited but the gravitational background is Minkowski space (or another non-degenerate solution of general relativity). We will see shortly, however, that the ground state we are using here represents a state in which even gravity is "unexcited" in the sense that it defines a completely degenerate geometry.

the diffeomorphism invariant measure singled out by the representation, the Ashtekar–Lewandowski measure [22].

Note also that the quantum theory should be invariant under SU(2)-rotations of the fields since a rotated triad does not give us a new metric. Holonomies are not gauge invariant in this sense, but as in lattice gauge theories we can use Wilson loops instead which are defined as traces of holonomies along a closed loop. Repeating the above construction only using Wilson loops results in gauge invariant states.

Since we used holonomies to construct our state space, their action can be obtained by multiplication and subsequent decomposition in the independent states. Fluxes, on the other hand, are built from the conjugate of connections and thus become derivative operators. Their action is most easy to understand for a flux with a surface S which is transversal to all curves used in constructing a given state. Since the value of a triad in a given point is conjugate to the connection in the same point but Poisson commutes with values of the connection in any other point, the flux operator will only notice *intersection points* of the surface with all the edges which will be summed over with individual contributions. The contributions of all the intersection points are the same if we count intersections with overlapping curves separately. In this way, acting with a flux operator on a state returns the same state multiplied with the intersection number between the surface of the flux and all the curves in the state. This immediately shows us the eigenvalues of flux operators which turn out to be *discrete*. Since the fluxes are the basic operators representing the triad from which geometric quantities like length, area and volume are constructed, it shows that geometry is discrete [27, 28, 29, 30]. The main part of the area spectrum for a given surface S (the one disregarding intersections of curves in the state) is

$$A(S) = \gamma \ell_{\mathrm{P}}^2 \sum_i \sqrt{j_i(j_i+1)} \tag{16}$$

where the sum is over all intersections of the surface S with curves in the state, and the SU(2)-labels j_i parameterize the multiplicity if curves overlap (without overlapping curves, all j_i are $\frac{1}{2}$). Thus, quantum geometry predicts that geometric spectra are discrete, and it also provides an explicit form. Note that the Planck length appears (which arises because the basic Poisson brackets (3) contain the gravitational constant while $\hbar$ enters by quantizing a derivative operator), but the scale of the discreteness is set by the Barbero–Immirzi parameter γ. While different γ lead to equivalent classical theories, the value of the parameter does matter in the quantum theory. If γ would be large the discreteness would be important already at large scales despite the smallness of the Planck length. Calculations from black hole entropy, however, show that γ must be smaller than one, its precise value being $\log(2)/\pi\sqrt{3}$ [5].

Thus, quantum geometry already fulfills one of our expectations of Sect. 2, namely that quantum gravity should predict a discreteness of geometry with a scale set roughly by the Planck length. Note that the use of holonomies in constructing the quantum theory, which was necessary for a well-defined formulation, is essential in obtaining the result about the discreteness. This fact has been overlooked in the Wheeler–DeWitt quantization which, consequently, does not show the discreteness.

4.2 Composite Operators

We now have a well-defined framework with a quantization of our basic quantities, holonomies and fluxes. Using them we can construct composite operators, e.g. geometric ones like area and volume or the constraint operator which governs the dynamics. Many of them have already been defined, but they are quite complicated. The volume operator, for instance, has been constructed and it has been shown to have a discrete spectrum [27, 29, 31]; determining all its eigenvalues, however, would require the diagonalization of arbitrarily large matrices. Since it plays an important role in constructing other operators, in particular the Hamiltonian constraint [32, 33], it makes explicit calculations in the whole theory complicated.

The constraint, for instance can be quantized by using a small Wilson loop along some loop α, which has the expansion $h_\alpha = 1 + A s_1^a s_2^b F_{ab}^i \tau_i$ where A is the coordinate area of the loop and s_1 and s_2 are tangent vectors to two of its edges, to quantize the curvature of the connection. The product of triads divided by the determinant appears to be problematic because the triad can be degenerate resulting in a vanishing determinant. However, one can make use of the classical identity [32]

$$\frac{E_i^a E_j^b \epsilon^{ijk}}{\sqrt{|\det E|}} = \frac{1}{4\pi\gamma G} \epsilon^{abc} \{A_c^k, V\} \,, \tag{17}$$

replace the connection components by holonomies h_s and use the volume operator to quantize this expression in a non-degenerate way. For the first part[10] of the constraint (4) this results in

$$\hat{H} = \sum_{v \in \mathcal{V}} \sum_{v(\Delta)=v} \epsilon^{IJK} \, \mathrm{tr}(h_{\alpha_{IJ}(\Delta)} h_{s_K(\Delta)} [h^{-1}_{s_K(\Delta)}, \hat{V}_v]) \tag{18}$$

where we sum over the set $\mathcal{V}$ of vertices of the graph belonging to the state we act on, and over all possible choices (up to diffeomorphisms) to form a tetrahedron Δ with a loop $\alpha_{IJ}(\Delta)$ sharing two sides with the graph and a third transversal curve $s_K(\Delta)$. The first holonomy along $\alpha_{IJ}(\Delta)$ quantizes the curvature components while $h_{s_K(\Delta)}$ together with the commutator quantizes the triad components.

[10] The remaining part of the constraint involving extrinsic curvature components can be obtained from the first part since the extrinsic curvature can be written as a Poisson bracket of the first part of the constraint with the volume [32].

A similar strategy can be used for matter Hamiltonians which usually also require to divide by the determinant of the triad at least in their kinetic terms. Here it is enough to cite some operators as examples to illustrate the general structure which will later be used in Sect. 6; for details we refer to [33]. For electromagnetism, for instance, we need to quantize

$$H_{\mathrm{Maxwell}} = \int_{\Sigma} \mathrm{d}^3x \, \frac{q_{ab}}{2Q^2\sqrt{\det q}} [\underline{E}^a \underline{E}^b + \underline{B}^a \underline{B}^b] \tag{19}$$

where $(\underline{A}_a, \underline{E}^a/Q^2)$ are the canonical fields of the electromagnetic sector with gauge group U(1) and coupling constant Q, related to the dimensionless fine structure constant by $\alpha_{\mathrm{EM}} = Q^2\hbar$. Furthermore, $\underline{B}^b$ is the magnetic field of the U(1) connection $\underline{A}$, i.e. the dual of its curvature.

Along the lines followed for the gravitational Hamiltonian we obtain the full electromagnetic Hamiltonian operator [33] (with a weight factor $w(v)$ depending on the graph)

$$\begin{aligned}\hat{H}_{\mathrm{Maxwell}} = \frac{1}{2\ell_{\mathrm{P}}^4 Q^2} \sum_{v\in\mathcal{V}} w(v) & \sum_{v(\Delta)=v(\Delta')=v} \mathrm{tr}\left(\tau_i\, h_{s_L(\Delta)} \left[h^{-1}_{s_L(\Delta)}, \sqrt{\hat{V}_v}\right]\right) \\ & \times \mathrm{tr}\left(\tau_i h_{s_P(\Delta')} \left[h^{-1}_{s_P(\Delta')}, \sqrt{\hat{V}_v}\right]\right) \\ & \times \epsilon^{JKL}\epsilon^{MNP}\left[\left(\mathrm{e}^{-\mathrm{i}\hat{\Phi}^B_{JK}} - 1\right)\left(\mathrm{e}^{-\mathrm{i}\hat{\Phi}'^B_{MN}} - 1\right) - \hat{\Phi}^E_{JK}\hat{\Phi}'^E_{MN})\right] .\end{aligned} \tag{20}$$

Let us emphasize the structure of the above regularized Hamiltonian. There is a common gravitational factor included in the SU(2) trace. The basic entities that quantize the electromagnetic part are the corresponding fluxes (as operators acting on a state for the electromagnetic field): one is associated with the magnetic field, which enters through a product of exponential flux factors $\exp(-\mathrm{i}\Phi^{(\prime)B})$ constructed from holonomies in Δ and Δ', respectively, while the other is related to the electric field, entering in a bilinear product of electric fluxes $\Phi^{(\prime)E}$.

Similarly, one can set up a theory for fermions which would be coupled to the gauge fields and to gravity. For a spin-$\frac{1}{2}$ field θ_A one obtains the kinetic part (with the Planck mass $m_{\mathrm{P}} = \hbar/\ell_{\mathrm{P}}$)

$$\begin{aligned}\hat{H}_{\mathrm{spin}-1/2} = -\frac{m_{\mathrm{P}}}{2\ell_{\mathrm{P}}^3} \sum_{v\in\mathcal{V}} \sum_{v(\Delta)=v} & \epsilon^{ijk}\epsilon^{IJK}\, \mathrm{tr}\left(\tau_i h_{s_I(\Delta)} \left[h^{-1}_{s_I(\Delta)}, \sqrt{\hat{V}_v)}\right]\right) \\ & \times \mathrm{tr}\left(\tau_j h_{s_J(\Delta)} \left[h^{-1}_{s_J(\Delta)}, \sqrt{\hat{V}_v)}\right]\right) \\ & \times \left[\left[(\tau_k h_{s_K(\Delta)}\theta)|_{s_K(\Delta)} - \theta|_v\right]_A \frac{\partial}{\partial\theta_A(v)} + h.c.\right] .\end{aligned} \tag{21}$$

All of these matter Hamiltonians are well-defined, bounded operators, which is remarkable since in quantum field theories on a classical background matter Hamiltonians usually have ultraviolet divergences. This can be interpreted as a natural cut-off implied by the discrete structure. Compared to the Wheeler–DeWitt quantization it is a huge progress that well-defined Hamiltonian constraint operators are available in the full theory. Not surprisingly, their action is very complicated for several reasons. The most obvious ones are the fact that Wilson loops necessary to quantize curvature components create many new curves in a state which is acted on, and that the volume operator is being used to quantize triad components. The first fact implies that complicated graphs are created, while the second one shows that even a single one of those contributions is difficult to analyze due to the unknown volume spectrum. And after determining the action of the constraint operator on states we still have to solve it, i.e. find its kernel. Furthermore, there are always several possible ways to quantize a classical Hamiltonian such that the ones we wrote down should be considered as possible choices which incorporate the main features.

The complicated nature should not come as a surprise, though. After all, we are dealing with a quantization of full general relativity without any simplifying assumptions. Even the classical equations are difficult to solve and to analyze if we do not assume symmetries or employ approximation schemes. Those simplifications are also available for quantum geometry, which is the subject of the rest of this article. Symmetries can be introduced at the level of states which can be rigorously defined as distributional, i.e. non-normalizable states (they cannot be ordinary states since the discrete structure would break any continuous symmetry). Approximations can be done in many ways, and different schemes are currently being worked out.

5 Loop Quantum Cosmology

Loop quantum cosmology aims to investigate quantum geometry in simplified situations which are obtained by implementing symmetries. In contrast to a Wheeler–DeWitt quantization and its minisuperspace models, there is now also a full theory available. It is then possible to perform all the steps of the quantization in a manner analogous to those in the full theory. In particular, one can be careful enough to respect all consistency conditions as, e.g., the use of holonomies.[11] There is a tighter relation between symmetric models and the full theory which goes beyond pure analogy. For instance, symmetric states can be defined rigorously [18, 34, 35] and the relation between operators is currently being investigated. In this section, as already in the previous one, we use intuitive ideas to describe the results.

[11] This sometimes requires to perform manipulations which seem more complicated than necessary or even unnatural from the point of view of a reduced model. However, all of this can be motivated from the full theory, and in fact exploiting simplifications which are not available in a full theory can always be misleading.

In addition to testing implications of the full theory in a simpler context, it is also possible to derive physical results. Fortunately, many interesting and realistic physical situations can be approximated by symmetric ones. This is true in particular for cosmology where one can assume the universe to be homogeneous and isotropic at large scales.

5.1 Symmetric States and Basic Operators

As seen before, the canonical fields of a theory of gravity are completely described by two numbers (depending on time) in an isotropic context. For Ashtekar's variables, isotropic connections and triads take the form

$$A_a^i(x)\mathrm{d}x^a = c\,\omega^i \quad , \quad E_i^a \frac{\partial}{\partial x^a} = p\,X_i \tag{22}$$

where ω^i are invariant 1-forms and X_i invariant vector fields. For a spatially flat configuration, $\omega^i = \mathrm{d}x^i$ are just coordinate differentials, while X_i are the derivatives; for non-zero spatial curvature the coordinate dependence is more complicated. In all isotropic models, c and p are functions just of time. Their relation to the isotropic variables used before is

$$c = \tfrac{1}{2}(k - \gamma\dot{a}) \quad , \quad |p| = a^2 \tag{23}$$

such that $\{c, p\} = (8\pi/3)\gamma G$. An important difference is that p can have both signs, corresponding to the two possible orientations of a triad. In a classical theory we would ultimately have to restrict to one sign since $p = 0$ represents a degenerate triad, and positive and negative signs are disconnected. But the situation can (and will) be different in a quantum theory.

We can now perform an analog of the construction of states in the full theory. The symmetry condition can be implemented by using only invariant connections (22) in holonomies as creation operators, i.e.

$$h_i(c) = \exp(c\tau_i) = \cos(c/2) + 2\tau_i \sin(c/2) \ .$$

Consequently, all the states we construct by acting on the ground state **1** will be functions of only the variable c. All the complication of the full states with an arbitrary number of curves has collapsed because of our symmetry assumption. The analog of the spin network basis, an orthonormal basis in the connection representation, is given by[12] [37]

$$\langle c|n\rangle = \frac{\exp(inc/2)}{\sqrt{2}\sin(c/2)} \tag{24}$$

for all integer n.

[12] A more careful analysis shows that the Hilbert space of loop quantum cosmology is not separable [36]. For our purposes, however, we can restrict to the separable subspace used here which is left invariant by our operators.

An analog of the flux operator is given by a quantization of the isotropic triad component p,

$$\hat{p}|n\rangle = \tfrac{1}{6}\gamma\ell_{\mathrm{P}}^2 n|n\rangle\ . \tag{25}$$

This immediately allows a number of observations: Its spectrum is discrete (n corresponds to the intersection number in the full theory) which results in a discrete geometry. The scale factor $\hat{a} = \sqrt{|\hat{p}|}$ also has a discrete spectrum which is very different from the Wheeler–DeWitt quantization where the scale factor is just a multiplication operator with a continuous spectrum. Thus, we obtain a different quantization with deviations being most important for small scale factors, close to the classical singularity. The quantization $\hat{p}$ also tells us that the sign of the "intersection number" n determines the orientation of space. There is only one state, $|0\rangle$, which is annihilated by $\hat{p}$; we identify it with the classical singularity.

We can also use $\hat{p}$ in order to obtain a quantization of the volume [38]; here we use the convention

$$V_{(|n|-1)/2} = \left(\tfrac{1}{6}\gamma\ell_{\mathrm{P}}^2\right)^{\frac{3}{2}} \sqrt{(|n|-1)|n|(|n|+1)} \tag{26}$$

for its eigenvalue on a state $|n\rangle$. Thus, we achieved our aims: We have a quantization of a model which simplifies the full theory. This can be seen from the simple nature of the states and the explicit form of the volume spectrum which is not available in the full theory. As we will see later, this also allows us to derive explicit composite operators [39, 37]. At the same time, we managed to preserve essential features of the full theory leading to a quantization different from[13] the Wheeler–DeWitt one which lacks a relation to a full theory.

5.2 Inverse Powers of the Scale Factor

We can now use the framework to perform further tests of essential aspects of quantum cosmology. In the Wheeler–DeWitt quantization we have seen that the inverse scale factor a^{-1} becomes an unbounded operator. Since its powers also appear in matter Hamiltonians, their quantizations will also be unbounded, reflecting the classical divergence. The situation in loop quantum cosmology looks even worse at first glance: $\hat{a}$ still contains zero in its spectrum, but now as a discrete point. Then its inverse does not even exist. There can now be two possibilities: It can be that we cannot get a quantization of the classically diverging a^{-1}, which would mean that there is no way to resolve the classical singularity. As the other possibility it can turn out that there are admissible quantizations of a^{-1} in the sense that they have the correct classical limit and are densely defined operators. The second possibility exists because, as noted earlier, there are usually several possibilities

[13] Both quantizations are in fact inequivalent because, e.g., the operator $\hat{a}$ has discrete spectrum in one and a continuous spectrum in the other quantization.

to construct a non-basic operator like a^{-1}. If the simplest one fails (looking for an inverse of $\hat{a}$), it does not mean that there is no quantization at all.

It turns out that the second possibility is realized [40], in a way special to quantum geometry. We can use the identity (17), which has been essential in quantizing Hamiltonians, in order to rewrite a^{-1} in a classically equivalent form which allows a well-defined quantization. In this way, again, we stay very close to the full theory, repeating only what can be done there, and at the same time obtain physically interesting results.

The reformulation can be written in a simple way for a symmetric context, e.g.,[14]

$$a^{-1} = \left(\frac{1}{2\pi\gamma G}\{c, |p|^{3/4}\}\right)^2 = \left(\frac{1}{3\pi\gamma G}\sum_i \mathrm{tr}(\tau_i h_i(c)\{h_i(c)^{-1}, \sqrt{V}\})\right)^2$$
$$= \left((2\pi\gamma G\, j(j+1)(2j+1))^{-1}\sum_i \mathrm{tr}_j(\tau_i h_i(c)\{h_i(c)^{-1}, \sqrt{V}\})\right)^2 \quad (27)$$

indicating in the last step that we can choose any SU(2)-representation when computing the trace without changing the classical expression (the trace without label is in the fundamental representation, $j = \frac{1}{2}$). Note that we only need a positive power of p at the right hand side which can easily be quantized. We just have to use holonomy operators and the volume operator, and turn the Poisson bracket into a commutator. This results in a well-defined operator which has eigenstates $|n\rangle$ and, for $j = \frac{1}{2}$, eigenvalues

$$\widehat{a^{-1}}|n\rangle = \frac{16}{\gamma^2 \ell_P^4}\left(\sqrt{V_{|n|/2}} - \sqrt{V_{|n|/2-1}}\right)^2 |n\rangle \quad (28)$$

in terms of the volume eigenvalues (26).

It has the following properties [40, 41]; see Fig. 1:

1. It is a *finite* operator with upper bound[15] $(a^{-1})_{\mathrm{max}} = \frac{32(2-\sqrt{2})}{3\ell_P}$ at a peak at $n = 2$. Now we can see that the situation is just as in the case of the hydrogen atom: The classically pathological behavior is cured by quantum effects which – purely for dimensional reasons – require $\hbar$ in the denominator (recall $\ell_P = \sqrt{8\pi G\hbar}$). A finite value for the upper bound is possible only with non-zero ℓ_P; in the classical limit $\ell_P \to 0$ we reobtain the classical divergence.

[14] One can easily see that there are many ways to rewrite a^{-1} in such a way. Essential features, however, are common to all these reformulations, for instance the fact that always the absolute value of p will appear in the Poisson bracket rather than p itself. We will come back to this issue shortly in the context of quantization ambiguities.

[15] We only use the general form of the upper bound. Its precise value depends on quantization ambiguities and is not important in this context.

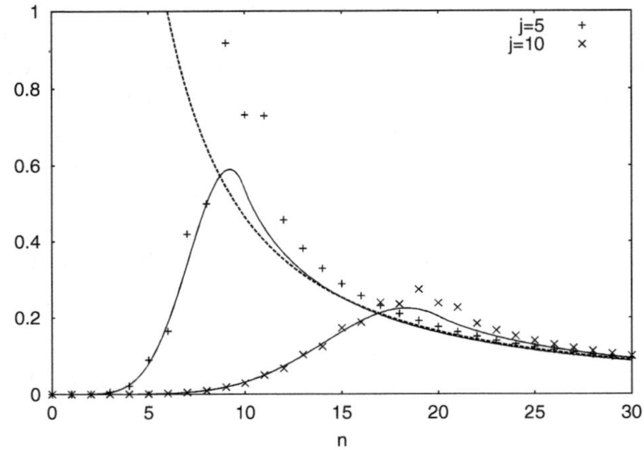

Fig. 1. Two examples for eigenvalues of inverse scale factor operators with $j = 5$ (+) and $j = 10$ (×), compared to the classical behavior (dashed) and the approximations (30) [41].

2. The first point demonstrates that the classical behavior is modified at small volume, but one can see that it is approached rapidly for volumes larger than the peak position. Thus, the quantization (28) has the correct classical limit and is perfectly admissible.
3. While the first two points verify our optimistic expectations, there is also an unexpected feature. The classical divergence is not just cut off at a finite value, the eigenvalues of the inverse scale factor drop off when we go to smaller volume and are exactly zero for $n = 0$ (where the eigenvalue of the scale factor is also zero). This feature, which will be important later, is explained by the fact that the right hand side of (27) also includes a factor of $\mathrm{sgn}(p)^2$ since the absolute value of p appears in the Poisson bracket. Strictly speaking, we can only quantize $\mathrm{sgn}(p)^2 a^{-1}$, not just a^{-1} itself. Classically, we cannot distinguish between both expressions – both are equally ill-defined for $a = 0$ and we would have to restrict to positive p. As it turns out, however, the expression with the sign does have well-defined quantizations, while the other one does not. Therefore, we have to use the sign when quantizing expressions involving inverse powers of a, and it is responsible for pushing the eigenvalue of the inverse scale factor at $n = 0$ to zero.
4. As already indicated in (27), we can rewrite the classical expression in many equivalent ways. Quantizations, however, will not necessarily be the same. In particular, using a higher representation $j \neq \frac{1}{2}$ in (27), the holonomies in a quantization will change n by amounts larger than one. In (28) we will then have volume eigenvalues not just with $n-1$ and $n+1$, but from $n - 2j$ to $n + 2j$ corresponding to the coupling rules of angular

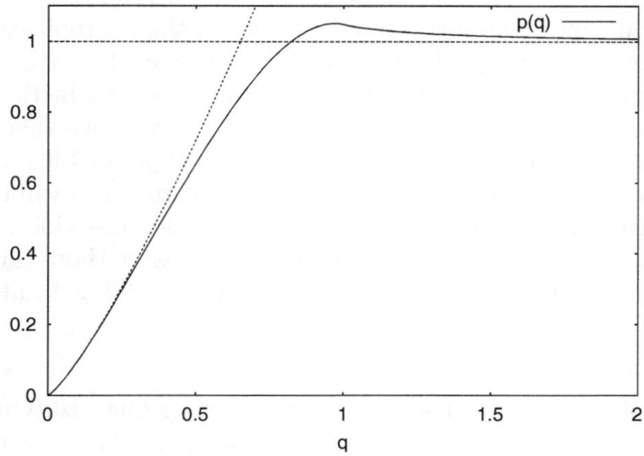

Fig. 2. The function $p(q)$ of (29), derived in [41]. For small q, $p(q)$ increases like $p(q) = \frac{12}{7}q^{5/4}(1 - q + O(q^2))$ (dashed).

momentum. Quantitative features depend on the particular value of j (or other quantization ambiguities), but qualitative aspects – in particular the ones in points 1 to 3 – *do not change*. Thus, the quantization is robust under ambiguities, but there can be small changes depending on which particular quantization is used. Such a freedom can also be exploited in a phenomenological analysis of some effects.

Let us make the last point more explicit. The exact formula for eigenvalues with a non-fundamental representation is quite complicated. It can, however, be approximated using a rather simple function [41]

$$p(q) = \tfrac{8}{77}\, q^{1/4}\left[7\left((q+1)^{11/4} - |q-1|^{11/4}\right) - 11q\left((q+1)^{7/4} - \mathrm{sgn}(q-1)|q-1|^{7/4}\right)\right] , \tag{29}$$

see Fig. 2, such that the eigenvalues of a quantization of a^{-m} with positive m are given by

$$(a^{-m})_n^{(j)} = V_{|n|/2}^{-m/3}\, p(|n|/2j)^{2m} \tag{30}$$

with the ambiguity parameter j. There are many other ambiguities which can change also the function p, but the one indicated by j is most important. It parameterizes the position of the peak in an inverse power of the scale factor, which roughly coincides with the boundary between classical behavior and quantum modifications.

Note also that (30) displays the observation that inverse powers of the scale factor annihilate the singular state $|0\rangle$, thanks to $\lim_{q\to 0}(q^{-1/4}p(q)) = 0$. This has important consequences for matter Hamiltonians: they usually consist of a kinetic term containing components of the inverse metric and a potential term containing metric components. In the isotropic case, a widely used example in cosmology is the scalar Hamiltonian (7). We have already discussed the divergence at the singularity of the classical kinetic term, unless $p_\phi = 0$. The potential term, on the other hand, vanishes there. When we quantize this expression, we have to use an inverse power (30) in the kinetic term; for the potential term we can just use volume eigenvalues. Now, the potential term still vanishes at $n = 0$, but so does the kinetic term after quantization. Similarly, one can check that any matter Hamiltonian $\hat{H}_{\mathrm{matter}}$ (assuming for simplicity the absence of curvature couplings) fulfills

$$\hat{H}_{\mathrm{matter}}|0\rangle = 0 \tag{31}$$

irrespective of the particular kind of matter and its quantization.

All these observations indicate that the quantum behavior is much better, less singular, than the classical one. The real test, however, can only come from studying the quantum evolution. An absence of singularities can be confirmed only if it is possible to extend the evolution through the singular boundary; the theory has to tell us what happens at the singularity and beyond.

5.3 Dynamics[16]

To study the dynamics of a theory we need its evolution equation which for gravity is given by the Hamiltonian constraint. In the Wheeler–DeWitt quantization we have seen that the constraint equation takes the form of an evolution equation after quantizing in a metric or triad representation and choosing an internal time a.

We can follow the same steps here if we first transform from the connection representation used so far in quantum geometry to a triad representation. This can be done straightforwardly since we already know the triad eigenstates $|n\rangle$. A state $|\psi\rangle$ can then be expanded in these eigenstates, $|\psi\rangle = \sum_n \psi_n(\phi)|n\rangle$ denoting possible matter degrees of freedom collectively by ϕ. The coefficients $\psi_n(\phi)$ in the expansion then define, as usually, the state in the triad representation. Since n denotes the eigenvalues of $\hat{p}$, it will now play the role of an internal time. Here we observe another difference to the Wheeler–DeWitt quantization: due to the discrete geometry, also time is discrete in an internal time picture.

The Wheeler–DeWitt quantization now proceeded by quantizing the gravitational momentum p_a by a differential operator as in quantum mechanics. An analogous step is not possible in quantum geometry; momenta here, i.e. connection components, have to be quantized using holonomies which do not act as differential operators. Instead, they act according to the SU(2) coupling rules, e.g.

[16] A brief summary of the results in this subsection can be found in [42].

$$\langle c|h_i(c)|n\rangle = \langle c|\cos(c/2) + 2\tau_i \sin(c/2)|n\rangle$$
$$= \tfrac{1}{2}(\langle c|n+1\rangle + \langle c|n-1\rangle) - \tfrac{1}{2}\mathrm{i}\tau_i(\langle c|n+1\rangle - \langle c|n-1\rangle)\,.$$

Thus, in a triad representation holonomies act by changing the label n in $\psi_n(\phi)$ by ± 1 since, e.g.,

$$(\sin(c/2)\psi)_n = -\tfrac{1}{2}\mathrm{i}\sum_n \psi_n(|n+1\rangle - |n-1\rangle) = \tfrac{1}{2}\mathrm{i}\sum_n (\psi_{n+1} - \psi_{n-1})|n\rangle\,.$$

The constraint operator contains several holonomy operators and also the volume operator. It leads to the constraint equation [37, 43]

$$\begin{aligned}
&(V_{|n+4|/2} - V_{|n+4|/2-1})\mathrm{e}^{\mathrm{i}k}\psi_{n+4}(\phi) - (2+\gamma^2k^2)(V_{|n|/2} - V_{|n|/2-1})\psi_n(\phi)\\
&+(V_{|n-4|/2} - V_{|n-4|/2-1})\mathrm{e}^{-\mathrm{i}k}\psi_{n-4}(\phi)\\
&= -\frac{8\pi}{3}G\gamma^3\ell_{\mathrm{P}}^2 \hat{H}_{\mathrm{matter}}(n)\psi_n(\phi)
\end{aligned} \tag{32}$$

which is a *difference equation* rather than a differential equation thanks to the discrete internal time. The parameter k again signifies the intrinsic curvature; for technical reasons the above equation has only been derived for the values $k = 0$ and $k = 1$, not for $k = -1$.

While the left hand side is very different from the Wheeler–DeWitt case, the right hand side looks similar. This is, however, only superficially so; for we have to use the quantizations of the preceding subsection for inverse metric components, in particular in the kinetic term.

We can eliminate the phase factors $\mathrm{e}^{\pm \mathrm{i}k}$ in (32) by using a wave function $\tilde{\psi}_n(\phi) := \mathrm{e}^{\mathrm{i}nk/4}\psi_n(\phi)$ which satisfies the same equation without the phase factors (of course, it is different from the original wave function only for $k = 1$). The phase factor can be thought of as representing rapid oscillations of the wave function caused by non-zero intrinsic curvature.

Large Volume Behavior

Since the Wheeler–DeWitt equation corresponds to a straightforward quantization of the model, it should at least approximately be valid when we are far away from the singularity, i.e. when the volume is large enough. To check that it is indeed reproduced we assume large volume, i.e. $n \gg 1$, and that the discrete wave function $\psi_n(\phi)$ does not display rapid oscillations at the Planck scale, i.e. from n to $n+1$, because this would indicate a significantly quantum behavior. We can thus interpolate the discrete wave function by a continuous one $\tilde{\psi}(p,\phi) = \tilde{\psi}_{n(p)}(\phi)$ with $n(p) = 6p/\gamma\ell_{\mathrm{P}}^2$ from (25). By our assumption of only mild oscillations, $\tilde{\psi}(p,\phi)$ can be assumed to be smooth with small higher order derivatives. We can then insert the smooth wave function in (32) and perform a Taylor expansion of $\tilde{\psi}_{n\pm 4}(\phi) = \tilde{\psi}(p(n) \pm \frac{2}{3}\gamma\ell_{\mathrm{P}}^2)$ in terms of $p/\gamma\ell_{\mathrm{P}}^2$. It is easy to check that this yields to leading order the equation

$$\frac{1}{2}\left(\frac{4}{9}\ell_{\rm P}^4\frac{\partial^2}{\partial p^2}-k\right)\tilde{\psi}(p,\phi)=-\frac{8\pi}{3}G\hat{H}_{\rm matter}(p)\tilde{\psi}(p,\phi)$$

which with $a = \sqrt{|p|}$ is the Wheeler–DeWitt equation (13) in the ordering given before [44]. Thus, indeed, at large volume the Wheeler–DeWitt equation is reproduced which demonstrates that the difference equation has the correct continuum limit at large volume. It also shows that the old Wheeler–DeWitt quantization, though not the fundamental evolution equation from the point of view of quantum geometry, can be used reliably as long as only situations are involved where the discreteness is not important. This includes many semiclassical situations, but not questions about the singularity.

When the volume is small, we are not allowed to do the Taylor expansions since n is of the order of one. There we expect important deviations between the difference equation and the approximate differential equation. This is close to the classical singularity, where we want corrections to occur since the Wheeler–DeWitt quantization cannot deal with the singularity problem.

Non-singular Evolution

To check the issue of the singularity in loop quantum cosmology we have to use the exact equation (32) without any approximations [45]. We start with initial values for $\psi_n(\phi)$ at large, positive n where we know that the behavior is close to the classical one. Then, we can evolve backwards using the evolution equation as a recurrence relation for $\psi_{n-4}(\phi)$ in terms of the initial values. In this way, we evolve toward the classical singularity and we will be able to see what happens there. The evolution is unproblematic as long as the coefficient $V_{|n-4|/2} - V_{|n-4|/2-1}$ of ψ_{n-4} in the evolution equation is non-zero. It is easy to check, however, that it can be zero, if and only if $n = 4$. When n is four, we are about to determine the value of the wave function at $n = 0$, i.e. right at the classical singularity, which is thus impossible. It seems that we are running into a singularity problem again: the evolution equation does not tell us the value $\psi_0(\phi)$ there.

A closer look confirms that there is *no* singularity. Let us first ignore the values $\psi_0(\phi)$ and try to evolve *through* the classical singularity. First there are no problems: for $\psi_{-1}(\phi)$ we only need $\psi_3(\phi)$ and $\psi_7(\phi)$ which we know in terms of our initial data. Similarly we can determine $\psi_{-2}(\phi)$ and $\psi_{-3}(\phi)$. When we come to $\psi_{-4}(\phi)$ it seems that we would need the unknown $\psi_0(\phi)$ which, fortunately, is not the case because $\psi_0(\phi)$ drops out of the evolution equation completely. It does not appear in the middle term on the left hand side because now, for $n = 0$, $V_{|n|/2} - V_{|n|/2-1} = 0$. Furthermore, we have seen as a general conclusion of loop quantizations that the matter Hamiltonian annihilates the singular state $|0\rangle$, which in the triad representation translates to $\hat{H}_{\rm matter}(n = 0) = 0$ independently of the kind of matter. Thus, $\psi_0(\phi)$ drops out completely and ψ_{-4} is determined solely by ψ_4. The further evolution to all negative n then proceeds without encountering any problems.

Intuitively, we obtain a branch of the universe at times "before" the classical singularity, which cannot be seen in the classical description nor in the Wheeler–DeWitt quantization. Note, however, that the classical space-time picture and the notion of time resolves around the singularity; the system can only be described by quantum geometry. The branch at negative times collapses to small volume, eventually reaching volume zero in the Planck regime. There, however, the evolution does not stop, but the universe bounces to enter the branch at positive time we observe. During the bounce, the universe "turns its inside out" in the sense that the orientation of space, given by $\mathrm{sgn}(p)$, changes.

In the discussion we ignored the fact that we could not determine $\psi_0(\phi)$ by using the evolution equation. Is it problematic that we do not know the values at the classical singularity? There is no problem at all because those values just decouple from values at non-zero n. Therefore, we can just choose them freely; they do not influence the behavior at positive volume. In particular, they cannot be determined in the above way from the initial data just because they are completely independent.

The decoupling of $\psi_0(\phi)$ was crucial in the way we evolved through the classical singularity. Had the values not decoupled completely, it would have been impossible to continue to all negative n. It could have happened that the lowest order coefficient is zero at some n, not allowing to determine ψ_{n-4}, but that this unknown value would not drop out when trying to determine lower ψ_n. In fact, this would have happened had we chosen a factor ordering different from the one implicitly assumed above. Thus, the requirement of a non-singular evolution selects the factor ordering in loop quantum cosmology which, in turn, fixes the factor ordering of the Wheeler–DeWitt equation (13) via the continuum limit. One can then re-check results of Wheeler–DeWitt quantum cosmology which are sensitive to the ordering [12] with the one we obtain here. It is not one of the orderings usually used for aesthetic reasons such that adaptations can be expected. An initial step of the analysis has been done in [43].

To summarize, the evolution equation of loop quantum cosmology allows us, for the first time, to push the evolution through the classical singularity. The theory tells us what happens beyond the classical singularity which means that there is no singularity at all. We already know that energy densities do not diverge in a loop quantization, and now we have seen that the evolution does not stop. Thus, none of the conditions for a singularity is satisfied.

Dynamical Initial Conditions

In the Wheeler–DeWitt quantization the singularity problem has been glossed over by imposing initial conditions at $a = 0$, which does have the advantage of selecting a unique state (up to norm) appropriate for the unique universe we observe. This issue appears now in a new light because $n = 0$ does not correspond to a "beginning" so that it does not make sense to choose initial conditions there. Still, $n = 0$ does play a special role, and in fact the behavior of the evolution equation at $n = 0$ *implies* conditions for a wave function [46]. The dynamical law and the issue of initial conditions are intertwined with each other and not separate as usually in physics. One object, the constraint equation, both governs the evolution and provides initial conditions. Due to the intimate relation with the dynamical law, initial conditions derived in this way are called *dynamical initial conditions.*

To see this we have to look again at the recurrence performed above. We noted that the constraint equation does not allow us to determine $\psi_0(\phi)$ from the initial data. We then just ignored the equation for $n = 4$ and went on to determine the values for negative n. The $n = 4$-equation, however, is part of the constraint equation and has to be fulfilled. Since ψ_0 drops out, it is a linear condition between ψ_4 and ψ_8 or, in a very implicit way, a linear condition for our initial data. If we only consider the gravitational part, i.e. the dependence on n, this is just what we need. Because the second order Wheeler–DeWitt equation is reproduced at large volume, we have a two-parameter freedom of choosing the initial values in such a way that the wave function oscillates only slowly at large volume. Then, one linear condition is enough to fix the wave function up to norm. When we also take into account the matter field, there is still more freedom since the dependence of the initial value on ϕ is not restricted by our condition. But the freedom is still reduced from two functions to one. Since we have simply coupled the scalar straightforwardly to gravity, its initial conditions remain independent. Further restrictions can only be expected from a more universal description. Note also that there are solutions with a wave length the size of the Planck length which are unrestricted (since the evolution equation only relates the wave function at n and $n \pm 4$). Their role is not understood so far, and progress can only be achieved after the measurement process or, in mathematical terms, the issue of the physical inner product is better understood.

In its spirit, the dynamical initial conditions are very different from the old proposals since they do not amount to prescribing a value of the wave function at $a = 0$. Still, they can be compared at least at an approximate level concerning implications for a wave function. They are quite similar to DeWitt's original proposal that the wave function vanishes at $a = 0$. The value at $a = 0$ itself would not be fixed, but quite generally the wave function has to approach zero when it reaches $n = 0$. In this sense, the dynamical initial conditions can be seen to provide a generalization of DeWitt's initial condition which does not lead to ill-posed initial value problems [47].

For the closed model with $k = 1$ we can also compare the implications with those of the tunnelling and the no-boundary proposals which have been defined only there. It turns out that the dynamical initial conditions are very close to the no-boundary proposal while they differ from the tunnelling one [43].

5.4 Phenomenology

So far, we have discussed mainly conceptional issues. Now that we know that loop quantum cosmology is able to provide a complete, non-singular description of a quantum universe we can also ask whether there are observational consequences. We already touched this issue by comparing with the older boundary proposals which have been argued to have different implications for the likelihood of inflation from the initial inflaton values they imply. However, the discussion has not come to a definite conclusion since the arguments rely on assumptions about Planck scale physics and also the interpretation of the wave function about which little is known. Loop quantum cosmology provides a more complete description and thus can add substantially to this discussion. An analysis analogous to the one in the Wheeler–DeWitt quantization has not been undertaken so far; instead a strategy has been used which works with an effective classical description implementing important quantum geometry effects and thus allows to sidestep interpretational problems of the wave function. It provides a general technique to study quantum effects in a phenomenological way which is currently being used in a variety of models.

Effective Friedmann Equation

The central idea is to isolate the most prominent effects of quantum geometry and transfer them into effective classical equations of motion, in the case of isotropic cosmology an effective Friedmann equation [48]. The most prominent effect we have seen is the cut-off observed for inverse powers of the scale factor (which can be thought of as a curvature cut-off). It is a non-perturbative effect and has the additional advantage that its reach can be extended into the semiclassical regime by choosing a large ambiguity parameter j.

In the equations for the isotropic model, inverse powers of the scale factor appear in the kinetic term of the matter Hamiltonian, e.g. (7) for a scalar. We have discussed in Sect. 2 that it is difficult to suppress this term by arranging the evolution of ϕ. Now we know, however, that quantum geometry provides a different suppression mechanism in the inverse scale factor operator. This has already played an important role in showing the absence of singularities since in fact the matter Hamiltonian vanishes for $n = 0$. Instead of a^{-3} we have to use a quantization of the inverse scale factor, e.g. in the form $\widehat{a^{-3}}$ whose eigenvalues (30) with $m = 3$ are bounded above. We can introduce the effect into the classical equations of motion by replacing $d(a) = a^{-3}$ with the bounded function $d_j(a) = a^{-3}p(3a^2/j\gamma\ell_{\mathrm{P}}^2)^6$ where $p(q)$ is defined in (29) and we choose a half-integer value for j. The effective scalar energy density then can be parameterized as

$$\rho_{\mathrm{eff}}(a) = \tfrac{1}{2}x\, a^{l(a)-3}\ell_{\mathrm{P}}^{-l(a)-3}\, p_\phi^2 + W(\phi)$$

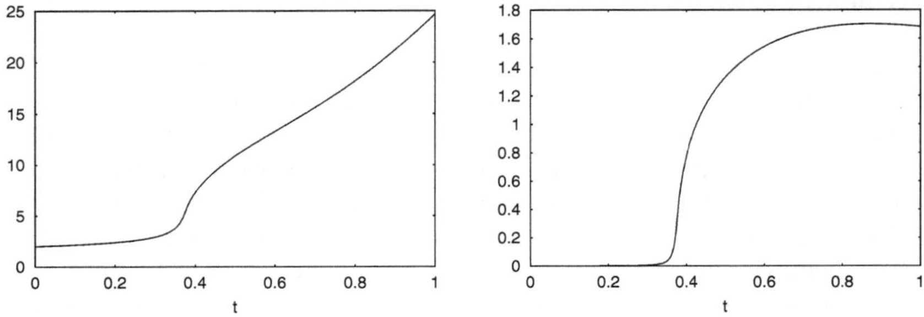

Fig. 3. Behavior of the scale factor (left) and the scalar field (right) during quantum geometry inflation (ending at $t \approx 0.4$ for $j = 100$) [43], both plotted in Planck units. The potential is just a mass term $8\pi GW(\phi) = 10^{-3}\hbar\phi^2/2$, and initial conditions for the numerical integration are $\phi_0 = 0$, $\sqrt{\kappa}\dot{\phi}_0 = 10^{-5}\ell_P^{-1}$ at $a_0 = 2\ell_P$.

with parameters x and l which also depend on quantization ambiguities, though not significantly. Note that ℓ_P appears in the denominator demonstrating that the effect is non-perturbative; it could not be obtained in a perturbative quantization in an expansion of G. With our function $p(q)$ we have $l = 12$ for very small a (but it decreases with increasing a), and it is usually larger than 3, even taking into account different quantization choices, thanks to the high power of $p(q)$ in $d_j(a)$. Thus, the effective equation of state parameter $w = -l/3$ in the parametrization (12) is smaller than -1; quantum geometry *predicts that the universe starts with an initial phase of inflation* [48]. It is a particular realization of super-inflation, but since w increases with a, there is no pole as would be the case with a constant w. Note that this does not require any special arrangements of the fields and their potentials, not even an introduction of a special inflaton field: any matter Hamiltonian acquires the modified kinetic term such that even a vanishing potential implies inflation. Inflation appears as a natural part of cosmological models in loop quantum cosmology. Moreover, the inflationary phase ends automatically once the expanding scale factor reaches the value $a \approx \sqrt{j\gamma/3}\,\ell_P$ where the modified density reaches its peak and starts to decrease (Fig. 3).

Inflation

Thus, inflation appears as a natural consequence, but it is less clear what role it can play. For an inflationary period responsible for structure formation it has to last long enough (in terms of e-foldings, i.e. a large ratio of the final a and the initial a) and to be very close to standard inflation, i.e. $w = -1$. The final scale factor for quantum geometry inflation is easy to find, $a \approx \sqrt{j\gamma/3}\,\ell_P$ as just discussed. The initial value, however, is more complicated. In a flat model, the inflationary period starts as close to $a = 0$ as we want, but at those small values the effective classical description must break down. Moreover, in a closed model the region close to $a = 0$ is classically forbidden[17] which also sets a lower limit for the initial a. All these issues depend more sensitively on the kind of matter added to the model and have not yet been analyzed systematically.

An alternative application of quantum geometry inflation can be seen in combination with standard inflation. We have discussed that the standard scenario requires a special potential and also special, very large initial values for the inflaton. For instance, for chaotic inflation with the potential $W(\phi) = \frac{1}{2}m\phi^2$ we need to start with $\phi_0 > m_{\mathrm{P}} = \hbar/\ell_{\mathrm{P}}$ which is huge compared to its own mass m. If we couple quantum geometry inflation with chaotic inflation, we would first observe an inflationary expansion at small volume which can stop at small a (i.e. j can be of the order one). During this phase also the evolution of the scalar is modified compared to the standard one since now $d_j(a)$ appears in the Hamiltonian equations of motion instead of a^{-3}. This leads to a differential equation

$$\ddot{\phi} = \frac{\mathrm{d}\log d_j(a)}{\mathrm{d}a}\dot{a}\dot{\phi} - a^3 d_j(a) W'(\phi)$$

for ϕ. For the always decreasing a^{-3} instead of $d_j(a)$ we obtain the previous equation (10) with the friction term. The modified $d_j(a)$, however, is increasing for small a such that we obtain a friction term with the opposite sign. This will require the inflaton to move up the potential, reaching large values even if it would start in $\phi(a = 0) = 0$ [43]; see Fig. 3.

5.5 Homogeneous Cosmology

The framework of loop quantum cosmology is available for all homogeneous, but in general anisotropic models [35]. When we require that the metric of a homogeneous model is diagonal, the volume operator simplifies again allowing an explicit analysis [49]. One obtains a more complicated evolution equation which is now a partial difference equation for three degrees of freedom, the three diagonal components of the metric. Nevertheless, the same mechanism for a removal of the classical singularity as in the isotropic case applies.

This is in particular important since it suggests an absence of singularities even in the full theory. It has been argued [50] that close to singularities points on a space-like slice decouple from each other such that the metric in each one is described by a particular homogeneous model, called Bianchi IX. If this is true and extends to the quantum theory, it would be enough to have a non-singular Bianchi IX model for singularity freedom of the full theory. Even though the classical evolution of the Bianchi IX model is very complicated and suspected to be chaotic [51], one can see that its loop quantization

[17] There is, however, a mechanism which leads to a small classically allowed region for small a including $a = 0$ even in a closed model [43]. This comes from a suppression of intrinsic curvature analogous to the cut-off of a^{-1} which would be a suppression of extrinsic curvature.

is singularity-free [52]. In fact, again the cut-off in the inverse scale factor leads to modified effective classical equations of motion which do not show the main indication for chaos. The Bianchi IX universe would still evolve in a complicated way, but its behavior simplifies once it reaches small volume. At this stage, a simple regular transition through the classical singularity occurs. This issue is currently being investigated in more detail. Also other homogeneous models provide a rich class of different systems which can be studied in a phenomenological way including quantum geometry modifications.

6 Quantum Gravity Phenomenology

Since quantum gravity is usually assumed to hold at scales near the Planck length $\ell_{\mathrm{P}} \sim 10^{-32}\mathrm{cm}$ or, equivalently, Planck energy $E_{\mathrm{P}} := \hbar/\ell_{\mathrm{P}} \sim 10^{18}\mathrm{GeV}$ experiments to probe such a regime were considered out of reach for most of the past. Recently, however, phenomena have been proposed which compensate for the tiny size of the Planck scale by a large number of small corrections adding up. These phenomena include in vacuo dispersion relations for gamma ray astrophysics [53, 54, 55], laser-interferometric limits on distance fluctuations [56, 57], neutrino oscillations [58], threshold shifts in ultra high energy cosmic ray physics [59, 60, 61, 62], CPT violation [63] and clock-comparison experiments in atomic physics [64]. They form the so called *quantum gravity phenomenology* [65].

The aim is to understand the imprint which the structure of space-time predicted by a specific theory of quantum gravity can have on matter propagation. Specifically, dispersion relations are expected to change due to a non-trivial microscopic structure (as in condensed matter physics where the dispersion relations deviate from the continuum approximation once atomic scales are reached). For particles with energy $E \ll E_{\mathrm{P}}$ and momentum $\boldsymbol{p}$ the following modified vacuum dispersion relations have been proposed [54]:

$$\boldsymbol{p}^2 = E^2\left(1 + \xi\, E/E_{\mathrm{P}} + \mathcal{O}\left((E/E_{\mathrm{P}})^2\right)\right) , \tag{33}$$

where $\xi \sim 1$ has been assumed, which still has to be verified in concrete realizations. Furthermore, this formula is based on a power series expansion which rests on the assumption that the momentum is analytic at $E = 0$ as a function of the energy. In general, the leading corrections can behave as $(E/E_{\mathrm{P}})^{\Upsilon+1}$, where $\Upsilon \geq 0$ is a positive real number.

Since the Planck energy is so large compared to that of particles which can be observed from Earth, the correction would be very tiny even if it is only of linear order. However, if a particle with the modified dispersion relation travels a long distance, the effects can become noticeable. For instance, while all photons travelling at the speed of light in Minkowski space would arrive at the same time if they had been emitted in a brief burst, (33) implies an energy dependent speed for particles with the modified dispersion relations. Compared to a photon travelling a distance L in Minkowski space, the retardation time is

$$\Delta t \approx \xi \, L \, E / E_{\mathrm{P}} \,. \tag{34}$$

If L is of a cosmological scale, the smallness of E/E_{P} can be compensated, thus bringing Δt close to possible observations. Candidates for suitable signals are Gamma Ray Bursts (GRB's), intense short bursts of energy around $E \sim 0.20\mathrm{MeV}$ that travel a cosmological distance $L \sim 10^{10}$ ly until they reach Earth. These values give $\Delta t \sim 0.01\mathrm{ms}$ which is only two orders of magnitude below the sensitivity δt for current observations of GRB's [66, 67] (for planned improvements see [68]). For the delay of two photons detected with an energy difference ΔE, the observational bound $E_{\mathrm{P}}/\xi \geq 4 \times 10^{16}$ GeV was established in [69] by identifying events having $\Delta E = 1$ TeV arriving to Earth from the active galaxy Markarian 421 within the time resolution $\Delta t = 280\,\mathrm{s}$ of the measurement. Moreover, GRB's also seem to generate Neutrino Bursts (NB) in the range $10^5 - 10^{10}$ GeV in the so-called fireball model [70, 71] which can be used for additional observations [72, 58, 73].

In summary, astrophysical observations of photons, neutrinos and also cosmic rays could make tests of quantum gravity effects possible, or at least restrict possible parameters in quantum gravity theories.

Within loop quantum gravity attention has focused on light [74, 75] and neutrino propagation [76]. Other approaches aimed at investigating similar quantum gravity effects include string theory [77], an open system approach [78], perturbative quantum gravity [79, 80] and non commutative geometry [81]. A common feature to all these approaches is that correction terms arise which break Lorentz symmetry. These studies overlap with a systematic analysis providing a general power counting renormalizable extension of the standard model that incorporates both Lorentz and CPT violations [82]. Progress in setting bounds to such symmetry violation has been reported in [55, 64, 83, 84, 85].

6.1 An Implementation in Loop Quantum Gravity

In order to implement the central idea, one needs states approximating a classical geometry at lengths much larger than the Planck length. The first proposed states of this type in loop quantum gravity were weave states [86]. Flat weave states $|W\rangle$ with characteristic length $\mathcal{L}$ were constructed as in Sect. 4 using collections of circles of Planck size radius (measured with the classical background geometry to be approximated) in random orientation. At distances $d \gg \mathcal{L}$ the continuous flat classical geometry is reproduced, while for distances $d \ll \mathcal{L}$ the discrete structure of space is manifest. The search for more realistic coherent states, which not only approximate the classical metric, but also its conjugate, the extrinsic curvature, is still ongoing [87, 88, 89, 90]. Current calculations have been done at a heuristic level by assuming simple properties of semiclassical states. The prize to pay is that

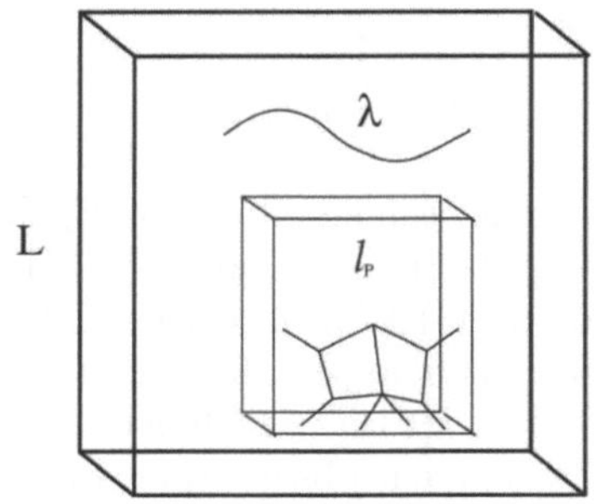

Fig. 4. Representation of the different scales of the problem. A coarse graining scale $\mathcal{L} \sim \lambda$ indicates Planck length features ℓ_P are minute as compared to matter scales. L here represents the size of a large piece of space.

some parameters, and even their order of magnitude, remain undetermined. Thus, these studies explore possible quantum gravity effects without using details of specific semiclassical states. Once properties of semiclassical states become better known, one can then check if the existing calculations have to be modified.

The setup requires to consider semiclassical states $|S\rangle$ for both gravity, the background for the propagation, and the propagating matter. One has to require that they are peaked at the classical configurations of interest with well defined expectation values such that there exists a coarse-grained expansion involving ratios of the relevant scales of the problem. Those are the Planck length ℓ_P, the characteristic length $\mathcal{L}$ and the matter wavelength λ satisfying $\ell_P \ll \mathcal{L} \leq \lambda$; see Fig. 4. The effective Hamiltonian is thus defined by

$$H^{\rm eff}_{\rm Matter} :=< S|\hat{H}_{\rm Matter}|S > . \tag{35}$$

Light

The full quantum Hamiltonian for the electromagnetic field is of the form (20). With the above assumptions about semiclassical states one can arrive at an effective electromagnetic Hamiltonian [75]

$$\begin{aligned} H^{\rm eff}_{\rm EM} = \frac{1}{Q^2}\int {\rm d}^3\boldsymbol{x} \Big[& \tfrac{1}{2}\left(1+\theta_7\,(\ell_P/\mathcal{L})^{2+2\Upsilon}\right)\left(\underline{\boldsymbol{B}}^2+\underline{\boldsymbol{E}}^2\right) \\ & +\theta_3\,\ell_P^2\left(\underline{\boldsymbol{B}}\cdot\nabla^2\underline{\boldsymbol{B}}+\underline{\boldsymbol{E}}\cdot\nabla^2\underline{\boldsymbol{E}}\right) \\ & +\theta_8\ell_P\Big(\underline{\boldsymbol{B}}\cdot(\nabla\times\underline{\boldsymbol{B}})+\underline{\boldsymbol{E}}\cdot(\nabla\times\underline{\boldsymbol{E}})\Big)+\cdots\Big] , \end{aligned} \tag{36}$$

up to order ℓ_P^2 and neglecting non linear terms. The coefficients θ_i have not yet been derived systematically; rather, the expression is to be understood as a collection of all the terms which can be expected. Precise values, and even the order of magnitude, can depend significantly on the explicit procedure followed to obtain the values from a semiclassical state. Moreover, as usual there are quantization ambiguities in the quantum Hamiltonian which influence the coefficients of the correction terms [41].

From the effective Hamiltonian (36) we obtain the equations of motion

$$A(\nabla \times \underline{\boldsymbol{B}}) - \frac{\partial \underline{\boldsymbol{E}}}{\partial t} + 2\ell_{\mathrm{P}}^2 \theta_3 \nabla^2 (\nabla \times \underline{\boldsymbol{B}}) - 2\theta_8 \ell_{\mathrm{P}} \nabla^2 \underline{\boldsymbol{B}} = 0, \tag{37}$$

$$A(\nabla \times \underline{\boldsymbol{E}}) + \frac{\partial \underline{\boldsymbol{B}}}{\partial t} + 2\ell_{\mathrm{P}}^2 \theta_3 \nabla^2 (\nabla \times \underline{\boldsymbol{E}}) - 2\theta_8 \ell_{\mathrm{P}} \nabla^2 \underline{\boldsymbol{E}} = 0, \tag{38}$$

where

$$A = 1 + \theta_7 \left(\ell_{\mathrm{P}}/\mathcal{L}\right)^{2+2\Upsilon} . \tag{39}$$

The above equations are supplemented by $\nabla \cdot \underline{\boldsymbol{B}} = 0$, together with the constraint $\nabla \cdot \underline{\boldsymbol{E}} = 0$, appropriate for vacuum.

Modifications in the Maxwell equations (37) and (38) imply a modified dispersion relation which, neglecting the non-linear part, can be derived by introducing the plane wave ansatz

$$\boldsymbol{E} = \boldsymbol{E}_0 \, \mathrm{e}^{\mathrm{i}(\boldsymbol{kx} - \omega t)}, \qquad \boldsymbol{B} = \boldsymbol{B}_0 \, \mathrm{e}^{\mathrm{i}(\boldsymbol{kx} - \omega t)}, \qquad k = |\boldsymbol{k}| \, . \tag{40}$$

The result is

$$\omega = k \left(1 + \theta_7 \left(\ell_{\mathrm{P}}/\mathcal{L}\right)^{2+2\Upsilon} - 2\,\theta_3 \left(k\ell_{\mathrm{P}}\right)^2 \pm 2\theta_8 \left(k\ell_{\mathrm{P}}\right)\right) \tag{41}$$

where the two signs of the last term correspond to the different polarizations of the photon. The speed of a photon becomes

$$v = \left.\frac{\mathrm{d}\omega}{\mathrm{d}k}\right|_{\mathcal{L}=1/k} = 1 \pm 4\,\theta_8 \left(k\ell_{\mathrm{P}}\right) - 6\theta_3 (k\ell_{\mathrm{P}})^2 + \theta_7 \left(k\ell_{\mathrm{P}}\right)^{2+2\Upsilon} + \cdots \tag{42}$$

The scale $\mathcal{L}$ has been estimated by its maximal value $1/k$. Clearly (42) is valid only for momenta satisfying $(\ell_{\mathrm{P}}\, k) \ll 1$.

There are also possible non-linear terms in the effective Maxwell equations [75]. They can become significant in strong magnetic fields, but the corrections obtained in the corresponding refraction indices are much smaller than similar effects in Quantum Electrodynamics. Nevertheless, quantum gravity corrections have distinct signatures: a main difference is that the speed of photons with polarization parallel to the plane formed by the background magnetic field and the direction of the wave is isotropic.

Spin-1/2 Particles

Similarly, one can derive an effective Hamiltonian for a spin-$\frac{1}{2}$ field of mass m [76]:

$$H^{\rm eff}_{{\rm spin}\,\frac{1}{2}} = \mathrm{i}\int \mathrm{d}^3x \Big\{\pi(\boldsymbol{x})\tau^d\partial_d\, \hat{A}\, \xi(\boldsymbol{x}) + \text{c.c.} + (4\mathcal{L})^{-1}\, \pi(\boldsymbol{x})\, \hat{C}\, \xi(\boldsymbol{x}) \qquad (43)$$

$$+ \frac{m}{2\hbar}[\xi^T(\boldsymbol{x})\, \sigma_2\, (\alpha + \beta\ell_{\rm P}\tau^a\partial_a)\, \xi(\boldsymbol{x}) + \pi^T(\boldsymbol{x})\, (\alpha + \beta\ell_{\rm P}\tau^a\partial_a)\, \sigma_2\pi(\boldsymbol{x})]\Big\},$$

where

$$\hat{A} = \left(1 + \kappa_1\, (\ell_{\rm P}/\mathcal{L})^{\Upsilon+1} + \kappa_2\, (\ell_{\rm P}/\mathcal{L})^{2\Upsilon+2} + \tfrac{1}{2}\,\kappa_3\, \ell_{\rm P}^2\ \nabla^2\right),$$

$$\hat{C} = \tfrac{1}{2}\,\kappa_7\, (\ell_{\rm P}/\mathcal{L})^{\Upsilon}\ \ell_{\rm P}^2\ \nabla^2$$

$$\alpha = 1 + \kappa_8\, (\ell_{\rm P}/\mathcal{L})^{\Upsilon+1}\quad , \qquad \beta = \kappa_9 + \kappa_{11}\, (\ell_{\rm P}/\mathcal{L})^{\Upsilon+1}, \qquad (44)$$

This leads to wave equations

$$\mathrm{i}\hbar\left[\frac{\partial}{\partial t} - \hat{A}\,\boldsymbol{\sigma}\cdot\nabla - \mathrm{i}\frac{\hat{C}}{2\mathcal{L}}\right]\xi(t,\boldsymbol{x}) + m\left(\alpha - \tfrac{1}{2}\mathrm{i}\beta\ell_{\rm P}\ \boldsymbol{\sigma}\cdot\nabla\right)\chi(t,\boldsymbol{x}) = 0 \qquad (45)$$

$$\mathrm{i}\hbar\left[\frac{\partial}{\partial t} + \hat{A}\,\boldsymbol{\sigma}\cdot\nabla + \mathrm{i}\frac{\hat{C}}{2\mathcal{L}}\right]\chi(t,\boldsymbol{x}) + m\left(\alpha - \tfrac{1}{2}\mathrm{i}\beta\ell_{\rm P}\ \boldsymbol{\sigma}\cdot\nabla\right)\xi(t,\boldsymbol{x}) = 0 \qquad (46)$$

with $\chi(t,\boldsymbol{x}) = \mathrm{i}\,\sigma_2\xi^*(t,\boldsymbol{x})$. As before, the dispersion relation can be obtained by inserting plane wave solutions, this time positive and negative energy solutions

$$W(\boldsymbol{p},h)\mathrm{e}^{\mp\mathrm{i}Et/\hbar\pm\mathrm{i}\boldsymbol{p}\cdot\boldsymbol{x}/\hbar} \qquad (47)$$

where $W(\boldsymbol{p},h)$ are helicity $(\boldsymbol{\sigma}\cdot\hat{p})$ eigenstates, with $h=\pm1$, so that

$$W(\boldsymbol{p},1) = \begin{pmatrix}\cos(\theta/2)\\ \mathrm{e}^{\mathrm{i}\phi}\,\sin(\theta/2)\end{pmatrix},\quad W(\boldsymbol{p},-1) = \begin{pmatrix}-\mathrm{e}^{-\mathrm{i}\phi}\,\sin(\theta/2)\\ \cos(\theta/2)\end{pmatrix}. \qquad (48)$$

For ultra-relativistic neutrinos ($p \gg m$) one obtains

$$\ell_{\rm P}E_\pm(p,\mathcal{L}) = p\ell_{\rm P} + \ell_{\rm P}m^2/2p \pm \tfrac{1}{2}\,(\ell_{\rm P}m)^2\,\kappa_9 - \tfrac{1}{2}\,\kappa_3\,(\ell_{\rm P}p)^3 \qquad (49)$$

$$+ (\ell_{\rm P}/\mathcal{L})^{\Upsilon+1}\left[\kappa_1p\ell_{\rm P} \mp \tfrac{1}{4}\,\kappa_7\,(\ell_{\rm P}p)^2\right] + (\ell_{\rm P}/\mathcal{L})^{2\Upsilon+2}\,\kappa_2\,p\ell_{\rm P}$$

and

$$v_\pm(p,\mathcal{L}) = 1 - \frac{m^2}{2p^2} - \tfrac{3}{2}\,\kappa_3\,(\ell_{\rm P}p)^2 + (\ell_{\rm P}/\mathcal{L})^{\Upsilon+1}\left(\kappa_1 \mp \tfrac{1}{2}\,\kappa_7\ell_{\rm P}p\right) + (\ell_{\rm P}/\mathcal{L})^{2\Upsilon+2}\ \kappa_2.$$

There are two physically interesting effects related to the dispersion relations just described for neutrinos. Namely neutrino oscillations for different flavors and time delay between neutrinos and photons coming from the same GRB. Estimates in this respect have been obtained in [76].

6.2 Summary

The phenomenological considerations described here are intended to give an idea of possible consequences of quantum gravity corrections. They start with assumptions about a state approximating a classical flat metric, a classical flat gravitational connection and a generic classical matter field, at scales larger than the coarse-grained characteristic length $\mathcal{L} \gg \ell_P$. Under these assumptions, modified dispersion relations can be expanded in the Planck length. In general, there are different types of corrections, which can have different dependence on, e.g., the helicity or the scale $\mathcal{L}$. This also includes the parameter Υ encoding our (current) ignorance of the scaling of the gravitational connection in a semiclassical expectation value.

The following motivations for the value of Υ have been made [75]: (i) $\Upsilon = 0$ can be understood as that the connection can not be probed below the coarse graining scale $\mathcal{L}$. The corresponding correction scales as $(k\ell_P)^2$. (ii) $\Upsilon = 1$ may be interpreted as the analog of a simple analysis [91], based on a saturation of Heisenberg's uncertainty relation inside a box of volume $\mathcal{L}^3$: $\Delta E \sim \ell_P/\mathcal{L}$, $\Delta A \sim \ell_P/\mathcal{L}^2$ and $\Delta E \Delta A \sim G\hbar/\mathcal{L}^3$. Then the correction behaves as $(k\ell_P)^4$. (iii) A value $\Upsilon = -\frac{1}{2}$ would lead to a helicity independent first order correction (i.e. $(k\ell_P)$); a negative value, however, is not allowed. Further fractional values have been obtained in [92] from a detailed proposal for coherent states in loop quantum gravity [87, 88]. From an observational point of view, lower order correction terms would certainly be preferable. Most of the evaluations so far have been done for first order terms, but recently also higher order corrections have been started to be compared with observations [61, 93].

7 Outlook

As discussed in this article, loop quantum gravity is at a stage where physical results are beginning to emerge which will eventually be confronted by observations. To obtain these results, as usually, approximation schemes have to be employed which capture the physically significant contributions of a full theory. In our applications we used the minisuperspace approximation to study cosmological models and a semiclassical approximation for the propagation of particles. We have to stress, however, that these approximation schemes are currently realized at different levels of precision, both having open issues to be filled in. Loop quantum cosmological models are based on symmetric states which have been explicitly constructed as distributional states in the full theory. There are no further assumptions besides the central one of symmetries. A partially open issue is the relation of symmetric operators to those of the full theory. A precise derivation of this relation will complete our understanding of the models and also of the full theory, but it is not expected to imply changes of the physical results since we know that they are robust under quantization ambiguities.

As for loop quantum gravity phenomenology its central ingredient are semiclassical states which are being investigated with different strategies leading to different proposals. The present explicit calculations are based on simple assumptions about semiclassical states which have to be probed in an eventual realization. Thus, there is not only a central simplifying assumption, semiclassicality, but also additional assumptions about its realization. These assumptions affect the presence as well as the magnitude of possible correction terms. It is not just the relation between phenomenology and observations, but also the one between phenomenology and the basic theory which has to be understood better.

Furthermore, there are important conceptual issues which are not yet completely understood. For instance, it would be essential to see the emergence of a classical space-time from semiclassical quantum states in order to study a particle moving in a state which approximates Minkowski space. A related issue is the fact that the discreteness of quantum geometry is supposed to lead to correction terms violating Lorentz symmetry. Such a violation, in turn, implies the existence of a distinguished time-like vector. An open conceptual issue is how such a distinguished vector can arise from the discrete formulation.[18] For this purpose one would need a distinguished rest frame which could be identified using the cosmic background radiation [74].

Future work will progress along several lines according to the different open problems. First, at a basic level, the conceptual issues will have to be understood better. In the case of quantum gravity phenomenology this will come as a consequence of additional insights into semiclassical states which are under investigation [87, 88, 89, 90]. This will also change the way how explicit calculations are implemented, and the precision of known results will be enhanced leading to a stronger confrontation with observations. Finally, there are many phenomenological effects which have not yet been investigated in the context of loop quantum gravity. Loop effects will lead to changes whose significance regarding observations has to be studied.

Already the present stage of developments proves that loop quantum gravity is a viable description of aspects of the real world. It offers natural solutions to problems, as e.g. the singularity problem, which in some cases have been open for decades and plagued all other theories developed so far. At the same time, sometimes surprising consequences emerged which lead to a coherent picture of a universe described by a discrete geometry. All this establishes the viability of loop quantum gravity, and we are beginning to test the theory also observationally.

[18] Models to understand modifications of the usual Lorentz symmetry have been developed in [94, 95].

Acknowledgements

We thank J. Alfaro, A. Ashtekar, H. Sahlmann and L.F. Urrutia for discussions. M.B. is grateful to the Universidad Autónoma Metropolitana Iztapalapa for hospitality during the completion of this article. The work of M.B. was supported in part by NSF grant PHY00-90091 and the Eberly research funds of Penn State as well as 40745-F CONACyT grant. H.A.M.T. acknowledges partial support from 40745-F CONACyT grant.

References

1. R. Arnowitt, S. Deser, and C. W. Misner: The Dynamics of General Relativity. In *Gravitation: An Introduction to Current Research*, ed by L. Witten (Wiley, New York 1962)
2. A. Ashtekar: Phys. Rev. D **36**, 1587 (1987)
3. J. F. Barbero G.: Phys. Rev. D **51**, 5507 (1995), [gr-qc/9410014]
4. A. Ashtekar, C. Beetle, O. Dreyer, S. Fairhurst, B. Krishnan, J. Lewandowski, and J. Wisniewski: Phys. Rev. Lett. **85** (2000) 3564, [gr-qc/0006006]; A. Ashtekar, C. Beetle, and J. Lewandowski: Class. Quantum Grav. **19**, 1195 (2002), [gr-qc/0111067]
5. A. Ashtekar, J. C. Baez, A. Corichi, and K. Krasnov: Phys. Rev. Lett. **80**, 904 (1998), [gr-qc/9710007]; A. Ashtekar, J. C. Baez, and K. Krasnov: Adv. Theor. Math. Phys. **4**, 1 (2001), [gr-qc/0005126]
6. A. Friedmann: Z. Phys. **10**, 377 (1922)
7. F. Lucchin and S. Matarrese: Phys. Lett. **B164**, 282 (1985); Phys. Rev. D **32**, 1316 (1985)
8. S. W. Hawking and G. F. R. Ellis: *The Large Scale Structure of Space-Time* (Cambridge University Press 1973)
9. D. L. Wiltshire: An introduction to quantum cosmology. In *Cosmology: The Physics of the Universe*, ed by B. Robson, N. Visvanathan, and W. S. Woolcock (World Scientific, Singapore 1996), pages 473–531, [gr-qc/0101003]
10. B. S. DeWitt: Phys. Rev. **160**, 1113 (1967)
11. P. G. Bergmann: Rev. Mod. Phys. **33**, 510 (1961); C. Rovelli: Phys. Rev. D **43**, 442 (1991)
12. N. Kontoleon and D. L. Wiltshire: Phys. Rev. D **59**, 063513 (1999), [gr-qc/9807075]
13. A. Vilenkin: Phys. Rev. D **30**, 509 (1984)
14. J. B. Hartle and S. W. Hawking: Phys. Rev. D **28**, 2960 (1983)
15. A. Ashtekar: *Lectures on non-perturbative canonical gravity* (World Scientific, Singapore 1991)
16. C. Rovelli: Liv. Rev. Relat. **1**, 1 (1998), http://www.livingreviews.org/Articles/Volume1/1998–1rovelli, [gr-qc/9710008].
17. T. Thiemann: *Introduction to Modern Canonical Quantum General Relativity*, [gr-qc/0110034]
18. M. Bojowald and H. A. Kastrup: Class. Quantum Grav. **17**, 3009 (2000), [hep-th/9907042]

19. R. Haag: *Local quantum physics: Fields, particles, algebras* (Springer, Berlin 1992)
20. H. Sahlmann: *Some Comments on the Representation Theory of the Algebra Underlying Loop Quantum Gravity*, [gr-qc/0207111]; *When Do Measures on the Space of Connections Support the Triad Operators of Loop Quantum Gravity?*, [gr-qc/0207112]; H. Sahlmann and T. Thiemann: *On the superselection theory of the Weyl algebra for diffeomorphism invariant quantum gauge theories*, [gr-qc/0302090; *Irreducibility of the Ashtekar–Isham–Lewandowski representation*, [gr-qc/0303074]; A. Okolow and J. Lewandowski: Class. Quantum Grav. **20**, 3543 (2003), [gr-qc/0302059]
21. A. Ashtekar, J. Lewandowski, D. Marolf, J. Mourão, and T. Thiemann: J. Math. Phys. **36**, 6456 (1995), [gr-qc/9504018]
22. A. Ashtekar and J. Lewandowski: J. Geom. Phys. **17**, 191 (1995), [hep-th/9412073]
23. A. Ashtekar and J. Lewandowski: J. Math. Phys. **36**, 2170 (1995)
24. A. Ashtekar and J. Lewandowski: *Background independent quantum gravity: A status report*, in preparation
25. C. Rovelli and L. Smolin: Nucl. Phys. **B331**, 80 (1990)
26. C. Rovelli and L. Smolin: Phys. Rev. D **52**, 5743 (1995)
27. C. Rovelli and L. Smolin: Nucl. Phys. **B442**, 593 (1995), [gr-qc/9411005], Erratum: Nucl. Phys. **B456**, 753 (1995)
28. A. Ashtekar and J. Lewandowski: Class. Quantum Grav. **14**, A55 (1997), [gr-qc/9602046]
29. A. Ashtekar and J. Lewandowski: Adv. Theor. Math. Phys. **1**, 388 (1997), [gr-qc/9711031]
30. T. Thiemann: J. Math. Phys. **39**, 3372 (1998), [gr-qc/9606092]
31. T. Thiemann: J. Math. Phys. **39**, 3347 (1998), [gr-qc/9606091]
32. T. Thiemann: Phys. Lett. B **380**, 257 (1996), [gr-qc/9606088]; Class. Quantum Grav. **15**, 839 (1998), [gr-qc/9606089
33. T. Thiemann: Class. Quantum Grav. **15**, 1281 (1998), [gr-qc/9705019]
34. M. Bojowald, *Quantum Geometry and Symmetry* (Shaker-Verlag, Aachen 2000)
35. M. Bojowald: Class. Quantum Grav. **17**, 1489 (2000), [gr-qc/9910103]
36. A. Ashtekar, M. Bojowald, and J. Lewandowski: Adv. Theor. Math. Phys., **7**, 233 (2003), [gr-qc/0304074]
37. M. Bojowald: Class. Quantum Grav. **19**, 2717 (2002), [gr-qc/0202077]
38. M. Bojowald: Class. Quantum Grav. **17**, 1509 (2000), [gr-qc/9910104]
39. M. Bojowald: Class. Quantum Grav. **18**, 1055 (2001), [gr-qc/0008052]
40. M. Bojowald: Phys. Rev. D **64**, 084018 (2001), [gr-qc/0105067]
41. M. Bojowald: Class. Quantum Grav. **19**, 5113 (2002), [gr-qc/0206053]
42. M. Bojowald: Gen. Rel. Grav. **35**, to appear (2003), [gr-qc/0305069]
43. M. Bojowald and K. Vandersloot: Phys. Rev. D **67**, 124023 (2003), [gr-qc/0303072]
44. M. Bojowald: Class. Quantum Grav. **18**, L109 (2001), [gr-qc/0105113]
45. M. Bojowald: Phys. Rev. Lett. **86**, 5227 (2001), [gr-qc/0102069]
46. M. Bojowald: Phys. Rev. Lett. **87**, 121301 (2001), [gr-qc/0104072]
47. M. Bojowald and F. Hinterleitner: Phys. Rev. D **66**, 104003 (2002), [gr-qc/0207038]
48. M. Bojowald: Phys. Rev. Lett. **89**, 261301 (2002), [gr-qc/0206054]
49. M. Bojowald: Class. Quantum Grav. **20**, 2595 (2003), [gr-qc/0303073]

50. V. A. Belinskii, I. M. Khalatnikov, and E. M. Lifschitz: Adv. Phys. **13**, 639 (1982)
51. D. Hobill, A. Burd, and A. Coley: *Deterministic chaos in general relativity* (Plenum Press, New York 1994)
52. M. Bojowald, G. Date, and K. Vandersloot: *Homogeneous loop quantum cosmology: The role of the spin connection*, [gr-qc/0311004]
53. P. Huet and M. Peskin: Nucl. Phys. B **434**, 3 (1995); J. Ellis, J. López, N. E. Mavromatos and D. V. Nanopoulos: Phys. Rev. D **53**, 3846 (1996)
54. G. Amelino-Camelia, J. Ellis, N. E. Mavromatos, D. V. Nanopoulos and S. Sarkar: Nature **393**, 763 (1998)
55. R. J. Gleiser and C. N. Kozameh: Phys. Rev. D **64**, 083007 (2001), [gr-qc/0102093]
56. G. Amelino-Camelia: Nature **398**, 216 (1999), [gr-qc/9808029]
57. Y. J. Ng and H. van Dam: Found. Phys. **30**, 795 (2000), [gr-qc/9906003]
58. R. Brustein, D. Eichler and S. Foffa: Phys. Rev. D **65**, 105006 (2002)
59. T. Kifune: Astrophys. J. Lett. **518**, L21 (1999), [astro-ph/9904164]
60. G. Amelino-Camelia and T. Piran: Phys. Rev. D **64**, 036005 (2001), [gr-qc/0008107]
61. J. Alfaro and G. Palma: Phys. Rev. D **65**, 103516 (2002), [hep-th/0111176; Phys. Rev. D **67**, 083003 (2003), [hep-th/0208193]
62. T. J. Konopka and S. A. Major: New Journal of Physics **4**, 57 (2002), [hep-ph/0201184]
63. J. R. Ellis, J. L. López, N. E. Mavromatos and D. V. Nanopoulos: Phys. Rev. D **53**, 3846 (1996), [hep-ph/9505340]
64. D. Sudarsky, L. F. Urrutia and H. Vucetich: Phys. Rev. Lett. **89**, 231301 (2002), [gr-qc/0204027]
65. J. Ellis, N. E. Mavromatos and D. V. Nanopoulos: Gen. Rel. Grav. **31**, 1257 (1999), [gr-qc/9905048]; G. Z. Adunas, E. Rodriguez-Milla and D. V. Ahluwalia: Phys. Lett. B **485**, 215 (2000), [gr-qc/0006021]; G. Amelino-Camelia: Lect. Notes Phys. **541**, 1–49 (2000), [gr-qc/9910089]
66. J. van Paradis et al.: Nature **386**, 686 (1997); M. L. Metzger et al.: Nature **387**, 878 (1997)
67. P. N. Bhat, G. J. Fishman, C. A. Meegan, R. B. Wilson, M. N. Brock and W. S. Paclesas: Nature **359**, 217 (1992)
68. P. Mészáros: Nucl. Phys. B (Proc. Suppl.) **80**, 63 (2000)
69. S. D. Biller et al.: Phys. Rev. Lett. **83**, 2108 (1999)
70. E. Waxman and J. Bahcall: Phys. Rev. Lett. **78**, 2292 (1997); E. Waxman: Nucl. Phys. B (Proc. Suppl.) **91**, 494 (2000); Nucl. Phys. B (Proc. Suppl.) **87**, 345 (2000)
71. M. Vietri: Phys. Rev. Lett. **80**, 3690 (1998)
72. M. Roy, H. J. Crawford and A. Trattner: *The prediction and detection of UHE Neutrino Bursts*, [astro-ph/9903231]
73. S. Choubey and S. F. King: Phys. Rev. D **67**, 073005 (2003), [hep-ph/0207260]
74. R. Gambini and J. Pullin: Phys. Rev. D **59**, 124021 (1999), [gr-qc/9809038]
75. J. Alfaro, H. A. Morales-Técotl, and L. F. Urrutia: Phys. Rev. D **65**, 103509 (2002), [hep-th/0108061]
76. J. Alfaro, H. A. Morales-Técotl, and L. F. Urrutia: Phys. Rev. Lett. **84**, 2318 (2000), [gr-qc/9909079]; Phys. Rev. D **66**, 124006 (2002), [hep-th/0208192].

77. N. E. Mavromatos, *The quest for quantum gravity: testing times for theories?*, [astro-ph/0004225]; J. Ellis, *Perspectives in High-Energy Physics*, JHEP Proceedings (2000), [hep-ph/0007161]; J. Ellis, *Testing fundamental physics with high-energy cosmic rays*, Nuovo Cim. C **24**, 483 (2001), [astro-ph/0010474]
78. F. Benatti and R. Floreanini: Phys. Rev. D **64**, 085015 (2001), [hep-ph/0105303]; Phys. Rev. D **62**, 125009 (2000), [hep-ph/0009283]
79. D. A. R. Dalvit, F. D. Mazzitelli, C. Molina-Paris: Phys. Rev. D **63**, 084023 (2001), [hep-th/0010229].
80. T. Padmanabhan: Phys. Rev. D **57**, 6206 (1998); K. Srinivasan, L. Sriramkumar and T. Padmanabhan: Phys. Rev. D **58**, 044009 (1998); S. Shankaranarayanan and T. Padmanabhan: Int. J. Mod. Phys. D **10**, 351 (2001)
81. G. Amelino-Camelia, T. Piran: Phys. Lett. **B497**, 265 (2001), [[hep-ph /0006210]
82. For recent reviews see V. A. Kostelecky: *Topics in Lorentz and CPT violation*, [hep-ph/0104227; R. Bluhm: *Probing the Planck scale in low-energy atomic physics*, [hep-ph/0111323], and references therein.
83. J. M. Carmona and J. L. Cortés: Phys. Lett. **B494**, 75 (2000), [hep-ph/0007057]
84. S. Liberati, T. Jacobson and D. Mattingly: *High-energy constraints on Lorentz symmetry violations*, hep-ph/0110094; T. Jacobson, S. Liberati and D. Mattingly, Phys. Rev. D **66**, 081302 (2002), [hep-ph/0112207].
85. R. C. Myers, M. Pospelov: Phys. Rev. Lett.**90**, 211601 (2003), [hep-ph/0301124].
86. A. Ashtekar, C. Rovelli and L. Smolin: Phys. Rev. Lett. **69**, 237 (1992), [hep-th/9203079]
87. T. Thiemann: Class. Quantum Grav. **18**, 2025 (2001), [hep-th/0005233]
88. H. Sahlmann, T. Thiemann, and O. Winkler: Nucl. Phys. B **606**, 401 (2001), [gr-qc/0102038]
89. M. Varadarajan: Phys. Rev. D **64**, 104003 (2001), [gr-qc/0104051]
90. A. Ashtekar and J. Lewandowski: Class. Quantum Grav. **18**, L117 (2001), [gr-qc/0107043]
91. For the analogous situation in electrodynamics see for example W. Heitler: *Quantum Theory of Radiation*, 3rd edn (Clarendon Press, Oxford, England 1954)
92. H. Sahlmann and T. Thiemann: *Towards the QFT on Curved Spacetime Limit of QGR. I: A General Scheme*, gr-qc/0207030; *II: A Concrete Implementation*, [gr-qc/0207031]
93. G. Amelino-Camelia: Int. J. Mod. Phys. D **12**, 1633 (2003), [gr-qc/0305057]
94. G. Amelino-Camelia: Int. J. Mod. Phys. D **11**, 35 (2002); J. Magueijo and L. Smolin: Phys. Rev. Lett. **88**, 190403 (2002), [hep-th/0112090]
95. T. Jacobson and D. Mattingly: Phys. Rev. D **63**, 041502(R) (2001)

Index

Lecture Notes in Physics

For information about Vols. 1–601
please contact your bookseller or Springer-Verlag
LNP Online archive: springerlink.com

Vol.602: T. Dauxois, S. Ruffo, E. Arimondo (Eds.), Dynamics and Thermodynamics of Systems with Long Range Interactions.

Vol.603: C. Noce, A. Vecchione, M. Cuoco, A. Romano (Eds.), Ruthenate and Rutheno-Cuprate Materials. Superconductivity, Magnetism and Quantum Phase.

Vol.604: J. Frauendiener, H. Friedrich (Eds.), The Conformal Structure of Space-Time: Geometry, Analysis, Numerics.

Vol.605: G. Ciccotti, M. Mareschal, P. Nielaba (Eds.), Bridging Time Scales: Molecular Simulations for the Next Decade.

Vol.606: J.-U. Sommer, G. Reiter (Eds.), Polymer Crystallization. Obervations, Concepts and Interpretations.

Vol.607: R. Guzzi (Ed.), Exploring the Atmosphere by Remote Sensing Techniques.

Vol.608: F. Courbin, D. Minniti (Eds.), Gravitational Lensing:An Astrophysical Tool.

Vol.609: T. Henning (Ed.), Astromineralogy.

Vol.610: M. Ristig, K. Gernoth (Eds.), Particle Scattering, X-Ray Diffraction, and Microstructure of Solids and Liquids.

Vol.611: A. Buchleitner, K. Hornberger (Eds.), Coherent Evolution in Noisy Environments.

Vol.612: L. Klein, (Ed.), Energy Conversion and Particle Acceleration in the Solar Corona.

Vol.613: K. Porsezian, V.C. Kuriakose (Eds.), Optical Solitons. Theoretical and Experimental Challenges.

Vol.614: E. Falgarone, T. Passot (Eds.), Turbulence and Magnetic Fields in Astrophysics.

Vol.615: J. Büchner, C.T. Dum, M. Scholer (Eds.), Space Plasma Simulation.

Vol.616: J. Trampetic, J. Wess (Eds.), Particle Physics in the New Millenium.

Vol.617: L. Fernández-Jambrina, L. M. González-Romero (Eds.), Current Trends in Relativistic Astrophysics, Theoretical, Numerical, Observational

Vol.618: M.D. Esposti, S. Graffi (Eds.), The Mathematical Aspects of Quantum Maps

Vol.619: H.M. Antia, A. Bhatnagar, P. Ulmschneider (Eds.), Lectures on Solar Physics

Vol.620: C. Fiolhais, F. Nogueira, M. Marques (Eds.), A Primer in Density Functional Theory

Vol.621: G. Rangarajan, M. Ding (Eds.), Processes with Long-Range Correlations

Vol.622: F. Benatti, R. Floreanini (Eds.), Irreversible Quantum Dynamics

Vol.623: M. Falcke, D. Malchow (Eds.), Understanding Calcium Dynamics, Experiments and Theory

Vol.624: T. Pöschel (Ed.), Granular Gas Dynamics

Vol.625: R. Pastor-Satorras, M. Rubi, A. Diaz-Guilera (Eds.), Statistical Mechanics of Complex Networks

Vol.626: G. Contopoulos, N. Voglis (Eds.), Galaxies and Chaos

Vol.627: S.G. Karshenboim, V.B. Smirnov (Eds.), Precision Physics of Simple Atomic Systems

Vol.628: R. Narayanan, D. Schwabe (Eds.), Interfacial Fluid Dynamics and Transport Processes

Vol.630: T. Brandes, S. Kettemann (Eds.), Anderson Localization and Its Ramifications

Vol.631: D. J. W. Giulini, C. Kiefer, C. Lämmerzahl (Eds.), Quantum Gravity, From Theory to Experimental Search

Vol.632: A. M. Greco (Ed.), Direct and Inverse Methods in Nonlinear Evolution Equations

Vol.633: H.-T. Elze (Ed.), Decoherence and Entropy in Complex Systems, Based on Selected Lectures from DICE 2002

Vol.634: R. Haberlandt, D. Michel, A. Pöppl, R. Stannarius (Eds.), Molecules in Interaction with Surfaces and Interfaces

Vol.635: D. Alloin, W. Gieren (Eds.), Stellar Candles for the Extragalactic Distance Scale

Vol.636: R. Livi, A. Vulpiani (Eds.), The Kolmogorov Legacy in Physics, A Century of Turbulence and Complexity

Vol.637: I. Müller, P. Strehlow, Rubber and Rubber Balloons, Paradigms of Thermodynamics

Vol.638: Y. Kosmann-Schwarzbach, B. Grammaticos, K.M. Tamizhmani (Eds.), Integrability of Nonlinear Systems

Vol.639: G. Ripka, Dual Superconductor Models of Color Confinement

Vol.640: M. Karttunen, I. Vattulainen, A. Lukkarinen (Eds.), Novel Methods in Soft Matter Simulations

Vol.641: A. Lalazissis, P. Ring, D. Vretenar (Eds.), Extended Density Functionals in Nuclear Structure Physics

Vol.642: W. Hergert, A. Ernst, M. Däne (Eds.), Computational Materials Science

Vol.643: F. Strocchi, Symmetry Breaking

Vol.644: B. Grammaticos, Y. Kosmann-Schwarzbach, T. Tamizhmani (Eds.) Discrete Integrable Systems

Vol.645: U. Schollwöck, J. Richter, D.J.J. Farnell, R.F. Bishop (Eds.), Quantum Magnetism

Vol.646: N. Bretón, J. L. Cervantes-Cota, M. Salgado (Eds.), The Early Universe and Observational Cosmology

MIX
Papier aus verantwortungsvollen Quellen
Paper from responsible sources
FSC® C105338

If you have any concerns about our products,
you can contact us on
ProductSafety@springernature.com

In case Publisher is established outside the EU,
the EU authorized representative is:
Springer Nature Customer Service Center GmbH
Europaplatz 3, 69115 Heidelberg, Germany

Printed by Libri Plureos GmbH
in Hamburg, Germany